Statistics and Computing

Series Editors:

J. Chambers

D. Hand

W. Härdle

For other titles published in this series, go to
www.springer.com/series/3022

Kenneth Lange

Numerical Analysis for Statisticians

Second Edition

Kenneth Lange
Departments of Biomathematics,
Human Genetics, and Statistics
David Geffen School of Medicine
University of California, Los Angeles
Le Conte Ave. 10833
Los Angeles, CA 90095-1766
USA
klange@ucla.edu

Series Editors:

ISBN 978-1-4614-2612-7 ISBN 978-1-4419-5945-4 (eBook)
DOI 10.1007/978-1-4419-5945-4
Springer New York Dordrecht Heidelberg London

Printed on acid-free paper

Springer is part of Springer Science+Business Media (www.springer.com)

Preface to the Second Edition

More than a decade has passed since the publication of the first edition of *Numerical Analysis for Statisticians.* During the interim, statistics rapidly evolved as a discipline. In particular, Markov chain Monte Carlo methods, data mining, and software resources all substantially improved. My own understanding of several subjects also matured. I accordingly set out to write a better book, trying to update topics and correct the errors and omissions of the first edition. The result is certainly a longer book. Whether it is a better book is probably best left to the judgment of readers.

One thing I learned from Springer's stable of reviewers is that there is no universally agreed-on environment for statistical computing. My own preference for coding algorithms in Fortran was quickly dismissed as unworkable. The suggestion of C or C++ might have garnered wider support, but no doubt Java devotees would have objected. In fact, most statisticians prefer higher-level environments such as SAS, R, or Matlab. Each of these environments has its advantages, but none is dominant. For computationally intensive tasks, interpreted languages are at a disadvantage, so compiled languages such as C, Fortran, and Java will survive. A more interesting question is which environment will adapt most quickly to parallel computing.

In any event, instructors will have to decide on their own computing policies. My choice of letting students submit homework in whatever language they want is popular, but unfortunately it promotes superficial grading and a tendency to proceed directly to output. There are many fine textbooks on computing that instructors can recommend as supplementary reading and spend class time covering. These are hardly substitutes for a theoretical treatment emphasizing mathematical motivation and derivations. However, students do need exposure to real computing and thoughtful numerical exercises. Mastery of theory is enhanced by the nitty gritty of coding, and in my experience many attractive algorithms simply fail the test of practicality. In response to users' suggestions, I have scattered many new numerical exercises throughout the text.

Some chapters of the text I hardly touched in revision. In contrast Chapters 11 through 16, on optimization, have been almost completely rewritten. There is now an entire chapter on the MM algorithm in addition to more comprehensive treatments of constrained optimization, convergence analysis, and penalized estimation. The new material reflects the focus of my recent teaching and to some extent my enduring research interests. An entire semester-long course could be constructed from these chapters alone. In this regard note also the additions to Chapter 5 on univariate optimization.

In response to criticisms of my over-reliance on the sweep operator in regression analysis, I have expanded Chapter 7 to include alternative methods of regression that are faster (Cholesky decomposition) or numerically more stable (Gram-Schmidt orthogonalization and the QR decomposition). Chapter 8 contains a new section on finding a single eigenvalue, and the entirely new Chapter 9 discusses the crucial singular value decomposition. Chapter 17 now contains an introduction to reproducing kernel Hilbert spaces. The importance of Hilbert spaces in computational statistics is destined to grow.

Chapter 22, on generating random deviates, has been expanded. Chapter 23, on independent Monte Carlo, and Chapter 24, on permutation testing and the bootstrap, now take a more practical approach that should enhance their value to readers. Nothing better illustrates the interplay between computational power and modern statistical inference than the bootstrap.

Chapter 25 contains new material on the EM algorithm in hidden Markov chains, on finding the equilibrium distribution of a continuous-time chain, and on stochastic simulation and τ-leaping. The latter topic comes up in the dynamical models of systems biology. Chapter 26, on Markov chain Monte Carlo (MCMC), features Gibbs and slice sampling and compares them on some fresh examples. Finally, the new Chapter 27 is devoted to advanced topics in MCMC, notably Markov random fields, reversible jump MCMC, and a brief investigation of rates of convergence in Gibbs sampling. The final sections on convergence combine subject matter from Hilbert spaces, orthogonal expansions, and the singular value decomposition into a single attractive package.

I have many people to thank in preparing this second edition. John Kimmel, my long-suffering editor, exhibited extraordinary patience in encouraging me to get on with this project. I was hosted by the Stanford Statistics Department during my 2006-2007 sabbatical year. It was a pleasure to learn from colleagues there such as Persi Diaconis, Brad Efron, Jerry Friedman, Trevor Hastie, and Rob Tibshirani. Among my UCLA colleagues I would like to single out Jan de Leeuw, Chiara Sabatti, Janet Sinsheimer, Eric Sobel, and Marc Suchard for their kind assistance and advice. Several postdocs and graduate students contributed as well. David Alexander, Kristin Ayers, Lara Bauman, John Ranola, Mary Sehl, Tongtong Wu, and the many other students in my graduate biomathematics classes helped enormously in spotting errors and forcing me to explain things more clearly. Finally, Hua Zhou contributed substantially to the content of Chapter 27 and taught several classes in my absence. To these people and many others not named, I owe a great debt. The faults of the book are, of course, solely mine. Some of these may be amended in future printings, so I hope that readers will send me corrections and suggested revisions at klange@ucla.edu.

I dedicated the first edition of *Numerical Analysis for Statisticians* to my late brother, Charles. Just before sending the final version of the second edition to Springer for editing, I read a copy of the UCLA Department of Mathematics' fall 2009 newsletter, *The Common Denominator.* I was delighted to find that it contained a tribute to Charles from a former student, Gordon Ellison. Here is what Mr. Ellison wrote:

> In about 1969 I enrolled in an Applied Mathematics class offered by UCLA extension and taught by Dr. Charles Lange. At that time I was writing a FORTRAN program using a Fourier series/Green's function solution for microelectronics applications. Dr. Lange's course included a portion on Green's functions, a subject I had studied rather extensively while pursuing my M.A. degree. On the last evening of the course, after I turned in my final exam, I asked Dr. Lange about a problem that slightly resembled my application, telling him that I would not ask for specific help, as I didn't have consultation money. We talked about the math and the simple example. It was the Memorial Day weekend and I went home and solved the little problem. On Monday Dr. Lange phoned me at home and explained that he had gone to the card catalog containing student information and obtained my phone number. During the phone conversation with him, he said he had a solution to the problem. We compared solutions and he told me I knew what I was doing and didn't need a consultant. Using the math that I learned from Dr. Lange, I discovered that my previous theory was not quite correct, and in some cases would have resulted in very large errors. I totally revised my code.
>
> A year or so later, I published a paper wherein I acknowledged Dr. Lange's assistance [1]. I am not exaggerating when I say that single problem launched a thirty-year career. I have related this story many times over the years and can almost feel the fear when considering what the outcome might have been had I not taken Dr. Lange's course and had he not taken the time to help me.
>
> The sad part is that in my youth I was not sufficiently clever or perhaps generous to find Dr. Lange and express my thanks in an appropriate manner. In the last several years I have performed several Internet searches trying to find him. I feel quite stupid that I never thought that he might still be at UCLA. For some reason, I had the idea that he was working at a local company and teaching his course as a side job.
>
> In late August of this year, I was once again telling my wife of my indebtedness to Dr. Lange, and that I have never found him. Then I realized that a Google search on *Charles Lange, PhD* might be productive, as it indeed was, but unfortunately I

only found an obituary. That was an incredibly sad day for me. My only consolation is that I can take the opportunity to tell other professional educators that those little things they do for even the least accomplished of their students sometimes have a very great and positive impact on their lives.

0.1 References

[1] Ellison GN (1973) The effect of some composite structures on the thermal resistance of substrates and integrated circuit chips. *IEEE Trans Electron Devices* ED-20:233–238

Preface to the First Edition

This book, like many books, was born in frustration. When in the fall of 1994 I set out to teach a second course in computational statistics to doctoral students at the University of Michigan, none of the existing texts seemed exactly right. On the one hand, the many decent, even inspiring, books on elementary computational statistics stress the nuts and bolts of using packaged programs and emphasize model interpretation more than numerical analysis. On the other hand, the many theoretical texts in numerical analysis almost entirely neglect the issues of most importance to statisticians. The closest book to my ideal was the classical text of Kennedy and Gentle [2]. More than a decade and a half after its publication, this book still has many valuable lessons to teach statisticians. However, upon reflecting on the rapid evolution of computational statistics, I decided that the time was ripe for an update.

The book you see before you represents a biased selection of those topics in theoretical numerical analysis most relevant to statistics. By intent this book is not a compendium of tried and trusted algorithms, is not a consumer's guide to existing statistical software, and is not an exposition of computer graphics or exploratory data analysis. My focus on principles of numerical analysis is intended to equip students to craft their own software and to understand the advantages and disadvantages of different numerical methods. Issues of numerical stability, accurate approximation, computational complexity, and mathematical modeling share the limelight and take precedence over philosophical questions of statistical inference. Accordingly, you must look elsewhere for a discussion of the merits of frequentist versus Bayesian inference. My attitude is that good data deserve inspection from a variety of perspectives. More often than not, these different perspectives reinforce and clarify rather than contradict one another.

Having declared a truce on issues of inference, let me add that I have little patience with the view that mathematics is irrelevant to statistics. While it is demeaning to statistics to view it simply as a branch of mathematics, it is also ridiculous to contend that statistics can prosper without the continued influx of new mathematical ideas. Nowhere is this more evident than in computational statistics. Statisticians need to realize that the tensions existing between statistics and mathematics mirror the tensions between other disciplines and mathematics. If physicists and economists can learn to live with mathematics, then so can statisticians. Theoreticians in any science will be attracted to mathematics and practitioners repelled. In the end, it really is just a matter of choosing the relevant parts of mathematics and ignoring the rest. Of course, the hard part is deciding what is irrelevant.

Each of the chapters of this book weaves a little mathematical tale with a statistical moral. My hope is to acquaint students with the main principles behind a numerical method without overwhelming them with detail. On first reading, this assertion may seem debatable, but you only have to delve a little more deeply to learn that many chapters have blossomed into full books written by better-informed authors. In the process of writing, I have had to educate myself about many topics. I am sure my ignorance shows, and to the experts I apologize. If there is anything fresh here, it is because my own struggles have made me more sensitive to the struggles of my classroom students. Students deserve to have logical answers to logical questions. I do not believe in pulling formulas out of thin air and expecting students to be impressed. Of course, this attitude reflects my mathematical bent and my willingness to slow the statistical discussion to attend to the mathematics.

The mathematics in this book is a mix of old and new. One of the charms of applying mathematics is that there is little guilt attached to resurrecting venerable subjects such as continued fractions. If you feel that I pay too much attention to these museum pieces, just move on to the next chapter. Note that although there is a logical progression tying certain chapters together—for instance, the chapters on optimization theory and the chapters on numerical integration—many chapters can be read as independent essays. At the opposite extreme of continued fractions, several chapters highlight recent statistical developments such as wavelets, the bootstrap, and Markov chain Monte Carlo methods. These modern topics were unthinkable to previous generations unacquainted with today's computers.

Any instructor contemplating a one-semester course based on this book will have to decide which chapters to cover and which to omit. It is difficult for me to provide sound advice because the task of writing is still so fresh in my mind. In reading the prepublication reviews of my second draft, I was struck by the reviewers' emphasis on the contents of Chapters 5, 7, 13, 14, 23, and 26. Instructors may want to cover material from Chapters 22 and 25 as a prelude to Chapters 23 and 26. Another option is to devote the entire semester to a single topic such as optimization theory. Finally, given the growing importance of computational statistics, a good case can be made for a two-semester course. This book contains adequate material for a rapidly paced yearlong course.

As with any textbook, the problems are nearly as important as the main text. Most problems merely serve to strengthen intellectual muscles strained by the introduction of new theory; some problems extend the theory in significant ways. The majority of any theoretical and typographical errors are apt to be found in the problems. I will be profoundly grateful to readers who draw to my attention errors anywhere in the book, no matter how small.

I have several people to thank for their generous help. Robert Jennrich taught me the rudiments of computational statistics many years ago. His influence pervades the book. Let me also thank the students in my graduate course at Michigan for enduring a mistake-ridden first draft. Ruzong Fan, in particular, checked and corrected many of the exercises. Michael Newton of the University of Wisconsin and Yingnian Wu of the University of Michigan taught from a corrected second draft. Their comments have been helpful in further revision. Robert Strawderman kindly brought to my attention Example 20.4.2, shared his notes on the bootstrap, and critically read Chapter 24. David Hunter prepared the index, drew several figures, and contributed substantially to the content of Chapter 22. Last of all, I thank John Kimmel of Springer for his patient encouragement and editorial advice.

This book is dedicated to the memory of my brother Charles. His close friend and colleague at UCLA, Nick Grossman, dedicated his recent book on celestial mechanics to Charles with the following farewell comments:

> His own work was notable for its devotion to real problems arising from the real world, for the beauty of the mathematics he invoked, and for the elegance of its exposition. Chuck died in summer, 1993, at the age of 51, leaving much undone. Many times since his death I have missed his counsel, and I know that this text would be far less imperfect if I could have asked him about a host of questions that vexed me. Reader, I hope that you have such a friend [1].

It is impossible for me to express my own regrets more poetically.

0.2 References

[1] Grossman N (1996) *The Sheer Joy of Celestial Mechanics.* Birkhäuser, Boston

[2] Kennedy WJ Jr, Gentle JE (1980) *Statistical Computing.* Marcel Dekker, New York

Contents

1

Recurrence Relations

1.1 Introduction

Recurrence relations are ubiquitous in computational statistics and probability. Devising good recurrence relations is both an art and a science. One general theme is the alpha and omega principle; namely, most recurrences are derived by considering either the first or last event in a chain of events. The following examples illustrate this principle and some other commonly employed techniques.

1.2 Binomial Coefficients

Let $\binom{n}{k}$ be the number of subsets of size k from a set of size n. Pascal's triangle is the recurrence scheme specified by

$$\binom{n+1}{k} = \binom{n}{k-1} + \binom{n}{k} \tag{1.1}$$

together with the boundary conditions $\binom{n}{0} = \binom{n}{n} = 1$. To derive (1.1) we take a set of size $n+1$ and divide it into a set of size n and a set of size 1. We can either choose $k-1$ elements from the n-set and combine them with the single element from the 1-set or choose all k elements from the n-set. The first choice can be made in $\binom{n}{k-1}$ ways and the second in $\binom{n}{k}$ ways.

As indicated by its name, we visualize Pascal's triangle as an infinite lower triangular matrix with n as row index and k as column index. The boundary values specify the first column and the diagonal as the constant 1. The recurrence proceeds row by row. If one desires only the binomial coefficients for a single final row, it is advantageous in coding Pascal's triangle to proceed from right to left along the current row. This minimizes computer storage by making it possible to overwrite safely the contents of the previous row with the contents of the current row. Pascal's triangle also avoids the danger of computer overflows caused by computing binomial coefficients via factorials.

1.3 Number of Partitions of a Set

Let B_n be the number of partitions of a set with n elements. By a partition we mean a division of the set into disjoint blocks. A partition induces an

K. Lange, *Numerical Analysis for Statisticians*, Statistics and Computing,
DOI 10.1007/978-1-4419-5945-4_1, © Springer Science+Business Media, LLC 2010

equivalence relation on the set in the sense that two elements are equivalent if and only if they belong to the same block. Two partitions are the same if and only if they induce the same equivalence relation.

Starting with $B_0 = 1$, the B_n satisfy the recurrence relation

$$B_{n+1} = \sum_{k=0}^{n} \binom{n}{k} B_{n-k} = \sum_{k=0}^{n} \binom{n}{k} B_k. \tag{1.2}$$

The reasoning leading to (1.2) is basically the same as in our last example. We divide our set with $n+1$ elements into an n-set and a 1-set. The 1-set can form a block by itself, and the n-set can be partitioned in B_n ways. Or we can choose $k \geq 1$ elements from the n-set in $\binom{n}{k}$ ways and form a block consisting of these elements and the single element from the 1-set. The remaining $n-k$ elements of the n-set can be partitioned in B_{n-k} ways.

1.4 Horner's Method

Suppose we desire to evaluate the polynomial

$$p(x) = a_0 x^n + a_1 x^{n-1} + \cdots + a_{n-1} x + a_n$$

for a particular value of x. If we proceed naively, then it takes $n-1$ multiplications to form the powers $x^k = x \cdot x^{k-1}$ for $2 \leq k \leq n$, n multiplications to multiply each power x^k by its coefficient a_{n-k}, and n additions to sum the resulting terms. This amounts to $3n-1$ operations in all. Horner's method exploits the fact that $p(x)$ can be expressed as

$$\begin{aligned} p(x) &= x(a_0 x^{n-1} + a_1 x^{n-2} + \cdots + a_{n-1}) + a_n \\ &= x b_{n-1}(x) + a_n. \end{aligned}$$

Since the polynomial $b_{n-1}(x)$ of degree $n-1$ can be similarly reduced, a complete recursive scheme for evaluating $p(x)$ is given by

$$\begin{aligned} b_0(x) &= a_0 \\ b_k(x) &= x b_{k-1}(x) + a_k, \quad k = 1, \ldots, n. \end{aligned} \tag{1.3}$$

This scheme, known as Horner's method, requires only n multiplications and n additions to compute $p(x) = b_n(x)$.

Interestingly enough, Horner's method can be modified to produce the derivative $p'(x)$ as well as $p(x)$. This modification is useful, for instance, in searching for a root of $p(x)$ by Newton's method. To discover the algorithm for evaluating $p'(x)$, we differentiate (1.3). This gives the amended Horner scheme

$$\begin{aligned} b_1'(x) &= b_0(x) \\ b_k'(x) &= xb_{k-1}'(x) + b_{k-1}(x), \quad k = 2, \ldots, n, \end{aligned}$$

requiring an additional $n-1$ multiplications and $n-1$ additions to compute $p'(x) = b_n'(x)$.

1.5 Sample Means and Variances

Consider a sequence $x_1, \ldots, x_n$ of n real numbers. After you have computed the sample mean and variance

$$\begin{aligned} \mu_n &= \frac{1}{n}\sum_{i=1}^n x_i \\ \sigma_n^2 &= \frac{1}{n}\sum_{i=1}^n (x_i - \mu_n)^2, \end{aligned}$$

suppose you are presented with a new observation x_{n+1}. It is possible to adjust the sample mean and variance without revisiting all of the previous observations. For example, it is obvious that

$$\mu_{n+1} = \frac{1}{n+1}(n\mu_n + x_{n+1}).$$

Because

$$\begin{aligned} (n+1)\sigma_{n+1}^2 &= \sum_{i=1}^{n+1}(x_i - \mu_{n+1})^2 \\ &= \sum_{i=1}^{n}(x_i - \mu_{n+1})^2 + (x_{n+1} - \mu_{n+1})^2 \\ &= \sum_{i=1}^{n}(x_i - \mu_n)^2 + n(\mu_{n+1} - \mu_n)^2 + (x_{n+1} - \mu_{n+1})^2 \end{aligned}$$

and

$$\begin{aligned} n(\mu_{n+1} - \mu_n)^2 &= n\Big(\mu_{n+1} - \frac{n+1}{n}\mu_{n+1} + \frac{1}{n}x_{n+1}\Big)^2 \\ &= \frac{1}{n}(x_{n+1} - \mu_{n+1})^2, \end{aligned}$$

it follows that

$$\sigma_{n+1}^2 = \frac{n}{n+1}\sigma_n^2 + \frac{1}{n}(x_{n+1} - \mu_{n+1})^2.$$

1.6 Expected Family Size

A married couple desires a family consisting of at least s sons and d daughters. At each birth the mother independently bears a son with probability p and a daughter with probability $q = 1-p$. They will quit having children when their objective is reached. Let N_{sd} be the random number of children born to them. Suppose we wish to compute the expected value $\mathrm{E}(N_{sd})$. Two cases are trivial. If either $s = 0$ or $d = 0$, then N_{sd} follows a negative binomial distribution. It follows that $\mathrm{E}(N_{0d}) = d/q$ and $\mathrm{E}(N_{s0}) = s/p$. When both s and d are positive, the distribution of N_{sd} is not so obvious. However, in this case we can condition on the outcome of the first birth and compute

$$\begin{aligned}\mathrm{E}(N_{sd}) &= p[1+\mathrm{E}(N_{s-1,d})] + q[1+\mathrm{E}(N_{s,d-1})] \\ &= 1 + p\,\mathrm{E}(N_{s-1,d}) + q\,\mathrm{E}(N_{s,d-1}).\end{aligned}$$

There are many variations on this idea. For instance, suppose we wish to compute the probability R_{sd} that the couple reaches its quota of s sons before its quota of d daughters. Then the R_{sd} satisfy the boundary conditions $R_{0d} = 1$ for $d > 0$ and $R_{s0} = 0$ for $s > 0$. When s and d are both positive, we have the recurrence relation

$$R_{sd} = pR_{s-1,d} + qR_{s,d-1}.$$

1.7 Poisson-Binomial Distribution

Let $X_1, \ldots, X_n$ be independent Bernoulli random variables with a possibly different success probability p_k for each X_k. The sum $S_n = \sum_{k=1}^n X_k$ is said to have a Poisson-binomial distribution. If all $p_k = p$, then S_n has a binomial distribution with n trials and success probability p. If each p_i is small, but the sum $\mu = \sum_{k=1}^n p_k$ is moderate in size, then S_n is approximately Poisson with mean μ. In many applications it is unnecessary to invoke this approximation because the exact distribution $q_n(i) = \Pr(S_n = i)$ can be calculated recursively by incrementing the number of summands n. Note first that $q_1(0) = 1 - p_1$ and $q_1(1) = p_1$. With these initial values, one can proceed inductively via

$$\begin{aligned}q_j(0) &= (1-p_j)q_{j-1}(0) \\ q_j(i) &= p_j q_{j-1}(i-1) + (1-p_j)q_{j-1}(i), \quad 1 \le i \le j-1 \\ q_j(j) &= p_j q_{j-1}(j-1)\end{aligned} \tag{1.4}$$

until reaching $j = n$. For the binomial distribution, this method can be superior to calculating each term directly by the standard formula

$$q_n(i) = \binom{n}{i} p^i (1-p)^{n-i}.$$

Just as with Pascal's triangle, it is preferable to proceed from right to left along a row. Furthermore, if the values $q_n(i)$ are only needed for a limited range $0 \leq i \leq k$, then the recurrence (1.4) can be carried out with this proviso.

1.8 A Multinomial Test Statistic

For relatively sparse multinomial data with known but unequal probabilities per category, it is useful to have alternatives to the classical chi-square test. For instance, the number of categories W_d with d or more observations can be a sensitive indicator of clustering. This statistic has mean $\lambda = \sum_{i=1}^m \mu_i$, where

$$\mu_i = \sum_{k=d}^{n} \binom{n}{k} p_i^k (1-p_i)^{n-k}$$

is the probability that the count N_i of category i satisfies $N_i \geq d$. Here we assume n trials, m categories, and a probability p_i attached to category i. If the variance of W_d is close to λ, then W_d follows an approximate Poisson distribution with mean λ [1, 3].

As a supplement to this approximation, it is possible to compute the distribution function $\Pr(W_d \leq j)$ recursively by adapting a technique of Sandell [5]. Once this is done, the p-value of an experimental result w_d can be recovered via $\Pr(W_d \geq w_d) = 1 - \Pr(W_d \leq w_d - 1)$. The recursive scheme can be organized by defining $t_{j,k,l}$ to be the probability that $W_d \leq j$ given k trials and l categories. The indices j, k, and l are confined to the ranges $0 \leq j \leq w_d - 1$, $0 \leq k \leq n$, and $1 \leq l \leq m$. The l categories implicit in $t_{j,k,l}$ refer to the first l of the overall m categories; the ith of these l categories is assigned the conditional probability $p_i/(p_1 + \cdots + p_l)$.

With these definitions in mind, note first the obvious initial values (a) $t_{0,k,1} = 1$ for $k < d$, (b) $t_{0,k,1} = 0$ for $k \geq d$, and (c) $t_{j,k,1} = 1$ for $j > 0$. Now beginning with $l = 1$, compute $t_{j,k,l}$ recursively by conditioning on how many observations fall in category l. Since at most $d-1$ observations can fall in category l without increasing W_d by 1, the recurrence relation for $j = 0$ is

$$\begin{aligned} & t_{0,k,l} \\ = & \sum_{i=0}^{\min\{d-1,k\}} \binom{k}{i} \left(\frac{p_l}{p_1 + \cdots + p_l}\right)^i \left(1 - \frac{p_l}{p_1 + \cdots + p_l}\right)^{k-i} t_{0,k-i,l-1}, \end{aligned}$$

and the recurrence relation for $j > 0$ is

$$\begin{aligned} t_{j,k,l} &= \sum_{i=0}^{\min\{d-1,k\}} \binom{k}{i} \left(\frac{p_l}{p_1+\cdots+p_l}\right)^i \left(1-\frac{p_l}{p_1+\cdots+p_l}\right)^{k-i} t_{j,k-i,l-1} \\ &\quad + \sum_{i=d}^{k} \binom{k}{i} \left(\frac{p_l}{p_1+\cdots+p_l}\right)^i \left(1-\frac{p_l}{p_1+\cdots+p_l}\right)^{k-i} t_{j-1,k-i,l-1}. \end{aligned}$$

These recurrence relations jointly permit replacing the matrix $(t_{j,k,l-1})$ by the matrix $(t_{j,k,l})$. At the end of this recursive scheme on $l = 2, \ldots, m$, we extract the desired probability $t_{w_d-1,n,m}$.

The binomial probabilities occurring in these formulas can be computed by our previous algorithm for the Poisson-binomial distribution. It is noteworthy that the Poisson-binomial recurrence increments the number of trials, whereas the recurrence for the distribution function of W_d increments the number of categories in the multinomial distribution.

1.9 An Unstable Recurrence

Not all recurrence relations are numerically stable. Henrici [2] dramatically illustrates this point using the integrals

$$y_n = \int_0^1 \frac{x^n}{x+a} dx. \tag{1.5}$$

The recurrence $y_n = 1/n - ay_{n-1}$ follows directly from the identity

$$\begin{aligned} \int_0^1 \frac{x^{n-1}(x+a-a)}{x+a} dx &= \int_0^1 x^{n-1} dx - a \int_0^1 \frac{x^{n-1}}{x+a} dx \\ &= \frac{1}{n} - a \int_0^1 \frac{x^{n-1}}{x+a} dx. \end{aligned}$$

In theory this recurrence furnishes a convenient method for calculating the y_n starting with the initial value $y_0 = \ln \frac{1+a}{a}$. Table 1.1 records the results of our computations in single precision when $a = 10$. It is clear that something has gone amiss. Computing in double precision only delays the onset of the instability.

We can diagnose the source of the problem by noting that for n moderately large most of the mass of the integral occurs near $x = 1$. Thus, to a good approximation

$$\begin{aligned} y_{n-1} &\approx \frac{1}{1+a} \int_0^1 x^{n-1} dx \\ &= \frac{1}{(1+a)n}. \end{aligned}$$

When a is large, the fraction $a/(1+a)$ in the difference

$$\begin{aligned} y_n &\approx \frac{1}{n} - a\frac{1}{(1+a)n} \\ &= \frac{1}{n}\Big(1 - \frac{a}{1+a}\Big) \end{aligned}$$

is close to 1. We lose precision whenever we subtract two numbers of the same sign and comparable magnitude. The moral here is that we must exercise caution in using recurrence relations involving subtraction. Fortunately, many recurrences in probability theory arise by conditioning arguments and consequently entail only addition and multiplication of nonnegative numbers.

TABLE 1.1. Computed Values of the Integral y_n

n	y_n	n	y_n
0	0.095310	5	0.012960
1	0.046898	6	0.037064
2	0.031020	7	-0.227781
3	0.023130	8	2.402806
4	0.018704	9	-23.916945

1.10 Quick Sort

Statisticians sort lists of numbers to compute sample quantiles and plot empirical distribution functions. It is a pleasant fact that the fastest sorting algorithm can be explained by a probabilistic argument [6]. At the heart of this argument is a recurrence relation specifying the average number of operations encountered in sorting n numbers. In this problem, we can explicitly solve the recurrence relation and estimate the rate of growth of its solution as a function of n. The recurrence relation is not so much an end in itself as a means to understanding the behavior of the sorting algorithm.

The quick sort algorithm is based on the idea of finding a splitting entry x_i of a sequence $x_1, \ldots, x_n$ of n distinct numbers in the sense that $x_j < x_i$ for $j < i$ and $x_j > x_i$ for $j > i$. In other words, a splitter x_i is already correctly ordered relative to the rest of the entries of the sequence. Finding a splitter reduces the computational complexity of sorting because it is easier to sort both of the subsequences $x_1, \ldots, x_{i-1}$ and $x_{i+1}, \ldots, x_n$ than it is to sort the original sequence. At this juncture, one can reasonably object that no splitter need exist, and even if one does, it may be difficult to locate. The quick sort algorithm avoids these difficulties by randomly selecting a splitting value and then slightly rearranging the sequence so that this splitting value occupies the correct splitting location.

In the background of quick sort is the probabilistic assumption that all $n!$ permutations of the n values are equally likely. The algorithm begins by randomly selecting one of the n values and moving it to the leftmost or first position of the sequence. Through a sequence of exchanges, this value is then promoted to its correct location. In the probabilistic setting adopted, the correct location of the splitter is uniformly distributed over the n positions of the sequence.

The promotion process works by exchanging or swapping entries to the right of the randomly chosen splitter x_1, which is kept in position 1 until a final swap. Let j be the current position of the sequence as we examine it from left to right. In the sequence up to position j, a candidate position i for the insertion of x_1 must satisfy the conditions $x_k < x_1$ for $1 < k \leq i$ and $x_k > x_1$ for $i < k \leq j$. Clearly, the choice $i = j$ works when $j = 1$ because then the set $\{k : 1 < k \leq i \text{ or } i < k \leq j\}$ is empty. Now suppose we examine position $j + 1$. If $x_{j+1} > x_1$, then we keep the current candidate position i. If $x_{j+1} < x_1$, then we swap x_{i+1} and x_{j+1} and replace i by $i+1$. In either case, the two required conditions imposed on i continue to obtain. Thus, we can inductively march from the left end to the right end of the sequence, carrying out a few swaps in the process, so that when $j = n$, the value i marks the correct position to insert x_1. Once this insertion is made, the subsequences $x_1, \ldots, x_{i-1}$ and $x_{i+1}, \ldots, x_n$ can be sorted separately by the same splitting procedure.

Now let e_n be the expected number of operations involved in quick sorting a sequence of n numbers. By convention $e_0 = 0$. If we base our analysis only on how many positions j must be examined at each stage and not on how many swaps are involved, then we can write the recurrence relation

$$\begin{aligned} e_n &= n - 1 + \frac{1}{n} \sum_{i=1}^{n} (e_{i-1} + e_{n-i}) \\ &= n - 1 + \frac{2}{n} \sum_{i=1}^{n} e_{i-1} \end{aligned} \tag{1.6}$$

by conditioning on the correct position i of the first splitter.

The recurrence relation (1.6) looks formidable, but a few algebraic maneuvers render it solvable. Multiplying equation (1.6) by n produces

$$ne_n = n(n-1) + 2 \sum_{i=1}^{n} e_{i-1}.$$

If we subtract from this the corresponding expression for $(n-1)e_{n-1}$, then we get

$$ne_n - (n-1)e_{n-1} = 2n - 2 + 2e_{n-1},$$

which can be rearranged to give

$$\frac{e_n}{n+1} = \frac{2(n-1)}{n(n+1)} + \frac{e_{n-1}}{n}. \tag{1.7}$$

Equation (1.7) can be iterated to yield

$$\begin{aligned}\frac{e_n}{n+1} &= 2\sum_{k=1}^{n} \frac{(k-1)}{k(k+1)} \\ &= 2\sum_{k=1}^{n} \Big(\frac{2}{k+1} - \frac{1}{k}\Big) \\ &= 2\sum_{k=1}^{n} \frac{1}{k} - \frac{4n}{n+1}.\end{aligned}$$

Because $\sum_{k=1}^{n} \frac{1}{k}$ approximates $\int_1^n \frac{1}{x}dx = \ln n$, it follows that

$$\lim_{n\to\infty} \frac{e_n}{2n\ln n} = 1.$$

Quick sort is indeed a very efficient algorithm on average. Press et al. [4] provide good computer code implementing it.

1.11 Problems

1. Let f_n be the number of subsets of $\{1, \ldots, n\}$ that do not contain two consecutive integers. Show that $f_1 = 2$, $f_2 = 3$, and $f_n = f_{n-1} + f_{n-2}$ for $n > 2$.

2. Suppose n, j, and $r_1, \ldots, r_j$ are positive integers with $n = r_1 + \cdots + r_j$ and with $r_1 \geq r_2 \geq \cdots \geq r_j \geq 1$. Such a decomposition is called a partition of n with largest part r_1. For example, $6 = 4 + 1 + 1$ is a partition of 6 into three parts with largest part 4. Let q_{nk} be the number of partitions of n with largest part k. Show that

$$q_{nk} = q_{n-1,k-1} + q_{n-k,k}.$$

3. In Horner's method suppose x_0 is a root of $p(x)$. Show that the numbers $b_k(x_0)$ produced by (1.3) yield the deflated polynomial

$$b_0(x_0)x^{n-1} + b_1(x_0)x^{n-2} + \cdots + b_{n-1}(x_0) = \frac{p(x)}{x - x_0}.$$

4. Prove that the characteristic polynomial $p_n(x) = \det(M - xI)$ of the $n \times n$ tridiagonal matrix

$$M = \begin{pmatrix} b_1 & c_2 & 0 & \cdots & 0 & 0 \\ a_2 & b_2 & c_3 & \cdots & 0 & 0 \\ \vdots & \vdots & \vdots & \ddots & \vdots & \vdots \\ 0 & 0 & 0 & \cdots & b_{n-1} & c_n \\ 0 & 0 & 0 & \cdots & a_n & b_n \end{pmatrix}$$

can be computed recursively by defining

$$\begin{aligned} p_0(x) &= 1 \\ p_1(x) &= (b_1 - x) \\ p_m(x) &= (b_m - x)p_{m-1}(x) - a_m c_m p_{m-2}(x), \quad m = 2, 3, \ldots, n. \end{aligned}$$

Why are the roots of $p_n(x)$ real in the symmetric case $a_m = c_m$ for all m? Devise a related recurrence to calculate the derivative $p_n'(x)$.

5. Give a recursive method for computing the second moments $\mathrm{E}(N_{sd}^2)$ in the family-planning model.

6. In the family-planning model, suppose the couple has an upper limit m on the number of children they can afford. Hence, they stop whenever they reach their goal of s sons and d daughters or m total children, whichever comes first. Let N_{sdm} now be their random number of children. Give a recursive method for computing $\mathrm{E}(N_{sdm})$.

7. In the family-planning model, suppose the husband and wife are both carriers of a recessive genetic disease. On average one quarter of their children will be afflicted. If the parents want at least s normal sons and at least d normal daughters, let T_{sd} be their random number of children. Give a recursive method for computing $\mathrm{E}(T_{sd})$.

8. Consider the multinomial model with m categories, n trials, and probability p_i attached to the ith category. Express the distribution function of the maximum number of counts $\max_i N_i$ observed in any category in terms of the distribution functions of the W_d. How can the algorithm for computing the distribution function of W_d be simplified to give an algorithm for computing a p-value of $\max_i N_i$?

9. Define the statistic U_d to be the number of categories i with $N_i < d$. Express the right-tail probability $\Pr(U_d \geq j)$ in terms of the distribution function of W_d. This gives a method for computing p-values of the statistic U_d. In some circumstances U_d has an approximate Poisson distribution. What do you conjecture about these circumstances?

10. Demonstrate that the integral y_n defined by equation (1.5) can be expanded in the infinite series

$$y_n = \sum_{k=0}^{\infty} \frac{(-1)^k}{(n+k+1)a^{k+1}}$$

when $a > 1$. This does provide a reasonably stable method of computing y_n for large a.

11. Show that the worst case of quick sort takes on the order of n^2 operations.

12. Let p be the probability that a randomly chosen permutation of n distinct numbers contains at least one pre-existing splitter. Show by an inclusion-exclusion argument that

$$\begin{aligned} p &= \sum_{i=1}^{n} \frac{(-1)^{i-1}}{i!} \\ &\approx 1 - e^{-1}. \end{aligned}$$

13. Continuing Problem 12, demonstrate that both the mean and variance of the number of pre-existing splitters equal 1.

1.12 References

[1] Barbour AD, Holst L, Janson S (1992) *Poisson Approximation.* Oxford University Press, Oxford

[2] Henrici P (1982) *Essentials of Numerical Analysis with Pocket Calculator Demonstrations.* Wiley, New York

[3] Kolchin VF, Sevast'yanov BA, Chistyakov VP (1978) *Random Allocations.* Winston, Washington DC

[4] Press WH, Teukolsky SA, Vetterling WT, Flannery BP (1992) *Numerical Recipes in Fortran: The Art of Scientific Computing,* 2nd ed. Cambridge University Press, Cambridge

[5] Sandell D (1991) Computing probabilities in a generalized birthday problem. *Math Scientist* 16:78–82

[6] Wilf HS (1986) *Algorithms and Complexity.* Prentice-Hall, New York

2

Power Series Expansions

2.1 Introduction

Power series expansions are old friends of all workers in the mathematical sciences [4, 5, 10]. This chapter emphasizes special techniques for handling and generating the power series encountered in computational statistics. Most expansions can be phrased in terms of recurrence relations. Logarithmic differentiation is one powerful device for developing recurrences. Our applications of logarithmic differentiation to problems such as the conversion between moments and cumulants illustrate some of the interesting possibilities.

Power series expansions are also available for many of the well-known distribution functions of statistics. Although such expansions are usually guaranteed to converge, roundoff error for an alternating series can be troublesome. Thus, either high-precision arithmetic should be used in expanding a distribution function, or the distribution function should be modified so that only positive terms are encountered in the series defining the modified function. Our expansions are coordinated with the discussion of special functions in *Numerical Recipes* [9]. We particularly stress connections among the various distribution functions.

2.2 Expansion of $P(s)^n$

Suppose $P(s) = \sum_{k=0}^{\infty} p_k s^k$ is a power series with $p_0 \neq 0$. If n is a positive integer, then the recurrence relation of J.C.P. Miller [5] permits one to compute the coefficients of $Q(s) = \sum_{k=0}^{\infty} q_k s^k = P(s)^n$ from those of $P(s)$. This clever formula is derived by differentiating $Q(s)$ and then multiplying the result by $P(s)$. By definition of $Q(s)$, this yields

$$P(s)Q'(s) = nP'(s)Q(s). \tag{2.1}$$

If we equate the coefficients of s^{k-1} on both sides of (2.1), then it follows that

$$\sum_{j=1}^{k} p_{k-j} j q_j = n \sum_{j=0}^{k-1} (k-j) p_{k-j} q_j,$$

K. Lange, *Numerical Analysis for Statisticians*, Statistics and Computing,
DOI 10.1007/978-1-4419-5945-4_2, © Springer Science+Business Media, LLC 2010

which can be solved for q_k in the form

$$q_k = \frac{1}{kp_0}\sum_{j=0}^{k-1}[n(k-j)-j]p_{k-j}q_j. \tag{2.2}$$

The obvious initial condition is $q_0 = p_0^n$. Sometimes it is more natural to compute q_k^*, where $q_k^*/k! = q_k$ and $p_k^*/k! = p_k$. Then the recurrence relation (2.2) can be rewritten as

$$q_k^* = \frac{1}{kp_0^*}\sum_{j=0}^{k-1}\binom{k}{j}[n(k-j)-j]p_{k-j}^*q_j^*. \tag{2.3}$$

2.2.1 *Application to Moments*

Suppose $X_1, \ldots, X_n$ are independent, identically distributed random variables. Let μ_k be the kth moment of X_1, and let ω_k be the kth moment of $S_n = \sum_{i=1}^n X_i$. Applying the recurrence (2.3) to the moment generating functions of X_1 and S_n gives

$$\omega_k = \frac{1}{k}\sum_{j=0}^{k-1}\binom{k}{j}[n(k-j)-j]\mu_{k-j}\omega_j.$$

As a concrete example, suppose $n = 10$ and X_1 has a uniform distribution on $[0, 1]$. Then $\mu_k = 1/(k+1)$. Table 2.1 records the first 10 moments ω_k of S_{10}.

TABLE 2.1. The Moments ω_k of the Sum of 10 Uniform Deviates

k	ω_k	k	ω_k
1	$.50000 \times 10^1$	6	$.24195 \times 10^5$
2	$.25833 \times 10^2$	7	$.14183 \times 10^6$
3	$.13750 \times 10^3$	8	$.84812 \times 10^6$
4	$.75100 \times 10^3$	9	$.51668 \times 10^7$
5	$.42167 \times 10^4$	10	$.32029 \times 10^8$

2.3 Expansion of $e^{P(s)}$

Again let $P(s)$ be a power series, and put $Q(s) = e^{P(s)}$ [8]. If one equates the coefficients of s^{k-1} in the obvious identity

$$Q'(s) = P'(s)Q(s),$$

then it follows that

$$q_k = \frac{1}{k}\sum_{j=0}^{k-1}(k-j)p_{k-j}q_j. \tag{2.4}$$

Clearly, $q_0 = e^{p_0}$. When $q_k^*/k! = q_k$ and $p_k^*/k! = p_k$, equation (2.4) becomes

$$q_k^* = \sum_{j=0}^{k-1}\binom{k-1}{j}p_{k-j}^*q_j^*. \tag{2.5}$$

2.3.1 Moments to Cumulants and Vice Versa

If $Q(s) = \sum_{k=0}^{\infty}\frac{m_k}{k!}s^k = e^{P(s)}$ is the moment generating function of a random variable, then $P(s) = \sum_{k=0}^{\infty}\frac{c_k}{k!}s^k$ is the corresponding cumulant generating function. Clearly, $m_0 = 1$ and $c_0 = 0$. The recurrence (2.5) can be rewritten as

$$m_k = \sum_{j=0}^{k-1}\binom{k-1}{j}c_{k-j}m_j.$$

From this one can deduce the equally useful recurrence

$$c_k = m_k - \sum_{j=1}^{k-1}\binom{k-1}{j}c_{k-j}m_j$$

converting moments to cumulants.

2.3.2 Compound Poisson Distributions

Consider a random sum $S_N = X_1 + \cdots + X_N$ of a random number N of independent, identically distributed random variables X_k. If N is independent of the X_k and has a Poisson distribution with mean λ, then S_N is said to have a compound Poisson distribution. If $R(s)$ is the common moment generating function of the X_k, then $Q(s) = e^{-\lambda+\lambda R(s)}$ is the moment generating function of S_N. Likewise, if the X_k assume only nonnegative integer values, and if $R(s)$ is their common probability generating function, then $Q(s) = e^{-\lambda+\lambda R(s)}$ is the probability generating function of S_N. Thus, the moments $\mathrm{E}(S_N^i)$ and probabilities $\Pr(S_N = i)$ can be recursively computed from the corresponding quantities for the X_k.

2.3.3 Evaluation of Hermite Polynomials

The Hermite polynomials $H_k(x)$ can be defined by the generating function

$$\sum_{k=0}^{\infty}\frac{H_k(x)}{k!}s^k = e^{xs-\frac{1}{2}s^2}.$$

From this definition it is clear that $H_0(x) = 1$ and $H_1(x) = x$. The recurrence (2.5) takes the form

$$H_k(x) = xH_{k-1}(x) - (k-1)H_{k-2}(x)$$

for $k \geq 2$. In general, any sequence of orthogonal polynomials can be generated by a linear, two-term recurrence relation. We will meet the Hermite polynomials later when we consider Gaussian quadrature and Edgeworth expansions.

2.4 Standard Normal Distribution Function

Consider the standard normal distribution

$$F(x) = \frac{1}{2} + \frac{1}{\sqrt{2\pi}} \int_0^x e^{-\frac{y^2}{2}} dy.$$

If we expand

$$e^{-\frac{y^2}{2}} = \sum_{n=0}^{\infty} \frac{(-1)^n y^{2n}}{2^n n!}$$

and integrate term by term, then it is clear that

$$F(x) = \frac{1}{2} + \frac{1}{\sqrt{2\pi}} \sum_{n=0}^{\infty} \frac{(-1)^n x^{2n+1}}{2^n(2n+1)n!}.$$

This is an alternating series that entails severe roundoff error even for x as small as 4.

To derive a more stable expansion, let

$$\begin{aligned} g(x) &= e^{\frac{x^2}{2}} \int_0^x e^{-\frac{y^2}{2}} dy \\ &= \sum_{n=0}^{\infty} c_n x^{2n+1}. \end{aligned}$$

By inspection, $g(x)$ satisfies the differential equation

$$g'(x) = xg(x) + 1. \tag{2.6}$$

Now $c_0 = 1$ because $g'(0) = 0g(0) + 1$. All subsequent coefficients are also positive. Indeed, equating coefficients of x^{2n} in (2.6) gives the recurrence relation

$$c_n = \frac{1}{2n+1} c_{n-1}.$$

Thus, the series for $g(x)$ converges stably for all $x > 0$. Since $g(x)$ is an odd function, only positive x need be considered. In evaluating

$$\begin{aligned} F(x) - \frac{1}{2} &= \frac{1}{\sqrt{2\pi}} e^{-\frac{x^2}{2}} g(x) \\ &= \sum_{n=0}^{\infty} a_n, \end{aligned}$$

we put $a_0 = \frac{1}{\sqrt{2\pi}} e^{-\frac{x^2}{2}} x$ and $a_n = a_{n-1} \frac{x^2}{2n+1}$. Then the partial sums $\sum_{i=0}^{n} a_i$ are well scaled, and $a_n = 0$ gives a machine independent test for the convergence of the series at its nth term.

2.5 Incomplete Gamma Function

The distribution function of a gamma random variable with parameters a and b is defined by

$$\begin{aligned} P(a, bx) &= \frac{1}{\Gamma(a)} \int_0^x b^a y^{a-1} e^{-by} dy \\ &= \frac{1}{\Gamma(a)} \int_0^{bx} z^{a-1} e^{-z} dz. \end{aligned} \tag{2.7}$$

We can expand $P(a, x)$ in a power series by repeated integration by parts. In fact,

$$\begin{aligned} P(a, x) &= \frac{x^a}{a\Gamma(a)} e^{-x} + \frac{1}{a\Gamma(a)} \int_0^x z^a e^{-z} dz \\ &= \frac{e^{-x} x^a}{\Gamma(a+1)} + P(a+1, x) \end{aligned}$$

leads to the stable series

$$P(a, x) = e^{-x} x^a \sum_{n=0}^{\infty} \frac{x^n}{\Gamma(a+n+1)}. \tag{2.8}$$

For the expansion (2.8) to be practical, we must have some method for evaluating the ordinary gamma function. One option is to iterate the functional identity

$$\ln \Gamma(a) = \ln \Gamma(a+1) - \ln a$$

until k is large enough so that $\ln \Gamma(a+k)$ is well approximated by Stirling's formula.

2.6 Incomplete Beta Function

For a and b positive, the incomplete beta function is defined by

$$I_x(a,b) = \frac{\Gamma(a+b)}{\Gamma(a)\Gamma(b)} \int_0^x y^{a-1}(1-y)^{b-1}dy.$$

Suppose we attempt to expand this distribution function in the form

$$\begin{aligned} I_x(a,b) &= x^a(1-x)^b \sum_{n=0}^{\infty} c_n x^n \\ &= \sum_{n=0}^{\infty} c_n x^{n+a}(1-x)^b. \end{aligned} \tag{2.9}$$

If we divide the derivative

$$\begin{aligned} \frac{d}{dx} I_x(a,b) &= \frac{\Gamma(a+b)}{\Gamma(a)\Gamma(b)} x^{a-1}(1-x)^{b-1} \\ &= \sum_{n=0}^{\infty} c_n[(n+a)(1-x) - bx]x^{n+a-1}(1-x)^{b-1} \end{aligned}$$

by $x^{a-1}(1-x)^{b-1}$, then it follows that

$$\frac{\Gamma(a+b)}{\Gamma(a)\Gamma(b)} = \sum_{n=0}^{\infty} c_n[(n+a)(1-x) - bx]x^n. \tag{2.10}$$

Equating the coefficients of x^n on both sides of (2.10) gives for $n = 0$

$$c_0 = \frac{\Gamma(a+b)}{a\Gamma(a)\Gamma(b)} = \frac{\Gamma(a+b)}{\Gamma(a+1)\Gamma(b)}$$

and for $n > 0$

$$c_n(n+a) - c_{n-1}(n-1+a+b) = 0,$$

which collapses to the recurrence relation

$$c_n = \frac{n-1+a+b}{n+a} c_{n-1}.$$

Therefore, all coefficients c_n are positive. The ratio test indicates that the power series (2.9) converges for $0 \le x < 1$. For x near 1, the symmetry relation $I_x(a,b) = 1 - I_{1-x}(b,a)$ can be employed to get a more quickly converging series.

2.7 Connections to Other Distributions

Evaluation of many classical distribution functions reduces to the cases already studied. Here are some examples.

2.7.1 Chi-square and Standard Normal

A chi-square random variable χ_n^2 with n degrees of freedom has a gamma distribution with parameters $a = n/2$ and $b = 1/2$. Hence, in terms of definition (2.7), we have $\Pr(\chi_n^2 \le x) = P(\frac{n}{2}, \frac{x}{2})$. If X has a standard normal distribution, then X^2 has a chi-square distribution with one degree of freedom. Obvious symmetry arguments therefore imply for $x \ge 0$ that $\Pr(X \le x) = \frac{1}{2} + \frac{1}{2}P(\frac{1}{2}, \frac{x^2}{2})$.

2.7.2 Poisson

The distribution function of a Poisson random variable X with mean λ can be expressed in terms of the incomplete gamma function (2.7) as

$$\Pr(X \le k-1) \quad = \quad 1 - P(k, \lambda).$$

The most illuminating proof of this result relies on constructing a Poisson process of unit intensity on $[0, \infty)$. In this framework $\Pr(X \le k-1)$ is the probability of $k-1$ or fewer random points on $[0, \lambda]$. Since the waiting time until the kth random point in the process follows a gamma distribution with parameters $a = k$ and $b = 1$, the probability of $k-1$ or fewer random points on $[0, \lambda]$ coincides with the probability $1 - P(k, \lambda)$ that the kth random point falls beyond λ.

2.7.3 Binomial and Negative Binomial

Let X be a binomially distributed random variable with n trials and success probability p. We can express the distribution function of X in terms of the incomplete beta function (2.9) as

$$\Pr(X \le k-1) \quad = \quad 1 - I_p(k, n-k+1). \tag{2.11}$$

To validate this expression, imagine distributing n points randomly on $[0, 1]$. The probability $\Pr(X \le k-1)$ is just the probability that $k-1$ or fewer of the random points occur on $[0, p]$. This latter probability is also the probability that the kth random point to the right of 0 falls on $[p, 1]$. But standard arguments from the theory of order statistics show that the kth random point to the right of 0 has beta density $n\binom{n-1}{k-1}y^{k-1}(1-y)^{n-k}$.

Alternatively, if we drop random points indefinitely on $[0, 1]$ and record the trial Y at which the kth point falls to the left of p, then Y follows a negative binomial distribution. By the preceding argument,

$$\Pr(Y > n) = \Pr(X \leq k-1),$$

which clearly entails $\Pr(Y \leq n) = I_p(k, n-k+1)$. If we focus on failures rather than total trials in the definition of the negative binomial, then the random variable $Z = Y - k$ is representative of this point of view. In this case, $\Pr(Z \leq m) = I_p(k, m+1)$.

2.7.4 F and Student's t

An $F_{m,n}$ random variable can be written as the ratio

$$F_{m,n} = \frac{n\chi_m^2}{m\chi_n^2}$$

of two independent chi-square random variables scaled to have unit means. Straightforward algebra gives

$$\begin{aligned}
\Pr(F_{m,n} \leq x) &= \Pr\left(\frac{\chi_m^2}{\chi_n^2} \leq \frac{mx}{n}\right) \\
&= \Pr\left(\frac{\chi_n^2}{\chi_m^2 + \chi_n^2} \geq \frac{n}{mx+n}\right).
\end{aligned}$$

If $p = n/2$ is an integer, then $\chi_n^2/2 = W_p$ is a gamma distributed random variable that can be interpreted as the waiting time until the pth random point in a Poisson process on $[0, \infty)$. Similarly, if $q = m/2$ is an integer, then $\chi_m^2/2 = W_q$ can be interpreted as the waiting time from the pth random point until the $(p+q)$th random point of the same Poisson process. In this setting, the ratio

$$\begin{aligned}
\frac{\chi_n^2}{\chi_m^2 + \chi_n^2} &= \frac{W_p}{W_q + W_p} \\
&\geq u
\end{aligned}$$

if and only if the waiting time until the pth point is a fraction u or greater of the waiting time until the $(p+q)$th point. Now conditional on the waiting time $W_p + W_q$ until random point $p+q$, the $p+q-1$ previous random points are uniformly and independently distributed on the interval $[0, W_p + W_q]$. It follows from equation (2.11) that

$$\begin{aligned}
\Pr\left(\frac{W_p}{W_q + W_p} \geq u\right) &= \sum_{j=0}^{p-1} \binom{p+q-1}{j} u^j (1-u)^{p+q-1-j} \\
&= 1 - I_u(p, p+q-1-p+1) \\
&= I_{1-u}(q, p).
\end{aligned}$$

In general, regardless of whether n or m is even, the identity

$$\Pr(F_{m,n} \leq x) \quad = \quad I_{\frac{mx}{mx+n}}\left(\frac{m}{2}, \frac{n}{2}\right) \tag{2.12}$$

holds, relating the F distribution to the incomplete beta function [1].

By definition a random variable t_n follows Student's t distribution with n degrees of freedom if it is symmetric around 0 and its square t_n^2 has an $F_{1,n}$ distribution. Therefore, according to equation (2.12),

$$\begin{aligned} \Pr(t_n \leq x) &= \frac{1}{2} + \frac{1}{2}\Pr(t_n^2 \leq x^2) \\ &= \frac{1}{2} + \frac{1}{2}\Pr(F_{1,n} \leq x^2) \\ &= \frac{1}{2} + \frac{1}{2} I_{\frac{x^2}{x^2+n}}\left(\frac{1}{2}, \frac{n}{2}\right) \end{aligned}$$

for $x \geq 0$.

2.7.5 Monotonic Transformations

Suppose X is a random variable with known distribution function $F(x)$ and $h(x)$ is a strictly increasing, continuous function. Then the random variable $h(X)$ has distribution function

$$\Pr[h(X) \leq x] \quad = \quad F[h^{-1}(x)],$$

where $h^{-1}(x)$ is the functional inverse of $h(x)$. If $h(x)$ is strictly decreasing and continuous, then

$$\begin{aligned} \Pr[h(X) < x] &= \Pr[X > h^{-1}(x)] \\ &= 1 - F[h^{-1}(x)]. \end{aligned}$$

Many common distributions fit this paradigm. For instance, if X is normal, then e^X is lognormal. If X is chi-square, then $1/X$, $1/\sqrt{X}$, and $\ln X$ are inverse chi-square, inverse chi, and log chi-square, respectively. If X has an $F_{m,n}$ distribution, then $\frac{1}{2}\ln X$ has Fisher's z distribution. Calculating any of these distributions therefore reduces to evaluating either an incomplete beta or an incomplete gamma function.

2.8 Problems

1. A symmetric random walk on the integer lattice points of R^k starts at the origin and at each epoch randomly chooses one of the $2k$ possible coordinate directions and takes a unit step in that direction [3]. If

u_{2n} is the probability that the walk returns to the origin at epoch $2n$, then one can show that

$$\sum_{n=0}^{\infty} \frac{u_{2n}}{(2n)!} x^{2n} = \Big[\sum_{n=0}^{\infty} \frac{1}{(2k)^{2n}(n!)^2} x^{2n}\Big]^k.$$

Derive a recurrence relation for computing u_{2n}, and implement it when $k = 2$. Check your numerical results against the exact formula

$$u_{2n} = \Big[\frac{1}{2^{2n}}\binom{2n}{n}\Big]^2.$$

Discuss possible sources of numerical error in using the recurrence relation.

2. Write recurrence relations for the Taylor coefficients of the functions $\left(\frac{1+s}{1-s}\right)^n$ and $\exp(\frac{1+s}{1-s})$.

3. Show that the coefficients of the exponential generating function

$$\sum_{n=0}^{\infty} \frac{B_n}{n!} s^n = e^{e^s - 1}$$

satisfy the recurrence relation (1.2) of Chapter 1. Check the initial condition $B_0 = 1$, and conclude that the coefficient B_n determines the number of partitions of a set with n elements.

4. Suppose the coefficients of a power series $\sum_{n=0}^{\infty} b_n x^n$ satisfy $b_n = p(n)$ for some polynomial p. Find a power series $\sum_{n=0}^{\infty} a_n x^n$ such that

$$p\left(x\frac{d}{dx}\right)\sum_{n=0}^{\infty} a_n x^n = \sum_{n=0}^{\infty} b_n x^n.$$

5. Show that $\sum_{n=1}^{m} n^2 = m(m+1)(2m+1)/6$ by evaluating

$$\left(x\frac{d}{dx}\right)^2 \sum_{n=0}^{m} x^n = \left(x\frac{d}{dx}\right)^2 \frac{x^{m+1}-1}{x-1}$$

at $x = 1$.

6. A family of discrete density functions $p_n(\theta)$ defined on $\{0, 1, \ldots\}$ and indexed by a parameter $\theta > 0$ is said to be a power series family if for all n

$$p_n(\theta) = \frac{c_n \theta^n}{g(\theta)}, \tag{2.13}$$

where $c_n \geq 0$, and where $g(\theta) = \sum_{k=0}^{\infty} c_k \theta^k$ is the appropriate normalizing constant. Show that the mean $\mu(\theta)$ and variance $\sigma^2(\theta)$ of the $p_n(\theta)$ reduce to

$$\begin{aligned}\mu(\theta) &= \frac{\theta g'(\theta)}{g(\theta)} \\ \sigma^2(\theta) &= \theta \mu'(\theta).\end{aligned}$$

7. Continuing Problem 6, suppose $X_1, \ldots, X_m$ is a random sample from the power series distribution (2.13). Show that $S_m = X_1 + \cdots + X_m$ follows a power series distribution with

$$\Pr(S_m = n) = \frac{a_{mn}\theta^n}{g(\theta)^m},$$

where a_{mn} is the coefficient of θ^n in $g(\theta)^m$. If $a_{mn} = 0$ for $n < 0$, then also prove that $a_{m,S_m - r}/a_{m,S_m}$ is an unbiased estimator of θ^r. This estimator is, in fact, the uniformly minimum variance, unbiased estimator of θ^r [6].

8. Suppose $f_n(x)$ and $F_n(x)$ represent, respectively, the density and distribution functions of a chi-square random variable with n degrees of freedom. The noncentral chi-square density [1] with noncentrality parameter 2λ and degrees of freedom n can be written as the Poisson mixture

$$f_{\lambda,n}(x) = \sum_{k=0}^{\infty} \frac{\lambda^k}{k!} e^{-\lambda} f_{n+2k}(x).$$

Show that

$$F_{n+2(k+1)}(x) = F_{n+2k}(x) - \frac{e^{-\frac{x}{2}} \left(\frac{x}{2}\right)^{\frac{n}{2}+k}}{\Gamma(\frac{n}{2}+k+1)}.$$

Hence, in evaluating the distribution function $F_{\lambda,n}(x)$ of $f_{\lambda,n}(x)$, it suffices to compute only the single incomplete gamma function $F_n(x)$. Prove the error estimate

$$\begin{aligned}0 &\leq F_{\lambda,n}(x) - \sum_{k=0}^{m} \frac{\lambda^k}{k!} e^{-\lambda} F_{n+2k}(x) \\ &\leq \left(1 - \sum_{k=0}^{m} \frac{\lambda^k}{k!} e^{-\lambda}\right) F_{n+2(m+1)}(x).\end{aligned}$$

9. To generalize Problem 8, consider the sum $S_N = \sum_{i=1}^{N} X_i$, where the summands X_i are independent, exponentially distributed random

variables with common mean $1/\nu$, and the number of summands N is a nonnegative, integer-valued random variable independent of the X_i. By definition, $S_N = 0$ when $N = 0$. If $\Pr(N = n) = p_n$, then show that

$$\begin{aligned}\Pr(S_N \le x) &= \sum_{n=0}^{\infty} p_n P(n, \nu x) \\ \mathrm{E}(e^{-\theta S_N}) &= \sum_{n=0}^{\infty} p_n \left(\frac{\nu}{\nu+\theta}\right)^n,\end{aligned}$$

where $P(n, x)$ is the incomplete gamma function. In view of Problem 8, how would you proceed in evaluating the distribution function $\Pr(S_N \le x)$? Finally, demonstrate that $\mathrm{E}(S_N) = \mathrm{E}(N)/\nu$ and $\mathrm{Var}(S_N) = [\mathrm{E}(N) + \mathrm{Var}(N)]/\nu^2$.

10. Prove the incomplete beta function identities

$$\begin{aligned}I_x(a,b) &= \frac{\Gamma(a+b)}{\Gamma(a+1)\Gamma(b)} x^a(1-x)^{b-1} + I_x(a+1, b-1), \quad b > 1 \\ I_x(a,b) &= \frac{\Gamma(a+b)}{\Gamma(a+1)\Gamma(b)} x^a(1-x)^{b} + I_x(a+1, b).\end{aligned}$$

These two relations form the basis of a widely used algorithm [7] for computing $I_x(a,b)$. (Hints: For the first, integrate by parts, and for the second, show that both sides have the same derivative.)

11. Suppose that Z has discrete density

$$\Pr(Z = j) = \binom{k+j-1}{j} p^k (1-p)^j,$$

where $k > 0$ and $0 < p < 1$. In other words, Z follows a negative binomial distribution counting failures, not total trials. Show that $\Pr(Z \le m) = I_p(k, m+1)$ regardless of whether k is an integer. (Hint: Use one of the identities of the previous problem.)

12. Let $X_{(k)}$ be the kth order statistic from a finite sequence $X_1, \ldots, X_n$ of independent, identically distributed random variables with common distribution function $F(x)$. Show that $X_{(k)}$ has distribution function $\Pr(X_{(k)} \le x) = I_{F(x)}(k, n-k+1)$.

13. Suppose the bivariate normal random vector $(X_1, X_2)^t$ has means $\mathrm{E}(X_i) = \mu_i$, variances $\mathrm{Var}(X_i) = \sigma_i^2$, and correlation ρ. Verify the decomposition

$$\begin{aligned}X_1 &= \sigma_1 |\rho|^{\frac{1}{2}} Y + \sigma_1 (1 - |\rho|)^{\frac{1}{2}} Z_1 + \mu_1 \\ X_2 &= \sigma_2 \operatorname{sgn}(\rho) |\rho|^{\frac{1}{2}} Y + \sigma_2 (1 - |\rho|)^{\frac{1}{2}} Z_2 + \mu_2,\end{aligned}$$

where Y, Z_1, and Z_2 are independent, standard normal random variables. Use this decomposition to deduce that

$$\begin{aligned}
&\Pr(X_1 \leq x_1, X_2 \leq x_2) \\
= \; & \frac{1}{\sqrt{2\pi}} \int_{-\infty}^{\infty} \Phi\left[\frac{x_1 - \mu_1 - \sigma_1 |\rho|^{\frac{1}{2}} y}{\sigma_1 (1 - |\rho|)^{\frac{1}{2}}}\right] \\
& \times \Phi\left[\frac{x_2 - \mu_2 - \sigma_2 \operatorname{sgn}(\rho) |\rho|^{\frac{1}{2}} y}{\sigma_2 (1 - |\rho|)^{\frac{1}{2}}}\right] e^{-\frac{y^2}{2}} dy,
\end{aligned}$$

where $\Phi(x)$ is the standard normal distribution [2].

2.9 References

[1] Bickel PJ, Doksum KA (1977) *Mathematical Statistics: Basic Ideas and Selected Topics.* Holden-Day, Oakland, CA

[2] Curnow RN, Dunnett CW (1962) The numerical evaluation of certain multivariate normal integrals. *Ann Math Stat* 33:571–579

[3] Feller W (1968) *An Introduction to Probability Theory and Its Applications, Volume 1*, 3rd ed. Wiley, New York

[4] Graham RL, Knuth DE, Patashnik O (1988) *Concrete Mathematics: A Foundation for Computer Science.* Addison-Wesley, Reading, MA

[5] Henrici P (1974) *Applied and Computational Complex Analysis, Volume 1.* Wiley, New York

[6] Lehmann EL (1991) *Theory of Point Estimation.* Wadsworth, Belmont, CA

[7] Majumder KL, Bhattacharjee GP (1973) Algorithm AS 63. The incomplete beta integral. *Appl Stat* 22:409–411

[8] Pourhamadi M (1984) Taylor expansion of $\exp(\sum_{k=0}^{\infty} a_k z^k)$ and some applications. *Amer Math Monthly* 91:303–307

[9] Press WH, Teukolsky SA, Vetterling WT, Flannery BP (1992) *Numerical Recipes in Fortran: The Art of Scientific Computing*, 2nd ed. Cambridge University Press, Cambridge

[10] Wilf HS (1990) *generatingfunctionology.* Academic Press, New York

3

Continued Fraction Expansions

3.1 Introduction

A continued fraction [2, 3, 4, 5] is a sequence of fractions

$$f_n = b_0 + \cfrac{a_1}{b_1 + \cfrac{a_2}{b_2 + \cfrac{a_3}{b_3 + \cdots + \cfrac{a_n}{b_n}}}} \tag{3.1}$$

formed from two sequences $a_1, a_2, \ldots$ and $b_0, b_1, \ldots$ of numbers. For typographical convenience, definition (3.1) is usually recast as

$$f_n = b_0 + \frac{a_1}{b_1+} \frac{a_2}{b_2+} \frac{a_3}{b_3+} \cdots \frac{a_n}{b_n}.$$

In many practical examples, the approximant f_n converges to a limit, which is typically written as

$$\lim_{n\to\infty} f_n = b_0 + \frac{a_1}{b_1+} \frac{a_2}{b_2+} \frac{a_3}{b_3+} \cdots.$$

Because the elements a_n and b_n of the two defining sequences can depend on a variable x, continued fractions offer an alternative to power series in expanding functions such as distribution functions. In fact, continued fractions can converge where power series diverge, and where both types of expansions converge, continued fractions often converge faster.

A lovely little example of a continued fraction is furnished by

$$\begin{aligned}
\sqrt{2} - 1 &= \frac{1}{2 + (\sqrt{2} - 1)} \\
&= \cfrac{1}{2 + \cfrac{1}{2 + (\sqrt{2} - 1)}} \\
&= \cfrac{1}{2 + \cfrac{1}{2 + \cfrac{1}{2 + (\sqrt{2} - 1)}}}.
\end{aligned}$$

K. Lange, *Numerical Analysis for Statisticians*, Statistics and Computing,
DOI 10.1007/978-1-4419-5945-4_3, © Springer Science+Business Media, LLC 2010

One can easily check numerically that the limit

$$\sqrt{2} = 1 + \frac{1}{2+}\frac{1}{2+}\frac{1}{2+}\cdots$$

is correct. It is harder to prove this analytically. For the sake of brevity, we will largely avoid questions of convergence. Readers interested in a full treatment of continued fractions can consult the references [2, 3, 5]. Problems 7 through 10 prove convergence when the sequences a_n and b_n are positive.

Before giving more examples, it is helpful to consider how we might go about evaluating the approximant f_n. One obvious possibility is to work from the bottom of the continued fraction (3.1) to the top. This obvious approach can be formalized by defining fractional linear transformations $t_0(x) = b_0 + x$ and $t_n(x) = a_n/(b_n + x)$ for $n > 0$. If the circle symbol $\circ$ denotes functional composition, then we take $x = 0$ and compute

$$\begin{aligned} t_n(0) &= \frac{a_n}{b_n} \\ t_{n-1} \circ t_n(0) &= \frac{a_{n-1}}{b_{n-1} + t_n(0)} \\ &\vdots \\ t_0 \circ t_1 \circ \cdots \circ t_n(0) &= f_n. \end{aligned}$$

This turns out to be a rather inflexible way to proceed because if we want the next approximant f_{n+1}, we are forced to start all over again. In 1655 J. Wallis [6] suggested an alternative strategy. (This is a venerable but often neglected subject.)

3.2 Wallis's Algorithm

According to Wallis,

$$t_0 \circ t_1 \circ \cdots \circ t_n(x) = \frac{A_{n-1}x + A_n}{B_{n-1}x + B_n} \tag{3.2}$$

for a certain pair of auxiliary sequences A_n and B_n. Taking $x = 0$ gives the approximant $f_n = A_n/B_n$. The sequences A_n and B_n satisfy the initial conditions

$$\begin{pmatrix} A_{-1} \\ B_{-1} \end{pmatrix} = \begin{pmatrix} 1 \\ 0 \end{pmatrix} \qquad \begin{pmatrix} A_0 \\ B_0 \end{pmatrix} = \begin{pmatrix} b_0 \\ 1 \end{pmatrix} \tag{3.3}$$

and for $n > 0$ the recurrence relation

$$\begin{pmatrix} A_n \\ B_n \end{pmatrix} = b_n \begin{pmatrix} A_{n-1} \\ B_{n-1} \end{pmatrix} + a_n \begin{pmatrix} A_{n-2} \\ B_{n-2} \end{pmatrix}. \tag{3.4}$$

From the initial conditions (3.3), it is clear that

$$t_0(x) = b_0 + x = \frac{A_{-1}x + A_0}{B_{-1}x + B_0}.$$

The general case of (3.2) is proved by induction. Suppose the formula is true for an arbitrary nonnegative integer n. Then the induction hypothesis and the recurrence relation (3.4) together imply

$$\begin{aligned}
t_0 \circ t_1 \circ \cdots \circ t_{n+1}(x) &= t_0 \circ t_1 \circ \cdots \circ t_n\Big(\frac{a_{n+1}}{b_{n+1} + x}\Big) \\
&= \frac{A_{n-1}\frac{a_{n+1}}{b_{n+1}+x} + A_n}{B_{n-1}\frac{a_{n+1}}{b_{n+1}+x} + B_n} \\
&= \frac{A_n x + (b_{n+1}A_n + a_{n+1}A_{n-1})}{B_n x + (b_{n+1}B_n + a_{n+1}B_{n-1})} \\
&= \frac{A_n x + A_{n+1}}{B_n x + B_{n+1}}.
\end{aligned}$$

3.3 Equivalence Transformations

The same continued fraction can be defined by more than one pair of sequences a_n and b_n. For instance, if $b_n \neq 0$ for all $n > 0$, then it is possible to concoct an equivalent continued fraction given by a pair of sequences a'_n and b'_n with $b'_0 = b_0$ and $b'_n = 1$ for all $n > 0$. This can be demonstrated most easily by defining the transformed auxiliary sequences

$$\begin{pmatrix} A'_n \\ B'_n \end{pmatrix} = \frac{1}{\prod_{k=1}^n b_k}\begin{pmatrix} A_n \\ B_n \end{pmatrix},$$

with the understanding that $\prod_{k=1}^n b_k = 1$ when $n = -1$ or 0. From this definition it follows that A'_n and B'_n satisfy the same initial conditions (3.3) as A_n and B_n. Furthermore, the recurrence relation (3.4) becomes

$$\begin{aligned}
\begin{pmatrix} A'_1 \\ B'_1 \end{pmatrix} &= \begin{pmatrix} A'_0 \\ B'_0 \end{pmatrix} + \frac{a_1}{b_1}\begin{pmatrix} A'_{-1} \\ B'_{-1} \end{pmatrix} \\
\begin{pmatrix} A'_n \\ B'_n \end{pmatrix} &= \begin{pmatrix} A'_{n-1} \\ B'_{n-1} \end{pmatrix} + \frac{a_n}{b_{n-1}b_n}\begin{pmatrix} A'_{n-2} \\ B'_{n-2} \end{pmatrix}, \qquad n > 1
\end{aligned}$$

after division by $\prod_{k=1}^n b_k$. Thus, the transformed auxiliary sequences correspond to the choice $a'_1 = a_1/b_1$ and $a'_n = a_n/(b_{n-1}b_n)$ for $n > 1$. By definition, the approximants $f'_n = A'_n/B'_n$ and $f_n = A_n/B_n$ coincide.

Faster convergence can often be achieved by taking the even part of a continued fraction. This is a new continued fraction whose approximant f'_n

equals the approximant f_{2n} of the original continued fraction. For the sake of simplicity, suppose that we start with a transformed continued fraction with all $b_n = 1$ for $n > 0$. We can then view the approximant f_n as the value at $x = 1$ of the iterated composition

$$s_1 \circ \cdots \circ s_n(x) \;=\; b_0 + \cfrac{a_1}{1 + \cfrac{a_2}{1 + \cfrac{a_3}{1 + \cdots + \cfrac{a_{n-1}}{1 + a_n x}}}}$$

of the n functions $s_1(x) = b_0 + a_1 x$ and $s_k(x) = 1/(1 + a_k x)$ for $2 \le k \le n$. If we compose these functions $r_n(x) = s_{2n-1} \circ s_{2n}(x)$ two by two, we get

$$\begin{aligned} r_1(x) &= b_0 + \frac{a_1}{1 + a_2 x}, \\ r_n(x) &= \cfrac{1}{1 + \cfrac{a_{2n-1}}{1 + a_{2n} x}} \\ &= 1 - \frac{a_{2n-1}}{1 + a_{2n-1} + a_{2n} x}, \qquad n > 1. \end{aligned}$$

The approximant f'_n of the new continued fraction is just

$$\begin{aligned} f'_n &= r_1 \circ \cdots \circ r_n(1) \\ &= b_0 + \cfrac{a_1}{1 + a_2 - \cfrac{a_2 a_3}{1 + a_3 + a_4 - \cdots - \cfrac{a_{2n-2} a_{2n-1}}{1 + a_{2n-1} + a_{2n}}}}. \end{aligned}$$

From this expansion of the even part, we read off the sequences $a'_1 = a_1$, $a'_n = -a_{2n-2}a_{2n-1}$ for $n > 1$, $b'_0 = b_0$, $b'_1 = 1 + a_2$, and $b'_n = 1 + a_{2n-1} + a_{2n}$ for $n > 1$.

3.4 Gauss's Expansion of Hypergeometric Functions

The hypergeometric function ${}_2F_1(a, b, c; x)$ is given by the power series

$$ {}_2F_1(a, b, c; x) \;=\; \sum_{n=0}^{\infty} \frac{a^{\overline{n}} b^{\overline{n}}}{c^{\overline{n}}} \frac{x^n}{n!}, \tag{3.5}$$

where $a^{\overline{n}}$, $b^{\overline{n}}$, and $c^{\overline{n}}$ are rising factorial powers defined by

$$a^{\overline{n}} \;=\; \begin{cases} 1 & n = 0 \\ a(a+1)\cdots(a+n-1) & n > 0, \end{cases}$$

and so forth. To avoid division by 0, the constant c in (3.5) should be neither 0 nor a negative integer. If either a or b is 0 or a negative integer, then the power series reduces to a polynomial. The binomial series

$$\begin{aligned}\frac{1}{(1-x)^a} &= \sum_{n=0}^{\infty}\binom{a+n-1}{n}x^n \\ &= \sum_{n=0}^{\infty} a^{\overline{n}}\frac{x^n}{n!} \\ &= {}_2F_1(a,1,1;x)\end{aligned}$$

and the incomplete beta function

$$\begin{aligned}I_x(a,b) &= \frac{\Gamma(a+b)}{\Gamma(a+1)\Gamma(b)}x^a(1-x)^b\sum_{n=0}^{\infty}\frac{(a+b)^{\overline{n}}}{(a+1)^{\overline{n}}}x^n \\ &= \frac{\Gamma(a+b)}{\Gamma(a+1)\Gamma(b)}x^a(1-x)^b\,{}_2F_1(a+b,1,a+1;x) \qquad (3.6)\end{aligned}$$

involve typical hypergeometric expansions. Straightforward application of the ratio test shows that the hypergeometric series (3.5) converges for all $|x| < 1$. As the binomial series makes clear, convergence can easily fail for $|x| \geq 1$.

In 1812 Gauss [1] described a method of converting ratios of hypergeometric functions into continued fractions. His point of departure was the simple identity

$$\frac{a^{\overline{n}}(b+1)^{\overline{n}}}{(c+1)^{\overline{n}}n!} - \frac{a^{\overline{n}}b^{\overline{n}}}{c^{\overline{n}}n!} = \frac{a(c-b)}{c(c+1)}\,\frac{(a+1)^{\overline{n-1}}(b+1)^{\overline{n-1}}}{(c+2)^{\overline{n-1}}(n-1)!}.$$

Multiplying this by x^n and summing on n yields the hypergeometric function identity

$$\begin{aligned}&{}_2F_1(a,b+1,c+1;x) - {}_2F_1(a,b,c;x) \\ &= \frac{a(c-b)x}{c(c+1)}\,{}_2F_1(a+1,b+1,c+2;x),\end{aligned}$$

which can be rewritten as

$$\begin{aligned}&\frac{{}_2F_1(a,b+1,c+1;x)}{{}_2F_1(a,b,c;x)} \\ &= \frac{1}{1-\frac{a(c-b)x}{c(c+1)}\cdot\frac{{}_2F_1(a+1,b+1,c+2;x)}{{}_2F_1(a,b+1,c+1;x)}}. \qquad (3.7)\end{aligned}$$

If in this derivation we interchange the roles of a and b and then replace b by $b+1$ and c by $c+1$, then we arrive at the similar identity

$$\begin{aligned}
&\frac{{}_2F_1(a+1,b+1,c+2;x)}{{}_2F_1(a,b+1,c+1;x)} \\
&= \frac{1}{1-\frac{(b+1)(c+1-a)x}{(c+1)(c+2)}\cdot\frac{{}_2F_1(a+1,b+2,c+3;x)}{{}_2F_1(a+1,b+1,c+2;x)}}.
\end{aligned} \tag{3.8}$$

Now the ratio (3.8) can be substituted for the ratio appearing in the denominator on the right of equation (3.7). Likewise, the ratio (3.7) with a, b, and c replaced by $a+1$, $b+1$, and $c+2$, respectively, can be substituted for the ratio appearing in the denominator on the right of equation (3.8). Alternating these successive substitutions produces Gauss's continued fraction expansion

$$\frac{{}_2F_1(a,b+1,c+1;x)}{{}_2F_1(a,b,c;x)} = \frac{1}{1+}\frac{d_1x}{1+}\frac{d_2x}{1+}\cdots \tag{3.9}$$

with

$$\begin{aligned}
d_{2n+1} &= -\frac{(a+n)(c-b+n)}{(c+2n)(c+2n+1)} \\
d_{2n+2} &= -\frac{(b+n+1)(c-a+n+1)}{(c+2n+1)(c+2n+2)}
\end{aligned}$$

for $n \geq 0$.

Gauss's expansion (3.9) is most useful when $b=0$, for then

$${}_2F_1(a,b,c;x) = 1.$$

For instance, the hypergeometric expansion of $(1-x)^{-a}$ has coefficients

$$\begin{aligned}
d_1 &= -a \\
d_{2n+1} &= -\frac{(a+n)}{2(2n+1)}, \qquad n \geq 1 \\
d_{2n+2} &= -\frac{(n+1-a)}{2(2n+1)}, \qquad n \geq 0.
\end{aligned}$$

In this example, note that the identity (3.7) continues to hold for $b=c=0$, provided ${}_2F_1(a,0,0;x)$ and the ratio $(c-b)/c$ are both interpreted as 1.

The hypergeometric function ${}_2F_1(a+b,1,a+1;x)$ determining the incomplete beta function (3.6) can be expanded with coefficients

$$\begin{aligned}
d_{2n+1} &= -\frac{(a+b+n)(a+n)}{(a+2n)(a+2n+1)} \\
d_{2n+2} &= -\frac{(n+1)(n+1-b)}{(a+2n+1)(a+2n+2)}
\end{aligned}$$

for $n \geq 0$. Press et al. [4] claim that this continued fraction expansion for the incomplete beta function is superior to the power series expansion (3.6) for all values of the argument x, provided one switches to the expansion of $I_{1-x}(b,a) = 1 - I_x(a,b)$ when $x > (a+1)/(a+b+2)$.

3.5 Expansion of the Incomplete Gamma Function

To expand the incomplete gamma function as a continued fraction, we take a detour and first examine the integral

$$J_x(a,b) = \frac{1}{\Gamma(a)} \int_0^\infty \frac{e^{-y} y^{a-1}}{(1+xy)^b} dy$$

for $a > 0$ and $x \geq 0$. This integral exhibits the surprising symmetry $J_x(a,b) = J_x(b,a)$. In fact, when both a and b are positive,

$$\begin{aligned} J_x(a,b) &= \frac{1}{\Gamma(a)} \int_0^\infty e^{-y} y^{a-1} \frac{1}{\Gamma(b)} \int_0^\infty e^{-z(1+xy)} z^{b-1} dz dy \\ &= \frac{1}{\Gamma(b)} \int_0^\infty e^{-z} z^{b-1} \frac{1}{\Gamma(a)} \int_0^\infty e^{-y(1+xz)} y^{a-1} dy dz \\ &= J_x(b,a). \end{aligned}$$

Because $J_x(a,0) = 1$ by definition of the gamma function, this symmetry relation yields $\lim_{a\to 0} J_x(a,b) = \lim_{a \to 0} J_x(b,a) = 1$. Thus, it is reasonable to define $J_x(0,b) = 1$ for $b > 0$.

To forge a connection to the incomplete gamma function, we consider $J_{x^{-1}}(1, 1-a)$. An obvious change of variables then implies

$$\begin{aligned} J_{x^{-1}}(1,1-a) &= \int_0^\infty e^{-y}\Big(1+\frac{y}{x}\Big)^{a-1} dy \\ &= x^{1-a} \int_0^\infty e^{-y}(x+y)^{a-1} dy \\ &= x^{1-a} e^x \int_x^\infty e^{-z} z^{a-1} dz. \end{aligned}$$

A final simple rearrangement gives

$$\frac{1}{\Gamma(a)} \int_0^x e^{-z} z^{a-1} dz = 1 - \frac{e^{-x} x^{a-1}}{\Gamma(a)} J_{x^{-1}}(1, 1-a). \tag{3.10}$$

The integral $J_x(a,b)$ also satisfies identities similar to equations (3.7) and (3.8) for the hypergeometric function. For instance,

$$J_x(a,b) = \frac{1}{\Gamma(a)} \int_0^\infty \frac{e^{-y} y^{a-1}(1+xy)}{(1+xy)^{b+1}} dy$$

$$\begin{aligned} &= J_x(a, b+1) + \frac{ax}{a\Gamma(a)} \int_0^\infty \frac{e^{-y} y^a}{(1+xy)^{b+1}} dy \\ &= J_x(a, b+1) + axJ_x(a+1, b+1) \end{aligned} \tag{3.11}$$

can be rearranged to give

$$\frac{J_x(a, b+1)}{J_x(a, b)} = \frac{1}{1 + ax\frac{J_x(a+1, b+1)}{J_x(a, b+1)}}. \tag{3.12}$$

Exploiting the symmetry $J_x(a, b) = J_x(b, a)$ when $b > 0$ or integrating by parts in general, we find

$$J_x(a, b+1) = J_x(a+1, b+1) + (b+1)xJ_x(a+1, b+2).$$

This in turn yields

$$\frac{J_x(a+1, b+1)}{J_x(a, b+1)} = \frac{1}{1 + (b+1)x\frac{J_x(a+1, b+2)}{J_x(a+1, b+1)}}. \tag{3.13}$$

Substituting equation (3.13) into equation (3.12) and vice versa in an alternating fashion leads to a continued fraction expansion of the form (3.9) for the ratio $J_x(a, b+1)/J_x(a, b)$. The coefficients of this expansion can be expressed as

$$\begin{aligned} d_{2n+1} &= a + n \\ d_{2n+2} &= b + n + 1 \end{aligned}$$

for $n \geq 0$. The special case $b = 0$ is important because $J_x(a, 0) = 1$.

If $a = 0$, then $J_x(0, b) = 1$, and it is advantageous to expand the continued fraction starting with identity (3.13) rather than identity (3.12). For example, the function $J_{x^{-1}}(1, 1-a)$ appearing in expression (3.10) for the incomplete gamma function can be expanded with coefficients

$$\begin{aligned} d_{2n+1} &= 1 - a + n \\ d_{2n+2} &= n + 1, \end{aligned}$$

provided we replace x in (3.9) by $1/x$, commence the continued fraction with identity (3.13), and take $a = 0$ and $b+1 = 1-a$. (See Problem 5.) Press et al. [4] recommend this continued fraction expansion for the incomplete gamma function on $x > a + 1$ and the previously discussed power series expansion on $x < a + 1$.

One subtle point in dealing with the case $a = 0$ is that we want the limiting value $J_x(0, b) = 1$ to hold for all b, not just for $b > 0$. To prove this slight extension, observe that iterating recurrence (3.11) leads to the representation

$$J_x(a,b) = J_x(a,b+n) + ax\sum_{k=1}^{n} p_k(x)J_x(a+k,b+n), \quad (3.14)$$

where the $p_k(x)$ are polynomials. If n is so large that $b+n>0$, then taking limits in (3.14) again yields $\lim_{a\to 0} J_x(a,b) = 1$.

3.6 Problems

1. Suppose a continued fraction has all $a_k = x$, $b_0 = 0$, and all remaining $b_k = 1-x$ for $|x| \neq 1$. Show that the nth approximant $f_n(x)$ satisfies

$$f_n(x) = \frac{x[1-(-x)^n]}{1-(-x)^{n+1}}.$$

Conclude that

$$\lim_{n\to\infty} f_n(x) = \begin{cases} x & |x|<1 \\ -1 & |x|>1. \end{cases}$$

Thus, the same continued fraction converges to two different analytic functions on two different domains.

2. Verify the identities

$$\begin{aligned} \ln(1-x) &= -x\,{}_2F_1(1,1,2;x) \\ \arctan(x) &= x\,{}_2F_1\Big(\frac{1}{2},1,\frac{3}{2};-x^2\Big) \\ \int_1^\infty \frac{e^{-xy}}{y^n}dy &= \frac{e^{-x}}{x}\int_0^\infty \frac{e^{-u}}{\left(1+\frac{u}{x}\right)^n}du. \end{aligned}$$

3. Find continued fraction expansions for each of the functions in the previous problem.

4. If ${}_1F_1(b,c;x) = \sum_{n=0}^\infty \frac{b^{\overline{n}}}{c^{\overline{n}}}\frac{x^n}{n!}$, then prove that

$${}_1F_1(b,c;x) = \lim_{a\to\infty} {}_2F_1\Big(a,b,c;\frac{x}{a}\Big). \quad (3.15)$$

Noting that $e^x = {}_1F_1(1,1;x)$, demonstrate that e^x has a continued fraction expansion given by the right-hand side of equation (3.9) with $d_{2n+1} = -(4n+2)^{-1}$ and $d_{2n+2} = -(4n+2)^{-1}$. (Hint: For the expansion of e^x, derive two recurrence relations by taking appropriate limits in equations (3.7) and (3.8).)

5. Check that the function $J_{x^{-1}}(1, 1-a)$ appearing in our discussion of the incomplete gamma function has the explicit expansion

$$J_{x^{-1}}(1,1-a) = x\Big(\frac{1}{x+}\,\frac{1-a}{1+}\,\frac{1}{x+}\,\frac{2-a}{1+}\,\frac{2}{x+}\cdots\Big).$$

Show that the even part of this continued fraction expansion amounts to

$$J_{x^{-1}}(1,1-a) = x\Big(\frac{1}{x+1-a-}\,\frac{1\cdot(1-a)}{x+3-a-}\,\frac{2\cdot(2-a)}{x+5-a-}\cdots\Big).$$

6. Lentz's method of evaluating the continued fraction (3.1) is based on using the ratios $C_n = A_n/A_{n-1}$ and $D_n = B_{n-1}/B_n$ and calculating f_n by $f_n = f_{n-1}C_nD_n$. This avoids underflows and overflows when the A_n or B_n tend to very small or very large values. Show that the ratios satisfy the recurrence relations

$$\begin{aligned} C_n &= b_n + \frac{a_n}{C_{n-1}} \\ D_n &= \frac{1}{b_n + a_n D_{n-1}}. \end{aligned}$$

7. Prove the determinant formulas

$$\det\begin{pmatrix} A_n & A_{n-1} \\ B_n & B_{n-1}\end{pmatrix} = (-1)^{n-1}\prod_{k=1}^{n} a_k \qquad (3.16)$$

$$\det\begin{pmatrix} A_{n+1} & A_{n-1} \\ B_{n+1} & B_{n-1}\end{pmatrix} = (-1)^{n-1}b_{n+1}\prod_{k=1}^{n} a_k \qquad (3.17)$$

in the notation of Wallis' algorithm.

8. Suppose the two sequences a_n and b_n generating a continued fraction have all elements nonnegative. Show that the approximants f_n satisfy $f_1 \geq f_3 \geq \cdots \geq f_{2n+1} \geq f_{2n} \geq \cdots \geq f_2 \geq f_0$. It follows that $\lim_{n\to\infty} f_{2n}$ and $\lim_{n\to\infty} f_{2n+1}$ exist, but unless further assumptions are made, there can be a gap between these two limits. (Hint: Use equation (3.17) from the previous problem to prove $f_{2n} \geq f_{2n-2}$ and $f_{2n+1} \leq f_{2n-1}$. Use equation (3.16) to prove $f_{2n+1} \geq f_{2n}$.)

9. Provided all $a_n \neq 0$, prove that the continued fraction (3.1) is also generated by the sequences $a_n' = 1$ and $b_n' = b_n \prod_{k=1}^n a_k^{(-1)^{n-k+1}}$.

10. Suppose that the sequences a_n and b_n are positive. The Stern-Stolz theorem [2] says that

$$\sum_{n=0}^{\infty} b_n \prod_{k=1}^{n} a_k^{(-1)^{n-k+1}} = \infty$$

is a necessary and sufficient condition for the convergence of the approximants f_n to the continued fraction (3.1). To prove the sufficiency of this condition, verify that:

(a) It is enough by the previous problem to take all $a_n = 1$.

(b) The approximants then satisfy

$$\begin{aligned} f_{2n+1} - f_{2n} &= \frac{A_{2n+1}}{B_{2n+1}} - \frac{A_{2n}}{B_{2n}} \\ &= \frac{A_{2n+1}B_{2n} - A_{2n}B_{2n+1}}{B_{2n}B_{2n+1}} \\ &= \frac{1}{B_{2n}B_{2n+1}} \end{aligned}$$

by the determinant formula (3.16).

(c) Because $B_n = b_n B_{n-1} + B_{n-2}$, the sequence B_n satisfies

$$\begin{aligned} B_{2n} &\geq B_0 \\ &= 1 \\ B_{2n+1} &\geq b_1. \end{aligned}$$

(d) The recurrence $B_n = b_n B_{n-1} + B_{n-2}$ and part (c) together imply

$$\begin{aligned} B_{2n} &\geq (b_{2n} + b_{2n-2} + \cdots + b_2)b_1 + 1 \\ B_{2n+1} &\geq b_{2n+1} + b_{2n-1} + \cdots + b_1; \end{aligned}$$

consequently, either $\lim_{n\to\infty} B_{2n} = \infty$, or $\lim_{n\to\infty} B_{2n+1} = \infty$.

(e) The sufficiency part of the theorem now follows from parts (b) and (d).

11. The Stieltjes function $F(x) = F(0)\int_0^\infty \frac{1}{1+xy}dG(y)$ plays an important role in the theoretical development of continued fractions [5]. Here $G(y)$ is an arbitrary probability distribution function concentrated on $[0,\infty)$ and $F(0) > 0$. In the region $\{x : x \neq 0,\ |\arg(x)| < \pi\}$ of the complex plane C excluding the negative real axis and 0, show that $F(x)$ has the following properties:

(a) $\frac{1}{x}F(\frac{1}{x}) = F(0)\int_0^\infty \frac{1}{x+y}dG(y)$,

(b) $F(x)$ is an analytic function,

(c) $\lim_{x\to 0} F(x) = F(0)$,

(d) The imaginary part of $F(x)$ satisfies

$$\text{Im}F(x) = \begin{cases} <0 & \text{Im}(x) > 0 \\ =0 & \text{Im}(x) = 0 \\ >0 & \text{Im}(x) < 0\,. \end{cases}$$

3.7 References

[1] Gauss CF (1812) Disquisitiones Generales circa Seriem Infinitam $1+\frac{\alpha\beta}{1\gamma}x+\frac{\alpha(\alpha+1)\beta(\beta+1)}{1\cdot 2\gamma(\gamma+1)}x^2+\frac{\alpha(\alpha+1)(\alpha+2)\beta(\beta+1)(\beta+2)}{1\cdot 2\cdot 3\gamma(\gamma+1)(\gamma+2)}x^3+$ etc. Pars prior, *Commentationes Societatis Regiae Scientiarium Gottingensis Recentiores* 2:1–46

[2] Jones WB, Thron WJ (1980) *Continued Fractions: Analytic Theory and Applications.* Volume 11 of *Encyclopedia of Mathematics and its Applications.* Addison-Wesley, Reading, MA

[3] Lorentzen L, Waadeland H (1992) *Continued Fractions with Applications.* North-Holland, Amsterdam

[4] Press WH, Teukolsky SA, Vetterling WT, Flannery BP (1992) *Numerical Recipes in Fortran: The Art of Scientific Computing*, 2nd ed. Cambridge University Press, Cambridge

[5] Wall HS (1948) *Analytic Theory of Continued Fractions.* Van Nostrand, New York

[6] Wallis J (1695) in *Opera Mathematica Volume 1.* Oxoniae e Theatro Shedoniano, reprinted by Georg Olms Verlag, Hildeshein, New York, 1972, p 355

4

Asymptotic Expansions

4.1 Introduction

Asymptotic analysis is a branch of mathematics dealing with the order of magnitude and limiting behavior of functions, particularly at boundary points of their domains of definition [1, 2, 4, 5, 7]. Consider, for instance, the function

$$f(x) = \frac{x^2+1}{x+1}.$$

It is obvious that $f(x)$ resembles the function x as $x \to \infty$. However, one can be more precise. The expansion

$$\begin{aligned} f(x) &= \frac{x^2+1}{x(1+\frac{1}{x})} \\ &= \Big(x+\frac{1}{x}\Big)\sum_{k=0}^{\infty}\Big(\frac{-1}{x}\Big)^k \\ &= x-1-2\sum_{k=1}^{\infty}\Big(\frac{-1}{x}\Big)^k \end{aligned}$$

indicates that $f(x)$ more closely resembles $x-1$ for large x. Furthermore, $f(x)-x+1$ behaves like $2/x$ for large x. We can refine the precision of the approximation by taking more terms in the infinite series. How far we continue in this and other problems is usually dictated by the application at hand.

4.2 Order Relations

Order relations are central to the development of asymptotic analysis. Suppose we have two functions $f(x)$ and $g(x)$ defined on a common interval I, which may extend to ∞ on the right or to $-\infty$ on the left. Let x_0 be either an internal point or a boundary point of I with $g(x) \neq 0$ for x close, but not equal, to x_0. Then the function $f(x)$ is said to be $O(g(x))$ if there exists a constant M such that $|f(x)| \leq M|g(x)|$ as $x \to x_0$. If $\lim_{x\to x_0} f(x)/g(x) = 0$, then $f(x)$ is said to be $o(g(x))$. Obviously, the relation $f(x) = o(g(x))$ implies the weaker relation $f(x) = O(g(x))$. Finally,

K. Lange, *Numerical Analysis for Statisticians*, Statistics and Computing,
DOI 10.1007/978-1-4419-5945-4_4, © Springer Science+Business Media, LLC 2010

if $\lim_{x\to x_0} f(x)/g(x) = 1$, then $f(x)$ is said to be asymptotic to $g(x)$. This is usually written $f(x) \asymp g(x)$. In many problems, the functions $f(x)$ and $g(x)$ are defined on the integers $\{1, 2, \ldots\}$ instead of on an interval I, and x_0 is taken as ∞.

For example, on $I = (1, \infty)$ one has $e^x = O(\sinh x)$ as $x \to \infty$ because

$$\begin{aligned} \frac{e^x}{\frac{e^x - e^{-x}}{2}} &= \frac{2}{1 - e^{-2x}} \\ &\le \frac{2}{1 - e^{-2}}. \end{aligned}$$

On $(0, \infty)$ one has $\sin^2 x = o(x)$ as $x \to 0$ because

$$\begin{aligned} \lim_{x\to 0} \frac{\sin^2 x}{x} &= \lim_{x\to 0} \sin x \lim_{x\to 0} \frac{\sin x}{x} \\ &= 0 \times 1. \end{aligned}$$

On $I = (0, \infty)$, our initial example can be rephrased as $(x^2+1)/(x+1) \asymp x$ as $x \to \infty$.

If $f(x)$ is bounded in a neighborhood of x_0, then we write $f(x) = O(1)$ as $x \to x_0$, and if $\lim_{x\to x_0} f(x) = 0$, we write $f(x) = o(1)$ as $x \to x_0$. The notation $f(x) = g(x) + O(h(x))$ means $f(x) - g(x) = O(h(x))$ and similarly for the o notation. For example,

$$\frac{x^2+1}{x+1} = x - 1 + O\Big(\frac{1}{x}\Big).$$

If $f(x)$ is differentiable at point x_0, then

$$f(x_0 + h) - f(x_0) = f'(x_0)h + o(h).$$

There are a host of miniature theorems dealing with order relations. Among these are

$$\begin{aligned} O(g) + O(g) &= O(g) \\ o(g) + o(g) &= o(g) \\ O(g_1)O(g_2) &= O(g_1 g_2) \\ o(g_1)O(g_2) &= o(g_1 g_2) \\ |O(g)|^\lambda &= O(|g|^\lambda), \quad \lambda > 0 \\ |o(g)|^\lambda &= o(|g|^\lambda), \quad \lambda > 0. \end{aligned}$$

4.3 Finite Taylor Expansions

One easy way of generating approximations to a function is via finite Taylor expansions. Suppose $f(x)$ has $n+1$ continuous derivatives near $x_0 = 0$. Then

$$f(x) = \sum_{k=0}^{n} \frac{1}{k!} f^{(k)}(0) x^k + O(x^{n+1})$$

as $x \to 0$. This order relation is validated by l'Hôpital's rule applied $n+1$ times to the quotient

$$\frac{f(x) - \sum_{k=0}^{n} \frac{1}{k!} f^{(k)}(0) x^k}{x^{n+1}}.$$

Of course, it is more informative to write the Taylor expansion with an explicit error term; for instance,

$$f(x) = \sum_{k=0}^{n} \frac{1}{k!} f^{(k)}(0) x^k + \frac{x^{n+1}}{n!} \int_0^1 f^{(n+1)}(tx)(1-t)^n dt. \quad (4.1)$$

This integral $\frac{x^{n+1}}{n!} \int_0^1 f^{(n+1)}(tx)(1-t)^n dt$ form of the remainder $R_n(x)$ after n terms can be derived by noting the recurrence relation

$$R_n(x) = -\frac{x^n}{n!} f^{(n)}(0) + R_{n-1}(x)$$

and the initial condition

$$R_0(x) = f(x) - f(0),$$

both of which follow from integration by parts. One virtue of formula (4.1) emerges when the derivatives of $f(x)$ satisfy $(-1)^k f^{(k)}(x) \geq 0$ for all $k > 0$. If this condition holds, then

$$\begin{aligned} 0 &\leq (-1)^{n+1} R_n(x) \\ &= \frac{x^{n+1}}{n!} \int_0^1 (-1)^{n+1} f^{(n+1)}(tx)(1-t)^n dt \\ &\leq \frac{x^{n+1}}{n!} (-1)^{n+1} f^{(n+1)}(0) \int_0^1 (1-t)^n dt \\ &= \frac{x^{n+1}}{(n+1)!} (-1)^{n+1} f^{(n+1)}(0) \end{aligned}$$

for any $x > 0$. In other words, the remainders $R_n(x)$ alternate in sign and are bounded in absolute value by the next term of the expansion. As an example, the function $f(x) = -\ln(1+x)$ satisfies the inequalities

$(-1)^k f^{(k)}(x) \geq 0$ and consequently also an infinity of Taylor expansion inequalities beginning with $0 \leq -\ln(1+x)+x \leq x^2/2$.

In large sample theory, finite Taylor expansions are invoked to justify asymptotic moment formulas for complicated random variables. The next proposition [6] is one species of a genus of results.

Proposition 4.3.1 *Let $X_1, X_2, \ldots$ be an i.i.d. sequence of random variables with common mean $\mathrm{E}(X_i) = \mu$ and variance $\mathrm{Var}(X_i) = \sigma^2$. Suppose that I is some interval with $\Pr(X_i \in I) = 1$ and that $m \geq 4$ is an even integer such that the first m moments of X_i exist. If $h(x)$ is any function whose mth derivative $h^{(m)}(x)$ is bounded on I, then the sample mean $A_n = \frac{1}{n}\sum_{i=1}^n X_i$ satisfies*

$$\mathrm{E}[h(A_n)] = h(\mu) + \frac{\sigma^2}{2n}h''(\mu) + O\Big(\frac{1}{n^2}\Big) \tag{4.2}$$

as $n \to \infty$. If $h(x)^2$ satisfies the same hypothesis as $h(x)$ with a possibly different m, then

$$\mathrm{Var}[h(A_n)] = \frac{\sigma^2}{n}h'(\mu)^2 + O\Big(\frac{1}{n^2}\Big) \tag{4.3}$$

as $n \to \infty$.

Proof: Let us begin by finding the order of magnitude of the kth moment μ_{nk} of the centered sum $S_n = \sum_{i=1}^n (X_i - \mu)$. We claim that μ_{nk} is a polynomial in n of degree $\lfloor \frac{k}{2} \rfloor$ or less, where $\lfloor \cdot \rfloor$ is the least integer function. This assertion is certainly true for $k \leq 2$ because $\mu_{n0} = 1$, $\mu_{n1} = 0$, and $\mu_{n2} = n\sigma^2$. The general case can be verified by letting c_j be the jth cumulant of $X_i - \mu$. Because a cumulant of a sum of independent random variables is the sum of the cumulants, nc_j is the jth cumulant of S_n. According to our analysis in Chapter 2, we can convert cumulants to moments via

$$\begin{aligned}
\mu_{nk} &= \sum_{j=0}^{k-1} \binom{k-1}{j} nc_{k-j}\mu_{nj} \\
&= \sum_{j=0}^{k-2} \binom{k-1}{j} nc_{k-j}\mu_{nj},
\end{aligned}$$

where the fact $c_1 = 0$ permits us to omit the last term in the sum. This formula and mathematical induction evidently imply that μ_{nk} is a polynomial in n whose degree satisfies

$$\begin{aligned}
\deg \mu_{nk} &\leq 1 + \max_{0 \leq j \leq k-2} \deg \mu_{nj} \\
&\leq 1 + \left\lfloor \frac{k-2}{2} \right\rfloor \\
&= \left\lfloor \frac{k}{2} \right\rfloor .
\end{aligned}$$

This calculation validates the claim.

Now consider the Taylor expansion

$$h(A_n) - \sum_{k=0}^{m-1} \frac{h^{(k)}(\mu)}{k!}(A_n - \mu)^k = \frac{h^{(m)}(\eta)}{m!}(A_n - \mu)^m \quad (4.4)$$

for η between A_n and μ. In view of the fact that $|h^{(m)}(A_n)| \leq b$ for some constant b and all possible values of A_n, taking expectations in equation (4.4) yields

$$\left| \mathrm{E}[h(A_n)] - \sum_{k=0}^{m-1} \frac{h^{(k)}(\mu)}{k!} \frac{\mu_{nk}}{n^k} \right| \leq \frac{b}{m!} \frac{\mu_{nm}}{n^m}. \quad (4.5)$$

Because μ_{nk} is a polynomial of degree at most $\lfloor k/2 \rfloor$ in n, the factor μ_{nk}/n^k is $O\left(n^{-k+\lfloor k/2 \rfloor}\right)$. This fact in conjunction with inequality (4.5) clearly gives the expansion (4.2).

If $h(x)^2$ satisfies the same hypothesis as $h(x)$, then

$$\mathrm{E}[h(A_n)^2] = h(\mu)^2 + \frac{\sigma^2}{2n} 2[h(\mu)h''(\mu) + h'(\mu)^2] + O\Big(\frac{1}{n^2}\Big).$$

Straightforward algebra now indicates that the difference

$$\mathrm{Var}[h(A_n)] = \mathrm{E}[h(A_n)^2] - \mathrm{E}[h(A_n)]^2$$

takes the form (4.3). ∎

The proposition is most easily applied if the X_i are bounded or $h(x)$ is a polynomial. For example, if the X_i are Bernoulli random variables with success probability p, then $h(A_n) = A_n(1 - A_n)$ is the maximum likelihood estimate of the Bernoulli variance $\sigma^2 = p(1-p)$. Proposition 4.3.1 implies

$$\begin{aligned} \mathrm{E}[A_n(1-A_n)] &= p(1-p) - \frac{p(1-p)2}{2n} \\ &= \Big(1 - \frac{1}{n}\Big)p(1-p) \\ \mathrm{Var}[A_n(1-A_n)] &= \frac{p(1-p)(1-2p)^2}{n} + O\Big(\frac{1}{n^2}\Big). \end{aligned}$$

The expression for the mean $\mathrm{E}[A_n(1-A_n)]$ is exact since the third and higher derivatives of $h(x) = x(1-x)$ vanish.

4.4 Expansions via Integration by Parts

Integration by parts often works well as a formal device for generating asymptotic expansions. Here are three examples.

4.4.1 Exponential Integral

Suppose Y has exponential density e^{-y} with unit mean. Given Y, let a point X be chosen uniformly from the interval $[0, Y]$. Then it is easy to show that X has density $E_1(x) = \int_x^\infty e^{-y}y^{-1}dy$ and distribution function $1 - e^{-x} + xE_1(x)$. To generate an asymptotic expansion of the exponential integral $E_1(x)$ as $x \to \infty$, one can repeatedly integrate by parts. This gives

$$\begin{aligned} E_1(x) &= -\frac{e^{-y}}{y}\bigg|_x^\infty - \int_x^\infty \frac{e^{-y}}{y^2}dy \\ &= \frac{e^{-x}}{x} + \frac{e^{-y}}{y^2}\bigg|_x^\infty + 2\int_x^\infty \frac{e^{-y}}{y^3}dy \\ &\vdots \\ &= e^{-x}\sum_{k=1}^n (-1)^{k-1}\frac{(k-1)!}{x^k} + (-1)^n n! \int_x^\infty \frac{e^{-y}}{y^{n+1}}dy. \end{aligned}$$

This is emphatically not a convergent series in powers of $1/x$. In fact, for any fixed x, we have $\lim_{k\to\infty} |(-1)^{(k-1)}(k-1)!/x^k| = \infty$.

Fortunately, the remainders $R_n(x) = (-1)^n n! \int_x^\infty e^{-y}y^{-n-1}dy$ alternate in sign and are bounded in absolute value by

$$\begin{aligned} |R_n(x)| &\leq \frac{n!}{x^{n+1}}\int_x^\infty e^{-y}dy \\ &= \frac{n!}{x^{n+1}}e^{-x}, \end{aligned}$$

the absolute value of the next term of the expansion. This suggests that we truncate the expansion when n is the largest integer with

$$\frac{\frac{n!}{x^{n+1}}e^{-x}}{\frac{(n-1)!}{x^n}e^{-x}} \leq 1.$$

In other words, we should choose $n \approx x$. If we include more terms, then the approximation degrades. This is in striking contrast to what happens with a convergent series.

Table 4.1 illustrates these remarks by tabulating a few representative values of the functions

$$\begin{aligned} I(x) &= xe^x E_1(x) \\ S_n(x) &= \sum_{k=1}^n (-1)^{k-1}\frac{(k-1)!}{x^{k-1}}. \end{aligned}$$

For larger values of x, the approximation noticeably improves. For instance, $I(10) = 0.91563$ while $S_{10}(10) = 0.91544$ and $I(100) = 0.99019 = S_4(100)$.

TABLE 4.1. Asymptotic Approximation of the Exponential Integral

x	$I(x)$	$S_1(x)$	$S_2(x)$	$S_3(x)$	$S_4(x)$	$S_5(x)$	$S_6(x)$
1	0.59634	1.0	0.0	2.0	-4.0		
2	0.72266	1.0	0.5	1.0	0.25	1.75	
3	0.78625	1.0	0.667	0.8999	0.6667	0.9626	0.4688
5	0.85212	1.0	0.8	0.88	0.8352	0.8736	0.8352

4.4.2 Incomplete Gamma Function

Repeated integration by parts of the right-tail probability of a gamma distributed random variable produces in the same manner

$$\begin{aligned} & \frac{1}{\Gamma(a)} \int_x^\infty y^{a-1} e^{-y} dy \\ = \;& x^a e^{-x} \sum_{k=1}^n \frac{1}{x^k \Gamma(a-k+1)} + \frac{1}{\Gamma(a-n)} \int_x^\infty y^{a-n-1} e^{-y} dy. \end{aligned}$$

If a is a positive integer, then the expansion stops at $n = a$ with remainder 0. Otherwise, if n is so large that $a - n - 1$ is negative, then the remainder satisfies

$$\left| \frac{1}{\Gamma(a-n)} \int_x^\infty y^{a-n-1} e^{-y} dy \right| \leq \left| \frac{1}{\Gamma(a-n)} \right| x^{a-n-1} e^{-x}.$$

Reasoning as above, we deduce that it is optimal to truncate the expansion when $|a-n|/x \approx 1$. The right-tail probability

$$\frac{1}{\sqrt{2\pi}} \int_x^\infty e^{-\frac{y^2}{2}} dy = \frac{1}{2\Gamma(\frac{1}{2})} \int_{\frac{x^2}{2}}^\infty z^{\frac{1}{2}-1} e^{-z} dz$$

of the standard normal random variable is covered by the special case $a = 1/2$ for $x > 0$; namely,

$$\frac{1}{\sqrt{2\pi}} \int_x^\infty e^{-\frac{y^2}{2}} dy = \frac{e^{-\frac{x^2}{2}}}{x\sqrt{2\pi}} \Big(1 - \frac{1}{x^2} + \frac{3}{x^4} - \frac{3 \cdot 5}{x^6} + \cdots \Big).$$

4.4.3 Laplace Transforms

The Laplace transform of a function $f(x)$ is defined by

$$\hat{f}(\lambda) = \int_0^\infty e^{-\lambda x} f(x) dx.$$

Repeated integration by parts yields

$$\begin{aligned}\hat{f}(\lambda) &= \frac{f(0)}{\lambda} + \frac{1}{\lambda}\int_0^\infty e^{-\lambda x} f'(x)dx \\ &\vdots \\ &= \sum_{k=0}^{n} \frac{f^{(k)}(0)}{\lambda^{k+1}} + \frac{1}{\lambda^{n+1}}\int_0^\infty e^{-\lambda x} f^{(n+1)}(x)dx,\end{aligned}$$

provided $f(x)$ is sufficiently well behaved that the required derivatives $f^{(k)}(0)$ and integrals $\int_0^\infty e^{-\lambda x}|f^{(k)}(x)|dx$ exist. The remainder satisfies

$$\lambda^{-n-1}\int_0^\infty e^{-\lambda x} f^{(n+1)}(x)dx = o(\lambda^{-n-1})$$

as $\lambda \to \infty$. Watson's lemma significantly generalizes this result [7].

4.5 General Definition of an Asymptotic Expansion

The previous examples suggest Poincaré's definition of an asymptotic expansion. Let $\phi_n(x)$ be a sequence of functions such that $\phi_{n+1}(x) = o(\phi_n(x))$ as $x \to x_0$. Then $\sum_{k=1}^\infty c_k\phi_k(x)$ is an asymptotic expansion for $f(x)$ if $f(x) = \sum_{k=1}^n c_k\phi_k(x) + o(\phi_n(x))$ holds as $x \to x_0$ for every $n \geq 1$. The constants c_n are uniquely determined by the limits

$$c_n = \lim_{x\to x_0} \frac{f(x) - \sum_{k=1}^{n-1} c_k\phi_k(x)}{\phi_n(x)}$$

taken recursively starting with $c_1 = \lim_{x\to x_0} f(x)/\phi_1(x)$. Implicit in this definition is the assumption that $\phi_n(x) \neq 0$ for x close, but not equal, to x_0.

4.6 Laplace's Method

Laplace's method gives asymptotic approximations for integrals

$$\int_c^d f(y)e^{-xg(y)}dy \tag{4.6}$$

depending on a parameter x as $x \to \infty$. Here the boundary points c and d can be finite or infinite. There are two cases of primary interest. If c is finite, and the minimum of $g(y)$ occurs at c, then the contributions to the integral around c dominate as $x \to \infty$. Without loss of generality, let us take $c = 0$

and $d = \infty$. (If d is finite, then we can extend the range of integration by defining $f(x) = 0$ to the right of d.) Now the supposition that the dominant contributions occur around 0 suggests that we can replace $f(y)$ by $f(0)$ and $g(y)$ by its first-order Taylor expansion $g(y) \approx g(0) + g'(0)y$. Making these substitutions leads us to conjecture that

$$\begin{aligned} \int_0^\infty f(y)e^{-xg(y)}\,dy &\asymp f(0)e^{-xg(0)} \int_0^\infty e^{-xyg'(0)}\,dy \\ &= \frac{f(0)e^{-xg(0)}}{xg'(0)}. \end{aligned} \tag{4.7}$$

In essence, we have reduced the integral to integration against the exponential density with mean $[xg'(0)]^{-1}$. As this mean approaches 0, the approximation becomes better and better. Under the weaker assumption that $f(y) \asymp ay^{b-1}$ as $y \to 0$ for $b > 0$, the integral (4.6) can be replaced by an integral involving a gamma density. In this situation,

$$\int_0^\infty f(y)e^{-xg(y)}\,dy \asymp \frac{a\Gamma(b)e^{-xg(0)}}{[xg'(0)]^b} \tag{4.8}$$

as $x \to \infty$.

The other case occurs when $g(y)$ assumes its minimum at an interior point, say 0, between, say, $c = -\infty$ and $d = \infty$. Now we replace $g(y)$ by its second-order Taylor expansion $g(y) = g(0) + \frac{1}{2}g''(0)y^2 + o(y^2)$. Assuming that the region around 0 dominates, we conjecture that

$$\begin{aligned} \int_{-\infty}^\infty f(y)e^{-xg(y)}\,dy &\asymp f(0)e^{-xg(0)} \int_{-\infty}^\infty e^{-\frac{xg''(0)y^2}{2}}\,dy \\ &= f(0)e^{-xg(0)}\sqrt{\frac{2\pi}{xg''(0)}}. \end{aligned} \tag{4.9}$$

In other words, we reduce the integral to integration against the normal density with mean 0 and variance $[xg''(0)]^{-1}$. As this variance approaches 0, the approximation improves.

The asymptotic equivalences (4.8) and (4.9) and their generalizations constitute Laplace's method. Before rigorously stating and proving the second of these conjectures, let us briefly consider some applications.

4.6.1 Moments of an Order Statistic

Our first application of Laplace's method involves a problem in order statistics. Let $X_1, \ldots, X_n$ be i.i.d. positive, random variables with common distribution function $F(x)$. We assume that $F(x) \asymp ax^b$ as $x \to 0$. Now consider the first order statistic $X_{(1)} = \min_{1 \le i \le n} X_i$. One can express the kth moment of $X_{(1)}$ in terms of its right-tail probability

$$\Pr(X_{(1)} > x) = [1 - F(x)]^n$$

as

$$\begin{aligned}
\mathrm{E}(X_{(1)}^k) &= k\int_0^\infty x^{k-1}[1-F(x)]^n dx \\
&= k\int_0^\infty x^{k-1}e^{n\ln[1-F(x)]}dx \\
&= \frac{k}{b}\int_0^\infty u^{\frac{k}{b}-1}e^{n\ln[1-F(u^{\frac{1}{b}})]}du,
\end{aligned}$$

where the last integral arises from the change of variable $u = x^b$. Now the function $g(u) = -\ln[1-F(u^{\frac{1}{b}})]$ has its minimum at $u = 0$, and an easy calculation invoking the assumption $F(x) \asymp ax^b$ yields $g(u) \asymp au$ as $u \to 0$. Hence, the first form (4.8) of Laplace's method implies

$$\mathrm{E}(X_{(1)}^k) \asymp \frac{k\Gamma(\frac{k}{b})}{b(na)^{\frac{k}{b}}}. \tag{4.10}$$

This asymptotic equivalence has an amusing consequence for a birthday problem. Suppose that people are selected one by one from a large crowd until two of the chosen people share a birthday. We would like to know how many people are selected on average before a match occurs. One way of conceptualizing this problem is to imagine drawing people at random times dictated by a Poisson process with unit intensity. The expected time until the first match then coincides with the expected number of people drawn [3]. Since the choice of a birthday from the available $n = 365$ days of the year is made independently for each random draw, we are in effect watching the evolution of n independent Poisson processes, each with intensity $1/n$.

Let X_i be the time when the second random point happens in the ith process. The time when the first birthday match occurs in the overall process is $X_{(1)} = \min_{1\le i\le n} X_i$. Now X_i has right-tail probability

$$\Pr(X_i > x) = \Big(1+\frac{x}{n}\Big)e^{-\frac{x}{n}}$$

because 0 or 1 random points must occur on $[0, x]$ in order for $X_i > x$. It follows that X_i has distribution function

$$\begin{aligned}
\Pr(X_i \le x) &= 1-\Big(1+\frac{x}{n}\Big)e^{-\frac{x}{n}} \\
&\asymp \frac{x^2}{2n^2},
\end{aligned}$$

and according to our calculation (4.10) with $a = 1/(2n^2)$, $b = 2$, and $k = 1$,

$$\mathrm{E}(X_{(1)}) \asymp \frac{\Gamma(\frac{1}{2})}{2(n\frac{1}{2n^2})^{\frac{1}{2}}} = \frac{1}{2}\sqrt{2\pi n}.$$

For $n = 365$ we get $\mathrm{E}(X_{(1)}) \approx 23.9$, a reasonably close approximation to the true value of 24.6.

4.6.2 Stirling's Formula

The behavior of the gamma function

$$\Gamma(\lambda) = \int_0^\infty y^{\lambda-1}e^{-y}dy$$

as $\lambda \to \infty$ can be ascertained by Laplace's method. If we define $z = y/\lambda$, then

$$\Gamma(\lambda+1) = \lambda^{\lambda+1}\int_0^\infty e^{-\lambda g(z)}dz$$

for the function $g(z) = z - \ln z$, which has its minimum at $z = 1$. Applying Laplace's second approximation (4.9) at $z = 1$ gives Stirling's asymptotic formula

$$\Gamma(\lambda+1) \asymp \sqrt{2\pi}\lambda^{\lambda+\frac{1}{2}}e^{-\lambda}$$

as $\lambda \to \infty$.

4.6.3 Posterior Expectations

In Bayesian calculations one is often confronted with the need to evaluate the posterior expectation

$$\frac{\int e^{h(\theta)}e^{l_n(\theta)+\pi(\theta)}d\theta}{\int e^{l_n(\theta)+\pi(\theta)}d\theta} \tag{4.11}$$

of some function $e^{h(\theta)}$ of the parameter θ. In formula (4.11), $\pi(\theta)$ is the logprior and $l_n(\theta)$ is the loglikelihood of n observations. If n is large and the observations are independent, then usually the logposterior $l_n(\theta)+\pi(\theta)$ is sharply peaked in the vicinity of the posterior mode $\hat{\theta}$.

In the spirit of Laplace's method, this suggests that the denominator in (4.11) can be approximated by

$$\begin{aligned}\int e^{l_n(\theta)+\pi(\theta)}d\theta &\approx e^{l_n(\hat{\theta})+\pi(\hat{\theta})}\int e^{\frac{1}{2}[l_n''(\hat{\theta})+\pi''(\hat{\theta})](\theta-\hat{\theta})^2}d\theta \\ &= e^{l_n(\hat{\theta})+\pi(\hat{\theta})}\sqrt{\frac{2\pi}{-[l_n''(\hat{\theta})+\pi''(\hat{\theta})]}}.\end{aligned}$$

If we also approximate the numerator of (4.11) by expanding the sum $h(\theta)+l_n(\theta)+\pi(\theta)$ around its maximum point $\tilde{\theta}$, then the ratio (4.11) can be approximated by

$$\begin{aligned}&\frac{\int e^{h(\theta)}e^{l_n(\theta)+\pi(\theta)}d\theta}{\int e^{l_n(\theta)+\pi(\theta)}d\theta} \\ &\approx e^{[h(\tilde{\theta})+l_n(\tilde{\theta})+\pi(\tilde{\theta})-l_n(\hat{\theta})-\pi(\hat{\theta})]}\sqrt{\frac{l_n''(\hat{\theta})+\pi''(\hat{\theta})}{h''(\tilde{\theta})+l_n''(\tilde{\theta})+\pi''(\tilde{\theta})}}.\end{aligned} \tag{4.12}$$

The major virtue of this approximation is that it substitutes optimization for integration. The approximation extends naturally to multidimensional settings, where the difficulty of integration is especially acute. Tierney and Kadane [8] provide a detailed analysis of the order of magnitude of the errors committed in using formula (4.12).

4.7 Validation of Laplace's Method

Here we undertake a formal proof of the second Laplace asymptotic formula (4.9). Proof of the first formula (4.7) is similar.

Proposition 4.7.1 *If the conditions*

(a) for every $\delta > 0$ there exists a $\rho > 0$ with $g(y) - g(0) \geq \rho$ for $|y| \geq \delta$,

(b) $g(y)$ is twice continuously differentiable in a neighborhood of 0 and $g''(0) > 0$,

(c) $f(y)$ is continuous in a neighborhood of 0 and $f(0) > 0$,

(d) the integral $\int_{-\infty}^{\infty} f(y)e^{-xg(y)}\,dy$ is absolutely convergent for $x \geq x_1$,

are satisfied, then the asymptotic relation (4.9) obtains.

Proof: By multiplying both sides of the asymptotic relation (4.9) by $e^{xg(0)}$, we can assume without loss of generality that $g(0) = 0$. Because $g(y)$ has its minimum at $y = 0$, l'Hôpital's rule implies $g(y) - \frac{1}{2}g''(0)y^2 = o(y^2)$ as $y \to 0$. Now let a small $\epsilon > 0$ be given, and choose $\delta > 0$ sufficiently small so that the inequalities

$$\begin{aligned}
(1-\epsilon)f(0) &\leq f(y) \\
&\leq (1+\epsilon)f(0) \\
\left|g(y) - \frac{1}{2}g''(0)y^2\right| &\leq \epsilon y^2
\end{aligned}$$

hold for $|y| \leq \delta$. Assumption (a) guarantees the existence of a $\rho > 0$ with $g(y) \geq \rho$ for $|y| \geq \delta$.

We next show that the contributions to the Laplace integral from the region $|y| \geq \delta$ are negligible as $x \to \infty$. Indeed, for $x \geq x_1$,

$$\begin{aligned}
\left|\int_{\delta}^{\infty} f(y)e^{-xg(y)}\,dy\right| &\leq \int_{\delta}^{\infty} |f(y)|e^{-(x-x_1)g(y)}e^{-x_1 g(y)}\,dy \\
&\leq e^{-(x-x_1)\rho}\int_{\delta}^{\infty} |f(y)|e^{-x_1 g(y)}\,dy \\
&= O(e^{-\rho x}).
\end{aligned}$$

Likewise, $\int_{-\infty}^{-\delta} f(y)e^{-xg(y)}\,dy = O(e^{-\rho x})$.

Owing to our choice of δ, the central portion of the integral satisfies

$$\int_{-\delta}^{\delta} f(y)e^{-xg(y)}\,dy \;\leq\; (1+\epsilon)f(0)\int_{-\delta}^{\delta} e^{-\frac{x}{2}[g''(0)-2\epsilon]y^2}\,dy.$$

Duplicating the above reasoning,

$$\int_{-\infty}^{-\delta} e^{-\frac{x}{2}[g''(0)-2\epsilon]y^2}\,dy + \int_{\delta}^{\infty} e^{-\frac{x}{2}[g''(0)-2\epsilon]y^2}\,dy \;=\; O(e^{-\omega x}),$$

where $\omega = \frac{1}{2}[g''(0) - 2\epsilon]\delta^2$. Thus,

$$\begin{aligned} & (1+\epsilon)f(0)\int_{-\delta}^{\delta} e^{-\frac{x}{2}[g''(0)-2\epsilon]y^2}\,dy \\ = \;& (1+\epsilon)f(0)\int_{-\infty}^{\infty} e^{-\frac{x}{2}[g''(0)-2\epsilon]y^2}\,dy + O(e^{-\omega x}) \\ = \;& (1+\epsilon)f(0)\sqrt{\frac{2\pi}{x[g''(0)-2\epsilon]}} + O(e^{-\omega x}). \end{aligned}$$

Assembling all of the relevant pieces, we now conclude that

$$\begin{aligned} \int_{-\infty}^{\infty} f(y)e^{-xg(y)}\,dy \;\leq\; & (1+\epsilon)f(0)\sqrt{\frac{2\pi}{x[g''(0)-2\epsilon]}} \\ & + O(e^{-\rho x}) + O(e^{-\omega x}). \end{aligned}$$

Hence,

$$\limsup_{x\to\infty} \sqrt{x}\int_{-\infty}^{\infty} f(y)e^{-xg(y)}\,dy \;\leq\; (1+\epsilon)f(0)\sqrt{\frac{2\pi}{[g''(0)-2\epsilon]}},$$

and sending $\epsilon \to 0$ produces

$$\limsup_{x\to\infty} \sqrt{x}\int_{-\infty}^{\infty} f(y)e^{-xg(y)}\,dy \;\leq\; f(0)\sqrt{\frac{2\pi}{g''(0)}}.$$

A similar argument gives

$$\liminf_{x\to\infty} \sqrt{x}\int_{-\infty}^{\infty} f(y)e^{-xg(y)}\,dy \;\geq\; f(0)\sqrt{\frac{2\pi}{g''(0)}}$$

and proves the proposition. ■

4.8 Problems

1. Prove the following order relations:

   a) $1-\cos^2 x = O(x^2)$ as $x \to 0$,

   b) $\ln x = o(x^\alpha)$ as $x \to \infty$ for any $\alpha > 0$,

   c) $\frac{x^2}{1+x^3} + \ln(1+x^2) = O(x^2)$ as $x \to 0$,

   d) $\frac{x^2}{1+x^3} + \ln(1+x^2) = O(\ln x)$ as $x \to \infty$.

2. Show that $f(x) \asymp g(x)$ as $x \to x_0$ does not entail the stronger relation $e^{f(x)} \asymp e^{g(x)}$ as $x \to x_0$. Argue that the condition $f(x) = g(x) + o(1)$ is sufficient to imply $e^{f(x)} \asymp e^{g(x)}$.

3. For two positive functions $f(x)$ and $g(x)$, prove that $f(x) \asymp g(x)$ as $x \to x_0$ implies $\ln f(x) = \ln g(x) + o(1)$ as $x \to x_0$. Hence, $\lim_{x\to x_0} \ln f(x) \neq 0$ entails $\ln f(x) \asymp \ln g(x)$ as $x \to x_0$.

4. Suppose in Proposition 4.3.1 we replace $h(A_n)$ by $h(c_n A_n)$, where the sequence of constants $c_n = 1 + an^{-1} + O(n^{-2})$. How does this change the right hand sides of the asymptotic expressions (4.2) and (4.3)?

5. Continuing Problem 4, derive asymptotic expressions for the mean and variance of $\Phi\left[(u - A_n)\sqrt{n/(n-1)}\right]$, where u is a constant, A_n is the sample mean of a sequence $X_1, \ldots, X_n$ of i.i.d. normal random variables with mean μ and variance 1, and $\Phi(x)$ is the standard normal distribution function. The statistic $\Phi\left[(u - A_n)\sqrt{n/(n-1)}\right]$ is the uniformly minimum variance unbiased estimator of the percentile $p = \Phi(X_i \leq u)$ [6].

6. Find an asymptotic expansion for $\int_x^\infty e^{-y^4} dy$ as $x \to \infty$.

7. Suppose that $0 < c < \infty$ and that $f(x)$ is bounded and continuous on $[0, c]$. If $f(c) \neq 0$, then show that

$$\int_0^c x^n f(x) dx \asymp \frac{c^{n+1}}{n} f(c)$$

   as $n \to \infty$.

8. Let $F(x)$ be a distribution function concentrated on $[0, \infty)$ with moments $m_k = \int_0^\infty y^k dF(y)$. For $x \geq 0$ define the Stieltjes function $f(x) = \int_0^\infty \frac{1}{1+xy} dF(y)$. Show that $\sum_{k=0}^\infty (-1)^k m_k x^k$ is an asymptotic expansion for $f(x)$ satisfying

$$f(x) - \sum_{k=0}^n (-1)^k m_k x^k = (-x)^{n+1} \int_0^\infty \frac{y^{n+1}}{1+xy} dF(y).$$

Argue, therefore, that the remainders of the expansion alternate in sign and are bounded in absolute value by the first omitted term.

9. Show that $\int_0^\infty \frac{e^{-y}}{1+xy}dy \asymp \frac{\ln x}{x}$ as $x \to \infty$. (Hints: Write

$$\int_0^\infty \frac{e^{-y}}{1+xy}dy = \frac{1}{x}\int_0^\infty \frac{d}{dy}\ln(1+xy)e^{-y}dy,$$

and use integration by parts and the dominated convergence theorem.)

10. Prove that

$$\int_0^{\frac{\pi}{2}} e^{-x\tan y}dy \asymp \frac{1}{x}$$

$$\int_{-\frac{\pi}{2}}^{\frac{\pi}{2}} (y+2)e^{-x\cos y}dy \asymp \frac{4}{x}$$

as $x \to \infty$.

11. For $0 < \lambda < 1$, demonstrate the asymptotic equivalence

$$\sum_{k=0}^{n} \binom{n}{k} k! n^{-k}\lambda^k \asymp \frac{1}{1-\lambda}$$

as $n \to \infty$. (Hint: Use the identity $k!n^{-k-1} = \int_0^\infty y^k e^{-ny}dy$.)

12. Demonstrate the asymptotic equivalence

$$\sum_{k=0}^{n} \binom{n}{k} k! n^{-k} \asymp \sqrt{\frac{\pi n}{2}}$$

as $n \to \infty$. (Hint: See Problem (11).)

13. The von Mises density

$$\frac{e^{\kappa\cos(y-\alpha)}}{2\pi I_0(\kappa)}, \quad -\pi < y \leq \pi,$$

is used to model random variation on a circle. Here α is a location parameter, $\kappa > 0$ is a concentration parameter, and the modified Bessel function $I_0(\kappa)$ is the normalizing constant

$$I_0(\kappa) = \frac{1}{2\pi}\int_{-\pi}^{\pi} e^{\kappa\cos y}dy.$$

Verify that Laplace's method yields

$$I_0(\kappa) \asymp \frac{e^\kappa}{\sqrt{2\pi\kappa}}$$

as $\kappa \to \infty$. For large κ it is clear that the von Mises distribution is approximately normal.

14. Let $\phi(x)$ and $\Phi(x)$ be the standard normal density and distribution functions. Demonstrate the bounds

$$\frac{x}{1+x^2}\phi(x) \ \leq \ 1-\Phi(x) \ \leq \ \frac{1}{x}\phi(x)$$

for $x > 0$. (Hints: Exploit the derivatives

$$\begin{aligned}\frac{d}{dx}e^{-x^2/2} &= -xe^{-x^2/2} \\ \frac{d}{dx}\left(x^{-1}e^{-x^2/2}\right) &= -\left(1+x^{-2}\right)e^{-x^2/2}\end{aligned}$$

and simple inequalities for the integral $1-\Phi(x)$.)

4.9 References

[1] Barndorff-Nielsen OE, Cox DR (1989) *Asymptotic Techniques for Use in Statistics.* Chapman & Hall, London

[2] Bender CM, Orszag SA (1978) *Advanced Mathematical Methods for Scientists and Engineers.* McGraw-Hill, New York

[3] Bloom G, Holst L, Sandell D (1994) *Problems and Snapshots from the World of Probability.* Springer, New York

[4] de Bruijn NG (1981) *Asymptotic Methods in Analysis.* Dover, New York

[5] Graham RL, Knuth DE, Patashnik O (1988) *Concrete Mathematics: A Foundation for Computer Science.* Addison-Wesley, Reading MA

[6] Lehmann EL (1991) *Theory of Point Estimation.* Wadsworth, Belmont, CA

[7] Murray JD (1984) *Asymptotic Analysis.* Springer, New York

[8] Tierney L, Kadane J (1986) Accurate approximations for posterior moments and marginal densities. *J Amer Stat Soc* 81:82–86

5

Solution of Nonlinear Equations

5.1 Introduction

Solving linear and nonlinear equations is a major preoccupation of applied mathematics and statistics. For nonlinear equations, closed-form solutions are the exception rather than the rule. Here we will concentrate on three simple techniques—bisection, functional iteration, and Newton's method—for solving equations in one variable. Insight into how these methods operate can be gained by a combination of theory and examples. Since functional iteration and Newton's method generalize to higher-dimensional problems, it is particularly important to develop intuition about their strengths and weaknesses. Equipped with this intuition, we can tackle harder problems with more confidence and understanding.

The last three sections of this chapter introduce the topics of minimization by golden section search and cubic interpolation. These two one-dimensional methods are also applicable as part of line-search algorithms in multidimensional optimization. They are paired here because they illustrate the tradeoffs in reliability and speed so commonly encountered in numerical analysis.

5.2 Bisection

Bisection is a simple, robust method of finding solutions to the equation $g(x) = 0$. In contrast to faster techniques such as Newton's method, no derivatives of $g(x)$ are required. Furthermore, under minimal assumptions on $g(x)$, bisection is guaranteed to converge to some root. Suppose that $g(x)$ is continuous, and an interval $[a, b]$ has been identified such that $g(a)$ and $g(b)$ are of opposite sign. If $g(x)$ is continuous, then the intermediate value theorem implies that $g(x)$ vanishes somewhere on $[a, b]$. Consider the midpoint $c = (a+b)/2$ of $[a, b]$. If $g(c) = 0$, then we are done. Otherwise, either $g(a)$ and $g(c)$ are of opposite sign, or $g(b)$ and $g(c)$ are of opposite sign. In the former case, the interval $[a, c]$ brackets a root; in the latter case,the interval $[c, b]$ does. In either case, we replace $[a, b]$ by the corresponding subinterval and continue. If we bisect $[a, b]$ a total of n times, then the final bracketing interval has length $2^{-n}(b - a)$. For n large enough, we can stop and approximate the bracketed root by the midpoint of the final bracketing

K. Lange, *Numerical Analysis for Statisticians*, Statistics and Computing,
DOI 10.1007/978-1-4419-5945-4_5, © Springer Science+Business Media, LLC 2010

interval. If we want to locate nearly all of the roots of $g(x)$ on $[a, b]$, then we can subdivide $[a, b]$ into many small adjacent intervals and apply bisection to each small interval in turn.

5.2.1 Computation of Quantiles by Bisection

Suppose we are given a continuous distribution function $F(x)$ and desire to find the α-quantile of $F(x)$. This amounts to solving the equation $g(x) = 0$ for $g(x) = F(x) - \alpha$. Bisection is applicable if we can find a bracketing interval to start the process. One strategy exploiting the monotonicity of $g(x)$ is to take an arbitrary initial point a and examine $g(a)$. If $g(a) < 0$, then we look for the first positive integer k with $g(a + k) > 0$. When this integer is found, the interval $[a + k - 1, a + k]$ brackets the α-quantile. If $g(a) > 0$, then we look for the first negative integer k such that $g(a+k) < 0$. In this case $[a + k, a + k + 1]$ brackets the α-quantile. Once a bracketing interval is found, bisection can begin. An obvious candidate for a is the mean. Instead of incrementing or decrementing by 1 in finding the initial bracketing interval, it usually is preferable to increment or decrement by the standard deviation of $F(x)$.

TABLE 5.1. Bracketing Intervals Given by Bisection

Iteration n	Interval	Iteration n	Interval
0	[1.291,2.582]	6	[1.997,2.017]
1	[1.936,2.582]	7	[2.007,2.017]
2	[1.936,2.259]	8	[2.012,2.017]
3	[1.936,2.098]	9	[2.015,2.017]
4	[1.936,2.017]	10	[2.015,2.016]
5	[1.977,2.017]	11	[2.015,2.015]

As a numerical example, consider the problem of calculating the .95–quantile of a t distribution with $n = 5$ degrees of freedom. A random variable with this distribution has mean 0 and standard deviation $\sqrt{n/(n-2)}$. Using the search tactic indicated above, we find an initial bracketing interval of $[1.291, 2.582]$. This and the subsequent bracketing intervals produced by bisection are noted in Table 5.1.

5.2.2 Shortest Confidence Interval

In forming a confidence interval or a Bayesian credible interval, it is natural to ask for the shortest interval $[a, b]$ with fixed content $H(b) - H(a) = \alpha$ for some probability distribution $H(x)$. This problem is not always well posed. To avoid logical difficulties, let us assume that $H(x)$ possesses a density $h(x)$ and ask for the region S_α of smallest Lebesgue measure $\mu(S_\alpha)$

satisfying $\int_{S_\alpha} h(x)dx = \alpha$. If $h(x)$ is unimodal, strictly increasing to the left of its mode, and strictly decreasing to the right of its mode, then S_α is a well-defined interval, and its Lebesgue measure is just its length.

In general, the reformulated problem makes sense in m-dimensional space R^m [1]. Its solution is given by

$$S_\alpha = \{x : h(x) \geq \lambda(\alpha)\}$$

for some number $\lambda(\alpha)$ depending on α. Such a number $\lambda(\alpha)$ exists if $\int_{\{x:h(x)=\lambda\}} h(x)dx = 0$ for all values of λ. In fact, if X is a random vector with density $h(x)$, then this condition guarantees that the right-tail probability

$$\Pr(h(X) \geq \lambda) = \int_{\{x:h(x)\geq\lambda\}} h(x)dx$$

is continuous and decreasing as a function of λ. In view of the intermediate value theorem, at least one λ must then qualify for each $\alpha \in (0,1)$.

The solution set S_α is unique, but only up to a set of Lebesgue measure 0. This can be checked by supposing that T also satisfies $\int_T h(x)dx = \alpha$. Subtracting this equation from the same equation for S_α yields

$$\int_{S_\alpha \setminus T} h(x)dx - \int_{T\setminus S_\alpha} h(x)dx = 0. \tag{5.1}$$

Because $h(x) \geq \lambda(\alpha)$ on $S_\alpha \setminus T$, it follows that

$$\int_{S_\alpha \setminus T} h(x)dx \geq \lambda(\alpha)\mu(S_\alpha \setminus T). \tag{5.2}$$

If $\mu(T \setminus S_\alpha) > 0$, it likewise follows that

$$\int_{T\setminus S_\alpha} h(x)dx < \lambda(\alpha)\mu(T \setminus S_\alpha). \tag{5.3}$$

Now if $\mu(T) \leq \mu(S_\alpha)$, then

$$\mu(T \setminus S_\alpha) \leq \mu(S_\alpha \setminus T). \tag{5.4}$$

The three inequalities (5.2), (5.3), and (5.4) are inconsistent with equality (5.1) unless $\mu(T \setminus S_\alpha) = 0$. But if $\mu(T \setminus S_\alpha) = 0$, then

$$\begin{aligned} \alpha &= \int_T h(x)dx \\ &= \int_{T\cap S_\alpha} h(x)dx \\ &< \int_{S_\alpha} h(x)dx \end{aligned}$$

unless $\mu(S_\alpha \setminus T) = 0$. Therefore, both $\mu(T \setminus S_\alpha)$ and $\mu(S_\alpha \setminus T)$ equal 0, and S_α and T differ by at most a set of measure 0.

As a concrete illustration of this principle, consider the problem of finding the shortest interval $[c, d]$ with a fixed probability $\int_c^d h(x)dx = \alpha$ for the gamma density $h(x) = \Gamma(a)^{-1}b^a x^{a-1}e^{-bx}$. Because

$$\frac{b^a}{\Gamma(a)} \int_c^d x^{a-1}e^{-bx}dx = \frac{1}{\Gamma(a)} \int_{bc}^{bd} z^{a-1}e^{-z}dz,$$

it suffices to take the scale constant $b = 1$. If $a \leq 1$, then the gamma density $h(x)$ is strictly decreasing in x, and the left endpoint of the shortest interval is given by $c = 0$. The right endpoint d can be found by bisection using our previously devised methods of evaluating the incomplete gamma function $P(a, x)$.

If the constant $a > 1$, $h(x)$ first increases and then decreases. Its modal value $\Gamma(a)^{-1}(a-1)^{a-1}e^{-(a-1)}$ occurs at $x = a-1$. One strategy for finding the shortest interval is to consider for each λ satisfying

$$0 < \lambda < \frac{1}{\Gamma(a)}(a-1)^{a-1}e^{-(a-1)}$$

the interval $[c_\lambda, d_\lambda]$ where $h(x) \geq \lambda$. The endpoints c_λ and d_λ are implicitly defined by $h(c_\lambda) = h(d_\lambda) = \lambda$ and can be found by bisection or Newton's method. Once $[c_\lambda, d_\lambda]$ is determined, the corresponding probability

$$\frac{1}{\Gamma(a)} \int_{c_\lambda}^{d_\lambda} x^{a-1}e^{-x}dx = P(a, d_\lambda) - P(a, c_\lambda)$$

can be expressed in terms of the incomplete gamma function. Thus, the original problem reduces to finding the particular λ satisfying

$$P(a, d_\lambda) - P(a, c_\lambda) = \alpha. \tag{5.5}$$

This λ can be straightforwardly computed by bisection. Note that this iterative process involves inner iterations to find c_λ and d_λ within each outer bisection iteration on λ.

Table 5.2 displays the endpoints c_λ and d_λ generated by the successive midpoints λ in a bisection scheme to find the particular λ satisfying equation (5.5) for $a = 2$ and $\alpha = 0.95$.

5.3 Functional Iteration

Suppose we are interested in finding a root of the equation $g(x) = 0$. If we let $f(x) = g(x) + x$, then this equation is trivially equivalent to the equation $x = f(x)$. In many examples, the iterates $x_n = f(x_{n-1})$ converge

TABLE 5.2. Bisection Iterates for the Shortest .95 Confidence Interval

Iteration n	c_λ	d_λ	**Iteration** n	c_λ	d_λ
1	0.2290	2.6943	9	0.0423	4.7669
2	0.1007	3.7064	10	0.0427	4.7559
3	0.0478	4.6198	11	0.0425	4.7614
4	0.0233	5.4844	12	0.0424	4.7642
5	0.0354	4.9831	13	0.0423	4.7656
6	0.0415	4.7893	14	0.0424	4.7649
7	0.0446	4.7019	15	0.0424	4.7652
8	0.0431	4.7449			

to a root of $g(x)$ starting from any point x_0 nearby. For obvious reasons, a root of $g(x)$ is said to be a fixed point of $f(x)$. Precise sufficient conditions for the existence of a unique fixed point of $f(x)$ and convergence to it are offered by the following proposition.

Proposition 5.3.1 *Suppose the function $f(x)$ defined on a closed interval I satisfies the conditions*

(a) $f(x) \in I$ whenever $x \in I$,

(b) $|f(y) - f(x)| \leq \lambda |y - x|$ for any two points x and y in I.

Then, provided the Lipschitz constant λ is in $[0, 1)$, $f(x)$ has a unique fixed point $x_\infty \in I$, and the functional iterates $x_n = f(x_{n-1})$ converge to x_∞ regardless of their starting point $x_0 \in I$. Furthermore, we have the precise error estimate

$$|x_n - x_\infty| \leq \frac{\lambda^n}{1-\lambda}|x_1 - x_0|. \tag{5.6}$$

Proof: The inequality

$$\begin{aligned} |x_{k+1} - x_k| &= |f(x_k) - f(x_{k-1})| \\ &\leq \lambda |x_k - x_{k-1}| \\ &\vdots \\ &\leq \lambda^k |x_1 - x_0| \end{aligned}$$

implies for $m > n$ the further inequality

$$\begin{aligned} |x_n - x_m| &\leq \sum_{k=n}^{m-1} |x_k - x_{k+1}| \\ &\leq \sum_{k=n}^{m-1} \lambda^k |x_1 - x_0| \end{aligned} \tag{5.7}$$

$$\leq \frac{\lambda^n}{1-\lambda}|x_1 - x_0|.$$

It follows from inequality (5.7) that x_n is a Cauchy sequence. Because the interval I is closed, the limit x_∞ of the sequence x_n exists in I. Invoking the continuity of $f(x)$ in the defining relation $x_n = f(x_{n-1})$ shows that x_∞ is a fixed point. Existence of a fixed point $y_\infty \neq x_\infty$ in I is incompatible with the inequality

$$\begin{aligned} |x_\infty - y_\infty| &= |f(x_\infty) - f(y_\infty)| \\ &\leq \lambda|x_\infty - y_\infty|. \end{aligned}$$

Finally, the explicit bound (5.6) follows from inequality (5.7) by sending m to ∞. ■

A function $f(x)$ having a Lipschitz constant $\lambda < 1$ is said to be contractive. In practice λ is taken to be any convenient upper bound of $|f'(x)|$ on the interval I. Such a choice is valid because of the mean value equality $f(x) - f(y) = f'(z)(x - y)$, where z is some number between x and y. In the vicinity of a fixed point x_∞ with $|f'(x_\infty)| < 1$, we can usually find a closed interval $I_d = [x_\infty - d, x_\infty + d]$ pertinent to the proposition. For instance, if $f(x)$ is continuously differentiable, then all sufficiently small, positive constants d yield $\lambda = \sup_{z \in I_d} |f'(z)| < 1$. Furthermore, $f(x)$ maps I_d into itself because

$$\begin{aligned} |f(x) - x_\infty| &= |f(x) - f(x_\infty)| \\ &\leq \lambda|x - x_\infty| \\ &\leq d \end{aligned}$$

for $x \in I_d$.

A fixed point x_∞ with $|f'(x_\infty)| < 1$ is said to be attractive. If x_∞ satisfies $f'(x_\infty) \in (-1, 0)$, then iterates $x_n = f(x_{n-1})$ converging to x_∞ eventually oscillate from side to side of x_∞. Convergence is eventually monotonic if $f'(x_\infty) \in (0, 1)$. If the inequality $|f'(x_\infty)| > 1$ holds, then the fixed point x_∞ is said to be repelling. Indeed, the mean value theorem implies in this situation that

$$\begin{aligned} |f(x) - x_\infty| &= |f'(z)(x - x_\infty)| \\ &> |x - x_\infty| \end{aligned}$$

for all x sufficiently close to x_∞. The case $|f'(x_\infty)| = 1$ is indeterminate and requires further investigation.

5.3.1 *Fractional Linear Transformations*

A fractional linear transformation $f(x) = (ax + b)/(cx + d)$ maps the complex plane into itself. For the sake of simplicity, let us assume that the complex constants a, b, c, and d are real and satisfy $ad - bc \neq 0$. Now consider the possibility of finding a real root of $x = f(x)$ by functional iteration. The solutions, if any, coincide with the two roots of the quadratic equation $cx^2 + (d - a)x - b = 0$. These roots can be expressed by the standard quadratic formula as

$$\begin{aligned} r_\pm &= \frac{-(d-a) \pm \sqrt{(d-a)^2 + 4bc}}{2c} \\ &= \frac{-(d-a) \pm \sqrt{(d+a)^2 - 4(ad-bc)}}{2c}. \end{aligned}$$

Both roots are purely real if and only if $(d + a)^2 \geq 4(ad - bc)$. Let us assume that this discriminant condition holds. Either root r is then locally attractive to the functional iterates, provided the derivative

$$\begin{aligned} f'(r) &= \frac{a}{cr+d} - \frac{(ar+b)c}{(cr+d)^2} \\ &= \frac{ad-bc}{(cr+d)^2} \end{aligned}$$

satisfies $|f'(r)| < 1$. It is locally repelling when $|f'(r)| > 1$.

Consider the product $(cr_+ + d)(cr_- + d) = ad - bc$. One of three things can happen. Either (a)

$$\begin{aligned} |cr_+ + d| &= |cr_- + d| \\ &= \sqrt{|ad-bc|}, \end{aligned}$$

or (b)

$$\begin{aligned} |cr_+ + d| &> \sqrt{|ad-bc|} \\ |cr_- + d| &< \sqrt{|ad-bc|}, \end{aligned}$$

or (c)

$$\begin{aligned} |cr_+ + d| &< \sqrt{|ad-bc|} \\ |cr_- + d| &> \sqrt{|ad-bc|}. \end{aligned}$$

Case (a) is indeterminate because $|f'(r_+)| = |f'(r_-)| = 1$. It turns out that functional iteration converges to the common root $r_+ = r_-$ when it exists [12]. Otherwise, case (a) leads to divergence unless the initial point is a root to begin with. In case (b) functional iteration converges locally to r_+ and diverges locally from r_-. This local behavior, in fact, holds globally [12]. In case (c), the opposite behavior relative to the two roots is observed. This analysis explains, for instance, why the continued fraction generated by the fractional linear transformation $f(x) = 1/(2+x)$ converges to $\sqrt{2}-1$ rather than to $-\sqrt{2} - 1$.

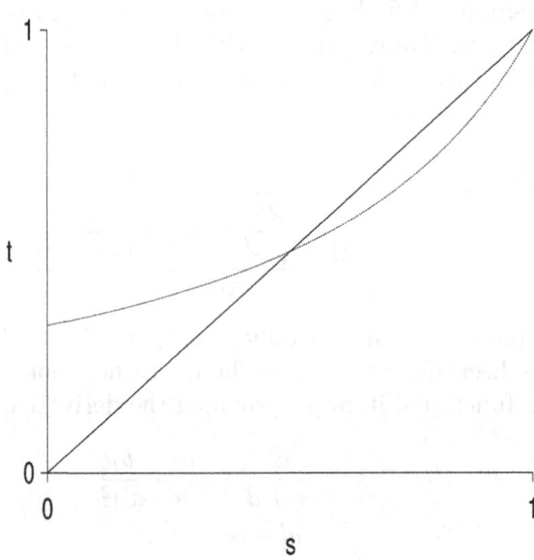

FIGURE 5.1. Intersection Points for a Supercritical Branching Process

5.3.2 *Extinction Probabilities by Functional Iteration*

In a branching process [3], particles reproduce independently at the end of each generation according to the same probabilistic law. Let p_k be the probability that a particle present at the current generation is replaced by k daughter particles at the next generation. Starting with a single particle at generation 0, we can ask for the probability s_∞ that the process eventually goes extinct. To characterize s_∞, we condition on the number of daughter particles k born to the initial particle. If extinction is to occur, then each line of descent emanating from a daughter particle must die out. If there are k daughter particles and consequently k lines of descent, then by independence of reproduction, all k lines of descent go extinct with probability s_∞^k. It follows that s_∞ satisfies the functional equation $s = \sum_{k=0}^\infty p_k s^k = P(s)$, where $P(s)$ is the generating function of the progeny distribution.

One can find the extinction probability by functional iteration starting at $s = 0$. Let s_n be the probability that extinction occurs in the branching process at or before generation n. Then $s_0 = 0$, $s_1 = p_0 = P(s_0)$, and, in general, $s_{n+1} = P(s_n)$. This recurrence relation can be deduced by again conditioning on the number of daughter particles in the first generation. If extinction is to occur at or before generation $n + 1$, then extinction must occur in n additional generations or sooner for each line of descent emanating from a daughter particle of the original particle.

On probabilistic grounds it is obvious that the sequence s_n increases monotonically to the extinction probability s_∞. To understand what is happening numerically, we need to know the number of fixed points of $s = P(s)$ and which fixed point is s_∞. Since $P''(s) = \sum_{k=2}^{\infty} k(k-1)p_k s^{k-2} \geq 0$, the curve $P(s)$ is convex. It starts at $P(0) = p_0 > 0$ above the diagonal line $t = s$. (Note that if $p_0 = 0$, then the process can never go extinct.) On the interval $[0, 1]$, the curve $P(s)$ and the diagonal line $t = s$ intersect in either one or two points. Figure 5.1 depicts the situation of two intersection points. The point $s = 1$ is certainly one intersection point because $P(1) = \sum_{k=0}^{\infty} p_k = 1$. There is a second intersection point to the left of $s = 1$ if and only if the slope of $P(s)$ at $s = 1$ is strictly greater than 1. The curve $P(s)$ then intersects $t = s$ at $s = 1$ from below. The slope $P'(1) = \sum_{k=0} kp_k$ equals the mean number of particles of the progeny distribution. Extinction is certain when the mean $P'(1) \leq 1$. When $P'(1) > 1$, the point $s = 1$ repels the iterates $s_n = P(s_{n-1})$. Hence, in this case the extinction probability is the smaller of the two fixed points of $s = P(s)$ on $[0, 1]$, and extinction is not certain.

TABLE 5.3. Functional Iteration for an Extinction Probability

Iteration n	Iterate s_n	Iteration n	Iterate s_n
0	0.000	10	0.847
1	0.498	20	0.873
2	0.647	30	0.878
3	0.719	40	0.879
4	0.761	50	0.880
5	0.788		

As a numerical example, consider the data of Lotka [6, 7] on the extinction of surnames among white males in the United States. Using 1920 census data, he computed the progeny generating function

$$\begin{aligned} P(s) &= .4982 + .2103s + .1270s^2 + .0730s^3 + .0418s^4 + .0241s^5 \\ &\quad + .0132s^6 + .0069s^7 + .0035s^8 + .0015s^9 + .0005s^{10}. \end{aligned}$$

Table 5.3 lists some representative functional iterates. Convergence to the correct extinction probability 0.880 is relatively slow.

5.4 Newton's Method

Newton's method can be motivated by the mean value theorem. Let x_{n-1} approximate the root x_∞ of the equation $g(x) = 0$. According to the mean value theorem,

$$\begin{aligned} g(x_{n-1}) &= g(x_{n-1}) - g(x_\infty) \\ &= g'(z)(x_{n-1} - x_\infty) \end{aligned}$$

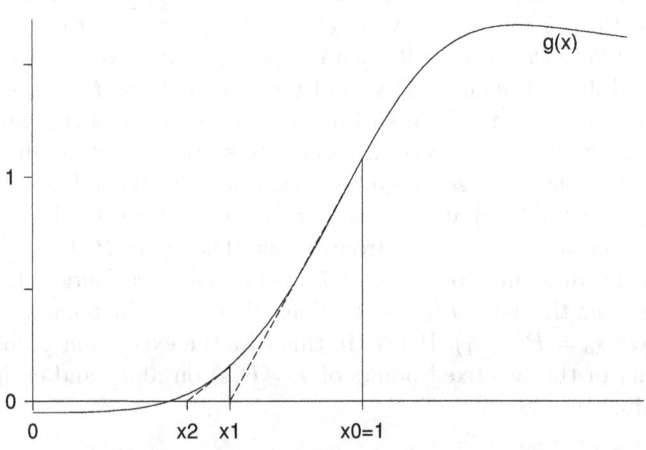

FIGURE 5.2. Newton's Method Applied to $g(x) = 1.95 - e^{-2/x} - 2e^{-x^4}$.

for some z on the interval between x_{n-1} and x_∞. If we substitute x_{n-1} for z and the next approximant x_n for x_∞, then this equality can be rearranged to provide the definition

$$x_n = x_{n-1} - \frac{g(x_{n-1})}{g'(x_{n-1})} \tag{5.8}$$

of Newton's method. From the perspective of functional iteration, Newton's method can be rephrased as $x_n = f(x_{n-1})$, where $f(x) = x - g(x)/g'(x)$. Figure 5.2 provides a geometric interpretation of Newton's method applied to a typical function $g(x)$, starting from $x_0 = 1$ and moving toward the unique root of $g(x) = 0$ on $(0, \infty)$. The iterate x_n is taken as the point of intersection of the x-axis and the tangent drawn through $[x_{n-1}, g(x_{n-1})]$. The method fails to converge if x_0 is chosen too far to the left or right.

The local convergence properties of Newton's method are determined by

$$\begin{aligned} f'(x_\infty) &= 1 - \frac{g'(x_\infty)}{g'(x_\infty)} + \frac{g(x_\infty)g''(x_\infty)}{g'(x_\infty)^2} \\ &= 0. \end{aligned}$$

If we let $e_n = x_n - x_\infty$ be the current error in approximating x_∞, then executing a second-order Taylor expansion around x_∞ yields

$$\begin{aligned} e_n &= f(x_{n-1}) - f(x_\infty) \\ &= f'(x_\infty)e_{n-1} + \frac{1}{2}f''(z)e_{n-1}^2 \end{aligned} \tag{5.9}$$

$$= \frac{1}{2}f''(z)e_{n-1}^2,$$

where z again lies between x_{n-1} and x_∞. Provided $f''(z)$ is continuous and x_0 is close enough to x_∞, the error representation (5.9) makes it clear that Newton's method converges and that

$$\lim_{n\to\infty} \frac{e_n}{e_{n-1}^2} = \frac{1}{2}f''(x_\infty).$$

This property is referred to as quadratic convergence. If an iteration function $f(x)$ satisfies $0 < |f'(x_\infty)| < 1$, then a first-order Taylor expansion implies $\lim_{n\to\infty} e_n/e_{n-1} = f'(x_\infty)$, which is referred to as linear convergence.

All else being equal, quadratic convergence is preferred to linear convergence. In practice, Newton's method can fail miserably if started too far from a desired root x_∞. Furthermore, it can be expensive to evaluate the derivative $g'(x)$. For these reasons, simpler, more robust methods such as bisection are often employed instead of Newton's method. The following two examples highlight favorable circumstances ensuring global convergence of Newton's method on a properly defined domain.

5.4.1 *Division without Dividing*

Forming the reciprocal of a number a is equivalent to solving for a root of the equation $g(x) = a - x^{-1}$. Newton's method (5.8) iterates according to

$$\begin{aligned} x_n &= x_{n-1} - \frac{a - x_{n-1}^{-1}}{x_{n-1}^{-2}} \\ &= x_{n-1}(2 - ax_{n-1}), \end{aligned}$$

which involves multiplication and subtraction but no division. If x_n is to be positive, then x_{n-1} must lie on the interval $(0, 2/a)$. If x_{n-1} does indeed reside there, then x_n will reside on the shorter interval $(0, 1/a)$ because the quadratic $x(2 - ax)$ attains its maximum of $1/a$ at $x = 1/a$. Furthermore, $x_n > x_{n-1}$ if and only if $2 - ax_{n-1} > 1$, and this latter inequality holds if and only if $x_{n-1} < 1/a$. Thus, starting on $(0, 1/a)$, the iterates x_n monotonically increase to their limit $1/a$. Starting on $[1/a, 2/a)$, the first iterate satisfies $x_1 \leq 1/a$, and subsequent iterates monotonically increase to $1/a$.

5.4.2 *Extinction Probabilities by Newton's Method*

Newton's method offers an alternative to functional iteration in computing the extinction probability s_∞ of a branching process. If $P(s)$ is the progeny generating function, then Newton's method starts with $x_0 = 0$ and iterates according to

$$x_n = x_{n-1} + \frac{P(x_{n-1}) - x_{n-1}}{1 - P'(x_{n-1})}. \tag{5.10}$$

Because extinction is certain when $P'(1) \leq 1$, we will make the contrary assumption $P'(1) > 1$. For such a supercritical process, $s_\infty < 1$. Because the curve $P(s)$ intersects the diagonal line $h(s) = s$ from above at s_∞, we infer that $P'(s_\infty) < 1$ in the supercritical case. This fact is important in avoiding division by 0 in the Newton's iterates (5.10).

It is useful to compare the sequence (5.10) to the sequence $s_n = P(s_{n-1})$ generated by functional iteration. Both schemes start at 0. We will show by induction that (a) $x_n \leq s_\infty$, (b) $x_{n-1} \leq x_n$, and (c) $s_n \leq x_n$ hold for all $n \geq 0$. Conditions (a) and (c) are true by definition when $n = 0$, while condition (b) is vacuous. In our inductive proof, we use the fact that condition (b) is logically equivalent to the condition $x_{n-1} \leq P(x_{n-1})$. With this in mind, suppose all three conditions hold for an arbitrary $n \geq 0$.

Because (a) is true for n and $P'(s)$ is increasing in s, the mean value theorem implies

$$\begin{aligned} s_\infty - P(x_n) &= P(s_\infty) - P(x_n) \\ &\geq P'(x_n)(s_\infty - x_n). \end{aligned}$$

Adding $x_n - s_\infty$ to this inequality leads to

$$x_n - P(x_n) \geq [1 - P'(x_n)](x_n - s_\infty),$$

which can be divided by $1 - P'(x_n)$ to yield

$$\frac{x_n - P(x_n)}{1 - P'(x_n)} \geq x_n - s_\infty.$$

Simple rearrangement gives the desired inequality (a) for $n + 1$.

Because $x_{n-1} \leq P(x_{n-1})$ and $x_{n-1} \leq x_n$ both hold by the induction hypothesis, it follows that the mean value theorem and definition (5.10) imply

$$\begin{aligned} P(x_n) - x_n &\geq P(x_{n-1}) + P'(x_{n-1})(x_n - x_{n-1}) - x_n \\ &= P(x_{n-1}) - x_{n-1} - [1 - P'(x_{n-1})](x_n - x_{n-1}) \\ &= P(x_{n-1}) - x_{n-1} - [1 - P'(x_{n-1})]\frac{P(x_{n-1}) - x_{n-1}}{1 - P'(x_{n-1})} \\ &= 0. \end{aligned}$$

This proves the alternate form $P(x_n) \geq x_n$ of (b) for $n + 1$.

To prove condition (c), we note that condition (b) implies

$$-P(x_n) \leq -x_n.$$

Multiplying this inequality by $P'(x_n)$ and then adding $P(x_n)$ yield

$$P(x_n)[1 - P'(x_n)] \;\leq\; P(x_n) - x_n P'(x_n).$$

Finally, dividing by $1 - P'(x_n)$ gives

$$\begin{aligned} P(x_n) &\leq \frac{P(x_n) - x_n P'(x_n)}{1 - P'(x_n)} \\ &= x_n + \frac{P(x_n) - x_n}{1 - P'(x_n)} \\ &= x_{n+1}. \end{aligned}$$

Since $s_{n+1} = P(s_n) \leq P(x_n) \leq x_{n+1}$, this completes the proof of (c).

TABLE 5.4. Newton's Method for an Extinction Probability

Iteration n	**Iterate** x_n	**Iteration** n	**Iterate** x_n
0	0.000	3	0.860
1	0.631	4	0.878
2	0.800	5	0.880

Application of Newton's method to the Lotka branching process data produces the iterates displayed in Table 5.4. Comparison of this table with Table 5.3 illustrates the much faster convergence of Newton's method. Properties (a), (b), and (c) are evident in these two tables. For those readers acquainted with multitype branching processes, it is noteworthy that all aspects of our comparison generalize if the differential $dP(\mathbf{1})$ of the vector of progeny generating functions is primitive and possesses a dominant eigenvalue strictly greater than 1.

5.5 Golden Section Search

We now turn to optimization of a function defined on an interval of the real line. Elementary calculus is replete with examples where optimization can be done analytically. Here we would like to focus on golden section search, a simple numerical algorithm for minimization. Golden section search brackets the minimum of a function much like bisection brackets the zero of a function. Golden section search is reliable and applies to any continuous function $f(x)$. Balanced against these strengths are its failure to generalize to higher dimensions and its relatively slow rate of convergence.

In golden section search we start with three points $a < b < c$ satisfying

$$f(b) \;<\; \min\{f(a), f(c)\}.$$

Numerical Recipes [8] discusses a general strategy for choosing an initial trio (a, b, c). If $f(x)$ has domain $[u, v]$ and $\lim_{x \to u} f(x) = \lim_{x \to v} f(x) = \infty$, then it is natural to choose a close to u, c close to v, and $b = \frac{1}{2}(u + v)$. To replace the bracketing interval (a, c) by a shorter interval, we choose $d \in (a, c)$ so that d belongs to the longer of the two intervals (a, b) and (b, c). Without loss of generality, suppose $b < d < c$. If $f(d) < f(b)$, then the three points $b < d < c$ bracket a minimum. If $f(d) > f(b)$, then the three points $a < b < d$ bracket a minimum. In the case of a tie, $f(d) = f(b)$, we choose $b < d < c$ when $f(c) < f(a)$ and $a < b < d$ when $f(a) < f(c)$.

These sensible rules do not address the problem of choosing d. Consider the fractional distances

$$\beta = \tfrac{b-a}{c-a}, \qquad \delta = \tfrac{d-b}{c-a}$$

along the interval (a, c). The next bracketing interval will have a fractional length of either $1 - \beta$ or $\beta + \delta$. To guard against the worst case, we should take $1 - \beta = \beta + \delta$. This determines $\delta = 1 - 2\beta$ and hence d. One could leave matters as they now stand, but the argument is taken one step further in golden section search. If we imagine repeatedly performing golden section search, then scale similarity is expected to set in eventually in the sense that

$$\beta = \frac{b-a}{c-a} = \frac{d-b}{c-b} = \frac{\delta}{1-\beta}.$$

Substituting $\delta = 1 - 2\beta$ in this identity and cross multiplying give the quadratic $\beta^2 - 3\beta + 1 = 0$ with solution

$$\beta = \frac{3 - \sqrt{5}}{2}$$

equal to the golden mean of ancient Greek mathematics. Following this reasoning, we should take $\delta = \sqrt{5} - 2 = 0.2361$.

Table 5.5 records the golden section search iterates for the binomial log-likelihood $f(x) = -7 \ln x - 3 \ln(1 - x)$ for the initial choices $a = 0.01$, $b = 0.50$, and $c = 0.99$. Convergence is slow but sure to the global minimum. In problems with multiple local minima, it is possible for golden section search to converge to a local minimum that is not the global minimum.

5.6 Minimization by Cubic Interpolation

Cubic interpolation offers a faster but less reliable method of minimization than golden section search. Suppose that the derivative $f'(x)$ is easy to compute and that we have evaluated $f(x)$ and $f'(x)$ at two points $a < b$.

TABLE 5.5. Golden Section Iterates for $f(x) = -7\ln x - 3\ln(1-x)$

Iteration	a	b	c	$f(b)$
0	0.01000	0.50000	0.99000	6.93147
1	0.31284	0.50000	0.99000	6.93147
2	0.50000	0.68716	0.99000	6.11251
3	0.50000	0.68716	0.80284	6.11251
4	0.61567	0.68716	0.80284	6.11251
5	0.61567	0.68716	0.73135	6.11251
10	0.69361	0.69759	0.70404	6.10878
15	0.69970	0.70006	0.70064	6.10864
20	0.69997	0.70000	0.70006	6.10864
25	0.70000	0.70000	0.70000	6.10864

In cubic interpolation, we match the values of a cubic polynomial $p(x)$ to these four values and minimize $p(x)$ on $[a, b]$. With luck this will lead to a lower value of $f(x)$. It simplifies matters to rescale the interval by setting $h(s) = f(a + sd)$ with $d = b - a$. Now $s = 0$ corresponds to a, and $s = 1$ corresponds to b. Furthermore, the chain rule implies $h'(s) = f'(a + sd)d$. Given these conventions, the theory of Hermite interpolation [4] suggests approximating $h(s)$ by the cubic polynomial

$$\begin{aligned} & p(s) \qquad\qquad (5.11) \\ = \; & (s-1)^2 h_0 + s^2 h_1 + s(s-1)[(s-1)(h_0' + 2h_0) + s(h_1' - 2h_1)] \\ = \; & (2h_0 + h_0' - 2h_1 + h_1')s^3 + (-3h_0 - 2h_0' + 3h_1 - h_1')s^2 + h_0's + h_0, \end{aligned}$$

where $h_0 = h(0)$, $h_0' = h'(0)$, $h_1 = h(1)$, and $h_1' = h'(1)$. One can readily verify that $p(0) = h_0$, $p'(0) = h_0'$, $p(1) = h_1$, and $p'(1) = h_1'$.

The local minima and maxima of $p(s)$ occur at the roots of the quadratic

$$\begin{aligned} p'(s) &= 3(2h_0 + h_0' - 2h_1 + h_1')s^2 + 2(-3h_0 - 2h_0' + 3h_1 - h_1')s + h_0' \\ &= c_2 s^2 + c_1 s + c_0. \end{aligned}$$

The discriminant $c_1^2 - 4c_2c_0$ determines whether $p'(s)$ attains the value 0. If $c_1^2 - 4c_2c_0 < 0$, then no real roots exist, and the minimum of $p(s)$ occurs at either 0 or 1. If $c_1^2 - 4c_2c_0 = 0$, then one double root exists, and if $c_1^2 - 4c_2c_0 > 0$, then two separate roots exist. The pertinent root of $p'(s) = 0$ is determined by the sign of the coefficient $2h_0 + h_0' - 2h_1 + h_1'$ of s^3 in $p(s)$. If this coefficient is positive, then the right root furnishes the local minimum, and if this coefficient is negative, then the left root furnishes the local minimum. The pertinent root r can be found by solving the quadratic equation $p'(s) = 0$. If r falls outside the interval $[0, 1]$, then the minimum of $p(s)$ occurs at either 0 or 1. Otherwise, $y = a + rd$ is a good candidate for an improved value of $f(x)$ on $[a, b]$.

Our previous function $f(x) = -7\ln x - 3\ln(1-x)$ is a good test case for assessing the performance of minimization by cubic interpolation. If we take $a = .0.10$ and $b = 0.99$, then the interpolating cubic has its minimum at b. This lack of progress shows the method at its worst. On the other hand, if we take $a = 0.50$ and $b = 0.75$, then the minimum of the interpolating cubic occurs at 0.6962, which is very close to the minimum of $f(x)$. Golden section search is incapable of converging this quickly. The best programs employ hybrid methods that balance reliability and speed.

5.7 Stopping Criteria

Deciding when to terminate an iterative method is more subtle than it might seem. In solving a nonlinear equation $g(x) = 0$, there are basically two tests. One can declare convergence when $|g(x_n)|$ is small or when x_n does not change much from one iteration to the next. Ideally, both tests should be satisfied. However, there are questions of scale. Our notion of small depends on the typical magnitudes of $g(x)$ and x, and stopping criteria should reflect these magnitudes [2]. Suppose $a > 0$ represents the typical magnitude of $g(x)$. Then a sensible criterion of the first kind is to stop when $|g(x_n)| < \epsilon a$ for $\epsilon > 0$ small. If $b > 0$ represents the typical magnitude of x, then a sensible criterion of the second kind is to stop when

$$|x_n - x_{n-1}| \quad \leq \quad \epsilon \max\{|x_n|, b\}. \tag{5.12}$$

To achieve p significant digits in the solution x_∞, take $\epsilon = 10^{-p}$.

When we optimize a function $f(x)$ with derivative $g(x) = f'(x)$, a third test comes into play. Now it is desirable for $f(x)$ to remain relatively constant near convergence. If $c > 0$ represents the typical magnitude of $f(x)$, then our final stopping criterion is

$$|f(x_n) - f(x_{n-1})| \quad \leq \quad \epsilon \max\{|f(x_n)|, c\}.$$

The second and third criteria generalize better than the first criterion to higher-dimensional problems because solutions often occur on boundaries or manifolds where the gradient $\nabla g(x)$ is not required to vanish. In higher dimensions, one should apply the criterion (5.12) to each coordinate of x. Choice of the typical magnitudes a, b, and c is problem specific, and some optimization programs leave this up to the discretion of the user. Often problems can be rescaled by an appropriate choice of units so that the choice $a = b = c = 1$ is reasonable. When in doubt about typical magnitudes, take this default and check whether the output of a preliminary computer run justifies the assumption.

5.8 Problems

1. Program any of the numerical examples discussed in this chapter and check the tabulated results.

2. Consider the quadratic function $x^2 - 2Ax + B$ whose coefficients A and B are independent, exponentially distributed random variables with common mean $1/\alpha$. The probability $p(\alpha)$ that both roots of this quadratic are real is given by the quantity

$$p(\alpha) \quad = \quad 1 - \sqrt{\pi\alpha}e^{\frac{\alpha}{4}}\left[1 - \Phi\left(\sqrt{\frac{\alpha}{2}}\right)\right],$$

where $\Phi(z)$ is the standard normal distribution function. Plot $p(\alpha)$ as a function of α. Find via bisection the value of α for which $p(\alpha) = \frac{1}{2}$.

3. Let $f(x)$ be a probability density and $g(x)$ a positive, measurable function. To minimize $\int_{S_\alpha} g(x)dx$ subject to $\int_{S_\alpha} f(x)dx = \alpha$, show that one should choose $\lambda(\alpha)$ and $S_\alpha = \{x : f(x)/g(x) > \lambda(\alpha)\}$ so that the constraint $\int_{S_\alpha} f(x)dx = \alpha$ is satisfied. If $f(x)$ and $g(x)$ are defined on an interval of the real line, and the ratio $f(x)/g(x)$ is increasing to the left and decreasing to the right of its mode, then S_α will be an interval.

4. To apply the Neyman-Pearson lemma of Problem 3, let $X_1, \ldots, X_n$ be a random sample from a normal distribution with mean μ and variance σ^2. The statistic

$$\frac{(n-1)S^2}{\sigma^2} \quad = \quad \frac{1}{\sigma^2}\sum_{i=1}^{n}(X_i - \bar{X})^2$$

is a pivot that follows a chi-square distribution with $n-1$ degrees of freedom. This pivot can be inverted to give a confidence interval for σ^2 of the form $(S^2/b, S^2/a)$. Design and implement an algorithm for computing the shortest confidence interval with a given confidence level. (Hint: As suggested in [5], use Problem 3 with $g(x) = x^{-2}$. You can check your results against the tables in [11].)

5. Show that the map $f(x) = \sqrt{2+x}$ is contractive on $[0, \infty)$. What is the smallest value of the Lipschitz constant? Identify the limit of the functional iterates $x_n = f(x_{n-1})$ from any starting point x_0.

6. Kepler's problem of celestial mechanics involves finding the eccentric anomaly E in terms of the mean anomaly M and the eccentricity $0 \le \epsilon < 1$ of an elliptical orbit. These three quantities are related by the equation $E = M + \epsilon \sin E$. Demonstrate that the corresponding function $f(E) = M + \epsilon \sin E$ is contractive on $(-\infty, \infty)$ with Lipschitz constant ϵ. Hence, the solution can be found by functional iteration.

7. Suppose $f(x) = -x^2 + x + \frac{1}{4}$. Prove that the iterates $x_n = f(x_{n-1})$ diverge if $x_0 < -\frac{1}{2}$ or $x_0 > \frac{3}{2}$, converge to $\frac{1}{2}$ if $-\frac{1}{2} < x_0 < \frac{3}{2}$, and converge to $-\frac{1}{2}$ if $x_0 = -\frac{1}{2}$ or $x_0 = \frac{3}{2}$.

8. For $0 \leq a \leq 4$, the function $f(x) = ax(1-x)$ maps the unit interval $[0, 1]$ onto itself. Show that:

   (a) The point 0 is a fixed point that is globally attractive when $a \leq 1$ and locally repelling when $a > 1$. Note that the rate of convergence to 0 is less than geometric when $a = 1$.

   (b) The point $1-a^{-1}$ is a fixed point for $a > 1$. It is locally attractive when $1 < a < 3$ and locally repelling when $3 < a \leq 4$.

   (c) For $1 < a \leq 2$, the fixed point $r = 1 - a^{-1}$ is globally attractive on $(0, 1)$. (Hint: Write $f(x) - r = (x - r)(1 - ax)$.)

   For $2 < a \leq 3$, the fixed point $1 - a^{-1}$ continues to be globally attractive on $(0, 1)$, but the proof of this fact is harder. For $a > 3$, the iterates $x_n = f(x_{n-1})$ no longer reliably converge. They periodically oscillate between several cluster points until at $a = 4$ they behave completely chaotically. See [10] for a nice intuitive discussion.

9. Functional iteration can often be accelerated. In searching for a fixed point of $x = f(x)$, consider the iteration scheme $x_n = f_\alpha(x_{n-1})$, where α is some constant and $f_\alpha(x) = (1-\alpha)x + \alpha f(x)$. Prove that any fixed point x_∞ of $f(x)$ is also a fixed point of $f_\alpha(x)$ and vice versa. Since $|f'_\alpha(x_\infty)|$ determines the rate of convergence of $x_n = f_\alpha(x_{n-1})$ to x_∞, find the α that minimizes $|f'_\alpha(x_\infty)|$ when $|f'(x_\infty)| < 1$. Unfortunately, neither x_∞ nor $f'(x_\infty)$ is typically known in advance.

10. The last problem is relevant to the branching process example of the text. Investigate numerically the behavior of the iterates

$$x_n = (1-\alpha)x_{n-1} + \alpha P(x_{n-1})$$

    for the choice $\alpha = 1/[1 - P'(0)]$ in the Lotka data. Is convergence to the extinction probability s_∞ faster than in ordinary functional iteration?

11. In the context of Problems 9 and 10, assume that $P'(1) > 1$. Show that the choice $\alpha = 1/[1-P'(0)]$ guarantees that the iterates increase monotonically from $x_0 = 0$ to the extinction probability $s_\infty < 1$.

12. Suppose the function $g(x)$ mapping a closed interval I into itself has a k-fold composition $f(x) = g \circ \cdots \circ g(x)$ satisfying the assumptions of Proposition 5.3.1. Prove that $g(x)$ has a unique fixed point.

13. What happens when you apply Newton's method to the functions

$$f(x) = \begin{cases} \sqrt{x} & x \geq 0 \\ -\sqrt{-x} & x < 0 \end{cases}$$

and $g(x) = \sqrt[3]{x}$?

14. Newton's method can be used to extract roots. Consider the function $g(x) = x^m - c$ for some integer $m > 1$ and $c > 0$. Show that Newton's method is defined by

$$x_n = x_{n-1}\left(1 - \frac{1}{m} + \frac{c}{mx_{n-1}^m}\right).$$

Prove that $x_n \geq c^{\frac{1}{m}}$ for all $x_{n-1} > 0$ and that $x_n \leq x_{n-1}$ whenever $x_{n-1} \geq c^{\frac{1}{m}}$. Thus, if $x_0 \geq c^{\frac{1}{m}}$, then x_n monotonically decreases to $c^{\frac{1}{m}}$. If $0 < x_0 < c^{\frac{1}{m}}$, then $x_1 > c^{\frac{1}{m}}$, but thereafter, x_n monotonically decreases to $c^{\frac{1}{m}}$.

15. Suppose the real-valued $f(x)$ satisfies $f'(x) > 0$ and $f''(x) > 0$ for all x in its domain (d, ∞). If $f(x) = 0$ has a root r, then demonstrate that r is unique and that Newton's method converges to r regardless of its starting point. Further, prove that x_n converges monotonically to r from above when $x_0 > r$ and that $x_1 > r$ when $x_0 < r$. How are these results pertinent to Problem 14? Cite at least one other example in the current chapter.

16. Problem 15 applies to polynomials $p(x)$ having only real roots. Suppose $p(x)$ is a polynomial of degree m with roots $r_1 < r_2 < \cdots < r_m$ and leading coefficient $c_m > 0$. Show that on the interval (r_m, ∞) the functions $p(x)$, $p'(x)$, and $p''(x)$ are all positive. Hence, if we seek r_m by Newton's method starting at $x_0 > r_m$, then the iterates x_n decrease monotonically to r_m. (Hint: According to Rolle's theorem, what can we say about the roots of $p'(x)$ and $p''(x)$?)

17. Suppose that the polynomial $p(x)$ has the known roots $r_1, \ldots, r_m$. Maehly's algorithm [9] attempts to extract one additional root r_{m+1} by iterating via

$$x_{n+1} = x_n - \frac{p(x_n)}{p'(x_n) - \sum_{k=1}^m \frac{p(x_n)}{(x_n - r_k)}}.$$

Show that this is just a disguised version of Newton's method. It has the virtue of being more numerically accurate than Newton's method applied to the deflated polynomial calculated from $p(x)$ by synthetic division. (Hint: Consider the polynomial $q(x) = p(x)\prod_{k=1}^m (x - r_k)^{-1}$.)

18. Apply Maehly's algorithm as sketched in Problem 17 to find the roots of the polynomial $p(x) = x^4 - 12x^3 + 47x^2 - 60x$.

19. In Example 5.4.1 suppose $x_0 = 1$ and $a \in (0, 2)$. Demonstrate that

$$\begin{aligned} x_n &= \frac{1-(1-a)^{2^n}}{a} \\ \Big|x_{n+1} - \frac{1}{a}\Big| &= a\Big|x_n - \frac{1}{a}\Big|^2. \end{aligned}$$

This shows very explicitly that x_n converges to $1/a$ at a quadratic rate.

20. In Problem 14 prove that

$$\begin{aligned} x_n &= \sqrt{c} + \frac{2\sqrt{c}}{\left[\left(1+\frac{2\sqrt{c}}{x_0-\sqrt{c}}\right)^{2^n} - 1\right]} \\ \Big|x_{n+1} - \sqrt{c}\Big| &\le \frac{1}{2\sqrt{c}}\Big|x_n - \sqrt{c}\Big|^2 \end{aligned}$$

when $m = 2$ and $x_0 > 0$. Thus, Newton's method converges at a quadratic rate. Use the first of these formulas or the iteration equation directly to show that $\lim_{n\to\infty} x_n = -\sqrt{c}$ for $x_0 < 0$.

21. Consider a function $f(x) = (x-r)^k g(x)$ with a root r of multiplicity k. If $g'(x)$ is continuous at r, and the Newton iterates x_n converge to r, then show that the iterates satisfy

$$\lim_{n\to\infty} \frac{|x_{n+1} - r|}{|x_n - r|} = 1 - \frac{1}{k}.$$

22. As an illustration of Problem 21, use Newton's method to extract a root of the polynomials $p_1(x) = x^2 - 1$ and $p_2(x) = x^2 - 2x + 1$ starting from $x_0 = 1$. Notice how much more slowly convergence occurs for $p_2(x)$ than for $p_1(x)$.

23. In minimization by cubic interpolation, show that the interpolating polynomial (5.11) is convex on the interval $[0, 1]$ if and only if

$$\tfrac{1}{3}h_1' + \tfrac{2}{3}h_0' \le h_1 - h_0 \le \tfrac{2}{3}h_1' + \tfrac{1}{3}h_0'. \qquad (5.13)$$

If $h_0' < 0$ and $h_1' > 0$, then argue further that $p(s)$ achieves its minimum on the open interval $(0, 1)$.

24. Consider finding a root of the equation $x^2 = 0$ by Newton's method starting from $x_0 = 1$. Show that it is impossible to satisfy the convergence criterion

$$|x_n - x_{n-1}| \le \epsilon \max\{|x_n|, |x_{n-1}|\}$$

for $\epsilon = 10^{-7}$ [2]. This example favors the alternative stopping rule (5.12).

5.9 References

[1] Box GEP, Tiao G (1973) *Bayesian Inference in Statistical Analysis.* Addison-Wesley, Reading, MA

[2] Dennis JE Jr, Schnabel RB (1996) *Numerical Methods for Unconstrained Optimization and Nonlinear Equations.* SIAM, Philadelphia

[3] Feller W (1968) *An Introduction to Probability Theory and Its Applications, Volume 1*, 3rd ed. Wiley, New York

[4] Henrici P (1982) *Essentials of Numerical Analysis with Pocket Calculator Demonstrations.* Wiley, New York

[5] Juola RC (1993) More on shortest confidence intervals. *Amer Statistician* 47:117–119

[6] Lotka AJ (1931) Population analysis—the extinction of families I. *J Wash Acad Sci* 21:377–380

[7] Lotka AJ (1931) Population analysis—the extinction of families II. *J Wash Acad Sci* 21:453–459

[8] Press WH, Teukolsky SA, Vetterling WT, Flannery BP (1992) *Numerical Recipes in Fortran: The Art of Scientific Computing*, 2nd ed. Cambridge University Press, Cambridge

[9] Stoer J, Bulirsch R (2002) *Introduction to Numerical Analysis*, 3rd ed. Springer, New York

[10] Strang G (1986) *Introduction to Applied Mathematics.* Wellesley–Cambridge Press, Wellesley, MA

[11] Tate RF, Klett GW (1969) Optimal confidence intervals for the variance of a normal distribution. *J Amer Stat Assoc* 54:674–682

[12] Wall HS (1948) *Analytic Theory of Continued Fractions.* Van Nostrand, New York

6

Vector and Matrix Norms

6.1 Introduction

In multidimensional calculus, vector and matrix norms quantify notions of topology and convergence [2, 4, 5, 6, 8, 12]. Because norms are also devices for deriving explicit bounds, theoretical developments in numerical analysis rely heavily on norms. They are particularly useful in establishing convergence and in estimating rates of convergence of iterative methods for solving linear and nonlinear equations. Norms also arise in almost every other branch of theoretical numerical analysis. Functional analysis, which deals with infinite-dimensional vector spaces, uses norms on functions.

6.2 Elementary Properties of Vector Norms

In our exposition of norms, we will assume a nodding familiarity with the Euclidean vector norm $||x||_2 = \sqrt{\sum_{i=1}^m x_i^2}$ in m-dimensional space R^m. This norm and others generalize the absolute value of a number on the real line. A norm on R^m is formally defined by four properties:

(a) $||x|| \geq 0$,

(b) $||x|| = 0$ if and only if $x = \mathbf{0}$,

(c) $||cx|| = |c| \cdot ||x||$ for every real number c,

(d) $||x + y|| \leq ||x|| + ||y||$.

In property (b), $\mathbf{0}$ is the vector with all m components 0. Property (d) is known as the triangle inequality. One immediate consequence of the triangle inequality is the further inequality $|\, ||x|| - ||y||\, | \leq ||x - y||$.

Two other simple but helpful norms are

$$\begin{aligned} ||x||_1 &= \sum_{i=1}^m |x_i| \\ ||x||_\infty &= \max_{1 \leq i \leq m} |x_i|. \end{aligned}$$

For each of the norms $||x||_p$, $p = 1$, 2, and ∞, a sequence of vectors x_n converges to a vector y if and only if each component sequence x_{ni} converges to y_i. Thus, all three norms give the same topology on R^m. The next proposition clarifies and generalizes this property.

K. Lange, *Numerical Analysis for Statisticians*, Statistics and Computing,
DOI 10.1007/978-1-4419-5945-4_6,

Proposition 6.2.1 *Let $||x||$ be any norm on R^m. Then there exist positive constants k_l and k_u such that $k_l||x||_1 \leq ||x|| \leq k_u||x||_1$ holds for all $x \in \mathsf{R}^m$.*

Proof: Let $e_1, \ldots, e_m$ be the standard basis vectors for R^m. Then the conditions (c) and (d) defining a norm indicate that $x = \sum_i x_i e_i$ satisfies

$$\begin{aligned} ||x|| &\leq \sum_i |x_i| \cdot ||e_i|| \\ &\leq \Big(\max_i ||e_i||\Big)||x||_1. \end{aligned}$$

This proves the upper bound with $k_u = \max_i ||e_i||$.

To establish the lower bound, we note that property (c) of a norm allows us to restrict attention to the set $S = \{x : ||x||_1 = 1\}$. Now the function $x \to ||x||$ is uniformly continuous on R^m because

$$\begin{aligned} \big|\,||x|| - ||y||\,\big| &\leq ||x - y|| \\ &\leq k_u||x - y||_1 \end{aligned}$$

follows from the upper bound just demonstrated. Since the set S is compact (closed and bounded), the function $x \to ||x||$ attains its lower bound k_l on S. Because of property (b), $k_l > 0$. ■

Proposition 6.2.1 immediately implies that $\sup_{x \neq \mathbf{0}} ||x||/||x||^{\dagger}$ is finite for every pair of norms $||x||$ and $||x||^{\dagger}$. For instance, it is straightforward to verify that

$$||x||_q \leq ||x||_p \tag{6.1}$$

$$||x||_p \leq m^{\frac{1}{p} - \frac{1}{q}}||x||_q \tag{6.2}$$

when p and q are chosen from $\{1, 2, \infty\}$ and $p < q$. These inequalities are sharp. Equality holds in (6.1) when $x = (1, 0, \ldots, 0)^t$, and equality holds in (6.2) when $x = (1, 1, \ldots, 1)^t$.

6.3 Elementary Properties of Matrix Norms

From one perspective an $m \times m$ matrix $A = (a_{ij})$ is simply a vector in R^{m^2}. Accordingly, we can define many norms involving A. However, it is profitable for a matrix norm also to be compatible with matrix multiplication. Thus, to the list of properties (a) through (d) for a vector norm, we add the requirement

(e) $||AB|| \leq ||A|| \cdot ||B||$

for any product of $m \times m$ matrices A and B. With this addition the Frobenius norm $||A||_F = \sqrt{\sum_{i=1}^m \sum_{j=1}^m a_{ij}^2} = \sqrt{\operatorname{tr}(AA^t)} = \sqrt{\operatorname{tr}(A^tA)}$ qualifies as a matrix norm. (Our reasons for writing $||A||_F$ rather than $||A||_2$ will soon be apparent.) Conditions (a) through (d) need no checking. Property (e) is verified by invoking the Cauchy-Schwarz inequality in

$$\begin{aligned}
||AB||_F^2 &= \sum_{i,j} \Big| \sum_k a_{ik} b_{kj} \Big|^2 \\
&\le \sum_{i,j} \Big(\sum_k a_{ik}^2 \Big) \Big(\sum_l b_{lj}^2 \Big) \\
&= \Big(\sum_{i,k} a_{ik}^2 \Big) \Big(\sum_{l,j} b_{lj}^2 \Big) \\
&= ||A||_F^2 ||B||_F^2.
\end{aligned}$$

Corresponding to any vector norm $||x||$ on R^m, there is an induced matrix norm $||A||$ on $m \times m$ matrices defined by

$$||A|| \quad = \quad \sup_{x \neq 0} \frac{||Ax||}{||x||} \quad = \quad \sup_{||x||=1} ||Ax||. \tag{6.3}$$

Using the same symbol for both the vector and inherited matrix norm ordinarily causes no confusion. All of the defining properties of a matrix norm are trivial to check for definition (6.3). For instance, consider property (e):

$$\begin{aligned}
||AB|| &= \sup_{||x||=1} ||ABx|| \\
&\le ||A|| \sup_{||x||=1} ||Bx|| \\
&= ||A|| \cdot ||B||.
\end{aligned}$$

Definition (6.3) also entails the equality $||I|| = 1$, where I is the $m \times m$ identity matrix. Because $||I||_F = \sqrt{m}$, the Frobenius norm $||A||_F$ and the induced matrix norm $||A||_2$ are definitely different. In infinite-dimensional settings, induced matrix norms are called operator norms.

In the following proposition, $\rho(C)$ denotes the absolute value of the dominant eigenvalue of the matrix C. This quantity is called the spectral radius of C.

Proposition 6.3.1 *If $A = (a_{ij})$ is an $m \times m$ matrix, then*

(a) $||A||_1 = \max_j \sum_i |a_{ij}|$,

(b) $||A||_2 = \sqrt{\rho(A^tA)}$, *which reduces to* $\rho(A)$ *if* A *is symmetric,*

(c) $||A||_2 = \max_{\{\|u\|_2=1, \|v\|_2=1\}} u^tAv$,

(d) $||A||_\infty = \max_i \sum_j |a_{ij}|$.

Proof: To prove (a) note that

$$\begin{aligned} ||A||_1 &= \sup_{||x||_1=1} \sum_i \Big| \sum_j a_{ij}x_j \Big| \\ &\leq \sup_{||x||_1=1} \sum_i \sum_j |a_{ij}| \cdot |x_j| \\ &= \sup_{||x||_1=1} \sum_j |x_j| \sum_i |a_{ij}| \\ &\leq \sup_{||x||_1=1} \Big(\sum_j |x_j|\Big)\Big(\max_k \sum_i |a_{ik}|\Big) \\ &= \max_k \sum_i |a_{ik}|. \end{aligned}$$

Equality holds throughout for the standard basis vector $x = e_k$ whose index k maximizes $\sum_i |a_{ik}|$.

To prove (b) choose an orthonormal basis of eigenvectors $u_1, \ldots, u_m$ for the symmetric matrix $A^t A$ with corresponding eigenvalues arranged so that $0 \leq \lambda_1 \leq \cdots \leq \lambda_m$. If $x = \sum_i c_i u_i$ is a unit vector, then $\sum_i c_i^2 = 1$, and

$$\begin{aligned} ||A||_2^2 &= \sup_{||x||_2=1} x^t A^t A x \\ &= \sum_i \lambda_i c_i^2 \\ &\leq \lambda_m. \end{aligned}$$

Equality is achieved when $c_m = 1$ and all other $c_i = 0$. If A is symmetric with eigenvalues μ_i arranged so that $|\mu_1| \leq \cdots \leq |\mu_m|$, then the u_i can be chosen to be the corresponding eigenvectors. In this case, clearly $\lambda_i = \mu_i^2$.

To prove (c) apply the Cauchy-Schwarz inequality and the definition of the matrix norm to the bilinear form $u^t A v$ with u and v unit vectors. This gives the inequality

$$u^t A v \;\leq\; \|u\|_2 \|Av\|_2 \;\leq\; \|u\|_2 \|A\|_2 \|v\|_2 \;=\; \|A\|_2.$$

Equality actually holds for a special choice of u and v. According to part (b), there is a unit vector v with $A^t A v = \|A\|_2^2 v$. If we let $w = Av$, then $\|w\|_2^2 = v^t A^t A v = \|A\|_2^2$. The unit vector $u = \|A\|_2^{-1} w$ now yields

$$u^t A v \;=\; \|A\|_2^{-1} v^t A^t A v \;=\; \|A\|_2.$$

To prove (d) note that

$$||A||_\infty \;=\; \sup_{||x||_\infty=1} \max_i \Big| \sum_j a_{ij} x_j \Big|$$

$$
\begin{aligned}
&\leq \sup_{\|x\|_\infty=1} \max_i \sum_j |a_{ij}| \Big(\max_k |x_k| \Big) \\
&= \max_i \sum_j |a_{ij}|.
\end{aligned}
$$

Equality holds throughout for

$$
x_j = \begin{cases} \frac{a_{kj}}{|a_{kj}|} & a_{kj} \neq 0 \\ 0 & a_{kj} = 0 \end{cases}
$$

if k is an index with maximum row sum $\sum_j |a_{kj}|$. ■

For theoretical purposes, it is convenient to consider vector and matrix norms defined over the complex vector space C^m. All of the properties studied so far generalize naturally to this setting. One needs to exercise a little care for the norm $\|x\|_2 = \sqrt{\sum_{i=1}^m |x_i|^2}$, where $|x_i|^2$ replaces x_i^2. This norm is induced by the complex inner product

$$
\langle x, y \rangle = \sum_{i=1}^m x_i y_i^*,
$$

with y_i^* denoting the complex conjugate of y_i. Proposition 6.3.1 (b) now refers to Hermitian matrices $A = (a_{ij}) = (a_{ji}^*) = A^*$ rather than to symmetric matrices. One of the advantages of extending norms to C^m is the following generalization of Proposition 6.3.1 (b) to arbitrary matrices.

Proposition 6.3.2 *The spectral radius $\rho(A)$ of a matrix A satisfies*

$$
\rho(A) \leq \|A\|
$$

for any induced matrix norm. Furthermore, for any A and $\epsilon > 0$, there exists some induced matrix norm with $\|A\| \leq \rho(A) + \epsilon$.

Proof: If λ is an eigenvalue of A with nontrivial eigenvector u, then the equality $\|Au\| = |\lambda| \cdot \|u\|$ for a vector norm entails the corresponding inequality $|\lambda| \leq \|A\|$ for the induced matrix norm.

Suppose A and $\epsilon > 0$ are given. There exists an invertible matrix S and an upper triangular matrix $T = (t_{ij})$ such that $A = STS^{-1}$. This fact follows directly from the Jordan canonical form or the Schur decomposition of A [7, 10, 8]. For $\delta > 0$ consider the diagonal matrix $D(\delta)$ whose ith diagonal entry is δ^{i-1}. It is straightforward to check that $[SD(\delta)]^{-1}ASD(\delta) = T(\delta)$ is upper triangular with entries $(t_{ij}\delta^{j-i})$ and consequently that the upper off-diagonal entries of $T(\delta)$ tend to 0 as $\delta \to 0$. It is also easy to check that $\|x\|_\delta = \|[SD(\delta)]^{-1}x\|_\infty$ defines a vector norm whose induced matrix norm is $\|A\|_\delta = \|[SD(\delta)]^{-1}ASD(\delta)\|_\infty = \|T(\delta)\|_\infty$. (See Problem 12.) According to Proposition 6.3.1 (c),

$$
\|A\|_\delta = \max_i \sum_j |t_{ij}|\delta^{j-i}.
$$

Because the eigenvalues of A coincide with the diagonal entries of T, we can take $\delta > 0$ so small that

$$\max_i \sum_j |t_{ij}|\delta^{j-i} \leq \rho(A) + \epsilon$$

Such a choice implies $||A||_\delta \leq \rho(A) + \epsilon$. ■

6.4 Norm Preserving Linear Transformations

An orthogonal matrix O satisfies the identity $OO^t = I$ [10]. From this definition, the further identities $O^tO = I$ and $(\det O)^2 = \det O \det O^t = 1$ follow. One can obviously divide the orthogonal matrices into rotations with $\det O = 1$ and reflections with $\det O = -1$. It is instructive to consider 2×2 matrices. A rotation through the angle θ has matrix representation

$$O = \begin{pmatrix} \cos\theta & -\sin\theta \\ \sin\theta & \cos\theta \end{pmatrix}.$$

One can derive this result by introducing the complex variable $z = x + iy$ and the complex exponential $e^{i\theta} = \cos\theta + i\sin\theta$. Rotation through the angle θ corresponds to the complex multiplication

$$e^{i\theta}z = \cos\theta\, x - \sin\theta\, y + i(\sin\theta\, x + \cos\theta\, y).$$

The reflection of a point across the line at angle $\frac{\theta}{2}$ with the x axis can be achieved by rotating the point through the angle $-\frac{\theta}{2}$, reflecting the result across the x axis by conjugation, and then rotating the reflected point back through the angle $\frac{\theta}{2}$. This corresponds to replacing $z = x + iy$ by the point

$$e^{i\theta}z^* = e^{i\theta/2}\Big(e^{-i\theta/2}z\Big)^*.$$

The matrix executing this reflection boils down to

$$O = \begin{pmatrix} \cos\theta & \sin\theta \\ \sin\theta & -\cos\theta \end{pmatrix}.$$

One can check that these prototype rotations and reflections satisfy the required properties $OO^t = I$ and $\det O = \pm 1$.

The set of orthogonal matrices forms a group under matrix multiplication. In other words, the product of two orthogonal matrices is orthogonal, and the inverse of an orthogonal matrix is orthogonal. The identity matrix is the unit of the group. The rotations constitute a subgroup of the orthogonal group, but the reflections do not since the product of two reflections is a rotation.

The trivial identity $(Ou)^tOv = u^tO^tOv = u^tv$ shows that an orthogonal transformation preserves inner products and Euclidean norms. Norm invariance has some profound consequences. For example, suppose λ is a real eigenvalue of O with real eigenvector v. Then $|\lambda|\|v\| = \|Ov\| = \|v\|$, and $|\lambda| = 1$. These considerations carry over to complex eigenvalues and force all eigenvalues of O to lie on the boundary of the unit circle. Norm invariance for vectors also leads to norm invariance for matrices. For the induced matrix norm $||A||_2$, invariance follows from Proposition 6.3.1 and the identities

$$\|OA\|_2^2 \;=\; \rho(A^tO^tOA) \;=\; \rho(A^tA) \;=\; \|A\|_2^2$$

and

$$\|AO\|_2 \;=\; \max_{\{\|u\|_2=1,\|v\|_2=1\}} u^tAOv \;=\; \max_{\{\|u\|_2=1,\|v\|_2=1\}} u^tAv \;=\; \|A\|_2.$$

If A has columns $a_1, \ldots, a_m$, then the Frobenius norm $||\cdot||_F$ satisfies

$$\|OA\|_F^2 \;=\; \sum_{k=1}^m \|Oa_k\|^2 \;=\; \sum_{k=1}^m \|a_k\|^2 \;=\; \|A\|_F^2.$$

A similar calculation via the rows of A demonstrates that $\|AO\|_F = \|A\|_F$.

Given a unit vector u, the Householder matrix $H = I - 2uu^t$ represents a reflection across the plane perpendicular to u. As we shall see later, Householder matrices play an important role in many areas of computational statistics. The calculation

$$HH^t \;=\; I - 4uu^t + 4u\|u\|_2^2u^t \;=\; I$$

verifies that H is orthogonal; H is also symmetric. Algebraically, it is clear that $Hv = v$ whenever v lies in the plane perpendicular to u. The vector u itself is taken into $Hu = u - 2u\|u\|_2^2 = -u$. These facts imply that H has one eigenvalue equal to -1 and all other eigenvalues equal to 1. Since the determinant of H is the product of its eigenvalues, H is indeed a reflection.

The outer product representation of a Householder matrix facilitates both theoretical understanding and practical numerical analysis. It is advantageous to write other matrices as sums of outer products. For instance, the spectral decomposition of a symmetric matrix A can be summarized by the formula

$$A \;=\; \sum_k \lambda_k v_k v_k^t,$$

where λ_k is a real eigenvalue with corresponding eigenvector v_k. The v_k form an orthonormal basis spanning the underlying space. An orthogonal projection matrix P can be represented by

$$P \;=\; \sum_k v_k v_k^t,$$

where the v_k form an orthonormal basis of the subspace fixed by P. This representation is not unique, but it does make it easy to check that $P^2 = P$ and $P^t = P$.

6.5 Iterative Solution of Linear Equations

Many numerical problems involve iterative schemes of the form

$$x_n = Bx_{n-1} + w \tag{6.4}$$

for solving the vector-matrix equation $(I - B)x = w$. Clearly, the map $f(x) = Bx + w$ satisfies

$$\begin{aligned} ||f(y) - f(x)|| &= ||B(y - x)|| \\ &\leq ||B|| \cdot ||y - x|| \end{aligned}$$

and therefore is contractive for a vector norm $||x||$ if $||B|| < 1$ holds for the induced matrix norm. If we substitute norms for absolute values in our convergence proof for one-dimensional functional iteration, then that proof generalizes to this vector setting, and we find that the iterates x_n converge to the unique solution x of $(I - B)x = w$. In light of the fact that w is arbitrary, it follows that $I - B$ is invertible. These facts are incorporated in the next proposition.

Proposition 6.5.1 *Let B be an arbitrary matrix with spectral radius $\rho(B)$. Then $\rho(B) < 1$ if and only if $||B|| < 1$ for some induced matrix norm. The inequality $||B|| < 1$ implies*

(a) $\lim_{n\to\infty} ||B^n|| = 0$,

(b) $(I - B)^{-1} = \sum_{n=0}^{\infty} B^n$,

(c) $\frac{1}{1+||B||} \leq ||(I - B)^{-1}|| \leq \frac{1}{1-||B||}$.

Proof: The first claim is an immediate consequence of Proposition 6.3.2. Assertion (a) follows from $||B^n|| \leq ||B||^n$. Assertion (b) follows if we let $x_0 = 0$ in the iteration scheme (6.4). Then $x_n = \sum_{i=0}^{n-1} B^i w$, and

$$\begin{aligned} (I - B)^{-1}w &= \lim_{n\to\infty} x_n \\ &= \lim_{n\to\infty} \sum_{i=0}^{n-1} B^i w. \end{aligned}$$

To prove the first inequality of assertion (c), note that taking norms in $I = (I - B)(I - B)^{-1}$ implies

$$\begin{aligned} 1 &\leq ||I - B|| \cdot ||(I - B)^{-1}|| \\ &\leq (1 + ||B||)||(I - B)^{-1}||. \end{aligned}$$

For the second inequality, use the identity $(I - B)^{-1} = I + B(I - B)^{-1}$. Taking norms now produces

$$||(I - B)^{-1}|| \leq 1 + ||B|| \cdot ||(I - B)^{-1}||,$$

which can be rearranged to give the desired result. ■

Linear iteration is especially useful in solving the equation $Ax = b$ for x when an approximation C to A^{-1} is known. If this is the case, then one can set $B = I - CA$ and $w = Cb$ and iterate via (6.4). Provided $||B|| < 1$, the unique fixed point of the scheme (6.4) satisfies $x = (I - CA)x + Cb$. If C^{-1} exists, then this is equivalent to $Ax = b$.

6.5.1 *Jacobi's Method*

Jacobi's method offers a typical example of this strategy. Suppose for the sake of simplicity that $A = (a_{ij})$ is strictly diagonally dominant in the sense that $|a_{ii}| > \sum_{j \neq i} |a_{ij}|$ holds for all rows i. Let D be the diagonal matrix with ith diagonal entry a_{ii}. Then the matrix $C = D^{-1}$ can be considered an approximate inverse A. The matrix $B = I - CA$ has diagonal elements $b_{ii} = 0$ and off-diagonal elements $b_{ij} = -a_{ij}/a_{ii}$. By definition

$$\begin{aligned} ||B||_\infty &= \max_i \sum_{j \neq i} |b_{ij}| \\ &< 1. \end{aligned}$$

This analysis has the side effect of showing that every strictly diagonally dominant matrix A is invertible.

6.5.2 *Landweber's Iteration Scheme*

In practice, the approximate inverse C can be rather crude. For instance, Landweber [9] suggests the surprising choice $C = \epsilon A^t$ for ϵ small and positive. Because $A^t A$ is positive definite, its eigenvalues can be arranged as $0 < \lambda_1 \leq \cdots \leq \lambda_m$. The eigenvalues of the symmetric matrix $I - \epsilon A^t A$ are then $1 - \epsilon\lambda_1, \ldots 1 - \epsilon\lambda_m$. As long as $1 - \epsilon\lambda_m > -1$, all eigenvalues of $I - \epsilon A^t A$ will occur on the interval $(-1, 1)$, which according to part (b) of Proposition 6.3.1 implies $||I - \epsilon A^t A||_2 < 1$. In other words, if $\epsilon < 2/\lambda_m = 2/||A||_2^2$, then linear iteration can be employed to solve $Ax = b$. Since finding the norm $||A||_2$ is cumbersome, one can replace it in bounding ϵ with more simply computed upper bounds. For instance, the inequalities $||A||_2 \leq ||A||_F$ and $||A||_2 \leq \sqrt{||A||_\infty ||A||_1}$ discussed in Problems 9 and 11 often serve well.

6.5.3 *Equilibrium Distribution of a Markov Chain*

A slightly different problem is to determine the equilibrium distribution of a finite state Markov chain. Recall that movement among the m states of a

Markov chain is governed by its $m \times m$ transition matrix $P = (p_{ij})$, whose entries are nonnegative and satisfy $\sum_j p_{ij} = 1$ for all i. A column vector x with nonnegative entries and norm $||x||_1 = \sum_i x_i = 1$ is said to be an equilibrium distribution for P provided $x^t P = x^t$, or equivalently $Qx = x$ for $Q = P^t$. Because the norm $||Q||_1 = 1$, one cannot immediately invoke the contraction mapping principle. However, if we restrict attention to the closed set $S = \{x : x_i \geq 0,\ i = 1, \ldots, m,\ \sum_i x_i = 1\}$, then we do get a contraction map under the hypothesis that some power Q^k has all entries positive [1]. Let $c > 0$ be the minimum entry of Q^k and $\mathbf{1}$ be the column vector of all 1's. The matrix $R = Q^k - c\mathbf{1}\mathbf{1}^t$ has all entries nonnegative and norm $||R||_1 < 1$.

Consider two vectors x and y from S. Owing to the fact $\mathbf{1}^t(x - y) = 0$, we get

$$\begin{aligned} ||Q^k x - Q^k y||_1 &= ||R(x-y)||_1 \\ &\leq ||R||_1 ||x - y||_1, \end{aligned}$$

and it follows that the map $x \to Q^k x$ is contractive on S with unique fixed point x_∞. Now for any $x \in S$,

$$\begin{aligned} Qx_\infty &= Q \lim_{n\to\infty} Q^{nk} x \\ &= \lim_{n\to\infty} Q^{nk} Q x \\ &= x_\infty. \end{aligned}$$

Thus, x_∞ is a fixed point of $x \to Qx$ as well. Because any integer n can be represented uniquely as $kl + r$ with $0 \leq r < k$, the inequality

$$\begin{aligned} ||Q^n x - x_\infty||_1 &= ||Q^{kl}(Q^r x - Q^r x_\infty)||_1 \\ &\leq ||R||_1^l ||Q^r x - Q^r x_\infty||_1 \end{aligned}$$

can be invoked to show that $\lim_{n\to\infty} Q^n x = x_\infty$ for all $x \in S$.

This method of finding the equilibrium distribution is termed the power method. One of its more interesting applications is to ranking internet nodes [3, 11]. Section 8.4 takes up the power method for more general matrices.

6.6 Condition Number of a Matrix

Consider the apparently innocuous matrix

$$A = \begin{pmatrix} 10 & 7 & 8 & 7 \\ 7 & 5 & 6 & 5 \\ 8 & 6 & 10 & 9 \\ 7 & 5 & 9 & 10 \end{pmatrix} \tag{6.5}$$

concocted by R. S. Wilson [2]. This matrix is symmetric and positive definite. The unique solution to the linear equation $Ax = b$ can be expressed as $x = A^{-1}b$. For the choice $b = (32, 23, 33, 31)^t$, we find $x = (1, 1, 1, 1)^t$. The slightly perturbed vector $b + \Delta b = (32.1, 22.9, 33.1, 30.9)^t$ leads to the violently perturbed solution $x + \Delta x = (9.2, -12.6, 4.5, -1.1)^t$. When we start with $b = (4, 3, 3, 1)^t$, then the solution of $Ax = b$ is $x = (1, -1, 1, -1)^t$. If we perturb A to $A + .01I$, then the solution of $(A + .01I)(x + \Delta x) = b$ is $x + \Delta x = (.59, -.32, .82, -.89)^t$. Thus, a relatively small change in A propagates to a large change in the solution of the linear equation.

One can gain insight into these disturbing patterns by defining the condition number of an invertible matrix. Consider a vector norm $||x||$ and its induced matrix norm $||A||$. If $Ax = b$ and $A(x + \Delta x) = b + \Delta b$, then by definition of the induced matrix norm,

$$\begin{aligned} ||b|| &\leq ||A|| \cdot ||x|| \\ ||\Delta x|| &\leq ||A^{-1}|| \cdot ||\Delta b||. \end{aligned}$$

Dividing the second of these inequalities by the first produces

$$\frac{||\Delta x||}{||x||} \leq \text{cond}(A)\frac{||\Delta b||}{||b||}, \tag{6.6}$$

where $\text{cond}(A) = ||A|| \cdot ||A^{-1}||$ is termed the condition number of the matrix A relative to the given norm. Inequality (6.6) is sharp. To achieve equality, one merely needs to choose x so that $||Ax|| = ||A|| \cdot ||x||$ and Δb so that $||A^{-1}\Delta b|| = ||A^{-1}|| \cdot ||\Delta b||$.

Now suppose $Ax = b$ and $(A + \Delta A)(x + \Delta x) = b$. It then follows from $\Delta x = -A^{-1}\Delta A(x + \Delta x)$ that

$$||\Delta x|| \leq ||A^{-1}|| \cdot ||\Delta A|| \cdot ||x + \Delta x||, \tag{6.7}$$

or equivalently

$$\frac{||\Delta x||}{||x + \Delta x||} \leq \text{cond}(A)\frac{||\Delta A||}{||A||}. \tag{6.8}$$

Inequality (6.8) is also sharp; see Problem 23.

A bound on the change $||\Delta x||/||x||$ is, perhaps, preferable to the bound (6.8). For $||\Delta A||$ small, one can argue that $x + \Delta x = (I + A^{-1}\Delta A)^{-1}x$ because

$$\begin{aligned} x &= A^{-1}b \\ &= A^{-1}(A + \Delta A)(x + \Delta x) \\ &= (I + A^{-1}\Delta A)(x + \Delta x). \end{aligned}$$

The identity $x + \Delta x = (I + A^{-1}\Delta A)^{-1}x$ in turn implies

$$||x + \Delta x|| \leq ||(I + A^{-1}\Delta A)^{-1}|| \cdot ||x||$$

$$
\begin{aligned}
&\leq \frac{||x||}{1 - ||A^{-1}\Delta A||} \\
&\leq \frac{||x||}{1 - ||A^{-1}|| \cdot ||\Delta A||}
\end{aligned}
$$

in view of part (c) of Proposition 6.5.1. Substituting this bound for $||x+\Delta x||$ in inequality (6.7) yields

$$
\begin{aligned}
\frac{||\Delta x||}{||x||} &\leq \frac{||A^{-1}|| \cdot ||\Delta A||}{1 - ||A^{-1}|| \cdot ||\Delta A||} \\
&= \operatorname{cond}(A)\frac{||\Delta A||}{(||A|| - \operatorname{cond}(A)||\Delta A||)}.
\end{aligned}
$$

The mysteries of the matrix (6.5) disappear when we compute its condition number $\operatorname{cond}_2(A)$ relative to the matrix norm $||A||_2$. Recalling part (b) of Proposition 6.3.1, it is clear that $\operatorname{cond}_2(A)$ is the ratio of the largest and smallest eigenvalues λ_4 and λ_1 of A. For the matrix (6.5), it turns out that $\lambda_1 = 0.01015$, $\lambda_4 = 30.2887$, and $\operatorname{cond}_2(A) = 2984$. We will learn later how to compute the dominant eigenvalues of A and A^{-1}. If A^{-1} is available, we can elect another more easily computed norm and calculate $\operatorname{cond}(A)$ relative to it.

6.7 Problems

1. Verify the vector norm inequalities (6.1) and (6.2) for p and q chosen from $\{1, 2, \infty\}$.

2. Show that $||x||_2^2 \leq ||x||_\infty ||x||_1 \leq \sqrt{m}||x||_2^2$ for any vector $x \in \mathsf{R}^m$.

3. Suppose x and y are vectors with

$$
\frac{||y - x||}{||x||} \leq c < 1
$$

for a norm $||\cdot||$. Demonstrate that

$$
\frac{||x - y||}{||y||} \leq \frac{c}{1 - c}.
$$

4. Let $x_1, \ldots, x_n$ be points in R^m. State and prove a necessary and sufficient condition under which the norm equality

$$
||x_1 + \cdots + x_n||_2 = ||x_1||_2 + \cdots + ||x_n||_2
$$

holds.

5. For $p \in (1, \infty)$ and $x \in \mathsf{R}^m$, define

$$\|x\|_p = \Big(\sum_{i=1}^m |x_i|^p\Big)^{1/p}.$$

Prove the identity

$$\|x\|_p = \sup_{\|y\|_q=1} y^t x,$$

where q satisfies $p^{-1} + q^{-1} = 1$. Use this second definition to show that $\|x\|_p$ gives a norm on R^m. (Hint: Look up and apply Hölder's inequality in verifying the equivalence of the two definitions.)

6. For the vector norm $\|x\|_p$ of the preceding problem, demonstrate that $\lim_{p\to\infty} \|x\|_p = \|x\|_\infty$.

7. Suppose T is a symmetric matrix. What further conditions on T guarantee that $\|x\| = \sqrt{|x^t T x|}$ is a vector norm?

8. Prove that $1 \le \|I\|$ and $\|A\|^{-1} \le \|A^{-1}\|$ for any matrix norm.

9. For an $m \times m$ matrix A, show that

$$\begin{aligned} \frac{1}{\sqrt{m}}\|A\|_1 &\le \|A\|_2 \le \sqrt{m}\|A\|_1 \\ \frac{1}{\sqrt{m}}\|A\|_\infty &\le \|A\|_2 \le \sqrt{m}\|A\|_\infty \\ \frac{1}{\sqrt{m}}\|A\|_F &\le \|A\|_2 \le \|A\|_F. \end{aligned}$$

(Hint: Use the vector norm inequalities (6.1) and (6.2), the matrix norm definition (6.3), and Proposition 6.3.1.)

10. Let A be an invertible $m \times m$ matrix. Demonstrate the formula

$$\|A^{-1}\| = \max_{v \neq \mathbf{0}} \frac{\|v\|}{\|Av\|}$$

for any vector norm $\|\cdot\|$ and its induced matrix norm.

11. Prove the inequality $\|A\|_2 \le \sqrt{\|A\|_\infty \|A\|_1}$. (Hint: If the dominant eigenvalue $\lambda \ge 0$ of $A^t A$ has eigenvector u, then bound $\lambda\|u\|_1$.)

12. Suppose $\|x\|$ is a vector norm and T is an invertible matrix. Show that $\|x\|^\dagger = \|Tx\|$ defines a vector norm whose induced matrix norm is $\|A\|^\dagger = \|TAT^{-1}\|$.

13. Define $\|A\| = \max_{i,j} |a_{ij}|$ for $A = (a_{ij})$. Show that this defines a vector norm but not a matrix norm on $m \times m$ matrices A.

14. Demonstrate that the outer product uv^t of two vectors in R^m has the matrix norms $\|uv^t\|_2 = \|uv^t\|_F = \|u\|_2\|v\|_2$.

15. Let P be an orthogonal projection. Prove that $I - P$ is an orthogonal projection and $I - 2P$ is an orthogonal transformation.

16. Demonstrate that a projection P satisfies the identity

$$\text{rank}(P) = \text{tr}(P).$$

17. Consider the linear map $f(x)$ from R^n to R^n with components

$$f_i(x) = \frac{1}{2}(x_i + x_{n-i+1}).$$

Prove that $f(x)$ is an orthogonal projection. Compute its rank, and identify a basis for its range.

18. Fix a positive integer n. Prove that a positive definite matrix M has a unique positive definite nth root R. (Hint: For uniqueness note that R commutes with M.)

19. Show that the matrix $\begin{pmatrix} 0 & 1 \\ 0 & 0 \end{pmatrix}$ has no square root even if one allows complex entries.

20. Let O_n be a sequence of orthogonal matrices. Show that there exists a subsequence O_{n_k} that converges to an orthogonal matrix. (Hint: Compute the norm $||O_n||_2$.)

21. Demonstrate that $\rho(A) = \lim_{n\to\infty} ||A^n||^{1/n}$ for any induced matrix norm. (Hints: $\rho(A^n)^{1/n} = \rho(A)$ and $[(\rho(A)+\epsilon)^{-1}A]^n \to 0$.)

22. Prove that the series $B_n = \sum_{k=0}^{n} \frac{A^k}{k!}$ converges. Its limit is the matrix exponential e^A.

23. Show that inequality (6.8) is sharp by choosing $w \neq \mathbf{0}$ so that

$$||A^{-1}w|| = ||A^{-1}|| \cdot ||w||.$$

Then take successively $\Delta x = -\beta A^{-1}w$, $x + \Delta x = w$, $\Delta A = \beta I$, and $b = (A + \Delta A)w$, where β is any nonzero number such that $A + \beta I$ is invertible.

24. Relative to any induced matrix norm, show that (a) $\text{cond}(A) \geq 1$, (b) $\text{cond}(A^{-1}) = \text{cond}(A)$, (c) $\text{cond}(AB) \leq \text{cond}(A)\,\text{cond}(B)$, and (d) $\text{cond}(cA) = \text{cond}(A)$ for any scalar $c \neq 0$. Also verify that if U is orthogonal, then $\text{cond}_2(U) = 1$ and

$$\text{cond}_2(A) = \text{cond}_2(AU) = \text{cond}_2(UA).$$

25. If $A + \Delta A$ is invertible, prove that

$$\frac{||(A+\Delta A)^{-1} - A^{-1}||}{||(A+\Delta A)^{-1}||} \leq \operatorname{cond}(A)\frac{||\Delta A||}{||A||}.$$

6.8 References

[1] Baldwin JT (1989) On Markov processes in elementary mathematics courses. *Amer Math Monthly* 96:147–153

[2] Ciarlet PG (1989) *Introduction to Numerical Linear Algebra and Optimization.* Cambridge University Press, Cambridge

[3] Eldén L (2007) *Matrix Methods in Data Mining and Pattern Recognition.* SIAM, Philadelphia

[4] Gill PE, Murray W, Wright MH (1991) *Numerical Linear Algebra and Optimization, Volume 1.* Addison-Wesley, Reading, MA

[5] Golub GH, Van Loan CF (1996) *Matrix Computations*, 3rd ed. Johns Hopkins University Press, Baltimore

[6] Hämmerlin G, Hoffmann KH (1991) *Numerical Mathematics.* Springer, New York

[7] Hoffman K, Kunze R (1971) *Linear Algebra*, 2nd ed. Prentice-Hall, Englewood Cliffs, NJ

[8] Isaacson E, Keller HB (1966) *Analysis of Numerical Methods.* Wiley, New York

[9] Landweber L (1951) An iteration formula for Fredholm integral equations of the first kind. *Amer J Math* 73:615–624

[10] Lang S (1971) *Linear Algebra*, 2nd ed. Addison-Wesley, Reading, MA

[11] Langville AN, Meyer CD (2006) *Google's PageRank and Beyond: The Science of Search Engine Rankings.* Princeton University Press, Princeton NJ

[12] Ortega JM (1990) *Numerical Analysis: A Second Course.* SIAM, Philadelphia

7

Linear Regression and Matrix Inversion

7.1 Introduction

Linear regression is the most commonly applied procedure in statistics. This fact alone underscores the importance of solving linear least squares problems quickly and reliably. In addition, iteratively reweighted least squares lies at the heart of a host of other optimization algorithms in statistics. The current chapter features four different methods for solving linear least squares problems: sweeping, Cholesky decomposition, the modified Gram-Schmidt procedure, and orthogonalization by Householder reflections. Later we take up solution by the singular value decomposition.

The sweep operator [1, 3, 6, 7, 8] is the workhorse of computational statistics. The matrices that appear in linear regression and multivariate analysis are almost invariably symmetric. Sweeping exploits this symmetry. Although there are faster and numerically more stable algorithms for inverting a matrix or solving a least-squares problem, no algorithm matches the conceptual simplicity and utility of sweeping. To highlight some of the typical quantities that sweeping calculates with surprising ease, we briefly review a few key ideas from linear regression and multivariate analysis.

A Cholesky decomposition furnishes a lower triangular square root of a positive definite matrix [2, 7, 10, 12, 13, 14, 15]. This turns out to be valuable because solving linear equations with triangular matrices is particularly simple. Cholesky decompositions also provide convenient parameterizations of covariance matrices. The complicated constraints required by positive definiteness melt away, leaving only positivity constraints on the diagonal of the Cholesky decomposition.

Gram-Schmidt orthogonalization, particularly in its modified form, offers a numerically more stable method of computing linear regression estimates than sweeping or Cholesky decomposition [2, 11, 13, 14, 15]. Although this is reason enough for introducing two of the major algorithms of matrix orthogonalization, we will meet further motivation in Chapter 11, where we discuss the computation of asymptotic standard errors of maximum likelihood estimates subject to linear constraints. Orthogonalization by Householder reflections leads to the same QR decomposition reached by Gram-Schmidt orthogonalization.

Woodbury's formula occupies a somewhat different niche than sweeping or matrix factorization [9, 10]. Many statistical models involve the inver-

K. Lange, *Numerical Analysis for Statisticians*, Statistics and Computing,
DOI 10.1007/978-1-4419-5945-4_7, © Springer Science+Business Media, LLC 2010

sion of matrices that are low-rank perturbations of matrices with known inverses. For instance, if D is an invertible diagonal matrix and u is a compatible column vector, the Sherman-Morrison formula [9, 10]

$$(D+uu^t)^{-1} = D^{-1} - \frac{1}{1+u^tD^{-1}u} D^{-1}uu^tD^{-1}$$

provides the inverse of the symmetric, rank-one perturbation $D+uu^t$ of D. Woodbury's formula generalizes the Sherman-Morrison formula. Both the original Sherman-Morrison formula and Woodbury's generalization permit straightforward computation of the determinant of the perturbed matrix from the determinant of the original matrix.

7.2 Motivation from Linear Regression

As motivation for studying least squares, we briefly review linear regression. The basic setup involves p independent observations that individually take the form

$$y_i = \sum_{j=1}^{q} x_{ij}\beta_j + u_i. \tag{7.1}$$

Here y_i depends linearly on the unknown parameters β_j through the known constants x_{ij}. The error u_i is assumed to be normally distributed with mean 0 and variance σ^2. If we collect the y_i into a $p \times 1$ observation vector y, the x_{ij} into a $p \times q$ design matrix X, the β_j into a $q \times 1$ parameter vector β, and the u_j into a $p \times 1$ error vector u, then the linear regression model can be rewritten in vector notation as $y = X\beta + u$. A maximum likelihood estimator $\hat{\beta}$ of β solves the normal equations $X^tX\beta = X^ty$. Throughout this chapter we assume that X has full rank. With this stipulation, the normal equations have the unique solution $\hat{\beta} = (X^tX)^{-1}X^ty$. This is also the least squares estimator of β even when the error vector u is nonnormal. In general, if u has uncorrelated components with common mean 0 and common variance σ^2, then the estimator $\hat{\beta}$ has mean and variance

$$\begin{aligned} \mathrm{E}(\hat{\beta}) &= \beta \\ \mathrm{Var}(\hat{\beta}) &= \sigma^2(X^tX)^{-1}. \end{aligned}$$

The difference $y-\hat{y} = y-X\hat{\beta}$ between the actual and predicted observations is termed the residual vector. Its Euclidean norm $\|y-\hat{y}\|_2^2$ squared, known as the residual sum of squares, is fundamentally important in inference. For example, σ^2 is usually estimated by $\|y-\hat{y}\|_2^2/(p-q)$. A single application of the sweep operator permits simultaneous computation of $\hat{\beta}$, $\mathrm{Var}(\hat{\beta})$, and $\|y-\hat{y}\|_2^2$.

In weighted least squares, one minimizes the criterion

$$f(\beta) = \sum_{i=1}^{p} w_i\Big(y_i - \sum_{j=1}^{q} x_{ij}\beta_j\Big)^2,$$

where the w_i are positive weights. This reduces to ordinary least squares if we substitute $\sqrt{w_i}y_i$ for y_i and $\sqrt{w_i}x_{ij}$ for x_{ij}. It is clear that any method for solving an ordinary least squares problem can be immediately adapted to solving a weighted least squares problem.

7.3 Motivation from Multivariate Analysis

A random vector $X \in \mathsf{R}^p$ with mean vector μ, covariance matrix Ω, and density

$$(2\pi)^{-\frac{p}{2}} \det(\Omega)^{-\frac{1}{2}} e^{-\frac{1}{2}(x-\mu)^t\Omega^{-1}(x-\mu)}$$

is said to follow a multivariate normal distribution. The sweep operator permits straightforward calculation of the quadratic form $(x-\mu)^t\Omega^{-1}(x-\mu)$ and the determinant of Ω. If we partition X and its mean and covariance so that

$$X = \begin{pmatrix} Y \\ Z \end{pmatrix}, \quad \mu = \begin{pmatrix} \mu_Y \\ \mu_Z \end{pmatrix}, \quad \Omega = \begin{pmatrix} \Omega_Y & \Omega_{YZ} \\ \Omega_{ZY} & \Omega_Z \end{pmatrix},$$

then conditional on the event $Y = y$, the subvector Z follows a multivariate normal density with conditional mean and variance

$$\begin{aligned} \mathrm{E}(Z \mid Y = y) &= \mu_Z + \Omega_{ZY}\Omega_Y^{-1}(y - \mu_Y) \\ \mathrm{Var}(Z \mid Y = y) &= \Omega_Z - \Omega_{ZY}\Omega_Y^{-1}\Omega_{YZ}. \end{aligned}$$

These quantities and the conditional density of Z given $Y = y$ can all be easily evaluated via the sweep operator.

7.4 Definition of the Sweep Operator

Suppose $A = (a_{ij})$ is an $m \times m$ symmetric matrix. Sweeping on the kth diagonal entry $a_{kk} \neq 0$ of A yields a new symmetric matrix $\hat{A} = (\hat{a}_{ij})$ with entries

$$\begin{aligned} \hat{a}_{kk} &= -\frac{1}{a_{kk}} \\ \hat{a}_{ik} &= \frac{a_{ik}}{a_{kk}} \\ \hat{a}_{kj} &= \frac{a_{kj}}{a_{kk}} \\ \hat{a}_{ij} &= a_{ij} - \frac{a_{ik}a_{kj}}{a_{kk}} \end{aligned}$$

for $i \neq k$ and $j \neq k$. Sweeping on the kth diagonal entry can be undone by inverse sweeping on the kth diagonal entry. Inverse sweeping sends the matrix $A = (a_{ij})$ into $\check{A} = (\check{a}_{ij})$ with entries

$$\begin{aligned}
\check{a}_{kk} &= -\frac{1}{a_{kk}} \\
\check{a}_{ik} &= -\frac{a_{ik}}{a_{kk}} \\
\check{a}_{kj} &= -\frac{a_{kj}}{a_{kk}} \\
\check{a}_{ij} &= a_{ij} - \frac{a_{ik}a_{kj}}{a_{kk}}
\end{aligned}$$

for $i \neq k$ and $j \neq k$. Because sweeping and inverse sweeping preserve symmetry, all operations can be carried out on either the lower or upper-triangular part of A alone. This saves both computation and storage. In practice, it is wise to carry out sweeping in double precision.

7.5 Properties of the Sweep Operator

We now develop the basic properties of the sweep operator following the exposition of Jennrich [6]. Readers familiar with Gaussian elimination or the simplex algorithm in linear programming have already been exposed to the major themes of pivoting and exchange [4]. Sweeping is a symmetrized version of Gauss-Jordan pivoting.

Proposition 7.5.1 *Let A be an $m \times m$ matrix and U and V be $p \times m$ matrices with columns $u_1, \ldots, u_m$ and $v_1, \ldots, v_m$, respectively. If $V = UA$ before sweeping on the kth diagonal entry of A, then $\hat{V} = \hat{U}\hat{A}$ after sweeping on the kth diagonal entry of A. Here sweeping sends A into $\hat{A}$, the matrix $\hat{U}$ coincides with U except for the exchange of column u_k for column v_k, and the matrix $\hat{V}$ coincides with V except for the exchange of column v_k for $-u_k$. The inverse sweep produces the same result except that u_k is exchanged for $-v_k$ and v_k is exchanged for u_k. Consequently, an inverse sweep undoes a sweep on the same diagonal entry and vice versa. An inverse sweep also coincides with a sweep cubed.*

Proof: By definition $v_{jl} = \sum_i u_{ji}a_{il}$ for all pairs j and l. After sweeping on a_{kk},

$$\begin{aligned}
\hat{v}_{jk} &= -u_{jk} \\
&= -\frac{1}{a_{kk}}\Big(v_{jk} - \sum_{i \neq k} u_{ji}a_{ik}\Big) \\
&= \hat{u}_{jk}\hat{a}_{kk} + \sum_{i \neq k} \hat{u}_{ji}\hat{a}_{ik}
\end{aligned}$$

$$= \sum_i \hat{u}_{ji}\hat{a}_{ik},$$

and for $l \neq k$,

$$\begin{aligned}
\hat{v}_{jl} &= v_{jl} \\
&= \sum_{i \neq k} u_{ji}a_{il} + u_{jk}a_{kl} \\
&= \sum_{i \neq k} u_{ji}a_{il} + \Big(v_{jk} - \sum_{i \neq k} u_{ji}a_{ik}\Big)\frac{a_{kl}}{a_{kk}} \\
&= \sum_{i \neq k} \hat{u}_{ji}\hat{a}_{il} + \hat{u}_{jk}\hat{a}_{kl} \\
&= \sum_i \hat{u}_{ji}\hat{a}_{il}.
\end{aligned}$$

Thus, $\hat{V} = \hat{U}\hat{A}$. Similar reasoning applies to an inverse sweep.

If a sweep is followed by an inverse sweep on the same diagonal entry, then the doubly transformed matrix $\check{\hat{A}}$ satisfies the equation $V = U\check{\hat{A}}$. Choosing square matrices U and V such that U is invertible allows one to write both A and $\check{\hat{A}}$ as $U^{-1}V$. Likewise, it is easy to check that the inverse and cube of a sweep transform U and V into exactly the same matrices $\check{U}$ and $\check{V}$. ∎

Performing a sequence of sweeps leads to the results stated in the next proposition.

Proposition 7.5.2 *Let the symmetric matrix A be partitioned as*

$$A = \begin{pmatrix} A_{11} & A_{12} \\ A_{21} & A_{22} \end{pmatrix}.$$

If possible, sweeping on the diagonal entries of A_{11} yields

$$\hat{A} = \begin{pmatrix} -A_{11}^{-1} & A_{11}^{-1}A_{12} \\ A_{21}A_{11}^{-1} & A_{22} - A_{21}A_{11}^{-1}A_{12} \end{pmatrix}. \tag{7.2}$$

In other words, sweeping on a matrix in block form conforms to the same rules as sweeping on the matrix entry by entry. Furthermore, if it is possible to sweep on a set of diagonal elements in more than one order, then the result is independent of the order chosen.

Proof: Applying Proposition 7.5.1 repeatedly in the equality

$$\begin{pmatrix} A_{11} & A_{12} \\ A_{21} & A_{22} \end{pmatrix} = \begin{pmatrix} I_{11} & \mathbf{0}_{12} \\ \mathbf{0}_{21} & I_{22} \end{pmatrix}\begin{pmatrix} A_{11} & A_{12} \\ A_{21} & A_{22} \end{pmatrix}$$

leads to

$$\begin{pmatrix} -I_{11} & A_{12} \\ \mathbf{0}_{21} & A_{22} \end{pmatrix} = \begin{pmatrix} A_{11} & \mathbf{0}_{12} \\ A_{21} & I_{22} \end{pmatrix}\begin{pmatrix} \hat{A}_{11} & \hat{A}_{12} \\ \hat{A}_{21} & \hat{A}_{22} \end{pmatrix},$$

where I_{11} and I_{22} are identity matrices and $\mathbf{0}_{12}$ and $\mathbf{0}_{21}$ are zero matrices. This implies

$$\begin{aligned} -I_{11} &= A_{11}\hat{A}_{11} \\ A_{12} &= A_{11}\hat{A}_{12} \\ \mathbf{0}_{21} &= A_{21}\hat{A}_{11} + \hat{A}_{21} \\ A_{22} &= A_{21}\hat{A}_{12} + \hat{A}_{22}. \end{aligned}$$

Solving these equations for the blocks of $\hat{A}$ yields the claimed results. ■

Sweeping is also a device for monitoring the positive definiteness of a matrix. In fact, sweeping is a more practical test than the classical criterion of Sylvester mentioned in Problem 13 of Chapter 8.

Proposition 7.5.3 *A symmetric matrix A is positive definite if and only if each diagonal entry can be swept in succession and is positive until it is swept. When a diagonal entry of a positive definite matrix A is swept, it becomes negative and remains negative thereafter. Furthermore, taking the product of the diagonal entries just before each is swept yields the determinant of A.*

Proof: The equivalence of the two conditions characterizing A is obvious if A is a 1×1 matrix. If A is $m \times m$, then suppose it has the form given in Proposition 7.5.2. Now the matrix identity

$$\begin{aligned} &\begin{pmatrix} A_{11} & \mathbf{0}_{12} \\ \mathbf{0}_{21} & A_{22} - A_{21}A_{11}^{-1}A_{12} \end{pmatrix} \\ &= \begin{pmatrix} I_{11} & \mathbf{0}_{12} \\ -A_{21}A_{11}^{-1} & I_{22} \end{pmatrix} \begin{pmatrix} A_{11} & A_{12} \\ A_{21} & A_{22} \end{pmatrix} \begin{pmatrix} I_{11} & -A_{11}^{-1}A_{12} \\ \mathbf{0}_{21} & I_{22} \end{pmatrix} \end{aligned} \tag{7.3}$$

shows that A is positive definite if and only if A_{11} and $A_{22} - A_{21}A_{11}^{-1}A_{12}$ are both positive definite. In view of equation (7.2) of Proposition 7.5.2, the equivalence of the sweeping condition and positive definiteness of A follows inductively from the same equivalence applied to the smaller matrices A_{11} and $A_{22} - A_{21}A_{11}^{-1}A_{12}$.

Once a diagonal entry of A has been swept, the diagonal entry forms part of the matrix $-A_{11}^{-1}$, which is negative definite. Hence, the swept diagonal entries must be negative. Finally, formula (7.3) shows that

$$\det A = \det(A_{11})\det(A_{22} - A_{21}A_{11}^{-1}A_{12}). \tag{7.4}$$

The validity of the asserted procedure for calculating $\det A$ now follows inductively since it is obviously true for a 1×1 matrix. ■

The determinant formula (7.4) is of independent interest. It does not depend on A being symmetric. Obviously, the analogous formula

$$\det A = \det(A_{22})\det(A_{11} - A_{12}A_{22}^{-1}A_{21}) \tag{7.5}$$

also holds.

7.6 Applications of Sweeping

The representation (7.2) is of paramount importance. For instance, in linear regression, suppose we construct the matrix

$$\begin{pmatrix} X^tX & X^ty \\ y^tX & y^ty \end{pmatrix} \tag{7.6}$$

and sweep on the diagonal entries of X^tX. Then the basic theoretical ingredients

$$\begin{aligned} &\begin{pmatrix} -(X^tX)^{-1} & (X^tX)^{-1}X^ty \\ y^tX(X^tX)^{-1} & y^ty - y^tX(X^tX)^{-1}X^ty \end{pmatrix} \\ = &\begin{pmatrix} -\frac{1}{\sigma^2}\operatorname{Var}(\hat{\beta}) & \hat{\beta} \\ \hat{\beta}^t & \|y-\hat{y}\|_2^2 \end{pmatrix} \end{aligned}$$

magically emerge.

When we construct the matrix

$$\begin{pmatrix} \Omega & x-\mu \\ x^t-\mu^t & 0 \end{pmatrix}$$

for the multivariate normal distribution and sweep on the diagonal entries of Ω, we get the quadratic form $-(x-\mu)^t\Omega^{-1}(x-\mu)$ in the lower-right block of the swept matrix. In the process we can also accumulate $\det\Omega$. To avoid underflows and overflows, it is better to compute $\ln\det\Omega$ by summing the logarithms of the diagonal entries as we sweep on them. If we partition X as $(Y^t, Z^t)^t$ and sweep on the upper-left block of

$$\begin{pmatrix} \Omega_Y & \Omega_{YZ} & \mu_Y - y \\ \Omega_{ZY} & \Omega_Z & \mu_Z \\ \mu_Y^t - y^t & \mu_Z^t & 0 \end{pmatrix},$$

then the conditional mean $\mathrm{E}(Z \mid Y = y) = \mu_Z + \Omega_{ZY}\Omega_Y^{-1}(y - \mu_Y)$ and conditional variance $\operatorname{Var}(Z \mid Y = y) = \Omega_Z - \Omega_{ZY}\Omega_Y^{-1}\Omega_{YZ}$ are immediately available.

7.7 Cholesky Decompositions

Let A be an $m \times m$ positive definite matrix. The Cholesky decomposition L of A is a lower-triangular matrix with positive diagonal entries that serves as an asymmetric square root of A. To prove that such a decomposition exists and is unique, we argue by induction. In the case of a positive scalar a, the Cholesky decomposition $l = \sqrt{a}$. For an $m \times m$ matrix $A = (a_{ij})$ with

$m > 1$, the square root condition $A = LL^t$ can be written in partitioned form as

$$\begin{pmatrix} a_{11} & a^t \\ a & A_{22} \end{pmatrix} = \begin{pmatrix} \ell_{11} & \mathbf{0}^t \\ \ell & L_{22} \end{pmatrix} \begin{pmatrix} \ell_{11} & \ell^t \\ \mathbf{0} & L_{22}^t \end{pmatrix}.$$

Equating the two sides block by block yields the identities

$$\begin{aligned} a_{11} &= \ell_{11}^2 \\ a &= \ell_{11}\ell \\ A_{22} &= \ell\ell^t + L_{22}L_{22}^t. \end{aligned}$$

Solving these equations gives

$$\begin{aligned} \ell_{11} &= \sqrt{a_{11}} \\ \ell &= \ell_{11}^{-1}a \\ L_{22}L_{22}^t &= A_{22} - \ell\ell^t \;=\; A_{22} - a_{11}^{-1}aa^t. \end{aligned}$$

Because $a_{11} > 0$, the values of ℓ_{11} and ℓ are uniquely determined. Proposition 7.5.2 implies that the matrix $A_{22} - a_{11}^{-1}aa^t$ is positive definite. Therefore in view of the induction hypothesis, L_{22} also exists and is unique.

The great virtue of this proof is that it is constructive and can be easily implemented in computer code. If we want $\det A$ as well as L, then we simply note that $\det A = (\det L)^2$ and that the determinant of L is the product of its diagonal entries. Another strength of the proof is that it shows that positive semidefinite matrices also possess Cholesky decompositions, though these are no longer unique. To adapt the above proof, we must examine the situation $a_{11} = 0$. In this case, we take $\ell_{11} = 0$. Fortunately, this does not conflict with the requirement $a = \ell_{11}\ell$ because the vector a also vanishes when a_{11} vanishes. Lack of uniqueness arises because ℓ is not fully determined, but certainly the specific choice $\ell = \mathbf{0}$ is viable. With this choice, $L_{22}L_{22}^t = A_{22}$, and induction again pushes the algorithm to completion. Problem 7 gives an example of nonuniqueness.

Regression analysis can be performed by finding the Cholesky factorization

$$(X, y)^t(X, y) = \begin{pmatrix} X^tX & X^ty \\ y^tX & y^ty \end{pmatrix} = \begin{pmatrix} L & \mathbf{0} \\ \ell^t & d \end{pmatrix} \begin{pmatrix} L^t & \ell \\ \mathbf{0}^t & d \end{pmatrix}.$$

Clearly L is the Cholesky decomposition of X^tX. The two equations

$$L\ell = X^ty, \quad L^t\beta = \ell,$$

which are equivalent to the normal equations, can be solved quickly in succession by forward and backward substitution for the estimated regression coefficients $\hat{\beta}$. Before commenting on this further, let us add that the equation $y^ty = \ell^t\ell + d^2$ implies that

$$d^2 = y^ty - \ell^t\ell = y^ty - y^tX(X^tX)^{-1}X^ty = \|y - \hat{y}\|_2^2.$$

Now let $L = (l_{ij})$ and $U = (u_{ij})$ be arbitrary $m \times m$ lower and upper triangular matrices with nonzero diagonal entries. To solve the equation $Lf = v$ by forward substitution, we take

$$\begin{aligned} f_1 &= l_{11}^{-1} v_1 \\ f_2 &= l_{22}^{-1}[v_2 - l_{21} f_1] \\ f_j &= l_{jj}^{-1}\Big[v_j - \sum_{k=1}^{j-1} l_{jk} f_k\Big], \quad j = 3, \ldots, m. \end{aligned}$$

To solve the equation $Ub = w$ by backward substitution, we take

$$\begin{aligned} b_m &= u_{mm}^{-1} w_m \\ b_{m-1} &= u_{m-1,m-1}^{-1}[w_{m-1} - u_{m-1,m} b_m] \\ b_j &= u_{jj}^{-1}\Big[w_j - \sum_{k=j+1}^{m} u_{jk} b_k\Big], \quad j = m-2, \ldots, 1. \end{aligned}$$

7.8 Gram-Schmidt Orthogonalization

The sweeping and Cholesky decomposition approaches to regression explicitly form the matrix X^tX. In contrast, orthogonalization methods operate directly on X. The QR decomposition represents a $p \times q$ matrix X with full column rank as the product QR of a $p \times q$ matrix Q with orthonormal columns and a $q \times q$ invertible upper-triangular matrix R. We will further assume without loss of generality that the diagonal entries of R are positive. When this is not the case, we multiply R by a diagonal matrix D whose diagonal entries are chosen from $\{-1, +1\}$ in such a way that DR has positive diagonal entries. The new representation $X = (QD)(DR)$ is a valid QR decomposition in the restricted sense.

Assuming for the moment that the QR decomposition exists, we can rephrase the normal equations $X^tX\beta = X^ty$ as

$$R^tQ^tQR\beta = R^tQ^ty.$$

Because R^t is invertible and Q^tQ equals the $q \times q$ identity matrix I_q, it follows that the normal equations reduce to $R\beta = Q^ty = r$. This system of equations can be solved by forward substitution for β. The equation

$$X^tX = R^tQ^tQR = R^tR$$

determines $L = R^t$ as the Cholesky decomposition of X^tX. Given that R is unique and invertible, the identity $Q = XR^{-1}$ determines Q.

Gram-Schmidt orthogonalization takes a collection of vectors such as the columns $x_1, \ldots, x_q$ of the design matrix X into an orthonormal collection of vectors $u_1, \ldots, u_q$ spanning the same column space. The algorithm begins by defining

$$u_1 = \frac{1}{\|x_1\|_2} x_1.$$

Given $u_1, \ldots, u_{k-1}$, the next unit vector u_k in the sequence is defined by dividing the column vector

$$v_k = x_k - \sum_{j=1}^{k-1} u_j^t x_k \, u_j \tag{7.7}$$

by its Euclidean norm. In other words, we subtract from x_k its projections onto each of the previously created u_j and normalize the result. A simple induction argument shows that the vectors $u_1, \ldots, u_k$ form an orthonormal basis of the subspace spanned by $x_1, \ldots, x_k$, assuming of course that these latter vectors are independent. The upper-triangular entries of the matrix R are given by the formulas $r_{jk} = u_j^t x_k$ for $1 \leq j < k$ and $r_{kk} = \|v_k\|_2$. This fact can be deduced by observing that component i of the vector identity (7.7) is just a disguised form of the matrix identity

$$x_{ik} = \sum_{j=1}^{k} u_{ij} r_{jk}$$

required by the QR decomposition with $Q = (u_{ij})$.

Computational experience has shown that the numerical stability of Gram-Schmidt orthogonalization can be improved by a simple device. In equation (7.7) we subtract from x_k all of its projections simultaneously. If the columns of X are nearly collinear, it is better to subtract off the projections sequentially. Thus, we let $v_k^{(1)} = x_k$ and sequentially compute

$$v_k^{(j+1)} = v_k^{(j)} - u_j^t v_k^{(j)} \, u_j, \quad r_{jk} = u_j^t v_k^{(j)}$$

until we reach $v_k = v_k^{(k)}$. As before, $r_{kk} = \|v_k\|_2$. With perfect arithmetic, the modified algorithm arrives at the same outcome as the previous algorithm. However, with imperfect arithmetic, the vectors $u_1, \ldots, u_q$ computed under the modified algorithm are more accurate and more nearly orthogonal.

In practice, the modified Gram-Schmidt method is applied to the partitioned matrix (X, y). All of the required information can be harvested from the extended QR decomposition

$$(X, y) = (Q, q) \begin{pmatrix} R & r \\ \mathbf{0} & d \end{pmatrix}.$$

If we premultiply $y = Qr + dq$ by Q^t, then we find that $Q^t y = r$, which is directly applicable in solving for β in $R\beta = Q^t y = r$. The predicted values $X\hat{\beta} = QRR^{-1}r = Qr$ and the residuals $y - X\hat{\beta} = y - Qr = dq$ are also immediately available. From the residuals, we get the residual sum of squares d^2 since q is a unit vector.

7.9 Orthogonalization by Householder Reflections

To construct the QR decomposition of the $p \times q$ matrix X, one can multiply it by a special sequence $H_1, \ldots, H_{q-1}$ of Householder reflections and arrive at the result

$$H_{q-1} \cdots H_2 H_1 X \quad = \quad \begin{pmatrix} R \\ \mathbf{0} \end{pmatrix}, \tag{7.8}$$

where R is a $q \times q$ upper triangular matrix with positive diagonal entries. If we let O be the $p \times p$ orthogonal matrix $H_{q-1} \cdots H_2 H_1$, then the residual sum of squares satisfies

$$\|y - X\beta\|_2^2 \;=\; \|Oy - OX\beta\|_2^2 \;=\; \left\|Oy - \begin{pmatrix} R \\ \mathbf{0} \end{pmatrix}\beta\right\|_2^2 \;=\; \left\|Oy - \begin{pmatrix} R\beta \\ \mathbf{0} \end{pmatrix}\right\|_2^2.$$

Putting $Oy = \begin{pmatrix} r_1 \\ r_2 \end{pmatrix}$, we recast this as

$$\|y - X\beta\|_2^2 \quad = \quad \|r_1 - R\beta\|_2^2 + \|r_2\|_2^2.$$

There is nothing we can do to diminish $\|r_2\|_2^2$, but we can eliminate the term $\|r_1 - R\beta\|_2^2$ by setting β equal to $\hat{\beta} = R^{-1}r_1$. The residual sum of squares then collapses to $\|r_2\|_2^2$. If we write

$$O \quad = \quad \begin{pmatrix} O_q \\ O_{p-q} \end{pmatrix}$$

with submatrices O_q and O_{p-q} having q and $p - q$ rows, respectively, then multiplying equation (7.8) by O^t produces the standard QR decomposition

$$X \quad = \quad (O_q^t, O_{p-q}^t)\begin{pmatrix} R \\ \mathbf{0} \end{pmatrix} \quad = \quad O_q^t R.$$

The key to determining the Householder reflections is the simple observation that for every pair of vectors v and w of equal Euclidean length, we can construct a Householder matrix $H = I - 2uu^t$ carrying v into w. Because the condition $Hv = w$ is equivalent to the condition

$$-2u^t v\, u \quad = \quad w - v,$$

the unit vector u must be a multiple of $w - v$. The choice

$$u = \frac{1}{\|w-v\|}(v-w)$$

gives

$$\begin{aligned} Hv &= v - \frac{2}{\|w-v\|^2}(v-w)(v-w)^t v \\ &= v + \frac{2\|v\|^2 - 2w^t v}{\|w-v\|^2}(w-v) \\ &= v + \frac{\|v\|^2 - 2w^t v + \|w\|^2}{\|w-v\|^2}(w-v) \\ &= v + (w-v) \\ &= w. \end{aligned}$$

Once H is determined, its action $Hx = x - 2u^t x u$ on any other vector x is trivial to compute.

Because we want R to be upper triangular, we choose $H = H_1$ to map the first column x of X to the vector $(\|x\|, 0, \ldots, 0)^t$. Except for the first entry, every entry of the vector u defining H is a positive multiple of the corresponding entry of x. The first entry u_1 is a positive multiple of $x_1 - \|x\|$. If x_1 is positive, we run the risk of losing significant digits in computing u_1. Application of the formula

$$x_1 - \|x\| = \frac{x_1^2 - \|x\|^2}{x_1 + \|x\|} = -\frac{x_2^2 + \cdots + x_p^2}{x_1 + \|x\|}$$

circumvents this pitfall. Once we multiply $H = H_1$ times the columns of X, we are left with a matrix of the form

$$H_1 X = \begin{pmatrix} r_{11} & r \\ \mathbf{0} & A \end{pmatrix}.$$

The columns of A are linearly independent because assuming otherwise renders the columns of $H_1 X$ linearly dependent. By induction we reduce A to an upper triangular matrix by multiplying it by a sequence of $\tilde{H}_{q-1}, \ldots, \tilde{H}_2$ Householder reflections. If we elevate each of the $(p-1) \times (p-1)$ matrices $\tilde{H}_i$ to

$$H_i = \begin{pmatrix} 1 & \mathbf{0}^t \\ \mathbf{0} & \tilde{H}_i \end{pmatrix},$$

then these matrices have the right dimension and leave the vector

$$(\|x\|, 0, \ldots, 0)^t$$

undisturbed.

7.10 Comparison of the Different Algorithms

The various algorithms can be compared on the basis of speed, accuracy, and flexibility. Table 7.1 gives the approximate number of floating point operations (flops) for each of the methods. Recall here that p is the number of cases and q is the number of regression coefficients. The various methods are listed in order of their numerical accuracy as rated by Seber and Lee [11], with sweeping the most prone to numerical error and the modified Gram-Schmidt procedure the least prone.

TABLE 7.1. Approximate Flop Counts for Different Regression Methods

Method	Flop Count
Sweeping	$pq^2 + q^3$
Cholesky Decomposition	$pq^2 + \frac{1}{3}q^3$
Householder Orthogonalization	$2pq^2 - \frac{2}{3}q^3$
Modified Gram-Schmidt	$2pq^2$

Although flops omit a great deal of behind-the-scenes processing in numerical computing, costs such as fetching array entries tend to be strongly correlated with flops [11]. For this reason, flops serve as a useful measure of computational speed. If p is large compared to q, then sweeping and Cholesky decomposition are approximately twice as fast as the other two methods. Their speed comes at a price of reduced accuracy, but using double precision mitigates most of the problems in practice. Despite its apparent inferiority to Cholesky decomposition, sweeping remains in common use because of its flexibility in stepwise regression and its advantages in dealing with the multivariate normal distribution.

7.11 Woodbury's Formula

Suppose A is an $m \times m$ matrix with known inverse A^{-1} and known determinant $\det A$. If U and V are $m \times n$ matrices of rank n, then $A+UV^t$ is a rank n perturbation of A. In many applications n is much smaller than m. If U has columns $u_1, \ldots, u_n$ and V has columns $v_1, \ldots, v_n$, then $A + UV^t$ can also be expressed as

$$A + UV^t = A + \sum_{i=1}^{n} u_i v_i^t.$$

Woodbury's formula amounts to

$$(A + UV^t)^{-1} = A^{-1} - A^{-1}U(I_n + V^t A^{-1}U)^{-1}V^t A^{-1}, \quad (7.9)$$

where I_n is the $n \times n$ identity matrix [9, 10]. Equation (7.9) is valuable because the smaller $n \times n$ matrix $I_n + V^t A^{-1} U$ is typically much easier to invert than the larger $m \times m$ matrix $A + UV^t$. When A is symmetric and $V = U$, Woodbury's formula is a consequence of sweeping the matrix

$$\begin{pmatrix} -A & U \\ U^t & I_n \end{pmatrix}$$

first on its upper-left block and then on its lower-right block and comparing the results to sweeping on these blocks in reverse order.

In solving the linear equation $(A + UV^t)x = b$, computing the whole inverse $(A+UV^t)^{-1}$ is unnecessary. Press et al. [10] recommend the following procedure. First compute the column vectors $z_1, \ldots, z_n$ of $Z = A^{-1}U$ by solving each linear equation $Az_i = u_i$. Then calculate $H = (I_n + V^t Z)^{-1}$. Finally, solve the linear equation $Ay = b$ for y. The solution to the linear equation $(A + UV^t)x = b$ can then be written as $x = y - ZHV^t y$.

If $A + UU^t$ is the covariance matrix of a multivariate normal random vector X, then to evaluate the density of X it is necessary to compute $\det(A + UU^t)$. (Observe that choosing $V = U$ preserves the symmetry of A.) The identity

$$\det(A + UV^t) = \det A \det(I_n + V^t A^{-1} U) \tag{7.10}$$

permits easy computation of $\det(A + UV^t)$. This identity also evidently implies that $A + UV^t$ is invertible if and only if $I_n + V^t A^{-1} U$ is invertible. To prove (7.10), we note that

$$\begin{aligned} \det(A + UV^t) &= \det A \det(I_m + A^{-1}UV^t) \\ &= \det A \det \begin{pmatrix} I_n & V^t \\ -A^{-1}U & I_m \end{pmatrix} \\ &= \det A \det(I_n + V^t A^{-1} U) \end{aligned}$$

follows directly from equations (7.4) and (7.5).

7.12 Problems

1. Consider the matrix

$$A = \frac{1}{3} \begin{pmatrix} 1 & -2 & -2 \\ -2 & 1 & -2 \\ -2 & -2 & 1 \end{pmatrix}.$$

Compute its inverse by sweeping. Determine whether A is positive definite based on the intermediate results of sweeping.

2. Calculate how many arithmetic operations it takes to compute one sweep of an $m \times m$ symmetric matrix A. If you calculate only the upper-triangular part of the result, how many operations do you save? Note that the revised sweeping scheme

$$\begin{aligned} \hat{a}_{kk} &= -\frac{1}{a_{kk}} \\ \hat{a}_{ik} &= -\hat{a}_{kk}a_{ik} \\ \hat{a}_{kj} &= -\hat{a}_{kk}a_{kj} \\ \hat{a}_{ij} &= a_{ij} - \hat{a}_{ik}a_{kj} \end{aligned}$$

for $i \neq k$ and $j \neq k$ is more efficient than the original sweeping scheme. How many operations does it take to compute A^{-1}, assuming all of the required diagonal sweeps are possible?

3. Suppose the positive definite matrix $A = (a_{ij})$ has inverse $B = (b_{ij})$. Show that $a_{ii}^{-1} \leq b_{ii}$ with equality if and only if $a_{ij} = a_{ji} = 0$ for all $j \neq i$. If A is an expected information matrix, what implications does this result have for maximum likelihood estimation in large samples?

4. The jackknife method of regression analysis can be implemented by replacing the linear regression matrix (7.6) by the matrix

$$\begin{pmatrix} X^t \\ I_p \\ y^t \end{pmatrix} \begin{pmatrix} X \; I_p \; y \end{pmatrix} = \begin{pmatrix} X^tX & X^t & X^ty \\ X & I_p & y \\ y^tX & y^t & y^ty \end{pmatrix},$$

sweeping on its upper-left block X^tX, and then sweeping on its $(q+k)$th diagonal entry for some k between 1 and p. Prove that this action yields the necessary ingredients for regression analysis omitting the kth observation y_k and the corresponding kth row of the $p \times q$ design matrix X [1]. (Hint: The additional sweep is equivalent to replacing the kth regression equation $y_k = \sum_{l=1}^q x_{kl}\beta_l + e_k$ by the regression equation $y_k = \sum_{l=1}^q x_{kl}\beta_l + \beta_{q+k} + e_k$ involving an additional parameter; the other regression equations are untouched. The parameter β_{q+k} can be adjusted to give a perfect fit to y_k. Hence, the estimates $\hat{\beta}_1, \ldots, \hat{\beta}_q$ after the additional sweep depend only on the observations y_i for $i \neq k$.)

5. Continuing Problem 4, let h_{kk} be the kth diagonal entry of the projection matrix $X(X^tX)^{-1}X^t$. If $\hat{y}_k$ is the predicted value of y_k and $\hat{y}_k^{-k}$ is the predicted value of y_k omitting this observation, then demonstrate that

$$y_k - \hat{y}_k^{-k} = \frac{y_k - \hat{y}_k}{1 - h_{kk}}.$$

(Hint: Apply the Sherman-Morrison-Woodbury formula.)

6. Find by hand the Cholesky decomposition of the matrix

$$A = \begin{pmatrix} 2 & -2 \\ -2 & 5 \end{pmatrix}.$$

7. Show that the matrices

$$B = \begin{pmatrix} 1 & 0 & 0 \\ 2 & 0 & 0 \\ 2 & 3 & 2 \end{pmatrix}, \quad C = \begin{pmatrix} 1 & 0 & 0 \\ 2 & 0 & 0 \\ 2 & 0 & \sqrt{13} \end{pmatrix}$$

are both valid Cholesky decompositions of the positive semidefinite matrix

$$A = \begin{pmatrix} 1 & 2 & 2 \\ 2 & 4 & 4 \\ 2 & 4 & 17 \end{pmatrix}.$$

8. Suppose the matrix $A = (a_{ij})$ is banded in the sense that $a_{ij} = 0$ when $|i-j| > d$. Prove that the Cholesky decomposition $B = (b_{ij})$ also satisfies the band condition $b_{ij} = 0$ when $|i-j| > d$.

9. How many arithmetic operations does it take to calculate the Cholesky decomposition L of an $m \times m$ positive definite matrix A? You may count a square root as a single operation. If the solution of the linear system $Ab = v$ is desired, one can successively solve $Lf = v$ and $L^t b = f$. Estimate the number of extra operations required to do this once L is obtained.

10. Find the QR decomposition of the matrix

$$X = \begin{pmatrix} 1 & 3 & 3 \\ 1 & 3 & 1 \\ 1 & 1 & 5 \\ 1 & 1 & 3 \end{pmatrix}$$

by the Gram-Schmidt process.

11. If $X = QR$ is the QR decomposition of X, then show that the projection matrix

$$X(X^t X)^{-1} X^t = QQ^t.$$

Show that $|\det X| = |\det R|$ when X is square and in general that $\det(X^t X) = (\det R)^2$.

12. Prove that (a) the product of two upper-triangular matrices is upper triangular, (b) the inverse of an upper-triangular matrix is upper triangular, (c) if the diagonal entries of an upper-triangular matrix

are positive, then the diagonal entries of its inverse are positive, and (d) if the diagonal entries of an upper-triangular matrix are unity, then the diagonal entries of its inverse are unity. Similar statements apply to lower-triangular matrices.

13. Demonstrate that an orthogonal upper-triangular matrix is diagonal.

14. Consider the $m \times m$ matrix

$$M = \begin{pmatrix} a & b & \cdots & b \\ b & a & \cdots & b \\ \vdots & \vdots & \vdots & \vdots \\ b & b & \cdots & a \end{pmatrix}.$$

If $\mathbf{1}$ is the column vector with all entries 1, then show that M has inverse and determinant

$$\begin{aligned} M^{-1} &= \frac{1}{a-b}\Big[I_m - \frac{b}{a+(m-1)b}\mathbf{1}\mathbf{1}^t\Big] \\ \det M &= (a-b)^{m-1}[a+(m-1)b]. \end{aligned}$$

15. Prove the slight generalization

$$(A+UDV^t)^{-1} = A^{-1} - A^{-1}U(D^{-1}+V^tA^{-1}U)^{-1}V^tA^{-1}$$

of the Woodbury formula (7.9) for compatible matrices A, U, D, and V.

16. Suppose A and B are invertible matrices of the same dimension. If $A^{-1}+B^{-1}$ is invertible, then show that $A+B$ is invertible, and conversely. Furthermore, prove the identities

$$(A^{-1}+B^{-1})^{-1} = A(A+B)^{-1}B = B(A+B)^{-1}A.$$

17. Let D be an $m \times m$ diagonal matrix with ith diagonal entry d_i. Show that the characteristic polynomial $p(\lambda)$ of the rank-one perturbation $D+uv^t$ of D reduces to

$$p(\lambda) = \prod_{i=1}^m(\lambda-d_i) - \sum_{i=1}^m u_iv_i\prod_{j\neq i}(\lambda-d_j).$$

Prove that d_i is an eigenvalue of $D+uv^t$ whenever $d_j = d_i$ for some $j \neq i$. If d_i is a unique diagonal entry, then show that it is an eigenvalue of $D+uv^t$ if and only if $u_iv_i = 0$. Finally, if none of the d_i is an eigenvalue, then demonstrate that the eigenvalues coincide with the roots of the rational function

$$r(\lambda) = 1 - \sum_{i=1}^m \frac{u_iv_i}{\lambda-d_i}.$$

18. Let A be an $m \times n$ matrix of full rank. One can easily show that

$$P = I_n - A^t(AA^t)^{-1}A$$

is the unique $n \times n$ orthogonal matrix projecting onto the null space of A; in other words, $P = P^t$, $P^2 = P$, and $Px = x$ if and only if $Ax = \mathbf{0}$. If b is a vector such that $b^tPb \neq 0$, then verify that the rank-one perturbation

$$Q = P - \frac{1}{b^tPb}Pbb^tP$$

is the unique orthogonal projection onto the null space of $\binom{A}{b^t}$. If $b^tPb = 0$, then verify that $Q = P$ serves as the orthogonal projector.

19. Let A be a square matrix of rank k and x and y vectors such that $x^tAy \neq 0$. Demonstrate that the matrix $A - (x^tAy)^{-1}Ayx^tA$ has rank $k-1$. More generally suppose that A is $m \times n$, B is $l \times m$, C is $l \times n$, and BAC^t is invertible. If A has rank k and $l \leq k$, then show that $A - AC^t(BAC^t)^{-1}BA$ has rank $k - l$. See the reference [5] for a history and applications of this problem.

7.13 References

[1] Dempster AP (1969) *Continuous Multivariate Analysis.* Addison-Wesley, Reading, MA

[2] Golub GH, Van Loan CF (1996) *Matrix Computations*, 3rd ed. Johns Hopkins University Press, Baltimore

[3] Goodnight JH (1979) A tutorial on the SWEEP operator. *Amer Statistician* 33:149–158

[4] Henrici P (1982) *Essentials of Numerical Analysis with Pocket Calculator Demonstrations.* Wiley, New York

[5] Hubert L, Meulman J, Heiser W (2000) Two purposes for matrix factorization: a historical appraisal. *SIAM Review* 42:68–82

[6] Jennrich RI (1977) Stepwise regression. *Statistical Methods for Digital Computers.* Enslein K, Ralston A, Wilf HS, editors, Wiley-Interscience, New York, pp 58–75

[7] Kennedy WJ Jr, Gentle JE (1980) *Statistical Computing.* Marcel Dekker, New York

[8] Little RJA, Rubin DB (1987) *Statistical Analysis with Missing Data.* Wiley, New York

[9] Miller KS (1987) *Some Eclectic Matrix Theory.* Robert E Krieger Publishing, Malabar, FL

[10] Press WH, Teukolsky SA, Vetterling WT, Flannery BP (1992) *Numerical Recipes in Fortran: The Art of Scientific Computing,* 2nd ed. Cambridge University Press, Cambridge

[11] Seber GAF, Lee AJ (2003) *Linear Regression Analysis,* 2nd ed. Wiley, Hoboken, NJ

[12] Stewart GW (1987) *Afternotes on Numerical Analysis.* SIAM, Philadelphia

[13] Strang G (1986) *Introduction to Applied Mathematics.* Wellesley-Cambridge Press, Wellesley, MA

[14] Thisted RA (1988) *Elements of Statistical Computing.* Chapman & Hall, New York

[15] Trefethen LN, Bau D III (1997) *Numerical Linear Algebra.* SIAM, Philadelphia

8

Eigenvalues and Eigenvectors

8.1 Introduction

Finding the eigenvalues and eigenvectors of a symmetric matrix is one of the basic tasks of computational statistics. For instance, in principal components analysis [13], a random m-vector X with covariance matrix Ω is postulated. As a symmetric matrix, Ω can be decomposed as

$$\Omega = UDU^t,$$

where D is the diagonal matrix of eigenvalues of Ω and U is the corresponding orthogonal matrix of eigenvectors. If the eigenvalues are distinct and ordered $0 \leq \lambda_1 < \lambda_2 < \ldots < \lambda_m$, then the columns $U_1, \ldots, U_m$ of U are unique up to sign. The random variables $V_j = U_j^t X$, $j = 1, \ldots, m$, have covariance matrix $U^t \Omega U = D$. These random variables are termed principal components. They are uncorrelated and increase in variance from the first, V_1, to the last, V_m.

Besides this classical application, there are other reasons for being interested in eigenvalues and eigenvectors. We have already seen how the dominant eigenvalue of a symmetric matrix Ω determines its norm $||\Omega||_2$. If Ω is the covariance matrix of a normally distributed random vector X with mean $\mathrm{E}(X) = \mu$, then the quadratic form and the determinant

$$\begin{aligned}
(x-\mu)^t \Omega^{-1} (x-\mu) &= [U^t(x-\mu)]^t D^{-1} U^t (x-\mu) \\
\det \Omega &= \prod_i \lambda_i
\end{aligned}$$

appearing in the density of X are trivial to calculate if Ω can be diagonalized explicitly. Understanding the eigenstructure of matrices also is crucial in proving convergence for maximum likelihood algorithms. This leads us to consider the Rayleigh quotient later in this chapter.

8.2 Jacobi's Method

Rather than survey the variety of methods for computing the eigenvalues and eigenvectors of a symmetric matrix Ω, we will focus on just one, the classical Jacobi method [1, 2, 4, 5, 8, 15, 16]. This is not necessarily the fastest method, but it does illustrate some useful ideas for proving convergence of iterative methods in general. One attractive feature of Jacobi's

K. Lange, *Numerical Analysis for Statisticians*, Statistics and Computing,
DOI 10.1007/978-1-4419-5945-4_8, © Springer Science+Business Media, LLC 2010

method is the ease with which it can be implemented on parallel computers. This fact suggests that it may regain its competitiveness on large-scale problems.

The idea of the Jacobi method is to gradually transform Ω to a diagonal matrix by a sequence of similarity transformations. Each similarity transformation involves a rotation designed to increase the sum of squares of the diagonal entries of the matrix currently similar to Ω. Section 6.4 reviews basic material on orthogonal transformations such as the rotation

$$R = \begin{pmatrix} \cos\theta & \sin\theta \\ -\sin\theta & \cos\theta \end{pmatrix} \tag{8.1}$$

in the plane. It is proved in Section 6.4 that two $m \times m$ matrices A and B related by the orthogonal similarity transformation $B = U^t AU$ satisfy the identity $||B||_F^2 = ||A||_F^2$. If we appeal to the circular permutation property of the matrix trace function, then it is also obvious that $\operatorname{tr}(B) = \operatorname{tr}(A)$.

Now consider the effect of a rotation involving row k and column l of the $m \times m$ matrix $A = (a_{ij})$. Without loss of generality, we take $k = 1$ and $l = 2$ and form the orthogonal matrix

$$U = \begin{pmatrix} R & \mathbf{0} \\ \mathbf{0}^t & I_{m-2} \end{pmatrix}.$$

The diagonal entry b_{ii} of $B = U^t AU$ equals a_{ii} when $i > 2$. The entries of the upper-left block of B are given by

$$\begin{aligned} b_{11} &= a_{11}\cos^2\theta - 2a_{12}\cos\theta\sin\theta + a_{22}\sin^2\theta \\ b_{12} &= (a_{11} - a_{22})\cos\theta\sin\theta + a_{12}(\cos^2\theta - \sin^2\theta) \\ b_{22} &= a_{11}\sin^2\theta + 2a_{12}\cos\theta\sin\theta + a_{22}\cos^2\theta. \end{aligned} \tag{8.2}$$

By virtue of the trigonometric identities

$$\begin{aligned} \cos^2\theta - \sin^2\theta &= \cos(2\theta) \\ \cos\theta\sin\theta &= \frac{1}{2}\sin(2\theta), \end{aligned}$$

it follows that

$$b_{12} = \frac{a_{11} - a_{22}}{2}\sin(2\theta) + a_{12}\cos(2\theta).$$

When $a_{22} - a_{11} \neq 0$, there is a unique $|\theta| < \pi/4$ such that

$$\tan(2\theta) = \frac{2a_{12}}{a_{22} - a_{11}}.$$

Making this choice of θ forces $b_{12} = 0$. When $a_{22} - a_{11} = 0$, the choice $\theta = \pi/4$ also gives $b_{12} = 0$.

Given $b_{12} = 0$, we infer from the first two formulas of (8.2) that

$$\begin{aligned} b_{11} &= a_{11}\cos^2\theta - 2a_{12}\cos\theta\sin\theta + a_{22}\sin^2\theta + b_{12}\tan\theta \\ &= a_{11} - a_{12}\tan\theta. \end{aligned} \tag{8.3}$$

The trace identity $b_{11} + b_{22} = a_{11} + a_{22}$ then yields the corresponding equality

$$b_{22} = a_{22} + a_{12}\tan\theta.$$

The two-dimensional version of the identity $||B||_F^2 = ||A||_F^2$ applies to the upper-left blocks of B and A. In other words, if $b_{12} = 0$, then

$$b_{11}^2 + b_{22}^2 = a_{11}^2 + 2a_{12}^2 + a_{22}^2.$$

In terms of the sums of squares of the diagonal entries of the matrices B and A, this translates into

$$\begin{aligned} S(B) &= \sum_{i=1}^m b_{ii}^2 \\ &= \sum_{i=1}^m a_{ii}^2 + 2a_{12}^2 \\ &= S(A) + 2a_{12}^2. \end{aligned}$$

Thus, choosing $b_{12} = 0$ forces $S(B) > S(A)$ whenever $a_{12} \neq 0$.

Beginning with a symmetric matrix Ω, Jacobi's method employs a sequence of rotations U_n as designed above to steadily decrease the sum of squares

$$||\Omega_n||_F^2 - S(\Omega_n) = ||\Omega||_F^2 - S(\Omega_n)$$

of the off-diagonal entries of the transformed matrices

$$\Omega_n = U_n^t \cdots U_1^t \Omega U_1 \cdots U_n.$$

For large n, approximate eigenvalues of Ω can be extracted from the diagonal of the nearly diagonal matrix Ω_n, and approximate eigenvectors can be extracted from the columns of the orthogonal matrix $O_n = U_1 \cdots U_n$.

In fact, there are several competing versions of Jacobi's method. The classical method selects the row i and column j giving the largest increase $2a_{ij}^2$ in the sum of squares $S(A)$ of the diagonal entries. The disadvantage of this strategy is that it necessitates searching through all off-diagonal entries of the current matrix A. A simpler strategy is to cycle through the off-diagonal entries according to some fixed schedule. In the threshold Jacobi method, this cyclic strategy is modified so that a rotation is undertaken only when the current off-diagonal entry a_{ij} is sufficiently large in absolute value. For purposes of theoretical exposition, it is simplest to adopt the classical strategy.

With this decision in mind, consider the sum of squares of the off-diagonal entries

$$\begin{aligned} L(A) &= \sum_k \sum_{l \neq k} a_{kl}^2 \\ &= ||A||_F^2 - S(A), \end{aligned}$$

and suppose that $B = U^t AU$, where U is a Jacobi rotation in the plane of i and j. Because a_{ij} is the largest off-diagonal entry in absolute value, $L(A) \leq m(m-1)a_{ij}^2$. Together with the relation $L(B) = L(A) - 2a_{ij}^2$, this implies

$$L(B) \leq \left[1 - \frac{2}{m(m-1)}\right] L(A).$$

Thus, the function $L(A)$ is driven to 0 at least as fast as the successive powers of $1-2/[m(m-1)]$. This clearly suggests the convergence of Jacobi's method to the diagonal matrix D.

Rigorously proving the convergence of Jacobi's method requires the next technical result.

Proposition 8.2.1 *Let x_n be a bounded sequence in R^p, and suppose*

$$\lim_{n \to \infty} ||x_{n+1} - x_n|| = 0$$

for some norm $||x||$. Then the collection T of cluster points of x_n is connected. If T is finite, it follows that T reduces to a single point and that $\lim_{n \to \infty} x_n = x_\infty$ exists.

Proof: It is straightforward to prove that T is a compact set. If it is disconnected, then it is contained in the union of two disjoint, open subsets S_1 and S_2 in such a way that neither $T \cap S_1$ nor $T \cap S_2$ is empty. The distance $d = \inf_{y \in T \cap S_1,\, z \in T \cap S_2} ||y - z||$ separating $T \cap S_1$ and $T \cap S_2$ must be positive; otherwise, there would be two sequences $y_n \in T \cap S_1$ and $z_n \in T \cap S_2$ with $||y_n - z_n|| < 1/n$. Because T is compact, there is a subsequence y_{n_k} of y_n that converges to a point of T. By passing to a subsubsequence if necessary, we can assume that z_{n_k} converges as well. The limits of these two convergent subsequences coincide. The fact that the common limit belongs to the open set S_1 and the boundary of S_2 or vice versa contradicts the disjointness of S_1 and S_2.

Now consider the sequence x_n in the statement of the proposition. For large enough n, we have $||x_{n+1} - x_n|| < d/4$. As the sequence x_n bounces back and forth between cluster points in S_1 and S_2, it must enter the closed set $W = \{y : \inf_{z \in T} ||y - z|| \geq d/4\}$ infinitely often. But this means that W contains a cluster point of x_n. Because W is disjoint from $T \cap S_1$ and

$T \cap S_2$, and these two sets are postulated to contain all of the cluster points of x_n, this contradiction implies that T is connected.

Since a finite set with more than one point is necessarily disconnected, T can be a finite set only if it consists of a single point. Finally, a bounded sequence with only a single cluster point has that point as its limit. ■

We are now in a position to prove convergence of Jacobi's method via the strategy of Michel Crouzeix [1]. Let us tackle eigenvalues first.

Proposition 8.2.2 *Suppose that Ω is an $m \times m$ symmetric matrix. The classical Jacobi strategy generates a sequence of rotations U_n and a sequence of similar matrices Ω_n related to Ω by*

$$\Omega_n = U_n^t \cdots U_1^t \Omega U_1 \cdots U_n.$$

With the rotations U_n chosen as described above, $\lim_{n\to\infty} \Omega_n$ exists and equals a diagonal matrix D whose entries are the eigenvalues of Ω in some order.

Proof: If Jacobi's method gives a diagonal matrix in a finite number of iterations, there is nothing to prove. Otherwise, let D_n be the diagonal part of Ω_n. We have already argued that the off-diagonal part $\Omega_n - D_n$ of Ω_n tends to the zero matrix $\mathbf{0}$. Because $||D_n||_F \leq ||\Omega_n||_F = ||\Omega||_F$, the sequence D_n is bounded in R^{m^2}. Let D_{n_k} be a convergent subsequence with limit D, not necessarily assumed to represent the eigenvalues of Ω. Owing to the similarity of the matrix Ω_{n_k} to Ω, we find

$$\begin{aligned} \det(D - \lambda I) &= \lim_{k\to\infty} \det(D_{n_k} - \lambda I) \\ &= \lim_{k\to\infty} \det(\Omega_{n_k} - \lambda I) \\ &= \det(\Omega - \lambda I). \end{aligned}$$

Thus, D possesses the same eigenvalues, counting multiplicities, as Ω. But the eigenvalues of D are just its diagonal entries.

To rule out more than one cluster point D, we apply Proposition 8.2.1, noting that there are only a finite number of permutations of the eigenvalues of Ω. According to equation (8.3) and its immediate sequel, if U_n is a rotation through an angle θ_n in the plane of entries i and j, then the diagonal entries of D_{n+1} and D_n satisfy

$$d_{n+1,kk} - d_{nkk} = \begin{cases} 0 & k \neq i,j \\ -\omega_{nij} \tan\theta_n & k = i \\ +\omega_{nij} \tan\theta_n & k = j, \end{cases}$$

where $\Omega_n = (\omega_{nkl})$. Because $|\theta_n| \leq \pi/4$ and $|\omega_{nij}| \leq ||\Omega_n - D_n||_F$, it follows that $\lim_{n\to\infty} ||D_{n+1} - D_n||_F = 0$. ■

Proposition 8.2.3 *If all of the eigenvalues λ_i of the matrix Ω are distinct, then the sequence $O_n = U_1 \cdots U_n$ of orthogonal matrices constructed in*

Proposition 8.2.2 converges to the matrix of eigenvectors of the limiting diagonal matrix D.

Proof: Mimicking the strategy of Proposition 8.2.2, we show that the sequence O_n is bounded, that it possesses only a finite number of cluster points, and that

$$\lim_{n\to\infty} ||O_{n+1} - O_n||_F = 0.$$

The sequence O_n is bounded because $||O_n||_F^2 = \operatorname{tr}(O_n^t O_n) = \operatorname{tr}(I)$. Suppose O_n has a convergent subsequence O_{n_k} with limit O. Then $D = O^t \Omega O$ holds because of Proposition 8.2.2. This implies that the columns of O are the orthonormal eigenvectors of Ω ordered consistently with the eigenvalues appearing in D. The eigenvectors are unique up to sign. Thus, O can be one of only 2^m possibilities.

To prove $\lim_{n\to\infty} ||O_{n+1} - O_n||_F = 0$, again suppose that U_n is a rotation through an angle θ_n in the plane of entries i and j. This angle is defined by $\tan(2\theta_n) = 2\omega_{nij}/(\omega_{njj} - \omega_{nii})$. Because Ω_n converges to D and the entries of D are presumed unique,

$$\begin{aligned}
|\omega_{njj} - \omega_{nii}| &> \frac{1}{2}\min_{k\neq l}|d_{kk} - d_{ll}| \\
&> 0
\end{aligned}$$

for all sufficiently large n. In view of the fact that $\lim_{n\to\infty} \omega_{nij} = 0$, this implies that $\lim_{n\to\infty} \theta_n = 0$, which in turn yields $\lim_{n\to\infty} U_n = I$. The inequality

$$\begin{aligned}
||O_{n+1} - O_n||_F &= ||O_n(U_{n+1} - I)||_F \\
&\leq ||O_n||_F ||U_{n+1} - I||_F
\end{aligned}$$

completes the proof. ∎

8.3 The Rayleigh Quotient

Sometimes it is helpful to characterize the eigenvalues and eigenvectors of an $m \times m$ symmetric matrix A in terms of the extrema of the Rayleigh quotient

$$R(x) = \frac{x^t A x}{x^t x}$$

defined for $x \neq \mathbf{0}$. This was the case, for example, in computing the norm $||A||_2$. To develop the properties of the Rayleigh quotient, let A have eigenvalues $\lambda_1 \leq \cdots \leq \lambda_m$ and corresponding orthonormal eigenvectors

$u_1, \ldots, u_m$. Because any x can be written as a unique linear combination $\sum_i c_i u_i$, the Rayleigh quotient can be expressed as

$$R(x) = \frac{\sum_i \lambda_i c_i^2}{\sum_i c_i^2}. \tag{8.4}$$

This representation clearly yields the inequality $R(x) \leq \lambda_m$ and the equality $R(u_m) = \lambda_m$. Hence, $R(x)$ is maximized by $x = u_m$ and correspondingly minimized by $x = u_1$. The Courant-Fischer theorem is a notable generalization of these results.

Proposition 8.3.1 (Courant-Fischer) *Let V_k be a k-dimensional subspace of R^m. Then*

$$\begin{aligned}\lambda_k &= \min_{V_k} \max_{x \in V_k,\, x \neq \mathbf{0}} R(x) \\ &= \max_{V_{m-k+1}} \min_{x \in V_{m-k+1},\, x \neq \mathbf{0}} R(x).\end{aligned}$$

The minimum in the first characterization of λ_k is attained for the subspace spanned by $u_1, \ldots, u_k$, and the maximum in the second characterization of λ_k is attained for the subspace spanned by $u_k, \ldots, u_m$.

Proof: If U_k is the subspace spanned by $u_1, \ldots, u_k$, then it is clear that

$$\lambda_k = \max_{x \in U_k,\, x \neq \mathbf{0}} R(x).$$

If V_k is an arbitrary subspace of dimension k, then there must be some nontrivial vector $x \in V_k$ orthogonal to $u_1, \ldots, u_{k-1}$. For this $x = \sum_{i=k}^m c_i u_i$, we find

$$\begin{aligned}R(x) &= \frac{\sum_{i=k}^m \lambda_i c_i^2}{\sum_{i=k}^m c_i^2} \\ &\geq \frac{\lambda_k \sum_{i=k}^m c_i^2}{\sum_{i=k}^m c_i^2} \\ &= \lambda_k.\end{aligned}$$

This proves that $\max_{x \in V_k,\, x \neq \mathbf{0}} R(x) \geq \lambda_k$. The second characterization of λ_k follows from the first characterization applied to $-A$, whose eigenvalues are $-\lambda_m \leq \cdots \leq -\lambda_1$. ■

The next proposition applies the Courant-Fischer theorem to the problem of estimating how much the eigenvalues of a symmetric matrix change under a symmetric perturbation of the matrix.

Proposition 8.3.2 *Let the $m \times m$ symmetric matrices A and $B = A + \Delta A$ have ordered eigenvalues $\lambda_1 \leq \cdots \leq \lambda_m$ and $\mu_1 \leq \cdots \leq \mu_m$, respectively. Then the inequality*

$$|\lambda_k - \mu_k| \leq ||\Delta A||_2$$

holds for all $k \in \{1, \ldots, m\}$.

Proof: Suppose that U_k is the subspace of R^m spanned by the eigenvectors $u_1, \ldots, u_k$ of A corresponding to $\lambda_1, \ldots, \lambda_k$. If V_k is an arbitrary subspace of dimension k, then the identity $R_B(x) = R_A(x) + R_{\Delta A}(x)$ and the Courant-Fischer theorem imply that

$$\begin{aligned}
\mu_k &= \min_{V_k} \max_{x \in V_k,\, x \neq \mathbf{0}} R_B(x) \\
&\leq \max_{x \in U_k,\, x \neq \mathbf{0}} R_B(x) \\
&\leq \lambda_k + \max_{x \in U_k,\, x \neq \mathbf{0}} R_{\Delta A}(x) \\
&\leq \lambda_k + \max_{x \neq \mathbf{0}} R_{\Delta A}(x) \\
&\leq \lambda_k + \|\Delta A\|_2.
\end{aligned}$$

If we reverse the roles of A and B, the inequality $\lambda_k \leq \mu_k + \|\Delta A\|_2$ follows similarly. ■

Finally, it is worth generalizing the above analysis to some nonsymmetric matrices. Suppose that A and B are symmetric matrices with B positive definite. An eigenvalue λ of $B^{-1}A$ satisfies $B^{-1}Ax = \lambda x$ for some $x \neq \mathbf{0}$. This identity is equivalent to the identity $Ax = \lambda Bx$. Taking the inner product of this latter identity with x suggests examining the generalized Rayleigh quotient [6]

$$R(x) = \frac{x^t Ax}{x^t Bx}. \tag{8.5}$$

For instance, it is easy to prove that the maximum and minimum eigenvalues of $B^{-1}A$ coincide with the maximum and minimum values of (8.5). Furthermore, the Courant-Fischer theorem carries over.

Another useful perspective on this subject is gained by noting that B has a symmetric square root $B^{1/2}$ defined in terms of its diagonalization $B = UDU^t$ by $B^{1/2} = UD^{1/2}U^t$. The eigenvalue equation $B^{-1}Ax = \lambda x$ is then equivalent to

$$B^{-\frac{1}{2}}AB^{-\frac{1}{2}}y = \lambda y$$

for $y = B^{1/2}x$. This lands us back in the territory of symmetric matrices and the ordinary Rayleigh quotient. The extended Courant-Fischer theorem now follows directly from the standard Courant-Fischer theorem.

8.4 Finding a Single Eigenvalue

In many applications, we seek either the largest or smallest eigenvalue rather than all of the eigenvalues of a matrix A. Let A be diagonalizable with eigenvalues $\lambda_1, \ldots, \lambda_m$ and corresponding eigenvectors $v_1, \ldots, v_m$. The power method finds the dominant eigenvector by taking the limit of the iterates

$$u_n \quad = \quad \frac{1}{\|Au_{n-1}\|_2} Au_{n-1}.$$

To see how this works, suppose $|\lambda_m| > |\lambda_i|$ for each $i < m$ and u_0 is a unit vector expressible as the linear combination $u_0 = c_1 v_1 + \cdots + c_m v_m$. We can rewrite

$$u_n \quad = \quad \frac{c_1 \left(\frac{\lambda_1}{\lambda_m}\right)^n v_1 + c_2 \left(\frac{\lambda_2}{\lambda_m}\right)^n v_2 + \cdots + c_m v_m}{\left\| c_1 \left(\frac{\lambda_1}{\lambda_m}\right)^n v_1 + c_2 \left(\frac{\lambda_2}{\lambda_m}\right)^n v_2 + \cdots + c_m v_m \right\|_2} \cdot \left(\frac{\lambda_m}{|\lambda_m|}\right)^n.$$

Assuming λ_m is real and $c_1 \neq 0$, it is obvious that

$$\lim_{n \to \infty} \left[u_n - \frac{c_m v_m}{\|c_m v_m\|_2} \cdot \left(\frac{\lambda_m}{|\lambda_m|}\right)^n \right] \quad = \quad 0.$$

In other words, u_n converges to a multiple of v_m or oscillates in sign. In either case, u_{2n} converges to a multiple of v_m. The rate of convergence depends on the ratio $\max_{1 \le i < m} |\lambda_i|/|\lambda_m|$. In theory, there is no guarantee that the coefficient c_m of an arbitrary initial vector u_0 is nonzero; in practice, choosing u_0 randomly overcomes this problem.

If we want to find the eigenvalue with smallest absolute value, then the inverse power method is at our disposal. In the inverse power method, we simply substitute A^{-1} for A in the power method. For other eigenvalues, neither method directly applies. However, suppose we have an approximate eigenvalue μ. The inverse matrix $(A - \mu I)^{-1}$ has eigenvalues $(\lambda_i - \mu)^{-1}$ with the same eigenvectors $v_1, \ldots, v_m$. This suggests substituting $(A - \mu I)^{-1}$ for A in the power method. The limit and the rate of convergence now depend on the magnitudes of the eigenvalues $(\lambda_i - \mu)^{-1}$. If we are fortunate enough to choose μ very close to a given λ_i, then the shifted inverse power method will converge very quickly to v_i.

The Rayleigh quotient iteration algorithm takes the shifted inverse power method one step further. Instead of relying on a static approximation μ to the dominant eigenvalue λ_m of a symmetric matrix A, it updates μ via the Rayleigh quotient $\mu_n = u_n^t A u_n$ from the unit vector

$$u_n \quad = \quad \frac{1}{\|(A - \mu_{n-1} I)^{-1} u_{n-1}\|_2} (A - \mu_{n-1} I)^{-1} u_{n-1}.$$

This algorithm converges at an extremely fast cubic rate; it works for the smallest eigenvalue if we start with a reasonable approximation to it.

Because the Rayleigh quotient algorithm carries no assurance of convergence to any particular eigenvalue, it is important to make a good initial guess of the desired eigenvalue. One approach is based on Gerschgorin's circle theorem.

Proposition 8.4.1 *Every eigenvalue λ of an $m \times m$ matrix $A = (a_{ij})$ satisfies at least one of the inequalities*

$$|\lambda - a_{ii}| \;\leq\; \sum_{j \neq i} |a_{ij}|.$$

Proof: Suppose v is an eigenvector corresponding to λ. For each component v_i of v, we have

$$(\lambda - a_{ii})v_i \;=\; \sum_{j \neq i} a_{ij} v_j.$$

If we select v_i so that $|v_i| \geq |v_j|$ for all j, then

$$|\lambda - a_{ii}| \;\leq\; \sum_{j \neq i} |a_{ij}| \frac{|v_j|}{|v_i|} \;\leq\; \sum_{j \neq i} |a_{ij}|.$$

Nothing in this argument requires A to be symmetric or λ to be real. ■

If A is symmetric and we want the largest eigenvalue λ_m, then one plausible guess of λ_m is $\mu = \max_i a_{ii}$. An even more conservative guess is

$$\mu \;=\; \max_{1 \leq i \leq m} \Big\{ a_{ii} + \sum_{j \neq i} |a_{ij}| \Big\}.$$

The references [10, 14] contain more material on Gerschgorin's circle theorem.

In addition to making a good initial guess, we need some way of monitoring the progress of the Rayleigh quotient algorithm. The next proposition furnishes a useful estimate.

Proposition 8.4.2 *Suppose that A is symmetric, u is a unit vector, and λ is the closest eigenvalue to μ. Then we have*

$$|\lambda - \mu| \;\leq\; \|Au - \mu u\|_2. \tag{8.6}$$

Proof: If A has spectral decomposition $A = \sum_i \lambda_i v_i v_i^t$ and u has orthogonal expansion $u = \sum_i v_i^t u \, v_i$, then

$$\begin{aligned}
\|Au - \mu u\|_2^2 &= \Big\| \sum_i (\lambda_i - \mu) v_i^t u \, v_i \Big\|_2^2 \\
&= \sum_i (\lambda_i - \mu)^2 (v_i^t u)^2 \\
&\geq \min_i (\lambda_i - \mu)^2 \sum_i (v_i^t u)^2 \\
&= \min_i (\lambda_i - \mu)^2.
\end{aligned}$$

TABLE 8.1. Iterations of the Rayleigh Quotient Algorithm

Iteration n	**Approximate Eigenvalue** μ_n	**Error Bound (8.6)**
0	15.000000000000000	5.8
1	16.201807549175964	1.1
2	16.313552340303307	1.2×10^{-2}
3	16.313567783095323	2.0×10^{-8}
4	16.313567783095326	2.1×10^{-15}

For a simple numerical example [12], consider the symmetric matrix

$$A = \begin{pmatrix} 3 & 2 & 1 \\ 2 & 8 & 3 \\ 1 & 3 & 15 \end{pmatrix}.$$

Table 8.1 displays the rapid progress of the Rayleigh quotient algorithm starting from $\mu_0 = 15$ and

$$u_0 = \frac{1}{\sqrt{3}}(1, 1, 1)^t.$$

Problem 8 explores an alternative algorithm of Hestenes and Karush [7] that converges more slowly but requires no matrix inversion. This algorithm is worth keeping in mind whenever the largest or smallest eigenvalue of a high-dimensional symmetric matrix is needed.

8.5 Problems

1. For symmetric matrices A and B, define $A \rhd 0$ to mean that A is positive semidefinite and $A \rhd B$ to mean that $A - B \rhd 0$. Show that $A \rhd B$ and $B \rhd C$ imply $A \rhd C$. Also show that $A \rhd B$ and $B \rhd A$ imply $A = B$. Thus, $\rhd$ induces a partial order on the set of symmetric matrices.

2. Find the eigenvalues and eigenvectors of the matrix

$$\Omega = \begin{pmatrix} 10 & 7 & 8 & 7 \\ 7 & 5 & 6 & 5 \\ 8 & 6 & 10 & 9 \\ 7 & 5 & 9 & 10 \end{pmatrix}$$

of R. S. Wilson by Jacobi's method. You may use the appropriate subroutine in Press et al. [11].

3. Find the eigenvalues and eigenvectors of the rotation matrix (8.1). Note that the eigenvalues are complex conjugates.

4. Find the eigenvalues and eigenvectors of the reflection matrix

$$\begin{pmatrix} \cos\theta & \sin\theta \\ \sin\theta & -\cos\theta \end{pmatrix}.$$

5. Suppose λ is an eigenvalue of the orthogonal matrix O with corresponding eigenvector v. Show that v has real entries only if $\lambda = \pm 1$.

6. A matrix A with real entries is said to be skew-symmetric if $A^t = -A$. Show that a skew-symmetric matrix has only imaginary eigenvalues.

7. Consider an $n \times n$ upper triangular matrix U with distinct nonzero diagonal entries. Let λ be its mth diagonal entry, and write

$$U = \begin{pmatrix} U_{11} & U_{12} & U_{13} \\ \mathbf{0} & \lambda & U_{23} \\ \mathbf{0} & \mathbf{0} & U_{33} \end{pmatrix}$$

in block form. Show that λ is an eigenvalue of U with eigenvector

$$w = \begin{pmatrix} v \\ -1 \\ \mathbf{0} \end{pmatrix}, \quad v = (U_{11} - \lambda I_{m-1})^{-1} U_{12},$$

where I_{m-1} is the $(m-1) \times (m-1)$ identity matrix.

8. Suppose the $m \times m$ symmetric matrix Ω has eigenvalues

$$\lambda_1 < \lambda_2 \le \cdots \le \lambda_{m-1} < \lambda_m.$$

The iterative scheme $x_{n+1} = (\Omega - \eta_n I)x_n$ can be used to approximate either λ_1 or λ_m [7]. Consider the criterion

$$\sigma_n = \frac{x_{n+1}^t \Omega x_{n+1}}{x_{n+1}^t x_{n+1}}.$$

Choosing η_n to maximize σ_n causes $\lim_{n\to\infty} \sigma_n = \lambda_m$, while choosing η_n to minimize σ_n causes $\lim_{n\to\infty} \sigma_n = \lambda_1$. If $\tau_k = x_n^t \Omega^k x_n$, then show that the extrema of σ_n as a function of η are given by the roots of the quadratic equation

$$0 = \det \begin{pmatrix} 1 & \eta & \eta^2 \\ \tau_0 & \tau_1 & \tau_2 \\ \tau_1 & \tau_2 & \tau_3 \end{pmatrix}.$$

9. Apply the algorithm of the previous problem to find the largest and smallest eigenvalues of the matrix in Problem 2.

10. Show that the extended Rayleigh quotient (8.5) has gradient

$$\frac{2[A - R(x)B]x}{x^t Bx}.$$

Argue that the eigenvalues and eigenvectors of $B^{-1}A$ are the stationary values and stationary points, respectively, of $R(x)$.

11. In Proposition 8.3.2 suppose the matrix ΔA is positive semidefinite. Prove that $\lambda_k \leq \mu_k$ for all k.

12. For m positive numbers $b_1 < \cdots < b_m$, define a matrix A with entries $a_{ij} = 1/(b_i + b_j)$. Show that A is positive definite with largest eigenvalue $\lambda_m \geq \frac{1}{2b_1}$ and smallest eigenvalue $\lambda_1 \leq \frac{1}{2b_m}$. (Hints: Use the identity

$$\frac{1}{b_i + b_j} = \int_0^\infty e^{-b_i x} e^{-b_j x} dx$$

and the Courant-Fischer theorem.)

13. Sylvester's criterion is a test for positive definiteness of a square matrix A. This test requires that every principal minor of A be positive. Recall that a principal minor is the determinant of an upper-left block of A. Use the Courant-Fischer theorem to prove that satisfaction of Sylvester's criterion is necessary and sufficient for positive definiteness [3].

14. Consider the $m \times m$ bordered symmetric matrix

$$A = \begin{pmatrix} B & c \\ c^t & d \end{pmatrix},$$

where B is symmetric, c is a vector, and d is a scalar. Let the eigenvalues of A be $\lambda_1 \leq \lambda_2 \leq \cdots \leq \lambda_m$ and the eigenvalues of B be $\mu_1 \leq \mu_2 \leq \cdots \leq \mu_{m-1}$. Use the Courant-Fischer theorem to prove the interlacing property $\lambda_i \leq \mu_i \leq \lambda_{i+1}$ for all i between 1 and $m-1$.

15. Let D be an $m \times m$ diagonal matrix with distinct and ordered diagonal entries $d_1 < d_2 < \cdots < d_m$. Also let v be a vector with nonzero components. Show that the eigenvalues λ_i of the rank-one perturbation $D \pm vv^t$ satisfy $d_i < \lambda_i < d_{i+1}$ for $i = 1, \ldots, m-1$ and $d_m < \lambda_m$ when the plus sign is taken and $\lambda_1 < d_1$ and $d_{i-1} < \lambda_i < d_i$ for $i = 2, \ldots, m$ when the minus sign is taken. Finally, demonstrate that the eigenvector corresponding to λ_i is a multiple of $(D - \lambda_i I)^{-1}v$. (Hint: This problem is a continuation of Problem 17 of Chapter 7.)

16. In the notation of Problem 1, show that two positive definite matrices A and B satisfy $A \rhd B$ if and only if they satisfy $B^{-1} \rhd A^{-1}$. If $A \rhd B$, then prove that $\det A \geq \det B$ and $\operatorname{tr} A \geq \operatorname{tr} B$.

17. Suppose the symmetric matrices A and B satisfy $A \rhd B$ in the notation of Problem 1. If the diagonal entries of A and B are equal, then demonstrate that $A = B$.

18. Let A and B be $m \times m$ symmetric matrices. Denote the smallest and largest eigenvalues of the convex combination $\alpha A + (1-\alpha)B$ by $\lambda_1[\alpha A + (1-\alpha)B]$ and $\lambda_m[\alpha A + (1-\alpha)B]$, respectively. For $\alpha \in [0,1]$, demonstrate that

$$\begin{aligned} \lambda_1[\alpha A + (1-\alpha)B] &\geq \alpha\lambda_1[A] + (1-\alpha)\lambda_1[B] \\ \lambda_m[\alpha A + (1-\alpha)B] &\leq \alpha\lambda_m[A] + (1-\alpha)\lambda_m[B]. \end{aligned}$$

19. Given the assumptions of the previous problem, show that the smallest and largest eigenvalues satisfy

$$\begin{aligned} \lambda_1[A+B] &\geq \lambda_1[A] + \lambda_1[B] \\ \lambda_m[A+B] &\leq \lambda_m[A] + \lambda_m[B]. \end{aligned}$$

20. Let A and B be positive semidefinite matrices of the same dimension. Show that the matrix $aA + bB$ is positive semidefinite for every pair of nonnegative scalars a and b. Thus, the set of positive semidefinite matrices is a convex cone.

21. One of the simplest ways of showing that a symmetric matrix is positive semidefinite is to show that it is the covariance matrix of a random vector. Use this insight to prove that if the symmetric matrices $A = (a_{ij})$ and $B = (b_{ij})$ are positive semidefinite, then the matrix $C = (c_{ij})$ with $c_{ij} = a_{ij}b_{ij}$ is also positive semidefinite [9]. (Hint: Take independent random vectors X and Y with covariance matrices A and B and form the random vector Z with components $Z_i = X_i Y_i$.)

22. Continuing Problem 21, suppose that the $n \times n$ symmetric matrices A and B have entries $a_{ij} = i(n-j+1)$ and $b_{ij} = \sum_{k=1}^{i} \sigma_k^2$ for $j \geq i$ and $\sigma_k^2 \geq 0$. Show that A and B are positive semidefinite [9]. (Hint: For A, consider the order statistics from a random sample of the uniform distribution on $[0,1]$.)

23. Use the Gerschgorin circle theorem to find four intervals containing

the eigenvalues of the matrix

$$A = \begin{bmatrix} 4 & \frac{1}{5} & -\frac{1}{10} & \frac{1}{10} \\ \frac{1}{5} & -1 & -\frac{1}{10} & \frac{1}{20} \\ -\frac{1}{10} & -\frac{1}{10} & 3 & \frac{1}{10} \\ \frac{1}{10} & \frac{1}{20} & \frac{1}{10} & -3 \end{bmatrix}.$$

Because these intervals do not overlap, each contains exactly one eigenvalue.

24. Suppose the rows of the square matrix $M = (m_{ij})$ obey the inequality $\sum_j |m_{ij}| < 1$. Demonstrate via the Gerschgorin circle theorem that all eigenvalues λ of M satisfy $|\lambda| < 1$.

25. Suppose the square matrix $M = (m_{ij})$ is diagonally dominant in the sense that $\sum_{j \neq i} |m_{ij}| < |m_{ii}|$ for every i. Apply the Gerschgorin circle theorem to prove that M is invertible.

8.6 References

[1] Ciarlet PG (1989) *Introduction to Numerical Linear Algebra and Optimization.* Cambridge University Press, Cambridge

[2] Demmel JW (1997) *Applied Numerical Linear Algebra.* SIAM, Philadelphia

[3] Gilbert GT (1991) Positive definite matrices and Sylvester's criterion. *Amer Math Monthly* 98:44–46

[4] Golub GH, Van Loan CF (1996) *Matrix Computations*, 3rd ed. Johns Hopkins University Press, Baltimore

[5] Hämmerlin G, Hoffmann KH (1991) *Numerical Mathematics.* Springer, New York

[6] Hestenes MR (1981) *Optimization Theory: The Finite Dimensional Case.* Robert E Krieger Publishing, Huntington, NY

[7] Hestenes MR, Karush WE (1951) A method of gradients for the calculation of the characteristic roots and vectors of a real symmetric matrix. *J Res Nat Bur Standards*, 47:471–478

[8] Isaacson E, Keller HB (1966) *Analysis of Numerical Methods.* Wiley, New York

[9] Olkin I (1985) A probabilistic proof of a theorem of Schur. *Amer Math Monthly* 92:50–51

[10] Ortega JM (1990) *Numerical Analysis: A Second Course.* SIAM, Philadelphia

[11] Press WH, Teukolsky SA, Vetterling WT, Flannery BP (1992) *Numerical Recipes in Fortran: The Art of Scientific Computing*, 2nd ed. Cambridge University Press, Cambridge

[12] Reiter C (1990) Easy algorithms for finding eigenvalues. *Math Magazine.* 63:173–178

[13] Rao CR (1973) *Linear Statistical Inference and its Applications*, 2nd ed. Wiley, New York

[14] Süli E, Mayers D (2003) *An Introduction to Numerical Analysis.* Cambridge University Press, Cambridge

[15] Thisted RA (1988) *Elements of Statistical Computing.* Chapman & Hall, New York

[16] Trefethen LN, Bau D III (1997) *Numerical Linear Algebra.* SIAM, Philadelphia

9

Singular Value Decomposition

9.1 Introduction

In many modern applications involving large data sets, statisticians are confronted with a large $m \times n$ matrix $X = (x_{ij})$ that encodes n features on each of m objects. For instance, in gene microarray studies x_{ij} represents the expression level of the ith gene under the jth experimental condition [13]. In information retrieval, x_{ij} represents the frequency of the jth word or term in the ith document [2]. The singular value decomposition (svd) captures the structure of such matrices. In many applications there are alternatives to the svd, but these are seldom as informative or as numerically accurate.

Most readers are well acquainted with the spectral theorem for symmetric matrices. This classical result states that an $m \times m$ symmetric matrix A can be written as $A = U\Sigma U^t$ for an orthogonal matrix U and a diagonal matrix Σ with diagonal entries σ_i. If U has columns $u_1, \ldots, u_m$, then the matrix product $A = U\Sigma U^t$ unfolds into the sum of outer products

$$A = \sum_{j=1}^{m} \sigma_j u_j u_j^t.$$

When $\sigma_j = 0$ for $j > k$, A has rank k and only the first k terms of the sum are relevant. The svd seeks to generalize the spectral theorem to nonsymmetric matrices. In this case there are two orthonormal sets of vectors $u_1, \ldots, u_k$ and $v_1, \ldots, v_k$ instead of one, and we write

$$A = \sum_{j=1}^{k} \sigma_j u_j v_j^t = U\Sigma V^t \tag{9.1}$$

for matrices U and V with orthonormal columns $u_1, \ldots, u_k$ and $v_1, \ldots, v_k$, respectively. Remarkably, it is always possible to find such a representation with nonnegative σ_j.

For some purposes, it is better to fill out the matrices U and V to full orthogonal matrices. If A is $m \times n$, then U is viewed as $m \times m$, Σ as $m \times n$,

K. Lange, *Numerical Analysis for Statisticians*, Statistics and Computing,
DOI 10.1007/978-1-4419-5945-4_9, © Springer Science+Business Media, LLC 2010

and V as $n \times n$. The svd then becomes

$$A = (u_1 \dots u_k \, u_{k+1} \dots u_m) \begin{pmatrix} \sigma_1 & \dots & 0 & 0 & \dots & 0 \\ \vdots & \ddots & \vdots & \vdots & \ddots & \vdots \\ 0 & \dots & \sigma_k & 0 & \dots & 0 \\ 0 & \dots & 0 & 0 & \dots & 0 \\ \vdots & \ddots & \vdots & \vdots & \ddots & \vdots \\ 0 & \dots & 0 & 0 & \dots & 0 \end{pmatrix} \begin{pmatrix} v_1^t \\ \vdots \\ v_k^t \\ v_{k+1}^t \\ \vdots \\ v_n^t \end{pmatrix},$$

assuming $k < \min\{m, n\}$. The scalars $\sigma_1, \dots, \sigma_k$ are said to be singular values and conventionally are listed in decreasing order. The vectors $u_1, \dots, u_k$ are known as left singular vectors and the vectors $v_1, \dots, v_k$ as right singular vectors. We will refer to any svd with U and V orthogonal matrices as full.

The numerical issues in constructing the svd are so complicated that outsiders get the impression that only experts should write code. Fortunately, the experts are kind enough to contribute open source code to cooperative efforts such as LAPACK [1]. It would be a mistake not to take advantage of these efforts. However, perfectly usable software can be written by novices for small-scale projects. Our discussion of Jacobi's method for constructing the svd targets such readers. Because it is accurate and receptive to parallelization, Jacobi's method remains in contention with other algorithms. In addition to these advantages, it is easy to explain.

There is a huge literature on the svd. The two books [9, 10] by Horn and Johnson provide an excellent overview of the mathematical theory. For all things numerical, the treatise of Golub and Van Loan [7] is the definitive source. Beginners will appreciate the more leisurely pace of the books [6, 15, 17]. At the same level as this text, the books [4, 16, 18] are also highly recommended.

9.2 Basic Properties of the SVD

Let us start with the crucial issue of existence following Karro and Li [12].

Proposition 9.2.1 *Every $m \times n$ matrix A has a singular value decomposition of the form (9.1) with positive diagonal entries for Σ.*

Proof: It suffices to prove that A can be represented as $U\Sigma V^t$ for full orthogonal matrices U and V. We proceed by induction on $\min\{m, n\}$. The cases $m = 1$ and $n = 1$ are trivial. In the first case, we write $A = \sigma 1 v^t$, and in the second case $A = \sigma u 1$. Thus, assume that $m > 1$ and $n > 1$. According to Proposition 11.2.1, the continuous function $(x, y) \mapsto x^t A y$ attains its maximum σ_1 on the compact set

$$S = \{(x, y) : x^t x = y^t y = 1\}$$

at some point $(x, y) = (u_1, v_1)$. Extend u_1 to an orthogonal matrix U with u_1 as its first column, and extend v_1 to an orthogonal matrix V with v_1 as its first column. The matrix $B = U^t AV$ has upper-left entry $\sigma_1 = u_1^t A v_1$. Let us show that $b_{k1} = b_{1k} = 0$ for the remaining entries in the first column and first row of B. To achieve this goal, let e_i denote the standard basis vector in R^m with 1 in position i and 0's elsewhere, and define the unit vector $x(\theta) = U[\cos(\theta)e_1 + \sin(\theta)e_k]$. Because the function $f(\theta) = x(\theta)^t A v_1$ is maximized by the choice $\theta = 0$, we have

$$0 \quad = \quad f'(0) \quad = \quad e_k^t U^t A v_1 \quad = \quad b_{k1}.$$

A similar argument with $y(\theta) = V[\cos(\theta)e_1 + \sin(\theta)e_k]$ gives $b_{1k} = 0$. Thus, B has the block diagonal form

$$B \quad = \quad U^t AV \quad = \quad \begin{pmatrix} \sigma_1 & \mathbf{0}^t \\ \mathbf{0} & C \end{pmatrix}.$$

One can now argue by induction that C has an svd. Since the set of orthogonal matrices is closed under multiplication and contains block matrices of the form

$$\begin{pmatrix} I & \mathbf{0}^t \\ \mathbf{0} & W \end{pmatrix}$$

for W orthogonal and I an identity matrix, it is clear that $A = UBV^t$ has an svd. Furthermore, the singular values are arranged in decreasing order owing to the definition of σ_1 and induction applied to C. ■

In view of the orthogonality of the bases $u_1, \ldots, u_k$ and $v_1, \ldots, v_k$, the singular value decomposition (9.1) leads to the formulas

$$\begin{aligned} A^t &= \sum_{j=1}^{k} \sigma_j v_j u_j^t \\ AA^t &= \sum_{j=1}^{k} \sigma_j^2 u_j u_j^t \\ A^t A &= \sum_{j=1}^{k} \sigma_j^2 v_j v_j^t. \end{aligned}$$

Hence, AA^t has nonzero eigenvalue σ_j^2 with corresponding eigenvector u_j, and $A^t A$ has nonzero eigenvalue σ_j^2 with corresponding eigenvector v_j. Although one cannot invert the $m \times n$ matrix A when $m \neq n$ or when A fails to be of full rank, there is a partial inverse that is important in practice.

Proposition 9.2.2 *The Moore-Penrose inverse*

$$A^- \quad = \quad \sum_{j=1}^{k} \sigma_j^{-1} v_j u_j^t$$

enjoys the properties

$$\begin{aligned} (AA^-)^t &= AA^- \\ (A^-A)^t &= A^-A \\ AA^-A &= A \\ A^-AA^- &= A^-. \end{aligned} \tag{9.2}$$

If A is square and invertible, then $A^- = A^{-1}$. If A has full column rank, then $A^- = (A^tA)^{-1}A^t$.

Proof: The first two properties in (9.2) are consequences of the representations

$$\begin{aligned} AA^- &= \sum_{j=1}^k u_j u_j^t \\ A^-A &= \sum_{j=1}^k v_j v_j^t. \end{aligned}$$

Multiplying these on the right by A and A^-, respectively, produces the last two properties in (9.2). If A is invertible, then multiplying $AA^-A = A$ on the right by A^{-1} shows that $AA^- = I$ and therefore that $A^- = A^{-1}$. Finally, if A has full column rank, then A^tA has inverse

$$(A^tA)^{-1} = \sum_{j=1}^k \sigma_j^{-2} v_j v_j^t.$$

Multiplying A^t on the left by $(A^tA)^{-1}$ yields A^-. ■

Recalling item (c) of Proposition 6.3.1 and the discussion of norm preserving transformations in Section 6.4, one can easily demonstrate that

$$\begin{aligned} \|A\|_2 &= \|\Sigma\|_2 = \max_i \sigma_i \\ \|A\|_F &= \|\Sigma\|_F = \Big(\sum_j \sigma_j^2\Big)^{1/2}. \end{aligned}$$

Norm preservation also permits a quick proof of the next proposition.

Proposition 9.2.3 *Suppose the matrix A has full svd $U\Sigma V^t$ with the diagonal entries σ_i of Σ appearing in decreasing order. The best rank-k approximation of A in the Frobenius norm is*

$$B = \sum_{j=1}^k \sigma_j u_j v_j^t. \tag{9.3}$$

Furthermore, $\|A - B\|_F = \big(\sum_{i>k} \sigma_i^2\big)^{1/2}$ and $\|A - B\|_2 = \sigma_{k+1}$.

Proof: Let B be an approximating matrix. Because

$$\|A - B\|_F = \|U\Sigma V^t - B\|_F = \|\Sigma - U^t BV\|_F, \qquad (9.4)$$

the best approximation is achieved by taking $C = U^t BV$ to be diagonal with its first k diagonal entries equal to σ_1 through σ_k and its remaining diagonal entries equal to 0. Multiplying C on the left by U and on the right by V^t shows that B has the form displayed in equation (9.3). ■

9.3 Applications

The svd is generally conceded to be the most accurate method of least squares estimation. As three of the next six examples illustrate, it offers other advantages in regression as well. The recent book [5] highlights applications in data mining.

9.3.1 Reduced Rank Regression

Occasionally in linear regression, the design matrix does not have full rank. When this occurs, there will be more than one parameter vector β that minimizes $\|y - X\beta\|_2^2$. The Moore-Penrose inverse picks out the solution with minimum norm. If X has full svd $X = \sum_j \sigma_j u_j v_j^t$ and $\beta = \sum_j \alpha_j v_j$, then the residual sum of squares is

$$\begin{aligned} \|y - X\beta\|_2^2 &= \Big\|\sum_j (u_j^t y - \sigma_j \alpha_j) u_j\Big\|_2^2 \\ &= \sum_j (u_j^t y - \sigma_j \alpha_j)^2. \end{aligned}$$

To minimize $\|y - X\beta\|_2^2$, we break the indices into two subsets. For those j with $\sigma_j > 0$, it is clear that we should set

$$\hat{\alpha}_j = \sigma_j^{-1} u_j^t y.$$

For those j with $\sigma_j = 0$, any value α_j will do. However, if we want a β of minimum norm, we should set $\hat{\alpha}_j = 0$. If k is the rank of X, then these choices are summarized in the equations

$$\hat{\beta} = \sum_{j=1}^k \sigma_j^{-1} u_j^t y v_j = X^- y = (X^t X)^- X^t y.$$

Problem 9 asks the reader to verify the identity $X^- = (X^t X)^- X^t$. The solution $\hat{\beta} = X^- y$ clearly generalizes the solution $\hat{\beta} = X^{-1} y$ when X is invertible.

9.3.2 Ridge Regression

The svd also unifies the theories of ridge regression and ordinary regression with X of full rank. In ridge regression, we minimize the penalized sum of squares

$$\begin{aligned} f_\lambda(\beta) &= \|y - X\beta\|_2^2 + \lambda\|\beta\|_2^2 \\ &= (y - X\beta)^t(y - X\beta) + \lambda\beta^t\beta. \end{aligned}$$

The case $\lambda = 0$ corresponds to ordinary regression and the case $\lambda > 0$ to ridge regression. The gradient of $f_\lambda(\beta)$ is

$$\nabla f_\lambda(\beta) = -2X^t(y - X\beta) + 2\lambda\beta.$$

Equating the gradient $\nabla f_\lambda(\beta)$ to the $\mathbf{0}$ vector yields the revised normal equations

$$(X^tX + \lambda I)\beta = X^ty$$

with solution

$$\hat{\beta} = (X^tX + \lambda I)^{-1}X^ty.$$

If we let X have svd $X = \sum_j \sigma_j u_j v_j^t$, then the ingredients

$$X^ty = \sum_j \sigma_j u_j^t y\, v_j, \quad X^tX + \lambda I = \sum_j (\sigma_j^2 + \lambda) v_j v_j^t$$

necessary for solving ridge regression are at our disposal. The parameter estimates and predicted values reduce to

$$\begin{aligned} \hat{\beta} &= \sum_j \frac{\sigma_j}{\sigma_j^2 + \lambda} u_j^t y\, v_j \\ \hat{y} &= \sum_j \frac{\sigma_j^2}{\sigma_j^2 + \lambda} u_j^t y\, u_j. \end{aligned}$$

The above expression for $\hat{\beta}$ makes it evident that increasing λ shrinks estimates toward the origin and that shrinkage is most pronounced in directions v_j corresponding to low singular values.

9.3.3 Polar Decomposition

The svd affords a simple construction of the polar decomposition of a square invertible matrix A. Any nonzero complex number z can be written as the product ru of a positive radius r and a unit vector u determined by its angle with the horizontal axis. By analogy, a polar decomposition represents $A = RO$ as the product of a positive definite matrix R and an orthogonal matrix O. The svd of $A = U\Sigma V^t$ is just one step away. Indeed, because U and V are square and all diagonal entries of Σ are positive, we merely set $R = U\Sigma U^t$ and $O = UV^t$. Clearly, R is positive definite, and O is orthogonal.

9.3.4 Image Compression

In digital imaging, a scene is recorded as an $m \times n$ matrix $A = (a_{ij})$ of intensities. The entry a_{ij} represents the brightness of the pixel (picture element) in row i and column j of the scene. Because m and n are often very large, storage of A is an issue. One way of minimizing storage is to approximate A by a low-rank matrix $B = (b_{ij})$. Proposition 9.2.3 tells how to compute the best rank-k approximation in the Frobenius norm. The choice of k is obviously a crucial consideration. One approach is to take the first k that makes the ratio $\|A - B\|_F / \|A\|_F$ sufficiently small. Once k is chosen, the singular values and singular vectors defining B are stored rather than B itself.

9.3.5 Principal Components

In principal components analysis, one simplifies a random vector or a random sample by projecting it onto a new coordinate system. The first principal component is the linear combination $v_1^t Y$ explaining the largest variance, the second principal component is the linear combination $v_2^t Y$ uncorrelated with $v_1^t Y$ explaining the largest remaining variance, and so forth. More precisely, suppose Y is a random vector with mean vector $\mathrm{E}(Y) = \mathbf{0}$ and variance matrix $\mathrm{Var}(Y)$. The first principal component $v_1^t Y$ is the linear combination that maximizes

$$\mathrm{Var}(v^t Y) \quad = \quad v^t \, \mathrm{Var}(Y) v$$

subject to the constraint $\|v\|_2 = 1$. By arguments very similar to those presented in Section 9.2, v_1 turns out to be the unit eigenvector associated with the largest eigenvalue of $\mathrm{Var}(Y)$. In general, v_i is the unit eigenvector of $\mathrm{Var}(Y)$ associated with the ith largest eigenvalue. Because these eigenvectors are orthogonal, the corresponding linear combinations $v_i^t Y$ are uncorrelated. The eigenvalue associated with v_i is the variance of $v_i^t Y$.

A random sample $x_1, \ldots, x_m$ is treated by considering its centered empirical distribution as a theoretical distribution. Centering is accomplished by subtracting the sample average $\bar{x} = \frac{1}{m} \sum_{j=1}^m x_j$ from each observation x_i. If we assume this has been done, then the sample variance can be expressed as $X^t X$ using

$$X \quad = \quad \frac{1}{\sqrt{m}} \begin{pmatrix} x_1^t \\ \vdots \\ x_m^t \end{pmatrix} \quad = \quad \sum_j \sigma_j u_j v_j^t.$$

The principal components can be recovered from the svd of X without computing $X^t X$. As we have seen many times already, $X^t X = \sum_j \sigma_j^2 v_j v_j^t$. Hence, the ith principal direction is given by the unit eigenvector v_i, and the variance of $v_i^t x_j$ over j is given by σ_i^2.

In regression problems with large numbers of nearly collinear predictors, statisticians sometimes discard the original predictors and regress instead on a subset of the first few principal components. Although this is an effective method of data reduction, it always raises questions of interpretation. One should also keep in mind that lower-order components can be as revealing as higher-order components [11].

9.3.6 *Total Least Squares*

In total least squares, the predictors as well as the observations are considered subject to error [14]. Minimizing the criterion $\|y - X\beta\|_2^2$ fails to capture errors in the predictors. Instead, we should fit a hyperplane to the vectors $z_i = (x_{i1}, \ldots, x_{in}, y_i)^t$. Now a hyperplane H is determined by its unit perpendicular r and its base point w, the point closest to the origin. In total least squares, r and w are estimated by minimizing the criterion $\sum_{i=1}^m \text{dist}(z_i, H)^2$, where

$$\text{dist}(z, H)^2 = \min\left\{\|u - z\|_2^2 : r^t u = r^t w\right\}$$

can be explicitly calculated by seeking a stationary point of the Lagrangian

$$\mathcal{L}(u, \lambda) = \frac{1}{2}\|u - z\|_2^2 - \lambda(r^t u - r^t w).$$

Setting the gradient of $\mathcal{L}(u, \lambda)$ with respect to u equal to $\mathbf{0}$ gives

$$u = z + \lambda r.$$

If we multiply this on the left by r^t and use $\|r\|_2 = 1$ and $r^t u = r^t w$, then we recover the Lagrange multiplier

$$\lambda = r^t(w - z)$$

and accordingly

$$\|u - z\|_2^2 = [r^t(z - w)]^2.$$

Hence, in total least squares we minimize the criterion $\sum_{i=1}^m [r^t(z_i - w)]^2$ with respect to r and w.

The identity

$$\sum_{i=1}^m [r^t(z_i - w)]^2 = \sum_{i=1}^m [r^t(z_i - \bar{z})]^2 + m[r^t(\bar{z} - w)]^2$$

involving the average $\bar{z}$ of the z_i shows that the total least squares criterion is minimized by taking $w = \bar{z}$ regardless of r. To find r, we introduce the matrix

$$M^t = (z_1 - \bar{z}, \ldots, z_m - \bar{z})$$

and minimize r^tM^tMr subject to $\|r\| = 1$. But this problem is solved by taking the svd $M = U\Sigma V^t$ of M and extracting the smallest singular value σ_m and corresponding right singular vector v_m. It is obvious that $r = v_m$ and the minimum of the squared distances is σ_m^2. Exceptions to this solution happen when the minimum singular vector is not unique. Another anomaly occurs when the last component of r equals 0. Because the point z_i is projected onto the point $u_i = z_i + \lambda r = z_i + r^t(\bar{z} - z_i)r$, in this rare situation the last component of u_i is y_i itself. Since this component serves as a predictor of y_i, all residuals are 0, and the fit is perfect.

9.4 Jacobi's Algorithm for the SVD

At this juncture, readers may want to review the presentation of Jacobi's method in Chapter 8. There we employed a two-sided algorithm that multiplies the current matrix on the left and right by Givens rotations. Here we employ a one-sided algorithm that multiplies the current matrix on the right by a Givens rotation. Since a rotation is orthogonal and the product of two orthogonal matrices is orthogonal, each stage of the algorithm operates on a matrix product AV, where V is orthogonal and A is the original $m \times n$ matrix whose svd we seek. The representation AV persists in the limit as the Givens rotations tend to the identity matrix. Once the product AV stabilizes with orthogonal columns, we normalize the nontrivial columns. The resulting matrix can be represented as $U\Sigma$, where Σ is an $m \times n$ diagonal matrix with nonnegative diagonal entries, and the columns of U are either orthogonal unit vectors or $\mathbf{0}$ vectors. The representation $AV = U\Sigma$ is equivalent to the svd $A = U\Sigma V^t$ if we ignore the irrelevant columns of U corresponding to the 0 singular values. In practice, one can keep track of V by applying the same sequence of Givens rotations to a second matrix that starts out as the $n \times n$ identity matrix.

This overall strategy is predicated on choosing the Givens rotations wisely. In the cyclic Jacobi strategy, each pair of columns of the target matrix is considered in turn. If B represents the current version of the target, then the entries of B^tB are just the inner products of the columns of B. To make columns i and j of the transformed target orthogonal, we choose the pertinent Givens rotation R based on the entries in positions (i,i), (i,j), and (j,j) of B^tB. The matrix R^tB^tBR then has a 0 in position (i,j). In the Jacobi svd algorithm, we never actually compute either B^tB or R^tB^tBR. Computation of the three inner products and BR suffices.

9.5 Problems

1. Suppose the matrix A has svd (9.1). Prove rigorously that A has rank k, range equal to the span of $\{u_1, \ldots, u_k\}$, and null space equal to the

orthogonal complement of $\{v_1, \ldots, v_k\}$. State and prove corresponding claims for A^t.

2. Show that the singular values of a matrix A are uniquely determined and that, when A is square and the singular values are distinct, the left and right singular vectors are uniquely determined up to sign.

3. Let A be an invertible $m \times m$ matrix with singular values $\sigma_1, \ldots, \sigma_m$. Prove that

$$\text{cond}_2(A) = \frac{\max_i \sigma_i}{\min_i \sigma_i}.$$

See Section 6.6 for the definition and application of condition numbers.

4. Calculate the Moore-Penrose inverse of the outer product uv^t when u and v are not necessarily unit vectors.

5. Demonstrate that the limits

$$A^- = \lim_{\lambda \downarrow 0} (A^t A + \lambda I)^{-1} A^t = \lim_{\lambda \downarrow 0} A^t (AA^t + \lambda I)^{-1}$$

provide alternative definitions of the Moore-Penrose inverse.

6. Let A be an $m \times n$ matrix and b an $m \times 1$ vector. Show that the following four statements are logically equivalent:

   (a) The equation $Ax = b$ has a solution x.

   (b) The vector b is a linear combination of the columns of A.

   (c) The partitioned matrix $(A\, b)$ has the same rank as A.

   (d) The identity $AA^- b = b$ holds.

7. Suppose the matrix A is symmetric. For any vector x demonstrate that $Ax = \mathbf{0}$ if and only if $A^- x = \mathbf{0}$.

8. Show that the Moore-Penrose inverse represents the only solution of the system of equations (9.2). (Hint: Suppose A_1^- and A_2^- are two solutions. To prove that $A_1^- = A_2^-$, first verify that $A_1^- A = A_2^- A$ and $AA_1^- = AA_2^-$.)

9. Prove the identity

$$A^- = (A^t A)^- A^t.$$

If the matrix A has full column rank, then prove the further identity

$$(AA^t)^- = A(A^t A)^{-2} A^t.$$

10. Courrieu's [3] algorithm for computing the Moore-Penrose inverse of a matrix M exploits the Cholesky decomposition L of the symmetric matrix M^tM. If M is $n \times n$ of rank r, then our development of the Cholesky decomposition in Section 7.7 shows that L can be chosen to have $n-r$ columns consisting entirely of 0's. Let K be the $n \times r$ matrix derived by deleting these columns. Demonstrate that $M^tM = KK^t$ and that

$$\begin{aligned} M^- &= (M^tM)^- M^t \\ (M^tM)^- &= K(K^tK)^{-2}K^t. \end{aligned}$$

(Hint: Apply the identities in Problem 9.)

11. Program and test Courrieu's algorithm described in Problem 10.

12. Find the closest rank-one approximation to the matrix

$$M = \begin{pmatrix} 1 & 1 \\ 1 & 1+\epsilon \end{pmatrix}$$

in the Frobenius norm $\|\cdot\|_F$. Here ϵ is a positive constant.

13. Find expressions for the singular values σ_1 and σ_2 of the matrix

$$M = \begin{pmatrix} a & b \\ c & d \end{pmatrix}.$$

14. Demonstrate that any matrix A is the limit of a sequence A_n of matrices of full rank.

15. Prove the inequality $\|A\|_F \leq \text{rank}(A)\|A\|_2$.

16. Suppose the square matrix A has full svd $A = U\Sigma V^t$. Verify the spectral decomposition

$$\begin{pmatrix} \mathbf{0} & A^t \\ A & \mathbf{0} \end{pmatrix} = \frac{1}{\sqrt{2}}\begin{pmatrix} V & V \\ U & -U \end{pmatrix}\begin{pmatrix} \Sigma & \mathbf{0} \\ \mathbf{0} & -\Sigma \end{pmatrix}\frac{1}{\sqrt{2}}\begin{pmatrix} V^t & U^t \\ V^t & -U^t \end{pmatrix}.$$

Thus, any algorithm producing a spectral decomposition will produce an svd of A without computing AA^t and A^tA.

17. Show that the map $A \mapsto \Sigma$ taking a matrix to its ordered singular values is continuous. (Hints: Suppose $\lim_{n\to\infty} A_n = A$. In view of the norm identity $\|A_n\|_F = \|\Sigma_n\|_F$, the matrices Σ_n are bounded. By passing to an appropriate subsequence and taking limits, one can prove that any cluster point of the Σ_n reduces to the unique matrix Σ appearing in the svd of A.)

18. The nuclear norm $\|A\|_*$ on $m \times n$ matrices is the analog of the ℓ_1 norm on vectors. If $A = (a_{ij})$ has ith singular value σ_i, then we want $\|A\|_*$ to equal $\sum_i \sigma_i$. Unfortunately, this definition does not suggest simple proofs of the norm properties. An alternative is to employ the function $f(A) = \sum_{i=1}^{\min\{m,n\}} a_{ii}$ in the definition

$$\|A\|_* = \max_{R,S} f(RAS) = \max_{R^\#, S^\#} \operatorname{tr}(R^\# A S^\#),$$

where R ranges over all $m \times m$ orthogonal matrices, S ranges over all $n \times n$ orthogonal matrices, $R^\#$ equals R after the deletion of its last $m - \min\{m, n\}$ rows, and $S^\#$ equals S after the deletion of its last $n - \min\{m, n\}$ columns. Demonstrate that

(a) $\|A\|_* \geq \sum_i \sigma_i$,
(b) $\|A\|_* = 0$ if and only if $A = \mathbf{0}$,
(c) $\|cA\|_* = |c| \, \|A\|_*$,
(d) $\|A + B\|_* \leq \|A\|_* + \|B\|_*$,
(e) $\|UAV\|_* = \|A\|_*$ for U and V orthogonal,
(f) $\|A\|_* \leq \sum_i \sigma_i$,
(g) $\|A\|_* = \operatorname{tr}\left(\sqrt{A^t A}\right) = \operatorname{tr}\left(\sqrt{AA^t}\right) = \|A^t\|_*$,
(h) $\|AB\|_* \leq \|A\|_* \|B\|_*$.

(Hints: For property (f) first show that $\|A\|_* = \|\Sigma\|_*$. For property (h) suppose A has svd UDV^t and B has svd YEZ^t. Write

$$f(RABS) = f(TDOEX) = \sum_{i=1}^{\min\{m,n\}} \sum_j \sum_k t_{ij} d_j o_{jk} e_k x_{ki},$$

where A has m rows, B has n columns, and $T = RU$, $O = V^t Y$, and $X = Z^t S$ are orthogonal matrices. Apply the Cauchy-Schwarz inequality to $|\sum_{i=1}^{\min\{m,n\}} t_{ij} x_{ki}|$, and use the fact that $|o_{ij}| \leq 1$.)

19. Show that ridge regression can be accomplished by conducting ordinary regression on an augmented data set. (Hint: Augment the design matrix X by a multiple of the identity matrix. Set the corresponding observations y_i equal to 0.)

20. In ridge regression, verify that the residual sum of squares $\|y - \hat{y}\|_2^2$ increases and $\|\hat{\beta}\|_2^2$ decreases as λ increases.

21. The quantity $\operatorname{df}(\lambda) = \operatorname{tr}[X(X^t X + \lambda I)^{-1} X^t]$ is termed the effective degrees of freedom in ridge regression. Show that

$$\operatorname{df}(\lambda) = \sum_j \frac{\sigma_j^2}{\sigma_j^2 + \lambda}.$$

22. Consider the problem of minimizing $\|y - X\beta\|_2^2$ subject to $\|\beta\|_2^2 \le \alpha^2$. If X has svd $U\Sigma V^t$, then prove in the ridge regression context that you should choose either $\lambda = 0$ or the $\lambda > 0$ satisfying

$$\sum_j \frac{\sigma_j^2}{(\sigma_j^2 + \lambda)^2}(u_j^t y)^2 = \alpha^2.$$

23. Find the polar decomposition of the matrix

$$M = \begin{pmatrix} a & -b \\ b & a \end{pmatrix}.$$

24. Demonstrate that the polar decomposition of an invertible matrix is unique [8]. (Hint: If $R_1O_1 = R_2O_2$ are two polar decompositions, then the matrix $R_2^{-1}R_1 = O_2O_1^t$ is similar to $R_2^{-1/2}R_1R_2^{-1/2}$ and therefore has positive eigenvalues.)

25. Show that an invertible matrix A satisfies the normality condition $AA^t = A^tA$ if and only if its polar decomposition $A = RO$ satisfies the commutivity relation $RO = OR$. (Hint: If $AA^t = A^tA$, then show that R^2 has the two symmetric square roots R and O^tRO.)

26. Given the data points (1,3), (3,1), (4,5), (5,7), and (7,4), find the prediction lines $y = ax + b$ determined by ordinary least squares and total least squares.

9.6 References

[1] Anderson E, Bai Z, Bischof C, Blackford LS, Demmel J, Dongarra JJ, Du Croz J, Hammarling S, Greenbaum A, McKenney A, Sorensen D (1999) *LAPACK Users' Guide*, 3rd ed. SIAM, Philadelphia

[2] Berry MW, Drmac Z, Jessup ER (1999) Matrices, vector spaces, and information retrieval. *SIAM Review* 41:335–362

[3] Courrieu P (2005) Fast computation of Moore-Penrose inverse matrices. *Neural Information Processing – Letters and Reviews* 8:25–29

[4] Demmel J (1997) *Applied Numerical Linear Algebra.* SIAM, Philadelphia

[5] Eldén L (2007) *Matrix Methods in Data Mining and Pattern Recognition.* SIAM, Philadelphia

[6] Gill PE, Murray W, Wright MH (1991) *Numerical Linear Algebra and Optimization, Volume 1.* Addison-Wesley, Redwood City, CA

[7] Golub GH, Van Loan CF (1996) *Matrix Computations*, 3rd ed. Johns Hopkins University Press, Baltimore

[8] Hausner A (1967) Uniqueness of the polar decomposition. *Amer Math Monthly* 74:303–304

[9] Horn RA, Johnson CR (1985) *Matrix Analysis.* Cambridge University Press, Cambridge

[10] Horn RA, Johnson CR (1991) *Topics in Matrix Analysis.* Cambridge University Press, Cambridge

[11] Jolliffe IT (1982). A note on the use of principal components in regression. *J Roy Stat Soc C* 31:300–303

[12] Karro J, Li C-K (1997) A unified elementary approach to canonical forms of matrices. *SIAM Review* 39:305–309

[13] McLachlan GJ, Do K-A, Ambroise C (2004) *Analyzing Microarray Gene Expression Data.* Wiley, Hoboken NJ

[14] Nievergelt Y (1994) Total least squares: state-of-the-art regression in numerical analysis. *SIAM Review* 36:258–264

[15] Press WH, Teukolsky SA, Vetterling WT, Flannery BP (1992) *Numerical Recipes in Fortran: The Art of Scientific Computing*, 2nd ed. Cambridge University Press, Cambridge

[16] Seber GAF, Lee AJ (2003) *Linear Regression Analysis,* 2nd ed. Wiley, Hoboken, NJ

[17] Strang G (2003) *Introduction to Linear Algebra*, 3rd ed. Wellesley-Cambridge Press, Wellesley, MA

[18] Trefethen LN, Bau D (1997) *Numerical Linear Algebra.* SIAM, Philadelphia

10

Splines

10.1 Introduction

Splines are used for interpolating functions. Before the advent of computer graphics, a draftsman would draw a smooth curve through a set of points plotted on graph paper by forcing a flexible strip to pass over the points. The strip, made of wood, metal, or plastic, typically was held in place by weights as the draftsman drew along its edge. Subject to passing through the interpolating points, the strip or spline would minimize its stress by straightening out as much as possible. Beyond the terminal points on the left and right, the spline would be straight.

Mathematical splines are idealizations of physical splines. For simplicity we will deal only with the most commonly used splines, cubic splines. These are piecewise cubic polynomials that interpolate a tabulated function $f(x)$ at certain data points $x_0 < x_1 < \cdots < x_n$ called nodes or knots. There are, of course, many ways of interpolating a function. For example, Lagrange's interpolation formula provides a polynomial $p(x)$ of degree n that agrees with $f(x)$ at the nodes. Unfortunately, interpolating polynomials can behave poorly even when fitted to slowly varying functions. (See Problem 1 for a discussion of the classical example of Runge.) Splines minimize average squared curvature and consequently perform better than interpolating polynomials.

The program of this chapter is to investigate a few basic properties of cubic splines, paying particular attention to issues of computing. We then show how splines can be employed in nonparametric regression. For a much fuller exposition of splines, readers can consult one or more of the books [1, 2, 3, 5, 8].

10.2 Definition and Basic Properties

We start with a formal definition of a spline.

Definition 10.2.1 *Let the values $f(x_i) = f_i$ of the function $f(x)$ be given at the points $x_0 < x_1 < \cdots < x_n$. A natural, cubic, interpolatory spline $s(x)$ is a function on the interval $[x_0, x_n]$ possessing the following properties:*

(a) $s(x)$ is a cubic polynomial on each node-to-node interval $[x_i, x_{i+1}]$,

(b) $s(x_i) = f_i$ at each node x_i,

K. Lange, *Numerical Analysis for Statisticians*, Statistics and Computing,
DOI 10.1007/978-1-4419-5945-4_10, © Springer Science+Business Media, LLC 2010

(c) *the second derivative $s''(x)$ exists and is continuous throughout the entire interval $[x_0, x_n]$,*

(d) *at the terminal nodes, $s''(x_0) = s''(x_n) = 0$.*

For brevity we simply call $s(x)$ a spline.

Existence and uniqueness follow immediately from this definition.

Proposition 10.2.1 *There is exactly one function $s(x)$ on $[x_0, x_n]$ satisfying the above properties.*

Proof: For notational convenience, let

$$\begin{aligned} h_i &= x_{i+1} - x_i \\ \sigma_i &= s''(x_i) \\ s_i(x) &= s(x), \quad x \in [x_i, x_{i+1}]. \end{aligned}$$

Note that the second derivatives σ_i are as yet unknown. Because $s_i(x)$ is a cubic polynomial, $s_i''(x)$ is a linear polynomial that can be expressed as

$$s_i''(x) = \sigma_i \frac{x_{i+1} - x}{h_i} + \sigma_{i+1} \frac{x - x_i}{h_i}. \tag{10.1}$$

The function $s''(x)$ pieced together in this fashion is clearly continuous on $[x_0, x_n]$. Integrating equation (10.1) twice gives

$$\begin{aligned} s_i(x) &= \frac{\sigma_i}{6h_i}(x_{i+1} - x)^3 + \frac{\sigma_{i+1}}{6h_i}(x - x_i)^3 \\ &\quad + c_1(x - x_i) + c_2(x_{i+1} - x). \end{aligned} \tag{10.2}$$

The constants of integration c_1 and c_2 can be determined from the interpolation conditions

$$\begin{aligned} f_i &= s_i(x_i) \\ &= \frac{\sigma_i}{6} h_i^2 + c_2 h_i \\ f_{i+1} &= s_i(x_{i+1}) \\ &= \frac{\sigma_{i+1}}{6} h_i^2 + c_1 h_i. \end{aligned}$$

Solving for c_1 and c_2 and substituting the results in equation (10.2) produce

$$\begin{aligned} s_i(x) &= \frac{\sigma_i}{6h_i}(x_{i+1} - x)^3 + \frac{\sigma_{i+1}}{6h_i}(x - x_i)^3 \\ &\quad + \Big(\frac{f_{i+1}}{h_i} - \frac{\sigma_{i+1} h_i}{6}\Big)(x - x_i) + \Big(\frac{f_i}{h_i} - \frac{\sigma_i h_i}{6}\Big)(x_{i+1} - x). \end{aligned} \tag{10.3}$$

Since it satisfies the interpolation conditions, $s(x)$ as defined by (10.3) is continuous on $[x_0, x_n]$.

Choosing the σ_i appropriately also will guarantee that $s'(x)$ is continuous. Differentiating equation (10.2) yields

$$\begin{aligned} s_i'(x) &= -\frac{\sigma_i}{2h_i}(x_{i+1}-x)^2 + \frac{\sigma_{i+1}}{2h_i}(x-x_i)^2 \\ &\quad + \frac{f_{i+1}-f_i}{h_i} - \frac{h_i}{6}(\sigma_{i+1}-\sigma_i). \end{aligned} \tag{10.4}$$

Continuity is achieved when $s_{i-1}'(x_i) = s_i'(x_i)$. In terms of equation (10.4), this is equivalent to

$$\frac{1}{6}h_{i-1}\sigma_{i-1}+\frac{1}{3}(h_{i-1}+h_i)\sigma_i+\frac{1}{6}h_i\sigma_{i+1} = \frac{f_{i+1}-f_i}{h_i}-\frac{f_i-f_{i-1}}{h_{i-1}}. \tag{10.5}$$

This is a system of $n-1$ equations for the $n+1$ unknowns $\sigma_0, \ldots, \sigma_n$. However, two of these unknowns, σ_0 and σ_n, are 0 by assumption. If by good fortune the $(n-1)\times(n-1)$ matrix of coefficients multiplying the remaining unknowns is invertible, then we can solve for $\sigma_1, \ldots, \sigma_{n-1}$ uniquely. Invertibility follows immediately from the strict diagonal dominance conditions

$$\begin{aligned} \frac{1}{3}(h_0+h_1) &> \frac{1}{6}h_1 \\ \frac{1}{3}(h_{i-1}+h_i) &> \frac{1}{6}h_{i-1}+\frac{1}{6}h_i \qquad i=2,\ldots,n-2 \\ \frac{1}{3}(h_{n-1}+h_n) &> \frac{1}{6}h_{n-1}. \end{aligned}$$

This completes the proof because, as already indicated, the coefficients $\sigma_1, \ldots, \sigma_{n-1}$ uniquely determine the spline $s(x)$. ∎

To solve for $\sigma_1, \ldots, \sigma_{n-1}$, one can use functional iteration as described in Chapter 6. In practice, it is better to exploit the fact that the matrix of coefficients is tridiagonal. To do so, define

$$d_i = \frac{6}{h_i}\Big(\frac{f_{i+1}-f_i}{h_i} - \frac{f_i-f_{i-1}}{h_{i-1}}\Big)$$

and rewrite the system (10.5) as

$$\frac{h_{i-1}}{h_i}\sigma_{i-1} + 2\Big(1+\frac{h_{i-1}}{h_i}\Big)\sigma_i + \sigma_{i+1} = d_i. \tag{10.6}$$

Now set

$$\sigma_{i-1} = \rho_i\sigma_i + \tau_i, \tag{10.7}$$

where ρ_i and τ_i are constants to be determined. In view of $\sigma_0 = 0$, we take $\rho_1 = \tau_1 = 0$. In general, substitution of equation (10.7) in equation (10.6) leads to

$$\sigma_i = -\frac{\sigma_{i+1}}{\frac{h_{i-1}}{h_i}\rho_i + 2(1+\frac{h_{i-1}}{h_i})} + \frac{d_i - \frac{h_{i-1}}{h_i}\tau_i}{\frac{h_{i-1}}{h_i}\rho_i + 2(1+\frac{h_{i-1}}{h_i})}.$$

This has the form of equation (10.7) and suggests computing

$$\begin{aligned}\rho_{i+1} &= -\frac{1}{\frac{h_{i-1}}{h_i}\rho_i + 2(1+\frac{h_{i-1}}{h_i})} \\ \tau_{i+1} &= \frac{d_i - \frac{h_{i-1}}{h_i}\tau_i}{\frac{h_{i-1}}{h_i}\rho_i + 2(1+\frac{h_{i-1}}{h_i})}\end{aligned}$$

recursively beginning at $i = 0$. Once the constants ρ_i and τ_i are available, then the σ_i can be computed in order from equation (10.7) beginning at $i = n$.

The next proposition validates the minimum curvature property of natural cubic splines.

Proposition 10.2.2 *Let $s(x)$ be the spline interpolating the function $f(x)$ at the nodes $x_0 < x_1 < \cdots < x_n$. If $g(x)$ is any other twice continuously differentiable function interpolating $f(x)$ at these nodes, then*

$$\int_{x_0}^{x_n} g''(x)^2 dx \;\geq\; \int_{x_0}^{x_n} s''(x)^2 dx, \tag{10.8}$$

with equality only if $g(x) = s(x)$ throughout $[x_0, x_n]$.

Proof: If $\int_{x_0}^{x_n} g''(x)^2 dx = \infty$, then there is nothing to prove. Therefore, assume the contrary and consider the identity

$$\begin{aligned}\int_{x_0}^{x_n}[g''(x) - s''(x)]^2 dx &= \int_{x_0}^{x_n} g''(x)^2 dx - 2\int_{x_0}^{x_n}[g''(x) - s''(x)]s''(x)dx \\ &\quad - \int_{x_0}^{x_n} s''(x)^2 dx. \end{aligned}\tag{10.9}$$

Let us prove that the second integral on the right-hand side of equation (10.9) vanishes. Decomposing this integral and integrating each piece by parts give

$$\begin{aligned}&\int_{x_0}^{x_n}[g''(x) - s''(x)]s''(x)dx \\ &= \sum_{i=0}^{n-1}\int_{x_i}^{x_{i+1}}[g''(x) - s''(x)]s''(x)dx \\ &= \sum_{i=0}^{n-1}\Big\{[g'(x) - s'(x)]s''(x)|_{x_i}^{x_{i+1}} - \int_{x_i}^{x_{i+1}}[g'(x) - s'(x)]s'''(x)dx\Big\}.\end{aligned}$$

Since $s(x)$ is a piecewise cubic polynomial, $s'''(x)$ equals some constant α_i on $[x_i, x_{i+1}]$. Thus, we find

$$\begin{aligned}\int_{x_i}^{x_{i+1}}[g'(x) - s'(x)]s'''(x)dx\} &= \alpha_i[g(x) - s(x)]_{x_i}^{x_{i+1}} \\ &= 0\end{aligned}$$

because $g(x)$ and $s(x)$ agree with $f(x)$ at each node. We are left with

$$\sum_{i=0}^{n-1} \Big\{ s''(x_{i+1})[g'(x_{i+1}) - s'(x_{i+1})] - s''(x_i)[g'(x_i) - s'(x_i)] \Big\},$$

which telescopes to

$$s''(x_n)[g'(x_n) - s'(x_n)] - s''(x_0)[g'(x_0) - s'(x_0)].$$

By assumption, $s''(x_0) = s''(x_n) = 0$.

This proves our contention about the vanishing of the second integral on the right-hand side of equation(10.9) and allows us to write

$$\int_{x_0}^{x_n} g''(x)^2 dx = \int_{x_0}^{x_n} s''(x)^2 dx + \int_{x_0}^{x_n} [g''(x) - s''(x)]^2 dx. \quad (10.10)$$

Inequality (10.8) is now obvious. If equality obtains in inequality (10.8), then the continuous function $g''(x) - s''(x)$ is identically 0. This implies that $s(x) = g(x) + c_0 + c_1 x$ for certain constants c_0 and c_1. Because $s(x)$ and $g(x)$ both interpolate $f(x)$ at x_0 and x_1, it follows that $c_0 = c_1 = 0$. ■

Note that the curvature of a function $g(x)$ is technically the function $\kappa(x) = g''(x)/[1 + g'(x)^2]^{\frac{3}{2}}$. For $|g'(x)| \ll 1$, we recover $g''(x)$. Proposition 10.2.2 should be interpreted in this light. The final proposition of this section provides bounds on the errors committed in spline approximation.

Proposition 10.2.3 *Suppose that $f(x)$ is twice continuously differentiable and $s(x)$ is the spline interpolating $f(x)$ at the nodes $x_0 < x_1 < \cdots < x_n$. If $h = \max_{0 \le i \le n-1}(x_{i+1} - x_i)$, then*

$$\begin{aligned} \max_{x_0 \le x \le x_n} |f(x) - s(x)| &\le h^{\frac{3}{2}} \Big[\int_{x_0}^{x_n} f''(y)^2 dy \Big]^{\frac{1}{2}} \\ \max_{x_0 \le x \le x_n} |f'(x) - s'(x)| &\le h^{\frac{1}{2}} \Big[\int_{x_0}^{x_n} f''(y)^2 dy \Big]^{\frac{1}{2}}. \end{aligned} \quad (10.11)$$

It follows that $s(x)$ and $s'(x)$ converge uniformly to $f(x)$ and $f'(x)$ as the mesh length h goes to 0.

Proof: Any $x \in [x_0, x_n]$ lies in some interval $[x_i, x_{i+1}]$. Because $f(t) - s(t)$ vanishes at x_i and x_{i+1}, Rolle's theorem indicates that $f'(z) - s'(z) = 0$ for some $z \in [x_i, x_{i+1}]$. Hence,

$$f'(x) - s'(x) = \int_z^x [f''(y) - s''(y)] dy,$$

and consequently the Cauchy-Schwarz inequality implies

$$|f'(x) - s'(x)| \le \Big\{ \int_z^x [f''(y) - s''(y)]^2 dy \Big\}^{\frac{1}{2}} \Big\{ \int_z^x 1^2 dy \Big\}^{\frac{1}{2}}$$

$$
\begin{aligned}
&= \left\{\int_z^x [f''(y)-s''(y)]^2 dy\right\}^{\frac{1}{2}} |x-z|^{\frac{1}{2}} \\
&\le \left\{\int_{x_0}^{x_n} [f''(y)-s''(y)]^2 dy\right\}^{\frac{1}{2}} h^{\frac{1}{2}}.
\end{aligned}
$$

In view of equation (10.10) with $g(x) = f(x)$, this gives the second inequality of (10.11).

To prove the first inequality, again let $x \in [x_i, x_{i+1}]$. Then

$$
\begin{aligned}
|f(x)-s(x)| &= \left|\int_{x_i}^x [f'(y)-s'(y)]dy\right| \\
&\le \int_{x_i}^x \max_{x_0\le z\le x_n} |f'(z)-s'(z)|dy \\
&\le h \max_{x_0\le z\le x_n} |f'(z)-s'(z)|.
\end{aligned}
$$

Substituting in this inequality the second inequality of (10.11) yields the first inequality of (10.11). ■

Better error bounds are available when $f(x)$ possesses more derivatives and the nodes are uniformly spaced [3]. For instance, if the fourth derivative $f^{(4)}(x)$ exists and is continuous, and the uniform spacing is h, then

$$
\max_{x_0\le x\le x_n} |f(x)-s(x)| \le \frac{h^4}{16} \max_{x_0\le x\le x_n} |f^{(4)}(x)|. \tag{10.12}
$$

10.3 Applications to Differentiation and Integration

Splines can be quite useful in numerical differentiation and integration. For example, equation (10.4) of Proposition 10.2.1 offers an accurate method of numerically differentiating $f(x)$ at any point $x \in [x_0, x_n]$. To integrate $f(x)$, we note that equation (10.3) implies

$$
\begin{aligned}
&\int_{x_i}^{x_{i+1}} s_i(x)dx \\
&= \frac{\sigma_i}{24}h_i^3 + \frac{\sigma_{i+1}}{24}h_i^3 + \left(\frac{f_{i+1}}{h_i} - \frac{\sigma_{i+1}h_i}{6}\right)\frac{h_i^2}{2} + \left(\frac{f_i}{h_i} - \frac{\sigma_i h_i}{6}\right)\frac{h_i^2}{2} \\
&= \frac{f_i + f_{i+1}}{2}h_i - \frac{\sigma_i + \sigma_{i+1}}{24}h_i^3.
\end{aligned}
$$

It follows that

$$
\int_{x_0}^{x_n} f(x)dx \approx \int_{x_0}^{x_n} s(x)dx = \sum_{i=0}^{n-1}\left[\frac{f_i + f_{i+1}}{2}h_i - \frac{\sigma_i + \sigma_{i+1}}{24}h_i^3\right].
$$

According to inequality (10.12), the error committed in the approximation $\int_{x_0}^{x_n} f(x)dx \approx \int_{x_0}^{x_n} s(x)ds$ is bounded above by

$$\frac{h^4}{16}(x_n - x_0) \max_{x_0 \le x \le x_n} |f^{(4)}(x)|$$

for nodes with uniform spacing h.

10.4 Application to Nonparametric Regression

In parametric regression, one minimizes a weighted sum of squares

$$\sum_{i=0}^{n} w_i[y_i - g(x_i)]^2 \tag{10.13}$$

over a particular class of functions $g(x)$, taking the observations y_i and the weights $w_i > 0$ as given. For instance, in polynomial regression, the relevant class consists of all polynomials of a certain degree d or less. In time series analysis, the class typically involves linear combinations of a finite number of sines and cosines. However, often there is no convincing rationale for restricting attention to a narrow class of candidate regression functions. This has prompted statisticians to look at wider classes of functions.

At first glance, some restriction on the smoothness of the regression functions seems desirable. This is a valuable insight, but one needs to exercise caution because there exist many infinitely differentiable functions reducing the weighted sum of squares (10.13) to 0. For example, the unique polynomial of degree n interpolating the observed values y_i at the points x_i achieves precisely this. Smoothness per se is insufficient. Control of the overall size of the derivatives of the regression function is also important. One criterion incorporating these competing aims is the convex combination

$$J_\alpha(g) = \alpha \sum_{i=0}^{n} w_i[y_i - g(x_i)]^2 + (1-\alpha)\int_{x_0}^{x_n} g''(x)^2 dx \tag{10.14}$$

for $0 < \alpha < 1$. Minimizing $J_\alpha(g)$ reaches a compromise between minimizing the weighted sum of squares and minimizing the average squared curvature of the regression function. For α near 1, the weighted sum of squares predominates. For α near 0, the average squared curvature takes precedence. One immediate consequence of Proposition 10.2.2 is that the class of relevant functions collapses to the class of splines. For if $g(x)$ is twice continuously differentiable, then the spline $s(x)$ that interpolates $g(x)$ at the nodes x_i contributes the same weighted sum of squares and a reduced integral term; in other words, $J_\alpha(s) \le J_\alpha(g)$.

To find the spline $s(x)$ minimizing J_α, we take the approach of de Boor [1] and extend the notation of Proposition 10.2.1. The system of $n-1$ equations displayed in (10.5) can be summarized by defining the vectors

$$\begin{aligned} \sigma &= (\sigma_1, \ldots, \sigma_{n-1})^t \\ f &= (f_0, \ldots, f_n)^t \\ &= [s(x_0), \ldots, s(x_n)]^t \\ y &= (y_0, \ldots, y_n)^t, \end{aligned}$$

the $(n-1) \times (n-1)$ tridiagonal matrix R with entries

$$r_{ij} = \frac{1}{6} \begin{cases} h_{i-1} & j = i-1 \\ 2(h_{i-1} + h_i) & j = i \\ h_i & j = i+1 \\ 0 & \text{otherwise,} \end{cases}$$

and the $(n-1) \times (n+1)$ tridiagonal matrix Q with entries

$$q_{ij} = \begin{cases} \frac{1}{h_{i-1}} & j = i-1 \\ -(\frac{1}{h_{i-1}} + \frac{1}{h_i}) & j = i \\ \frac{1}{h_i} & j = i+1 \\ 0 & \text{otherwise.} \end{cases}$$

In this notation, the system of equations (10.5) is expressed as $R\sigma = Qf$. If we also let W be the diagonal matrix with ith diagonal entry w_i, then the weighted sum of squares (10.13) becomes $(y-f)^t W (y-f)$.

The integral contribution to $J_\alpha(s)$ can be represented in matrix notation by observing that equation (10.1) implies

$$\begin{aligned} & \int_{x_i}^{x_{i+1}} s''(x)^2 dx \\ = & \frac{1}{h_i^2} \int_{x_i}^{x_{i+1}} [\sigma_i(x_{i+1} - x) + \sigma_{i+1}(x - x_i)]^2 dx \\ = & \frac{1}{h_i^2} \frac{\sigma_i^2 h_i^3}{3} + \frac{2\sigma_i \sigma_{i+1}}{h_i^2} \int_{x_i}^{x_{i+1}} (x_{i+1} - x)(x - x_i) dx + \frac{1}{h_i^2} \frac{\sigma_{i+1}^2 h_i^3}{3} \\ = & \frac{h_i}{3} \sigma_i^2 + 2h_i \sigma_i \sigma_{i+1} \int_0^1 (1-z) z dz + \frac{h_i}{3} \sigma_{i+1}^2 \\ = & \frac{h_i}{3} (\sigma_i^2 + \sigma_i \sigma_{i+1} + \sigma_{i+1}^2). \end{aligned}$$

Taking into account $\sigma_0 = \sigma_n = 0$, we infer that

$$\int_{x_0}^{x_n} s''(x)^2 dx = \frac{1}{3} \sum_{i=0}^{n-1} h_i (\sigma_i^2 + \sigma_i \sigma_{i+1} + \sigma_{i+1}^2)$$

$$\begin{aligned} &= \frac{1}{6}\sum_{i=1}^{n-1}[h_{i-1}\sigma_{i-1}\sigma_i + 2\sigma_i^2(h_{i-1}+h_i) + h_i\sigma_i\sigma_{i+1}] \\ &= \sigma^t R\sigma. \end{aligned}$$

This shows that the symmetric, invertible matrix R is positive definite. Furthermore, because $\sigma = R^{-1}Qf$, the criterion $J_\alpha(s)$ reduces to

$$\begin{aligned} J_\alpha(s) &= \alpha(y-f)^t W(y-f) + (1-\alpha)\sigma^t R\sigma \\ &= \alpha(y-f)^t W(y-f) + (1-\alpha)f^t Q^t R^{-1}Qf. \end{aligned} \tag{10.15}$$

Based on the identity (10.15), it is possible to minimize $J_\alpha(s)$ as a function of f. At the minimum point of $J_\alpha(s)$, its gradient with respect to f satisfies

$$-2\alpha W(y-f) + 2(1-\alpha)Q^t R^{-1}Qf = \mathbf{0}. \tag{10.16}$$

Solving for the optimal f yields

$$\hat{f} = [\alpha W + (1-\alpha)Q^t R^{-1}Q]^{-1}\alpha W y.$$

Alternatively, equation (10.16) can be rewritten as

$$-2\alpha W(y-f) + 2(1-\alpha)Q^t\sigma = \mathbf{0}. \tag{10.17}$$

Thus, the optimal σ determines the optimal f through

$$y - \hat{f} = \Big(\frac{1-\alpha}{\alpha}\Big)W^{-1}Q^t\hat{\sigma}. \tag{10.18}$$

Multiplying equation (10.17) by QW^{-1} gives

$$-2\alpha Qy + 2\alpha R\sigma + 2(1-\alpha)QW^{-1}Q^t\sigma = \mathbf{0}$$

with solution

$$\hat{\sigma} = [\alpha R + (1-\alpha)QW^{-1}Q^t]^{-1}\alpha Q y.$$

This solution for $\hat{\sigma}$ has the advantage that the positive definite matrix $\alpha R + (1-\alpha)QW^{-1}Q^t$ inherits a banding pattern from R and Q. Solving linear equations involving banded matrices is more efficiently accomplished via their Cholesky decompositions than via the sweep operator; see Problem 8 of Chapter 7. Once $\hat{\sigma}$ is available, equation (10.18) not only determines $\hat{f}$ but also determines the weighted residual sum of squares

$$(y-\hat{f})^t W(y-\hat{f}) = \Big(\frac{1-\alpha}{\alpha}\Big)^2\hat{\sigma}^t QW^{-1}Q^t\hat{\sigma}.$$

10.5 Problems

1. Consider the function $f(x) = (1+25x^2)^{-1}$ on $[-1,1]$. Runge's example [4] involves fitting an interpolating polynomial $p_n(x)$ to $f(x)$ at $n+1$ equally spaced nodes

$$\begin{aligned} x_i &= -1+ih \quad i=0,1,\ldots,n, \\ h &= \frac{2}{n} \end{aligned}$$

using Lagrange's formula

$$p_n(x) = \sum_{i=0}^{n} f(x_i) \prod_{j \neq i} \frac{(x-x_j)}{(x_i-x_j)}.$$

Compare the fit of $p_n(x)$ to $f(x)$ to that of the natural, cubic, interpolating spline $s_n(x)$. In this comparison pay particular attention to the point $-1+n^{-1}$ as n increases. Please feel free to use relevant subroutines from [6] to carry out your computations.

2. Show that Proposition 10.2.1 remains valid if the condition

(d*) $s'(x_0) = f'(x_0)$ and $s'(x_n) = f'(x_n)$

replaces condition (d) in the definition of a cubic spline. Show that Proposition 10.2.2 also carries over if $g(x)$ as well as $s(x)$ satisfies (d*).

3. For nodes $x_0 < x_1 < \cdots < x_n$ and function values $f_i = f(x_i)$, develop a quadratic interpolating spline $s(x)$ satisfying the conditions:

(a) $s(x)$ is a quadratic polynomial on each interval $[x_i, x_{i+1}]$,

(b) $s(x_i) = f_i$ at each node x_i,

(c) the first derivative $s'(x)$ exists and is continuous throughout the entire interval $[x_0, x_n]$.

To simplify your theory, write

$$s(x) = a_i + b_i(x-x_i) + c_i(x-x_i)(x-x_{i+1})$$

for $x \in [x_i, x_{i+1}]$. Derive explicit expressions for the a_i and b_i from property (b). Using property (c), prove that

$$c_i = \frac{b_i - b_{i-1}}{x_{i+1}-x_i} - c_{i-1}\frac{x_i - x_{i-1}}{x_{i+1}-x_i}$$

for $i = 1, \ldots, n-1$. What additional information do you require to completely determine the spline?

4. Given the nodes $x_0 < x_1 < \cdots < x_n$, let V be the vector space of functions that are twice continuously differentiable at each node x_i and cubic polynomials on $(-\infty, x_0)$, (x_n, ∞), and each of the intervals (x_i, x_{i+1}). Show that any function $s(x) \in V$ can be uniquely represented as

$$s(x) = a_0 + a_1 x + a_2 x^2 + a_3 x^3 + \sum_{i=0}^{n} c_i (x - x_i)_+^3,$$

where $(x - x_i)_+^3$ is 0 for $x \leq x_i$ and $(x - x_i)^3$ otherwise. Conclude that this vector space has dimension $n + 5$.

5. Continuing Problem 4, consider the vector subspace $W \subset V$ whose functions are linear on $(-\infty, x_0)$ and (x_n, ∞). Prove that W is just the subspace of natural cubic splines and has dimension $n + 1$.

6. Let $s(x)$ be the natural cubic spline interpolating the function $f(x)$ at the three equally spaced nodes $x_0 < x_1 < x_2$. Explicitly evaluate the integral $\int_{x_0}^{x_2} s(x)dx$ and the derivatives $s'(x_i)$ in terms of the spacing $h = x_2 - x_1 = x_1 - x_0$ and the function values $f_i = f(x_i)$.

7. In the spline model for nonparametric regression, show that the positive definite matrix $\alpha R + (1 - \alpha)QW^{-1}Q^t$ is banded. How many subdiagonals display nonzero entries?

8. Continuing Problems 4 and 5, let

$$f_i(x) = a_{i0} + a_{i1}x + \sum_{j=0}^{n} c_{ij}(x - x_j)_+^3, \qquad i = 0, \ldots, n$$

be a basis of the vector space W. If $s(x) = \sum_{i=0}^{n} \beta_i f_i(x)$, then show that

$$\int_{x_0}^{x_n} s''(x)t''(x)dx = 6 \sum_{i=0}^{n} \sum_{j=0}^{n} \beta_i c_{ij} t(x_j)$$

for any $t(x) \in V$. (Hints: Integrate by parts as in Proposition 10.2.2, and use the fact that $\sum_{j=0}^{n} c_{ij} = 0$ by virtue of Problem 5.)

9. Mindful of Problems 4, 5, and 8, let $s(x) = \sum_{i=0}^{n} \beta_i f_i(x) \in W$ be the spline minimizing the functional $J_\alpha(g)$ defined in equation (10.14). Prove that $s(x)$ satisfies

$$\begin{aligned} 0 &= -\alpha \sum_{j=0}^{n} w_j [y_j - s(x_j)] t(x_j) + (1 - \alpha) \int_{x_0}^{x_n} s''(x)t''(x)dx \\ &= -\alpha \sum_{j=0}^{n} w_j \Big[y_j - \sum_{i=0}^{n} \beta_i f_i(x_j) \Big] t(x_j) + 6(1 - \alpha) \sum_{i=0}^{n} \sum_{j=0}^{n} \beta_i c_{ij} t(x_j) \end{aligned}$$

for any function $t(x) \in W$. Because the constants $t(x_j)$ are arbitrary, demonstrate that this provides the system of linear equations

$$\alpha w_j y_j = \alpha w_j \sum_{i=0}^{n} \beta_i f_i(x_j) + 6(1-\alpha) \sum_{i=0}^{n} \beta_i c_{ij}$$

determining the β_j. Summarize this system of equations as the single vector equation $\alpha W y = \alpha W F^t \beta + 6(1-\alpha) C^t \beta$ by defining appropriate matrices. Because the symmetric matrix FC^t has entry

$$\sum_{k=0}^{n} f_i(x_k) c_{jk} = \int_{x_0}^{x_n} f_i''(x) f_j''(x) dx \qquad (10.19)$$

in row i and column j, argue finally that the solution

$$\hat{\beta} = [\alpha F W F^t + 6(1-\alpha) F C^t]^{-1} \alpha F W y \qquad (10.20)$$

is well defined. This approach to minimizing $J_\alpha(g)$ can exploit any of several different bases for W [2]. (Hint: Adopting the usual calculus of variations tactic, evaluate the derivative $J'_\alpha(s + \epsilon t)|_{\epsilon=0}$. Equality (10.19) follows from Problem 8.)

10. It is possible to give a Bayesian interpretation to the spline solution (10.20) of Problem 9 [7]. Suppose in the notation of Problem 9 that

$$y_j = \sum_{i=0}^{n} \beta_i f_i(x_j) + \frac{1}{\sqrt{w_j}} \epsilon_j,$$

where the errors ϵ_j are independent, univariate normals with common mean 0 and common variance σ^2. Assuming that β has a multivariate normal prior with mean **0** and covariance

$$\sigma^2 \Omega = \sigma^2 [6(1-\alpha)\alpha^{-1} F C^t]^{-1},$$

demonstrate that the posterior mean of β is given by (10.20).

10.6 References

[1] de Boor C (1978) *A Practical Guide to Splines.* Springer, New York

[2] Eubank R (1990) *Smoothing Splines and Nonparametric Regression.* Marcel Dekker, New York

[3] Hämmerlin G, Hoffmann KH (1991) *Numerical Mathematics.* Springer, New York

[4] Isaacson E, Keller HB (1966) *Analysis of Numerical Methods.* Wiley, New York

[5] Powell MJD (1981) *Approximation Theory and Methods.* Cambridge University Press, Cambridge

[6] Press WH, Teukolsky SA, Vetterling WT, Flannery BP (1992) *Numerical Recipes in Fortran: The Art of Scientific Computing*, 2nd ed. Cambridge University Press, Cambridge

[7] Silverman BW (1985) Some aspects of the spline smoothing approach to non-parametric regression curve fitting (with discussion). *J Roy Stat Soc B* 47:1–52

[8] Wahba G (1990) *Spline Functions for Observational Data.* CBMS–NSF Regional Conference Series, SIAM, Philadelphia

11

Optimization Theory

11.1 Introduction

This chapter summarizes a handful of basic principles that permit the exact solution of many optimization problems. Misled by the beautiful examples of elementary calculus, students are disappointed when they cannot solve optimization problems analytically. More experienced scholars know that exact solutions are the exception rather than the rule. However, they cherish these exceptions because they form the basis of most iteration schemes in optimization.

Readers are expected to be comfortable with univariate optimization techniques commonly taught in elementary calculus. As the chapter proceeds, the difficulty of the presented theory escalates. Each new complication forces a climb to a new plateau. To adjust to the thinner atmosphere in preparation for the next ascent, take your time in understanding the specific examples and keep in mind how many centuries it took for the pioneers to scale the present height. In the compass of a short chapter, it is impossible to present all proofs. Fortunately, many texts [2, 5, 6, 12, 16, 17, 22] lay out all details.

We first consider multivariate optimization without constraints. The theory builds on previous experience and is largely a matter of exchanging scalar notation for vector notation. Thus, first derivatives become differentials and gradients, second derivatives become second differentials or Hessians, and positivity becomes positive definiteness. We then move on to optimization with equality constraints and meet those magical constructs called Lagrange multipliers. We next consider optimization problems with inequality constraints. Extension of the Lagrange multiplier rule to these problems by Karush, Kuhn, and Tucker ranks as one of the great triumphs of 20th century applied mathematics. The chapter concludes with a brief survey of block relaxation methods. Here exact methods are applied to successive subsets of the parameters of an objective function.

Mastery of the material in this chapter pays enormous dividends. Much of modern estimation theory is driven by optimization. Despite the advances of the past few centuries, the best methods for many large-scale problems are still a matter of debate. Although harder to teach, framing problems as optimization exercises is also vitally important. Without a proper grounding in optimization theory, many algorithms for solving problems lack a theoretical basis and appear hopelessly ad hoc.

K. Lange, *Numerical Analysis for Statisticians*, Statistics and Computing,
DOI 10.1007/978-1-4419-5945-4_11,

11.2 Unconstrained Optimization

In this section we discuss optimization of a continuous function $f(x)$ defined on R^m. Because we can always substitute $-f(x)$ for $f(x)$, we consider, without loss of generality, minimization rather than maximization. In this broad setting, $f(x)$ may be unbounded. For instance, the function $f(x) = x$ is unbounded on the real line. However, if we require the domain of $f(x)$ to be a compact set, then $f(x)$ is bounded and attains its minimum. Recall that a set is compact if and only if it is closed and bounded. According to the Bolzano-Weierstrass theorem [16], every sequence x_n defined on a compact set C has a subsequence x_{n_k} converging to a point $x \in C$.

Proposition 11.2.1 (Weierstrass) *A continuous function $f(x)$ defined on a compact set is bounded below and attains its minimum.*

Proof: Let $b = \inf_{x\in C} f(x)$. If $f(x)$ is unbounded below, then $b = -\infty$. By definition of b, there exists a sequence $x_n \in C$ with $\lim_{n\to\infty} f(x_n) = b$. Extract from this sequence a convergent subsequence x_{n_k} with limit $y \in C$. In view of the continuity of $f(x)$, we have $\lim_{k\to\infty} f(x_{n_k}) = f(y)$. Hence, the value $b = f(y)$ is finite and attained by $f(x)$ at the point y. ■

Weierstrass's theorem does not guarantee that a continuous function $f(x)$ possesses a minimum on an open domain $U \subset \mathsf{R}^m$ such as R^m itself. To achieve this desirable result, we must impose on $f(x)$ a property called coerciveness. When $U = \mathsf{R}^m$, coerciveness simply means

$$\lim_{\|x\|_2\to\infty} f(x) \quad = \quad \infty,$$

where as usual $\|\cdot\|_2$ denotes the standard Euclidean norm. This is not an adequate definition for a more general open domain U. We also want $f(x)$ to approach ∞ as x approaches the boundary of U. Hence, $f(x)$ is said to be coercive if for every constant c the set $\{x \in U : f(x) \leq c\}$ is compact. When $f(x)$ is coercive in this sense, Weierstrass's theorem remains valid on U because it applies to the smaller domain $\{x \in U : f(x) \leq f(y)\}$ for any $y \in U$.

Example 11.2.1 *The Fundamental Theorem of Algebra*

Consider a polynomial $p(z) = c_n z^n + c_{n-1}z^{n-1} + \cdots + c_0$ in the complex variable z with $c_n \neq 0$. The fundamental theorem of algebra says that $p(z)$ has a root. d'Alembert suggested an interesting optimization proof of this fact [5]. We begin by observing that if we identify a complex number with an ordered pair of real numbers, then the domain of the real-valued function $|p(z)|$ is R^2. The identity

$$|p(z)| \quad = \quad |z|^n \left| c_n + \frac{c_{n-1}}{z} + \cdots + \frac{c_0}{z^n} \right|$$

shows that $|p(z)|$ is coercive. According to the amended version of Proposition 11.2.1, $|p(z)|$ attains its minimum at some point y. Expanding $p(z)$ around y gives a polynomial

$$q(z) \quad = \quad p(z+y) \quad = \quad b_n z^n + b_{n-1}z^{n-1} + \cdots + b_0$$

with the same degree as $p(z)$. Furthermore, the minimum of $|q(z)|$ occurs at $z = 0$. Suppose $b_1 = \cdots = b_{k-1} = 0$ and $b_k \neq 0$. For some angle $\theta \in [0, 2\pi)$, the scaled complex exponential

$$u \quad = \quad \left|\frac{b_0}{b_k}\right|^{1/k} e^{i\theta/k}$$

is a root of the equation $b_k u^k + b_0 = 0$. The function $f(t) = |q(tu)|$ clearly satisfies

$$\begin{aligned} f(t) &= |b_k t^k u^k + b_0| + o(t^k) \quad = \quad |b_0(1-t^k)| + o(t^k) \\ f(t) &\geq |b_0| \end{aligned}$$

for t small and positive. These two conditions are compatible only if $b_0 = 0$. Hence, the minimum of $|q(z)| = |p(z+y)|$ is 0. ■

We now turn to differentiable functions. Suppose $f(x)$ is differentiable on the open set U. Its differential $df(x)$ is the row vector of partial derivatives of $f(x)$. Its gradient $\nabla f(x)$ is the transpose of $df(x)$. For instance, the differential and gradient of the linear function $f(x) = b^t x$ are

$$df(x) \quad = \quad b^t, \quad \nabla f(x) \quad = \quad b.$$

For a quadratic function $f(x) = \frac{1}{2}x^t Ax$, the componentwise calculation

$$\frac{\partial}{\partial x_k} f(x) \quad = \quad \frac{1}{2}\sum_i x_i a_{ik} + \frac{1}{2}\sum_j a_{kj} x_j$$

shows that

$$df(x) \quad = \quad x^t A, \quad \nabla f(x) \quad = \quad Ax$$

for a symmetric matrix $A = (a_{ij})$. As a rule, revert to componentwise differentiation whenever you harbor any doubts about the correctness of a displayed differential.

Most readers will recall the following necessary condition for a minimum.

Proposition 11.2.2 (Fermat) *Suppose a differentiable function $f(x)$ has a local minimum at the point y of the open set U. Then $\nabla f(x)$ vanishes at y.*

Proof: For any vector v, the function $g(s) = f(y+sv)$ has a local minimum at $s = 0$. Hence, $g'(0)$ vanishes. According to the chain rule, the inner product $g'(0) = df(y)v$ also vanishes. This can only happen for all v if $\nabla f(y) = \mathbf{0}$. ■

A point y satisfying the first-order condition $\nabla f(y) = \mathbf{0}$ is called a stationary point. When $f(x)$ is twice continuously differentiable, there is a second-order necessary condition for a local minimum as well. A slight strengthening of the second-order condition gives a sufficient condition for a stationary point to be a local minimum. These results depend on the symmetric matrix $d^2f(x) = \nabla^2 f(x)$ of second partial derivatives of $f(x)$; this matrix is called the second differential or Hessian of $f(x)$.

Proposition 11.2.3 *Suppose a twice continuously differentiable function $f(x)$ has a local minimum at the point y of the open set U. Then $d^2f(x)$ is positive semidefinite at y. Conversely, if y is a stationary point and $d^2f(y)$ is positive definite, then y is a local minimum.*

Proof: Consider again the function $g(s) = f(y + sv)$ of the scalar s. Expanding $g(s)$ in a second-order Taylor series around 0 yields

$$\begin{aligned} g(s) &= g(0) + g'(0)s + \frac{1}{2}g''(r)s^2 \\ &= f(y) + df(y)vs + \frac{1}{2}v^t d^2f(z)vs^2 \end{aligned}$$

for some r between 0 and s and $z = y+rv$. According to Fermat's principle, the linear terms $g'(0)s$ and $df(y)vs$ vanish. By continuity it follows that

$$\begin{aligned} \frac{1}{2}v^t d^2f(y)v &= \frac{1}{2}g''(0) \\ &= \lim_{s\to 0} \frac{g(s) - g(0)}{s^2} \\ &\geq 0. \end{aligned}$$

Hence, the quadratic form $v^t d^2f(y)v$ is nonnegative, and the corresponding matrix $d^2f(y)$ is positive semidefinite.

Conversely, if y is a stationary point and $d^2f(y)$ is positive definite, then suppose x_n is a sequence of points that converges to y and satisfies $f(x_n) < f(y)$. Let $z_n = y + r_n(x_n - y)$ be the corresponding point of evaluation of $d^2f(x)$ in the second-order Taylor expansion of $f(x_n)$ around $f(y)$. By passing to a subsequence if necessary, we may assume that the normalized sequence $v_n = (y - x_n)/\|y - x_n\|_2$ converges to a unit vector v. Taking limits along the sequence x_n produces

$$\begin{aligned} 0 &\geq \lim_{n\to\infty} \frac{f(x_n) - f(y)}{\|y - x_n\|_2^2} \\ &= \lim_{n\to\infty} \frac{1}{2}v_n^t d^2f(z_n)v_n \end{aligned}$$

$$= \frac{1}{2}v^t d^2 f(y)v.$$

This inequality contradicts the positive definiteness of $d^2 f(y)$. Hence, there is no such sequence x_n, and y represents a local minimum of $f(x)$. ■

Example 11.2.2 *Minimum of a Positive Definite Quadratic Function*

The quadratic function $f(x) = \frac{1}{2}x^t Ax + b^t x + c$ has gradient

$$\nabla f(x) = Ax + b$$

for A symmetric. Assuming that A is also invertible, the sole stationary point of $f(x)$ is $-A^{-1}b$. When A is positive definite, $f(x)$ is coercive, and this point furnishes the minimum of $f(x)$. Coerciveness follows from an examination of the behavior of $x^t Ax$ over the sphere $\{x : \|x\|_2 = 1\}$. Because the sphere is compact and A is positive definite, the minimum d of $x^t Ax$ is positive. The Cauchy-Schwarz inequality therefore gives

$$f(x) \geq \frac{d}{2}\|x\|_2^2 - \|b\|_2\|x\|_2 + c.$$

This lower bound demonstrates that $\lim_{\|x\|_2 \to \infty} f(x) = \infty$. ■

Example 11.2.3 *Maximum Likelihood for the Multivariate Normal*

The sample mean and sample variance

$$\bar{y} = \frac{1}{k}\sum_{j=1}^{k} y_j$$

$$S = \frac{1}{k}\sum_{j=1}^{k}(y_j - \bar{y})(y_j - \bar{y})^t$$

are also the maximum likelihood estimates of the theoretical mean μ and theoretical variance Ω of a random sample $y_1, \ldots, y_k$ from a multivariate normal distribution. (For a review of the multivariate normal, see the Appendix.) To prove this fact, we first note that maximizing the loglikelihood function

$$-\frac{k}{2}\ln\det\Omega - \frac{1}{2}\sum_{j=1}^{k}(y_j - \mu)^t\Omega^{-1}(y_j - \mu)$$

$$= -\frac{k}{2}\ln\det\Omega - \frac{k}{2}\mu^t\Omega^{-1}\mu + \Big(\sum_{j=1}^{k} y_j\Big)^t\Omega^{-1}\mu - \frac{1}{2}\sum_{j=1}^{k} y_j^t\Omega^{-1}y_j$$

$$= -\frac{k}{2}\ln\det\Omega - \frac{1}{2}\operatorname{tr}\Big[\Omega^{-1}\sum_{j=1}^{k}(y_j - \mu)(y_j - \mu)^t\Big]$$

constitutes a special case of the previous example with $A = k\Omega^{-1}$ and $b = -\Omega^{-1}\sum_{j=1}^{k} y_j$. This leads to the same estimate $\hat{\mu} = \bar{y}$ regardless of the value of Ω.

To estimate Ω, we exploit the Cholesky decompositions $\Omega = LL^t$ and $S = MM^t$ under the assumption that both Ω and S are invertible. In view of the identities $\Omega^{-1} = (L^{-1})^t L^{-1}$ and $\det \Omega = (\det L)^2$, the loglikelihood becomes

$$\begin{aligned} & k \ln \det L^{-1} - \frac{k}{2} \operatorname{tr}\left[(L^{-1})^t L^{-1} M M^t\right] \\ = \; & k \ln \det (L^{-1} M) - \frac{k}{2} \operatorname{tr}\left[(L^{-1}M)(L^{-1}M)^t\right] - k \ln \det M \end{aligned}$$

using the cyclic permutation property of the matrix trace function. Because products and inverses of lower triangular matrices are lower triangular, the matrix $R = L^{-1}M$ ranges over the set of lower triangular matrices with positive diagonal entries as L ranges over the same set. This permits us to reparameterize and estimate $R = (r_{ij})$ instead of L. Up to an irrelevant additive constant, the loglikelihood reduces to

$$k \ln \det R - \frac{k}{2} \operatorname{tr}(RR^t) \;=\; k \sum_i \ln r_{ii} - \frac{k}{2} \sum_i \sum_{j=1}^{i} r_{ij}^2 .$$

Clearly, this function is maximized by taking $r_{ij} = 0$ for $j \neq i$. The term $g(r_{ii}) = -k \ln r_{ii} + \frac{k}{2} r_{ii}^2$ is convex and coercive on the interval $(0, \infty)$. Differentiation of $g(r_{ii})$ shows that it is minimized by taking $r_{ii} = 1$. In other words, the maximum likelihood estimator $\hat{R}$ is the identity matrix I. This implies that $\hat{L} = M$ and consequently that $\hat{\Omega} = S$. ■

11.3 Optimization with Equality Constraints

The subject of Lagrange multipliers has a strong geometric flavor. Our proof of the multiplier rule will rely on tangent vectors and directions of steepest ascent and descent. Suppose the objective function $f(x)$ to be minimized is continuously differentiable and defined on R^n. The gradient direction $\nabla f(x) = df(x)^t$ is the direction of steepest ascent of $f(x)$ near the point x. We can motivate this fact by considering the linear approximation

$$f(x + su) \;=\; f(x) + s\,df(x)u + o(s)$$

for a unit vector u and a scalar s. The error term $o(s)$ becomes negligible compared to s as s approaches 0. The inner product $df(x)u$ in this approximation is greatest for the unit vector

$$u \;=\; \frac{1}{\|\nabla f(x)\|_2} \nabla f(x).$$

Thus, $\nabla f(x)$ points locally in the direction of steepest ascent of $f(x)$. Similarly, $-\nabla f(x)$ points locally in the direction of steepest descent.

The classical theory of Lagrange multipliers, which is all we consider in this section, is limited to equality constraints. These are defined by the conditions $g_i(x) = 0$ for continuously differentiable functions $g_1(x), \ldots, g_m(x)$. A point obeying the constraints is said to be feasible. A tangent vector (or direction) w at the feasible point x satisfies $dg_i(x)w = 0$ for all i. Of course, if the constraint surface is curved, we must interpret the tangent vectors as specifying directions of infinitesimal movement. From the perpendicularity relation $dg_i(x)w = 0$, it follows that the set of tangent vectors is the orthogonal complement $S^\perp(x)$ of the vector subspace $S(x)$ spanned by the $\nabla g_i(x)$. To avoid degeneracies, the vectors $\nabla g_i(x)$ must be linearly independent. Figure 11.1 depicts level curves $g(x) = c$ and gradients $\nabla g(x)$ for the function $\sin(x)\cos(y)$ over the square $[0, \pi] \times [-\frac{\pi}{2}, \frac{\pi}{2}]$. Tangent vectors are parallel to the level curves (contours) and perpendicular to the gradients (arrows).

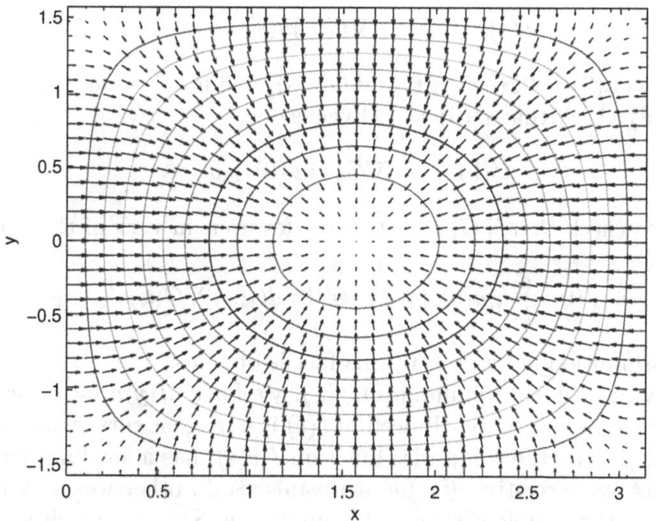

FIGURE 11.1. Level Curves and Steepest Ascent Directions for $\sin(x)\cos(y)$

These geometric insights can be made rigorous by introducing the notion of a tangent curve at the point x. This is simply a differentiable curve $v(s)$ having a neighborhood of the scalar 0 as its domain and satisfying $v(0) = x$ and $g_i[v(s)] = 0$ for all i and all s sufficiently close to 0. If we apply the chain rule to the composite function $g_i[v(s)] = 0$, then the identity $dg_i(x)v'(0) = 0$ emerges. The vector $w = v'(0)$ is said to be a tangent vector at x. Conversely, if w satisfies $dg_i(x)w = 0$ for all i, then we can construct a tangent curve at x with tangent vector w. This application of the implicit function theorem [16] requires some notation. Let $G(x)$ be the vector-valued function with ith component $g_i(x)$. The differential $dG(x)$ is the Jacobi matrix whose ith row is the differential $dg_i(x)$. In agreement with our earlier notation, $\nabla G(x)$ is the transpose of $dG(x)$.

Now consider the relationship

$$h(u, s) \quad = \quad G[x + \nabla G(x)u + sw] \quad = \quad \mathbf{0}.$$

Applying the chain rule to the function $h(u, s)$ gives

$$\begin{aligned} d_u h(\mathbf{0}, 0) &= dG[x + \nabla G(x)u + sw]\nabla G(x)\Big|_{(u,s)=(\mathbf{0},0)} \\ &= dG(x)\nabla G(x). \end{aligned}$$

Since $G(x) = \mathbf{0}$ and $dG(x)\nabla G(x)$ is invertible when $dG(x)$ has full row rank, the implicit function theorem implies that we can solve for u as a function of s in a neighborhood of 0. If we denote the resulting continuously differentiable function by $u(s)$, then our tangent curve is

$$v(s) \quad = \quad x + \nabla G(x)u(s) + sw.$$

By definition $u(0) = \mathbf{0}$, $v(0) = x$, and $G[v(s)] = \mathbf{0}$ for all s close to 0. Thus, we need only check that $v'(0) = w$. Because

$$v'(0) \quad = \quad \nabla G(x)u'(0) + w,$$

it suffices to check that $\nabla G(x)u'(0) = \mathbf{0}$. However, in view of the equality

$$0 \quad = \quad u'(0)^t \frac{d}{ds} h[u(0), 0] \quad = \quad u'(0)^t dG(x)[\nabla G(x)u'(0) + w]$$

and the assumption $dG(x)w = \mathbf{0}$, this fact is obvious.

If y provides a local minimum of $f(x)$, we have $df(y)w = 0$ for every tangent direction $w \in S^{\perp}(y)$. Indeed, if $v(s)$ is a tangent curve with tangent vector w at y, then the composite function $f[v(s)]$ has a local minimum at $s = 0$. Thus, its derivative $df(y)w$ at 0 vanishes. In other words, $\nabla f(y)$ is a member of the double orthogonal complement $S^{\perp\perp}(y)$, which we know from linear algebra equals $S(y)$. This enables us to write

$$\nabla f(y) \quad = \quad -\sum_{i=1}^{m} \lambda_i \nabla g_i(y)$$

for properly chosen constants $\lambda_1, \ldots, \lambda_m$. Alternatively, the Lagrangian function

$$\mathcal{L}(x, \omega) \quad = \quad f(x) + \sum_{i=1}^{m} \omega_i g_i(x) \tag{11.1}$$

has a stationary point at (y, λ). In this regard, note that

$$\frac{\partial}{\partial \omega_i}\mathcal{L}(y, \lambda) = 0$$

is equivalent to the constraint condition $g_i(y) = 0$.

In summary, we have proved the first part of the next proposition. In the statement and proof of the proposition, first and second differentials are with respect to x, holding λ fixed. For this reason we will suppress the dependence of the Lagrangian function on the Lagrange multipliers.

Proposition 11.3.1 (Lagrange) *Suppose the continuously differentiable function $f(x)$ has a local minimum at the feasible point y and that the constraint functions $g_1(x), \ldots, g_m(x)$ are continuously differentiable with linearly independent gradient vectors $\nabla g_i(y)$ at y. Then there exists a multiplier vector λ such that (y, λ) is a stationary point of the Lagrangian (11.1). Furthermore, if $f(x)$ and all $g_i(x)$ are twice continuously differentiable, then $v^t d^2\mathcal{L}(y)v \geq 0$ for every tangent vector v at y. Conversely, if (y, λ) is a stationary point of the Lagrangian and $v^t d^2\mathcal{L}(y)v > 0$ for every nontrivial tangent vector v at y, then y represents a local minimum of $f(x)$ subject to the constraints.*

Proof: Let $u(s)$ be a tangent curve at y with tangent vector $u'(0) = v$. Because the function $h(s) = \mathcal{L}[u(s)]$ has a minimum at $s = 0$, its second derivative

$$\begin{aligned} h''(0) &= d\mathcal{L}[u(0)]u''(0) + u'(0)^t d^2\mathcal{L}[u(0)]u'(0) \\ &= v^t d^2\mathcal{L}(y)v \end{aligned}$$

must be nonnegative.

Conversely, suppose that (y, λ) is a stationary point and that the quadratic form $v^t d^2\mathcal{L}(y)v$ is positive for every nontrivial tangent vector v. If y is not a local minimum, then there exists a feasible sequence of points x_n converging to y with $f(x_n) < f(y)$ for every n. By passing to a subsequence if necessary, we can assume that the unit vectors

$$v_n = \frac{1}{\|x_n - y\|_2}(x_n - y)$$

converge to a unit vector v. In view of the limit

$$0 = \lim_{n\to\infty} \frac{g_i(x_n) - g_i(y)}{\|x_n - y\|_2} = dg_i(y)v,$$

v is a tangent vector. Because $\mathcal{L}(x) = f(x)$ for any feasible point x, and y is a stationary point of $\mathcal{L}(x)$, we have

$$0 \geq \lim_{n\to\infty} \frac{f(x_n) - f(y)}{\|x_n - y\|_2^2}$$

$$\begin{aligned} &= \lim_{n\to\infty} \frac{\mathcal{L}(x_n) - \mathcal{L}(y)}{\|x_n - y\|_2^2} \\ &= \lim_{n\to\infty} \frac{1}{2} v^t d^2\mathcal{L}(y)v. \end{aligned}$$

But this contradicts the assumed positivity of $v^t d^2\mathcal{L}(y)v$, and therefore y represents a local minimum. ■

Example 11.3.1 *Estimation of Multinomial Probabilities*

Consider a multinomial experiment with n trials and observed outcomes $n_1, \ldots, n_m$ over m categories. The maximum likelihood estimate of the probability p_i of category i is $\hat{p}_i = n_i/n$. To demonstrate this fact, let

$$L(p) = \binom{n}{n_1 \ldots n_m} \prod_{i=1}^m p_i^{n_i}$$

denote the likelihood. If $n_i = 0$ for some i, then we interpret $p_i^{n_i}$ as 1 even when $p_i = 0$. This convention makes it clear that we can increase $L(p)$ by replacing p_i by 0 and p_j by $p_j/(1-p_i)$ for $j \neq i$. Thus, for purposes of maximum likelihood estimation, we can assume that all $n_i > 0$. Given this assumption, $L(p)$ tends to 0 when any p_i tends to 0, and $-\ln L(p)$ is coercive. It follows that we can further restrict our attention to the interior region where all $p_i > 0$ and maximize the loglikelihood $\ln L(p)$ subject to the equality constraint $\sum_{i=1}^m p_i = 1$. To find the minimum of $-\ln L(p)$, we look for a stationary point of the Lagrangian

$$= -\ln \binom{n}{n_1 \ldots n_m} - \sum_{i=1}^m n_i \ln p_i + \lambda\Big(\sum_{i=1}^m p_i - 1\Big).$$

Setting the partial derivative of $\mathcal{L}(p)$ with respect to p_i equal to 0 gives the equation

$$\frac{n_i}{p_i} = \lambda.$$

These m equations are satisfied subject to the constraint by taking $\lambda = n$ and $p_i = n_i/n$. Not only is the Lagrange multiplier rule of Proposition 11.3.1 true at $\hat{p}$, but the sufficient condition for a minimum also holds. In fact, the entries

$$\frac{\partial^2}{\partial p_i \partial p_j}\mathcal{L}(p) = -\frac{\partial^2}{\partial p_i \partial p_j}\ln L(p) = \begin{cases} \frac{n_i}{p_i^2} & i = j \\ 0 & i \neq j \end{cases}$$

of the Hessian $d^2\mathcal{L}(p)$ force it to be positive definite. ■

Example 11.3.2 *A Counterexample to Sufficiency*

The Lagrange multiplier rule by itself is not sufficient for a point to furnish a minimum. For example, consider the function $f(x) = x_1^3 - x_2$ subject to the constraint $g(x) = -x_2 = 0$. The Lagrange multiplier rule

$$\nabla f(\mathbf{0}) = \begin{pmatrix} 0 \\ -1 \end{pmatrix} = \nabla g(\mathbf{0})$$

holds, but the origin $\mathbf{0}$ fails to minimize $f(x)$. Indeed, the one-dimensional slice $x_1 \mapsto f(x_1, 0)$ has a saddle point at $x_1 = 0$. ■

Example 11.3.3 *Quadratic Programming with Equality Constraints*

Minimizing a quadratic function

$$q(x) = \frac{1}{2}x^t Ax + b^t x + c$$

on R^n subject to the m linear equality constraints

$$v_i^t x = d_i$$

is one of the most important problems in nonlinear programming. Here the symmetric matrix A is assumed positive definite. The constraints can be re-expressed as $Vx = d$ by defining V to be the $m \times n$ matrix with ith row v_i^t and d to be the column vector with ith entry d_i.

To minimize $q(x)$ subject to the constraints, we introduce the Lagrangian

$$\begin{aligned} \mathcal{L}(x, \lambda) &= \frac{1}{2}x^t Ax + b^t x + c + \sum_{i=1}^m \lambda_i [v_i^t x - d_i] \\ &= \frac{1}{2}x^t Ax + b^t x + c + \lambda^t (Vx - d). \end{aligned}$$

A stationary point of $\mathcal{L}(x, \lambda)$ is determined by the equations

$$\begin{aligned} b + Ax + V^t \lambda &= \mathbf{0} \\ Vx &= d, \end{aligned}$$

whose formal solution amounts to

$$\begin{pmatrix} x \\ \lambda \end{pmatrix} = \begin{pmatrix} A & V^t \\ V & \mathbf{0} \end{pmatrix}^{-1} \begin{pmatrix} -b \\ d \end{pmatrix}.$$

The next proposition shows that the indicated matrix inverse exists. Because the second differential $d^2 q(x) = A$ is positive definite, the second-order sufficient condition of Proposition 11.3.1 ensures that the calculated point provides a local minimum. In light of our subsequent remarks about strict convexity, this local minimum is a global minimum. ■

Proposition 11.3.2 *Let A be an $n \times n$ positive definite matrix and V be an $m \times n$ matrix. Then the matrix*

$$M = \begin{pmatrix} A & V^t \\ V & \mathbf{0} \end{pmatrix}$$

has inverse

$$M^{-1} = \begin{pmatrix} A^{-1} - A^{-1}V^t(VA^{-1}V^t)^{-1}VA^{-1} & A^{-1}V^t(VA^{-1}V^t)^{-1} \\ (VA^{-1}V^t)^{-1}VA^{-1} & -(VA^{-1}V^t)^{-1} \end{pmatrix}$$

if and only if V has linearly independent rows $v_1^t, \ldots, v_m^t$.

Proof: We first show that M is invertible with the specified inverse if and only if $(VA^{-1}V^t)^{-1}$ exists. If

$$M^{-1} = \begin{pmatrix} B & C \\ D & E \end{pmatrix},$$

then the identity

$$\begin{pmatrix} A & V^t \\ V & \mathbf{0} \end{pmatrix} \begin{pmatrix} B & C \\ D & E \end{pmatrix} = \begin{pmatrix} I_n & \mathbf{0} \\ \mathbf{0} & I_m \end{pmatrix}$$

implies that $VC = I_m$ and $AC + V^tE = \mathbf{0}$. Multiplying the last equality by VA^{-1} gives $I_m = -VA^{-1}V^tE$. Thus, $(VA^{-1}V^t)^{-1}$ exists. Conversely, if $(VA^{-1}V^t)^{-1}$ exists, then one can check by direct multiplication that M has the claimed inverse.

If $(VA^{-1}V^t)^{-1}$ exists, then V must have full row rank m. Conversely, if V has full row rank m, take any nontrivial $u \in \mathsf{R}^m$. Then the fact

$$u^tV = u_1v_1^t + \cdots u_mv_m^t \neq \mathbf{0}^t$$

and the positive definiteness of A imply $u^tVA^{-1}V^tu > 0$. Thus, $VA^{-1}V^t$ is also positive definite and invertible. ■

It is noteworthy that the matrix M of Proposition 11.3.2 can be inverted by sweeping on its diagonal entries. Indeed, sweeping on the diagonal entries of A takes

$$\begin{pmatrix} A & V^t \\ V & \mathbf{0} \end{pmatrix} \longrightarrow \begin{pmatrix} -A^{-1} & A^{-1}V^t \\ VA^{-1} & -VA^{-1}V^t \end{pmatrix}.$$

Sweeping is now possible for the remaining diagonal entries of M since $VA^{-1}V^t$ is positive definite.

Example 11.3.4 *Smallest Matrix Subject to Secant Conditions*

In some situations covered by Example 11.3.3, the answer can be radically simplified. Consider the problem of minimizing the Frobenius norm of a matrix M subject to the linear constraints $Mu_i = v_i$ for $i = 1, \ldots, q$. It is helpful to rewrite the constraints in matrix form as $MU = V$ for $U = (u_1, \ldots, u_q)$ and $V = (v_1, \ldots, v_q)$. Provided U has full column rank q, the minimum of the squared norm $\|M\|_F^2$ subject to the constraints is attained by the choice $M = V(U^tU)^{-1}U^t$. We can prove this assertion by taking the partial derivative of the Lagrangian

$$\begin{aligned}\mathcal{L} &= \frac{1}{2}\|M\|_F^2 + \sum_i \sum_k \lambda_{ik}\Big(\sum_j m_{ij}u_{jk} - v_{ik}\Big) \\ &= \frac{1}{2}\sum_i \sum_j m_{ij}^2 + \sum_i \sum_k \lambda_{ik}\Big(\sum_j m_{ij}u_{jk} - v_{ik}\Big)\end{aligned}$$

with respect to m_{ij} and equating it to 0. This gives the Lagrange multiplier equation

$$0 = m_{ij} + \sum_k \lambda_{ik}u_{jk},$$

which we collectively express in matrix notation as $\mathbf{0} = M + \Lambda U^t$. This equation and the constraint equation $MU = V$ uniquely determine the minimum of the objective function. Indeed, straightforward substitution shows that $M = V(U^tU)^{-1}U^t$ and $\Lambda = -V(U^tU)^{-1}$ constitute the solution. This result will come in handy later when we discuss accelerating the MM algorithm. ■

11.4 Optimization with Inequality Constraints

In the current section, we study the problem of minimizing an objective function $f(x)$ subject to the mixed constraints

$$\begin{aligned}g_i(x) &= 0, \qquad 1 \le i \le p \\ h_j(x) &\le 0, \qquad 1 \le j \le q.\end{aligned}$$

All of these functions share some open set $U \subset \mathsf{R}^n$ as their domain. In addition to the equality constraints defined by the $g_i(x)$, we now have inequality constraints defined by the $h_j(x)$. A constraint $h_j(x)$ is active at the feasible point x provided $h_j(x) = 0$; it is inactive if $h_j(x) < 0$. In general, we will assume that the feasible region is nonempty. The case $p = 0$ of no equality constraints and the case $q = 0$ of no inequality constraints are both allowed.

In exploring solutions to the above constrained minimization problem, we will meet a generalization of the Lagrange multiplier rule fashioned independently by Karush and later by Kuhn and Tucker. Under fairly weak regularity conditions, the rule holds at all extrema. In contrast to this necessary condition, the sufficient condition for an extremum involves second derivatives.

In our derivation of the Lagrange multiplier rule for equality constraints, we required that the gradients $\nabla g_i(y)$ be linearly independent at a local minimum y of $f(x)$. Without this condition, we could not generate a full set of tangent curves at y. In the presence of inequality constraints, the situation is more complicated. To avoid redundant constraints, not only do we need linear independence of the gradients of the equality constraints but also a restriction on the active inequality constraints. The simplest one to check is the Kuhn-Tucker [15] condition requiring linear independence of the set of vectors consisting of the gradients $\nabla g_i(y)$ plus the gradients $\nabla h_j(y)$ of the active inequality constraints. The Kuhn-Tucker condition implies the weaker Mangasarian-Fromovitz constraint qualification. This condition [19] requires that the gradients $\nabla g_i(y)$ be linearly independent and that there exists a vector v with $dg_i(y)v = 0$ for all i and $dh_j(y)v < 0$ for all inequality constraints active at y. The vector v is a tangent vector in the sense that infinitesimal motion from y along v stays within the feasible region.

Proposition 11.4.1 *Suppose that the objective function $f(x)$ of the constrained optimization problem just described has a local minimum at the feasible point y. If $f(x)$ and the constraint functions are continuously differentiable near y, and the Mangasarian-Fromovitz constraint qualification holds at y, then there exist Lagrange multipliers $\lambda_1, \ldots, \lambda_p$ and $\mu_1, \ldots, \mu_q$ such that*

$$\nabla f(y) + \sum_{i=1}^{p} \lambda_i \nabla g_i(y) + \sum_{j=1}^{q} \mu_j \nabla h_j(y) = \mathbf{0}. \tag{11.2}$$

Moreover, each of the multipliers μ_j is nonnegative, and $\mu_j = 0$ whenever $h_j(y) < 0$. The restriction $\mu_j h_j(y) = 0$ is called complementary slackness.

Proof: See the article by McShane [21] or Proposition 4.2.1 of [16]. ■

Example 11.4.1 *The Cauchy-Schwarz Inequality*

For a given vector $v \neq \mathbf{0}$, consider minimizing the linear function $f(x) = v^t x$ subject to the inequality constraint $h_1(x) = \|x\|_2^2 - 1 \leq 0$. The minimum exists since $f(x)$ is continuous and the feasible region is compact. Furthermore, the Mangasarian-Fromovitz constraint qualification is true at any point $x \neq \mathbf{0}$. According to the Lagrange multiplier rule of Proposition 11.4.1, there exist a nonnegative scalar μ and a feasible vector y such that

$$v + 2\mu y = \mathbf{0}.$$

This equation forces μ to be positive. The complementary slackness condition $\mu(\|y\|_2^2 - 1) = 0$ holds only if $\|y\|_2 = 1$. It follows that

$$\mu = \frac{1}{2}\|v\|_2, \qquad y = -\frac{1}{\|v\|_2}v,$$

and $f(x)$ has minimum $-\|v\|_2$. If we substitute $-x$ for x in the inequality $v^t x \geq -\|v\|_2$, then the standard Cauchy-Schwarz inequality $v^t x \leq \|v\|_2$ for $\|x\|_2 \leq 1$ appears. Equality is attained when x is a unit vector in the direction v. ■

Example 11.4.2 *Application to Another Inequality*

Let us demonstrate the inequality

$$\frac{x_1^2 + x_2^2}{4} \leq e^{x_1 + x_2 - 2}$$

subject to the constraints $x_1 \geq 0$ and $x_2 \geq 0$ [8]. It suffices to show that the minimum of

$$f(x_1, x_2) = -(x_1^2 + x_2^2)e^{-x_1 - x_2}$$

is $-4e^{-2}$. According to Proposition 11.4.1 with $h_1(x_1, x_2) = -x_1$ and $h_2(x_1, x_2) = -x_2$, a minimum point necessarily satisfies

$$\begin{aligned} -\frac{\partial}{\partial y_1} f(y_1, y_2) &= (2y_1 - y_1^2 - y_2^2)e^{-y_1 - y_2} \\ &= -\mu_1 \\ -\frac{\partial}{\partial y_2} f(y_1, y_2) &= (2y_2 - y_1^2 - y_2^2)e^{-y_1 - y_2} \\ &= -\mu_2, \end{aligned}$$

where the multipliers μ_1 and μ_2 are nonnegative and satisfy $\mu_1 y_1 = 0$ and $\mu_2 y_2 = 0$. In this problem, the Mangasarian-Fromovitz constraint qualification is trivial to check using the vector $v = \mathbf{1}$. If neither y_1 nor y_2 vanishes, then μ_1 and μ_2 both vanish and

$$2y_1 - y_1^2 - y_2^2 = 2y_2 - y_1^2 - y_2^2 = 0.$$

This forces $y_1 = y_2$ and $2y_1 - 2y_1^2 = 0$. It follows that $y_1 = y_2 = 1$, where $f(1, 1) = -2e^{-2}$. We can immediately eliminate the origin $\mathbf{0}$ from contention because $f(0, 0) = 0$. If $y_1 = 0$ and $y_2 > 0$, then $\mu_2 = 0$ and $2y_2 - y_2^2 = 0$. This implies that $y_2 = 2$ and $(0, 2)$ is a candidate minimum point. By symmetry, $(2, 0)$ is also a candidate minimum point. At these two boundary points, $f(2, 0) = f(0, 2) = -4e^{-2}$, and this verifies the claimed minimum value. ■

As with equality-constrained problems, we also have a sufficient condition for a local minimum.

Proposition 11.4.2 *Suppose that the objective function $f(x)$ of the constrained optimization problem satisfies the multiplier rule (11.2) at the point y. Let $f(x)$ and the constraint functions be twice continuously differentiable, and let $\mathcal{L}(x)$ be the Lagrangian*

$$\mathcal{L}(x) = f(x) + \sum_{i=1}^{p} \lambda_i g_i(x) + \sum_{j=1}^{q} \mu_j h_j(x).$$

If $v^t d^2\mathcal{L}(y)v > 0$ for every vector $v \neq \mathbf{0}$ satisfying $dg_i(y)v = 0$ and $dh_j(y)v \leq 0$ for all active constraints, then y provides a local minimum of $f(x)$.

Proof: See Proposition 4.5.1 of [16]. ■

Example 11.4.3 *Minimum of a Linear Reciprocal Function*

Consider minimizing the function $f(x) = \sum_{i=1}^{n} c_i x_i^{-1}$ subject to the linear inequality constraint $\sum_{i=1}^{n} a_i x_i \leq b$. Here all indicated variables and parameters are positive. It is obvious that the inequality constraints $x_i \geq 0$ are inactive at the minimum. Differentiating the Lagrangian

$$\mathcal{L}(x) = \sum_{i=1}^{n} c_i x_i^{-1} + \mu\left(\sum_{i=1}^{n} a_i x_i - b\right)$$

gives the multiplier equations

$$-\frac{c_i}{y_i^2} + \mu a_i = 0.$$

It follows that $\mu > 0$, that the constraint is active, and that

$$\begin{aligned} y_i &= \sqrt{\frac{c_i}{\mu a_i}}, \qquad 1 \leq i \leq n \\ \mu &= \left(\frac{1}{b}\sum_{i=1}^{n}\sqrt{a_i c_i}\right)^2. \end{aligned} \tag{11.3}$$

The second differential $d^2\mathcal{L}(y)$ is diagonal with ith diagonal entry $2c_i/y_i^3$. This matrix is certainly positive definite, and Proposition 11.4.2 confirms that the stationary point (11.3) provides the minimum of $f(x)$ subject to the constraint. ■

11.5 Convexity

Optimization theory is much simpler for convex functions than for ordinary functions. Here we briefly summarize some of the relevant facts without proof. A fuller treatment can be found in the references [4, 16, 23].

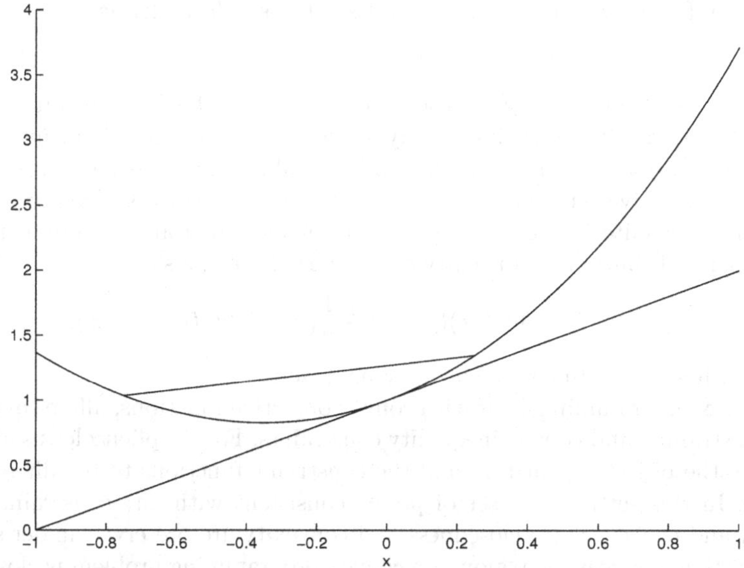

FIGURE 11.2. Plot of the Convex Function $e^x + x^2$

A set $S \subset \mathsf{R}^m$ is said to be convex if the line segment between any two points x and y of S lies entirely within S. Formally, this means that whenever $x, y \in S$ and $\alpha \in [0,1]$, the point $z = \alpha x + (1-\alpha)y \in S$. In general, any convex combination $\sum_{i=1}^n \alpha_i x_i$ of points $x_1, \ldots, x_n$ in S must also reside in S. Here, the coefficients α_i must be nonnegative and sum to 1. It is easy to concoct examples of convex sets. For example, every interval on the real line is convex; every ball in R^n, either open or closed, is convex; and every multidimensional rectangle, either open or closed, is convex.

Convex functions are defined on convex sets. A real-valued function $f(x)$ defined on a convex set S is convex provided

$$f[\alpha x + (1-\alpha)y] \;\; \leq \;\; \alpha f(x) + (1-\alpha)f(y) \tag{11.4}$$

for all $x, y \in S$ and $\alpha \in [0,1]$. Figure 11.2 depicts how in one dimension definition (11.4) requires the chord connecting two points on the curve $f(x)$ to lie above the curve. If strict inequality holds in (11.4) for every $x \neq y$ and $\alpha \in (0,1)$, then $f(x)$ is said to be strictly convex. One can prove by induction that inequality (11.4) extends to Jensen's inequality

$$f\Big(\sum_{i=1}^n \alpha_i x_i\Big) \;\; \leq \;\; \sum_{i=1}^n \alpha_i f(x_i)$$

for any convex combination of points from S. A concave function satisfies the reverse of inequality (11.4).

Figure 11.2 also illustrates how a tangent line to a convex curve lies below the curve. This supporting hyperplane inequality property characterizes convex differentiable functions. It can be expressed formally as

$$f(y) \geq f(x) + df(x)(y - x) \tag{11.5}$$

for all pairs of points (x, y) in the domain of $f(x)$. One can demonstrate that $f(x)$ is strictly convex if and only if strict inequality holds in inequality (11.5) when $y \neq x$. If $f(x)$ is twice differentiable, then it is convex when $d^2 f(x)$ is positive semidefinite for all x; it is strictly convex when $d^2 f(x)$ is positive definite for all x. These conclusions follow from the supporting hyperplane definition of convexity and the Taylor expansion

$$f(y) = f(x) + df(x)(y - x) + \frac{1}{2}(y - x)^t d^2 f(z)(y - x),$$

where z lies on the line segment between x and y.

Convex programming deals with convex objective functions, affine equality constraints, and convex inequality constraints. For simplicity let us also require the objective function and the constraint functions to be differentiable. In this setting, the set of points consistent with any constraint is closed and convex. Since closedness and convexity are preserved under set intersection, the feasible region of a convex programming problem is closed and convex. Convexity is important because a local minimum of a convex function on a convex domain is also a global minimum. If the objective function is strictly convex, then a local minimum is the unique global minimum. Finally, in convex programming, satisfaction of the multiplier rule (11.2) is sufficient for a minimum. As we have already pointed out, the multiplier rule is necessary under an appropriate constraint qualification.

11.6 Block Relaxation

In some problems it pays to update only a subset of the parameters at a time. Block relaxation divides the parameters into disjoint blocks and cycles through the blocks, updating only those parameters within a single block at each stage of a cycle [7]. Block relaxation is most successful when these updates are exact. When each block consists of a single parameter, block relaxation is called cyclic coordinate descent or cyclic coordinate ascent.

Example 11.6.1 *Sinkhorn's Algorithm*

Let $M = (m_{ij})$ be a rectangular matrix with positive entries. Sinkhorn's theorem [24] says that there exist two diagonal matrices A and B with positive diagonal entries a_i and b_j such that the matrix AMB has prescribed row and column sums. Let $r_i > 0$ be the ith row sum and $c_j > 0$ the jth column sum. Because AMB has entry $a_i m_{ij} b_j$ at the intersection of row i and column j, the constraints are

$$\sum_i a_i m_{ij} b_j = c_j, \quad \sum_j a_i m_{ij} b_j = r_i.$$

For these constraints to be consistent, we must have

$$\sum_i r_i = \sum_i \sum_j a_i m_{ij} b_j = \sum_j c_j.$$

Given this assumption, we now sketch a method for finding A and B. Consider minimizing the smooth function [13]

$$f(A,B) = -\sum_i r_i \ln a_i - \sum_j c_j \ln b_j + \sum_i \sum_j a_i m_{ij} b_j.$$

If any a_i or b_j approaches 0, then $f(a,b)$ tends to ∞. In view of this fact, the minimum occurs in a region where the parameters a_i and b_j are uniformly bounded below by a positive constant. Within this region, it follows that $a_i m_{ij} b_j$ tends to ∞ if either a_i or b_j tends to ∞. Hence, the minimum of $f(A,B)$ exists. At the minimum, Fermat's principle requires

$$\begin{aligned} \frac{\partial}{\partial a_i} f(A,B) &= -\frac{r_i}{a_i} + \sum_j m_{ij} b_j = 0 \\ \frac{\partial}{\partial b_j} f(A,B) &= -\frac{c_j}{b_j} + \sum_i a_i m_{ij} = 0. \end{aligned}$$

These equations are just a disguised form of Sinkhorn's constraints.

The direct attempt to solve the stationarity equations is almost immediately thwarted. It is much easier to minimize $f(A,B)$ with respect to A for B fixed or vice versa. If we fix B, then rearranging the first stationarity equation gives

$$a_i = \frac{r_i}{\sum_j m_{ij} b_j}.$$

Similarly, if we fix A, then rearranging the second stationarity equation yields

$$b_j = \frac{c_j}{\sum_i a_i m_{ij}}.$$

Sinkhorn's block relaxation algorithm [24] alternates the updates of A and B. ■

Example 11.6.2 *Poisson Sports Model*

Consider a simplified version of a model proposed by Maher [18] for a sports contest between two teams in which the number of points scored by team i against team j follows a Poisson process with intensity $e^{o_i - d_j}$, where o_i is an "offensive strength" parameter for team i and d_j is a "defensive strength" parameter for team j. (See Section 12.10 for a brief description of Poisson processes.) If t_{ij} is the length of time that i plays j and p_{ij} is the number of points that i scores against j, then the corresponding Poisson loglikelihood function is

$$\ell_{ij}(\theta) = p_{ij}(o_i - d_j) + p_{ij} \ln t_{ij} - t_{ij} e^{o_i - d_j} - \ln p_{ij}!, \qquad (11.6)$$

where $\theta = (o, d)$ is the parameter vector. Note that the parameters should satisfy a linear constraint such as $d_1 = 0$ in order for the model to be identifiable; otherwise, it is clearly possible to add the same constant to each o_i and d_j without altering the likelihood. We make two simplifying assumptions. First, the outcomes of the different games are independent. Second, each team's point total within a single game is independent of its opponent's point total. The second assumption is more suspect than the first since it implies that a team's offensive and defensive performances are somehow unrelated to one another; nonetheless, the model gives an interesting first approximation to reality. Under these assumptions, the full data loglikelihood is obtained by summing $\ell_{ij}(\theta)$ over all pairs (i, j). Setting the partial derivatives of the loglikelihood equal to zero leads to the equations

$$e^{-d_j} = \frac{\sum_i p_{ij}}{\sum_i t_{ij} e^{o_i}} \quad \text{and} \quad e^{o_i} = \frac{\sum_j p_{ij}}{\sum_j t_{ij} e^{-d_j}}$$

satisfied by the maximum likelihood estimate $(\hat{o}, \hat{d})$.

These equations do not admit a closed-form solution, so we turn to block relaxation [7]. If we fix the o_i, then we can solve for the d_j and vice versa in the form

$$d_j = -\ln\left(\frac{\sum_i p_{ij}}{\sum_i t_{ij} e^{o_i}}\right) \quad \text{and} \quad o_i = \ln\left(\frac{\sum_j p_{ij}}{\sum_j t_{ij} e^{-d_j}}\right).$$

Block relaxation consists in alternating the updates of the defensive and offensive parameters with the proviso that d_1 is fixed at 0.

Table 11.1 summarizes our application of the Poisson sports model to the results of the 2002–2003 regular season of the National Basketball Association. In these data, t_{ij} is measured in minutes. A regular game lasts 48 minutes, and each overtime period, if necessary, adds five minutes. Thus, team i is expected to score $48e^{\hat{o}_i - \hat{d}_j}$ points against team j when the two teams meet and do not tie. Team i is ranked higher than team j if $\hat{o}_i - \hat{d}_j > \hat{o}_j - \hat{d}_i$, which is equivalent to the condition $\hat{o}_i + \hat{d}_i > \hat{o}_j + \hat{d}_j$.

It is worth emphasizing some of the virtues of the model. First, the ranking of the 29 NBA teams on the basis of the estimated sums $\hat{o}_i + \hat{d}_i$ for the 2002-2003 regular season is not perfectly consistent with their cumulative wins; strength of schedule and margins of victory are reflected in the model. Second, the model gives the point-spread function for a particular game as the difference of two independent Poisson random variables.

Team	$\hat{o}_i + \hat{d}_i$	Wins	Team	$\hat{o}_i + \hat{d}_i$	Wins
Cleveland	-0.0994	17	Phoenix	0.0166	44
Denver	-0.0845	17	New Orleans	0.0169	47
Toronto	-0.0647	24	Philadelphia	0.0187	48
Miami	-0.0581	25	Houston	0.0205	43
Chicago	-0.0544	30	Minnesota	0.0259	51
Atlanta	-0.0402	35	LA Lakers	0.0277	50
LA Clippers	-0.0355	27	Indiana	0.0296	48
Memphis	-0.0255	28	Utah	0.0299	47
New York	-0.0164	37	Portland	0.0320	50
Washington	-0.0153	37	Detroit	0.0336	50
Boston	-0.0077	44	New Jersey	0.0481	49
Golden State	-0.0051	38	San Antonio	0.0611	60
Orlando	-0.0039	42	Sacramento	0.0686	59
Milwaukee	-0.0027	42	Dallas	0.0804	60
Seattle	0.0039	40			

TABLE 11.1. Ranking of all 29 NBA teams on the basis of the 2002-2003 regular season according to their estimated offensive strength plus defensive strength. Each team played 82 games.

Third, one can easily amend the model to rank individual players rather than teams by assigning to each player an offensive and defensive intensity parameter. If each game is divided into time segments punctuated by substitutions, then the block relaxation algorithm can be adapted to estimate the assigned player intensities. This might provide a rational basis for salary negotiations that takes into account subtle differences between players not reflected in traditional sports statistics. ■

Example 11.6.3 *K-Means Clustering*

In k-means clustering we must divide n points $x_1, \ldots, x_n$ in R^m into k clusters. Each cluster C_j is characterized by a cluster center μ_j. The best clustering of the points minimizes the criterion

$$f(\mu, C) = \sum_{j=1}^{k} \sum_{x_i \in C_j} \|x_i - \mu_j\|_2^2,$$

where μ is the matrix whose columns are the μ_j and C is the collection of clusters. Because this mixed continuous-discrete optimization problem has no obvious analytic solution, block relaxation is attractive. If we hold the clusters fixed, then it is clear from Example 11.2.3 that we should set

$$\mu_j = \frac{1}{|C_j|} \sum_{x_i \in C_j} x_i.$$

Similarly, it is clear that if we hold the cluster centers fixed, then we should assign point x_i to the cluster C_j minimizing $\|x_i - \mu_j\|_2$. Block relaxation, known as Lloyd's algorithm in this context, alternates cluster center redefinition and cluster membership reassignment. It is simple and effective. The initial cluster centers can be chosen randomly from the n data points. The evidence suggests that this should be done in a biased manner that spreads the centers out [1]. Changing the objective function to

$$g(\mu, C) = \sum_{j=1}^{k} \sum_{x_i \in C_j} \|x_i - \mu_j\|_1$$

makes it more resistant to outliers. The recentering step is now solved by replacing means by medians in each coordinate. This takes a little more computation but is usually worth the effort. ■

Example 11.6.4 *Canonical Correlations*

Consider a random vector Z partitioned into a subvector X of predictors and a subvector Y of responses. The most elementary form of canonical correlation analysis seeks two linear combinations a^tX and b^tY that are maximally correlated [20]. If we partition the variance matrix of Z into blocks

$$\mathrm{Var}(Z) = \begin{pmatrix} \Sigma_{11} & \Sigma_{12} \\ \Sigma_{21} & \Sigma_{22} \end{pmatrix}$$

consistent with X and Y, then the two linear combinations maximize the covariance $a^t\Sigma_{12}b$ subject to the variance constraints

$$a^t\Sigma_{11}a = b^t\Sigma_{22}b = 1.$$

This constrained maximization problem is an ideal candidate for block relaxation. Problem 28 sketches another method for finding a and b that exploits the singular value decomposition.

For fixed b we can easily find the best a. Introduce the Lagrangian

$$\mathcal{L}(a) = a^t\Sigma_{12}b - \frac{\lambda}{2}\Big(a^t\Sigma_{11}a - 1\Big),$$

and equate its gradient

$$\nabla\mathcal{L}(a) = \Sigma_{12}b - \lambda\Sigma_{11}a$$

to $\mathbf{0}$. This gives the minimum point

$$a = \frac{1}{\lambda}\Sigma_{11}^{-1}\Sigma_{12}b,$$

assuming the submatrix Σ_{11} is positive definite. Inserting this value into the constraint $a^t\Sigma_{11}a = 1$ allows us to solve for the Lagrange multiplier λ and hence pin down a as

$$a = \frac{1}{\sqrt{b^t\Sigma_{21}\Sigma_{11}^{-1}\Sigma_{12}b}}\Sigma_{11}^{-1}\Sigma_{12}b.$$

Because the second differential $d^2\mathcal{L} = -\lambda\Sigma_{11}$ is negative definite, a represents the maximum. Likewise, fixing a and optimizing over b gives the update

$$b = \frac{1}{\sqrt{a^t\Sigma_{12}\Sigma_{22}^{-1}\Sigma_{21}a}}\Sigma_{22}^{-1}\Sigma_{21}a.$$

TABLE 11.2. Iterates in Canonical Correlation Estimation

n	$\mathbf{a_{n1}}$	$\mathbf{a_{n2}}$	$\mathbf{b_{n1}}$	$\mathbf{b_{n2}}$
0	1.000000	1.000000	1.000000	1.000000
1	0.553047	0.520658	0.504588	0.538164
2	0.552159	0.521554	0.504509	0.538242
3	0.552155	0.521558	0.504509	0.538242
4	0.552155	0.521558	0.504509	0.538242

As a toy example consider the correlation matrix

$$\text{Var}(Z) = \begin{pmatrix} 1 & 0.7346 & 0.7108 & 0.7040 \\ 0.7346 & 1 & 0.6932 & 0.7086 \\ 0.7108 & 0.6932 & 1 & 0.8392 \\ 0.7040 & 0.7086 & 0.8392 & 1 \end{pmatrix}$$

with unit variances on its diagonal. Table 11.2 shows the first few iterates of block relaxation starting from $a = b = \mathbf{1}$. Convergence is exceptionally quick. ■

Example 11.6.5 *Iterative Proportional Fitting*

Our last example of block relaxation is taken from the contingency table literature [3, 9]. Consider a three-way contingency table with two-way interactions. If the three factors are indexed by i, j, and k and have r, s, and t levels, respectively, then a loglinear model for the observed data y_{ijk} is defined by an exponentially parameterized mean

$$\mu_{ijk} = e^{\lambda+\lambda_i^1+\lambda_j^2+\lambda_k^3+\lambda_{ij}^{12}+\lambda_{ik}^{13}+\lambda_{jk}^{23}}$$

for each cell ijk. To ensure that all parameters are identifiable, we make the usual assumption that a parameter set summed over one of its indices yields 0. For instance, $\lambda^1_{\cdot} = \sum_i \lambda^1_i = 0$ and $\lambda^{12}_{i\cdot} = \sum_j \lambda^{12}_{ij} = 0$. The overall effect λ is permitted to be nonzero.

If we postulate independent Poisson distributions for the random variables Y_{ijk} underlying the observed values y_{ijk}, then the loglikelihood is

$$L = \sum_i \sum_j \sum_k (y_{ijk} \ln \mu_{ijk} - \mu_{ijk}). \tag{11.7}$$

Maximizing L with respect to λ can be accomplished by setting

$$\begin{aligned} \frac{\partial}{\partial \lambda} L &= \sum_i \sum_j \sum_k (y_{ijk} - \mu_{ijk}) \\ &= 0. \end{aligned}$$

This tells us that whatever the other parameters are, λ should be adjusted so that $\mu_{\cdots} = y_{\cdots} = m$ is the total sample size. (Here again the dot convention signifies summation over the lost index.) In other words, if $\mu_{ijk} = e^{\lambda}\omega_{ijk}$, then λ is chosen so that $e^{\lambda} = m/\omega_{\cdots}$. With this proviso, the loglikelihood becomes

$$\begin{aligned} L &= \sum_i \sum_j \sum_k y_{ijk} \ln \frac{m\omega_{ijk}}{\omega_{\cdots}} - m \\ &= \sum_i \sum_j \sum_k y_{ijk} \ln \frac{\omega_{ijk}}{\omega_{\cdots}} + m \ln m - m, \end{aligned}$$

which is up to an irrelevant constant just the loglikelihood of a multinomial distribution with probability $\omega_{ijk}/\omega_{\cdots}$ attached to cell ijk. Thus, for purposes of maximum likelihood estimation, we might as well stick with the Poisson sampling model.

Unfortunately, no closed-form solution to the Poisson likelihood equations exists satisfying the complicated linear constraints. The resolution of this dilemma lies in refusing to update all of the parameters simultaneously. Suppose that we consider only the parameters λ, λ^1_i, λ^2_j, and λ^{12}_{ij} pertinent to the first two factors. If in equation (11.7) we let

$$\begin{aligned} \mu_{ij} &= e^{\lambda + \lambda^1_i + \lambda^2_j + \lambda^{12}_{ij}} \\ \alpha_{ijk} &= e^{\lambda^3_k + \lambda^{13}_{ik} + \lambda^{23}_{jk}}, \end{aligned}$$

then setting

$$\begin{aligned} \frac{\partial}{\partial \lambda^{12}_{ij}} L &= \sum_k (y_{ijk} - \mu_{ijk}) \\ &= y_{ij\cdot} - \mu_{ij\cdot} \\ &= y_{ij\cdot} - \mu_{ij}\alpha_{ij\cdot} \\ &= 0 \end{aligned}$$

leads to $\mu_{ij} = y_{ij.}/\alpha_{ij.}$. The constraint $\sum_k (y_{ijk} - \mu_{ijk}) = 0$ implies that the other partial derivatives

$$\begin{aligned}\frac{\partial}{\partial \lambda} L &= y_{...} - \mu_{...} \\ \frac{\partial}{\partial \lambda_i^1} L &= y_{i..} - \mu_{i..} \\ \frac{\partial}{\partial \lambda_j^2} L &= y_{.j.} - \mu_{.j.}\end{aligned}$$

vanish as well. This stationary point of the loglikelihood is also a stationary point of the Lagrangian with all Lagrange multipliers equal to 0.

Of course, we still must nail down λ, λ_i^1, λ_j^2, and λ_{ij}^{12}. The choice

$$\lambda_{ij}^{12} = \ln\left(\frac{y_{ij.}}{\alpha_{ij.}}\right) - \lambda - \lambda_i^1 - \lambda_j^2$$

certainly guarantees $\mu_{ij} = y_{ij.}/\alpha_{ij.}$. One can check that the further choices

$$\begin{aligned}\lambda &= \frac{1}{rs}\sum_i \sum_j \ln \mu_{ij} \\ \lambda_i^1 &= \frac{1}{s}\sum_j \ln \mu_{ij} - \lambda \\ \lambda_j^2 &= \frac{1}{r}\sum_i \ln \mu_{ij} - \lambda\end{aligned}$$

satisfy the relevant equality constraints $\lambda_{\cdot}^1 = 0$, $\lambda_{\cdot}^2 = 0$, $\lambda_{\cdot j}^{12} = 0$, and $\lambda_{i.}^{12} = 0$.

At the second stage, the parameter set $\{\lambda, \lambda_i^1, \lambda_k^3, \lambda_{ik}^{13}\}$ is updated, holding the remaining parameters fixed. At the third stage, the parameter set $\{\lambda, \lambda_j^2, \lambda_k^3, \lambda_{jk}^{23}\}$ is updated, holding the remaining parameters fixed. These three successive stages constitute one iteration of the iterative proportional fitting algorithm. Each stage either leaves all parameters unchanged or increases the loglikelihood. In this example, the parameter blocks are not disjoint. ■

11.7 Problems

1. Which of the following functions is coercive on its domain?

   (a) $f(x) = x + 1/x$ on $(0, \infty)$

   (b) $f(x) = x - \ln x$ on $(0, \infty)$

   (c) $f(x) = x_1^2 + x_2^2 - 2x_1x_2$ on R^2

(d) $f(x) = x_1^4 + x_2^4 - 3x_1x_2$ on R^2

(e) $f(x) = x_1^2 + x_2^2 + x_3^2 - \sin(x_1x_2x_3)$ on R^3.

Give convincing reasons in each case.

2. Consider a polynomial $p(x)$ in n variables $x_1, \ldots, x_n$. Suppose that $p(x) = \sum_{i=1}^n c_i x_i^{2m}$ + lower-order terms, where all $c_i > 0$ and where a lower-order term is a product $bx_1^{m_1} \cdots x_n^{m_n}$ with $\sum_{i=1}^n m_i < 2m$. Prove rigorously that $p(x)$ is coercive on R^n. (Hint: Consider inequality (12.8) of Chapter 12.)

3. Demonstrate that $h(x) + k(x)$ is coercive on R^n if $k(x)$ is convex and $h(x)$ satisfies $\lim_{\|x\|_2 \to \infty} \|x\|_2^{-1} h(x) = \infty$. (Hint: Assume $k(x)$ is differentiable and apply the supporting hyperplane definition of convexity.)

4. In some problems it is helpful to broaden the notion of coerciveness. Consider a continuous function $f : \mathsf{R}^n \mapsto \mathsf{R}$ such that the limit

$$c = \lim_{r \to \infty} \inf_{\{\|x\| \geq r\}} f(x)$$

exists. The value of c can be finite or ∞ but not $-\infty$. Now let y be any point with $f(y) < c$. Show that the set $S_y = \{x : f(x) \leq f(y)\}$ is compact and that $f(x)$ attains its global minimum on S_y. The particular function

$$g(x) = \frac{x_1 + 2x_2}{1 + x_1^2 + x_2^2}$$

furnishes an example when $n = 2$. Demonstrate that the limit c equals 0. What is the minimum value and minimum point of $g(x)$? (Hint: What is the minimum value of $g(x)$ on the circle $\{x : \|x\| = r\}$?)

5. Find the minima of the functions

$$\begin{aligned} f(x) &= x \ln x \\ g(x) &= x - \ln x \\ h(x) &= x + \frac{1}{x} \end{aligned}$$

on $(0, \infty)$. Prove rigorously that your solutions are indeed the minima.

6. Find all of the stationary points of the function

$$f(x) = x_1^2 x_2 e^{-x_1^2 - x_2^2}$$

in R^2. Classify each point as either a local minimum, a local maximum, or a saddle point.

7. Demonstrate that the function $x_1^2+x_2^2(1-x_1)^3$ has a unique stationary point in R^2, which is a local minimum but not a global minimum. Can this occur for a continuously differentiable function with domain R?

8. Find the triangle of greatest area with fixed perimeter p. (Hint: Recall that a triangle with sides a, b, and c has area $\sqrt{s(s-a)(s-b)(s-c)}$, where $s = (a+b+c)/2 = p/2$.)

9. The equation

$$\sum_{i=1}^{n} a_i x_i^2 = c$$

defines an ellipse in R^n whenever all $a_i > 0$. The problem of Apollonius is to find the closest point on the ellipse from an external point y [5]. Demonstrate that the solution has coordinates

$$x_i = \frac{y_i}{1+\lambda a_i},$$

where λ is chosen to satisfy

$$\sum_{i=1}^{n} a_i \left(\frac{y_i}{1+\lambda a_i}\right)^2 = c.$$

Show how you can adapt this solution to solve the problem with the more general ellipse $(x-z)^t A(x-z) = c$ for A a positive definite matrix.

10. Use the techniques of this chapter to show that the minimum of the quadratic form $x^t A x$ subject to $\|x\|_2^2 = 1$ coincides with the smallest eigenvalue of the symmetric matrix A. The minimum point furnishes the corresponding eigenvector. Note that you will have to use the Lagrange multiplier rule for nonlinear constraints.

11. Find a minimum of $f(x) = x_1^2 + x_2^2$ subject to the inequality constraints $h_1(x) = -2x_1 - x_2 + 10 \le 0$ and $h_2(x) = -x_1 \le 0$. Prove that it is the global minimum.

12. Minimize the function $f(x) = e^{-(x_1+x_2)}$ subject to the constraints $h_1(x) = e^{x_1} + e^{x_2} - 20 \le 0$ and $h_2(x) = -x_1 \le 0$ on R^2.

13. Find the minimum and maximum of the function $f(x) = x_1 + x_2$ over the subset of R^2 defined by the constraints $h_i(x) \le 0$ for

$$\begin{aligned} h_1(x) &= -x_1 \\ h_2(x) &= -x_2 \\ h_3(x) &= 1 - x_1 x_2. \end{aligned}$$

14. Consider the problem of minimizing $f(x) = (x_1 + 1)^2 + x_2^2$ subject to the inequality constraint $h(x) = -x_1^3 + x_2^2 \leq 0$ on R^2. Solve the problem by sketching the feasible region and using a little geometry. Show that the Lagrange multiplier rule fails, and explain why.

15. Consider the inequality constraint functions

$$\begin{aligned} h_1(x) &= -x_1 \\ h_2(x) &= -x_2 \\ h_3(x) &= x_1^2 + 4x_2^2 - 4 \\ h_4(x) &= (x_1 - 2)^2 + x_2^2 - 5. \end{aligned}$$

Show that the Kuhn-Tucker constraint qualification fails but the Mangasarian-Fromovitz constraint qualification succeeds at the point $x = (0, 1)^t$. For the inequality constraint functions

$$\begin{aligned} h_1(x) &= x_1^2 - x_2 \\ h_2(x) &= -3x_1^2 + x_2, \end{aligned}$$

show that both constraint qualifications fail at the point $x = (0, 0)^t$ [10].

16. For $p > 1$ define the norm $\|x\|_p$ on R^n satisfying $\|x\|_p^p = \sum_{i=1}^n |x_i|^p$. For a fixed vector z, maximize $f(x) = z^t x$ subject to $\|x\|_p^p \leq 1$. Deduce Hölder's inequality $|z^t x| \leq \|x\|_p \|z\|_q$ for q defined by the equation $p^{-1} + q^{-1} = 1$.

17. If A is a matrix and y is a compatible vector, then $Ay \geq \mathbf{0}$ means that all entries of the vector Ay are nonnegative. Farkas's lemma says that $x^t y \geq 0$ for all vectors y with $Ay \geq \mathbf{0}$ if and only if x is a nonnegative linear combination of the rows of A. Prove Farkas's lemma assuming that the rows of A are linearly independent.

18. Establish the inequality

$$\Big(\prod_{i=1}^n a_i\Big)^{\frac{1}{n}} \leq \frac{1}{n}\sum_{i=1}^n a_i$$

between the geometric and arithmetic mean of n positive numbers. Verify that equality holds if and only if all $a_i = a$. (Hints: Replace a_i by e^{x_i} and appeal to convexity.)

19. Demonstrate that the Kullback-Leibler (cross-entropy) distance

$$f(x) = x_1 \ln \frac{x_1}{x_2} + x_2 - x_1$$

is convex on the set $\{x = (x_1, x_2) : x_1 > 0, x_2 > 0\}$.

20. In Sinkhorn's theorem, suppose the matrix M is square. Show that some entries of M can be 0 as long as some positive power M^p of M has all entries positive.

21. Let $M = (m_{ij})$ be a nontrivial $m \times n$ matrix. The dominant part of the singular value decomposition (svd) of M is an outer product matrix λuv^t with $\lambda > 0$ and u and v unit vectors. This outer product minimizes

$$\|M - \lambda uv^t\|_F^2 = \sum_i \sum_j (m_{ij} - \lambda u_i v_j)^2.$$

One can use alternating least squares to find λuv^t [11]. In the first step of the algorithm, one fixes v and estimates $w = \lambda u$ by least squares. Show that w has components

$$w_i = \sum_j m_{ij} v_j.$$

Once w is available, we set

$$\lambda = \|w\|_2, \qquad u = \frac{1}{\|w\|_2} w.$$

What are the corresponding updates for v and λ when you fix u? To find the next outer product in the svd, form the matrix $M - \lambda uv^t$ and repeat the process. Program and test this algorithm.

22. Suppose A is a symmetric matrix and B is a positive definite matrix of the same dimension. Formulate cyclic coordinate descent and ascent algorithms for minimizing and maximizing the Rayleigh quotient

$$R(x) = \frac{x^t A x}{x^t B x}$$

over the set $x \neq \mathbf{0}$. See Section 8.3 for connections to eigenvalue extraction.

23. For a positive definite matrix A, consider minimizing the quadratic function $f(x) = \frac{1}{2} x^t A x + b^t x + c$ subject to the constraints $x_i \geq 0$ for all i. Show that the cyclic coordinate descent updates are

$$\hat{x}_i = \max\Big\{0, x_i - a_{ii}^{-1}\Big[\sum_j a_{ij} x_j + b_i\Big]\Big\}.$$

If we impose the additional constraint $\sum_i x_i = 1$, the problem is harder. One line of attack is to minimize the penalized function

$$f_\mu(x) = f(x) + \frac{\mu}{2}\Big(\sum_i x_i - 1\Big)^2$$

for a large positive constant μ. One can show that the minimum of $f_\mu(x)$ tends to the constrained minimum of $f(x)$ as μ tends to ∞ [2]. Accepting this result, demonstrate that cyclic coordinate descent for $f_\mu(x)$ has updates

$$\hat{x}_i = \max\Big\{0, x_i - (a_{ii}+\mu)^{-1}\Big[\sum_j a_{ij}x_j + b_i + \mu\Big(\sum_j x_j - 1\Big)\Big]\Big\}.$$

Program this second algorithm and test it for the choices

$$A = \begin{pmatrix} 2 & 1 \\ 1 & 1 \end{pmatrix}, \quad b = \begin{pmatrix} 1 \\ 0 \end{pmatrix}.$$

Start with $\mu = 1$ and double it every time you update the full vector x. Do the iterates converge to the minimum of $f(x)$ subject to all constraints?

24. Program and test any one of the four examples 11.6.1, 11.6.2, 11.6.3, or 11.6.4.

25. Program and test a k-medians clustering algorithm and concoct an example where it differs from k-means clustering.

26. Consider the coronary disease data [9, 14] displayed in the three-way contingency Table 11.3. Using iterative proportional fitting, find the maximum likelihood estimates for the loglinear model with first-order interactions. Perform a chi-square test to decide whether this model fits the data better than the model postulating independence of the three factors.

27. As noted in the text, the loglinear model for categorical data can be interpreted as assuming independent Poisson distributions for the various categories with category i having mean $\mu_i(\theta) = e^{l_i^t\theta}$, where l_i is a vector whose entries are 0's or 1's. Calculate the observed information $-d^2L(\theta) = \sum_i e^{l_i^t\theta} l_i l_i^t$ in this circumstance, and deduce that it is positive semidefinite. In the presence of linear constraints on θ, show that any maximum likelihood estimate of θ is necessarily unique provided the vector subspace of possible θ is included in the linear span of the l_i.

28. In Example 11.6.4, make the change of variables $c = \Sigma_{11}^{1/2}a$ and $d = \Sigma_{22}^{1/2}b$. In the new variables show that one must maximize $c^t\Omega d$ subject to $\|c\|_2 = \|d\|_2 = 1$, where $\Omega = \Sigma_{11}^{-1/2}\Sigma_{12}\Sigma_{22}^{-1/2}$. The first singular vectors $c = u_1$ and $d = v_1$ of the svd $\Omega = \sum_k \sigma_k u_k v_k^t$ provide the solution. Hence, $a = \Sigma_{11}^{-1/2}u_1$ and $b = \Sigma_{22}^{-1/2}v_1$. The advantage of this approach is that one can now define higher-order canonical correlation vectors from the remaining singular vectors.

TABLE 11.3. Coronary Disease Data

Disease Status	Cholesterol Level	Blood Pressure 1	2	3	4	Total
Coronary	1	2	3	3	4	12
	2	3	2	1	3	9
	3	8	11	6	6	31
	4	7	12	11	11	41
Total		20	28	21	24	93
No Coronary	1	117	121	47	22	307
	2	85	98	43	20	246
	3	119	209	68	43	439
	4	67	99	46	33	245
Total		388	527	204	118	1237

11.8 References

[1] Arthur D, Vassilvitskii S (2007): k-means++: the advantages of careful seeding. 2007 Symposium on Discrete Algorithms (SODA)

[2] Beltrami EJ (1970) *An Algorithmic Approach to Nonlinear Analysis and Optimization.* Academic Press, New York

[3] Bishop YMM, Feinberg SE, Holland PW (1975) *Discrete Multivariate Analysis: Theory and Practice.* MIT Press, Cambridge, MA

[4] Boyd S, Vandenberghe L (2004) *Convex Optimization.* Cambridge University Press, Cambridge

[5] Brinkhuis J, Tikhomirov V (2005) *Optimization: Insights and Applications.* Princeton University Press, Princeton, NJ

[6] Ciarlet PG (1989) *Introduction to Numerical Linear Algebra and Optimization.* Cambridge University Press, Cambridge

[7] de Leeuw J (1994) Block relaxation algorithms in statistics. in *Information Systems and Data Analysis*, Bock HH, Lenski W, Richter MM, Springer, Berlin

[8] de Souza PN, Silva J-N (2001) *Berkeley Problems in Mathematics*, 2nd ed. Springer, New York

[9] Everitt BS (1977) *The Analysis of Contingency Tables.* Chapman & Hall, London

[10] Forsgren A, Gill PE, Wright MH (2002) Interior point methods for nonlinear optimization. *SIAM Review* 44:523–597

[11] Gabriel KR, Zamir S (1979) Lower rank approximation of matrices by least squares with any choice of weights. *Technometrics* 21:489–498

[12] Hestenes MR (1981) *Optimization Theory: The Finite Dimensional Case.* Robert E Krieger Publishing, Huntington, NY

[13] Kosowsky JJ, Yuille AL (1994) The invisible hand algorithm: solving the assignment problem with statistical physics. *Neural Networks* 7:477–490

[14] Ku HH, Kullback S (1974) Log-linear models in contingency table analysis. *Biometrics* 10:452–458

[15] Kuhn HW, Tucker AW (1951) Nonlinear programming. in *Proceedings of the Second Berkeley Symposium on Mathematical Statistics and Probability.* University of California Press, Berkeley, pp 481-492

[16] Lange, K (2004) *Optimization.* Springer, New York

[17] Luenberger DG (1984) *Linear and Nonlinear Programming,* 2nd ed. Addison-Wesley, Reading, MA

[18] Maher MJ (1982), Modelling association football scores. *Statistica Neerlandica* 36:109–118

[19] Mangasarian OL, Fromovitz S (1967) The Fritz John necessary optimality conditions in the presence of equality and inequality constraints. *J Math Anal Appl* 17:37–47

[20] Mardia KV, Kent JT, Bibby JM (1979) *Multivariate Analysis.* Academic, New York

[21] McShane EJ (1973) The Lagrange multiplier rule. *Amer Math Monthly* 80:922–925

[22] Peressini AL, Sullivan FE, Uhl JJ Jr (1988) *The Mathematics of Nonlinear Programming.* Springer, New York

[23] Ruszczynski A (2006) *Nonlinear Optimization.* Princeton University Press. Princeton, NJ

[24] Sinkhorn R (1967) Diagonal equivalence to matrices with prescribed row and column sums. *Amer Math Monthly* 74:402–405

12

The MM Algorithm

12.1 Introduction

Most practical optimization problems defy exact solution. In the current chapter we discuss an optimization method that relies heavily on convexity arguments and is particularly useful in high-dimensional problems such as image reconstruction [27]. This iterative method is called the MM algorithm. One of the virtues of the MM acronym is that it does double duty. In minimization problems, the first M stands for majorize and the second M for minimize. In maximization problems, the first M stands for minorize and the second M for maximize. When it is successful, the MM algorithm substitutes a simple optimization problem for a difficult optimization problem. Simplicity can be attained by (a) avoiding large matrix inversions, (b) linearizing an optimization problem, (c) separating the variables of an optimization problem, (d) dealing with equality and inequality constraints gracefully, and (e) turning a nondifferentiable problem into a smooth problem. In simplifying the original problem, we pay the price of iteration or iteration with a slower rate of convergence.

Statisticians have vigorously developed a special case of the MM algorithm called the EM algorithm, which revolves around notions of missing data [6, 32]. We present the EM algorithm in the next chapter. We prefer to present the MM algorithm first because of its greater generality, its more obvious connection to convexity, and its weaker reliance on difficult statistical principles. Readers eager to learn the history of the MM algorithm and see more applications can consult the survey articles [1, 4, 12, 27, 17, 40]. The book by Steele [38] is an especially good introduction to the art and science of inequalities.

12.2 Philosophy of the MM Algorithm

A function $g(x \mid x_n)$ is said to majorize a function $f(x)$ at x_n provided

$$\begin{aligned} f(x_n) &= g(x_n \mid x_n) \\ f(x) &\leq g(x \mid x_n), \qquad x \neq x_n. \end{aligned} \tag{12.1}$$

In other words, the surface $x \mapsto g(x \mid x_n)$ lies above the surface $f(x)$ and is tangent to it at the point $x = x_n$. Here x_n represents the current iterate in a search of the surface $f(x)$. Figure 12.1 provides a simple one-dimensional example of majorization.

K. Lange, *Numerical Analysis for Statisticians*, Statistics and Computing,
DOI 10.1007/978-1-4419-5945-4_12,

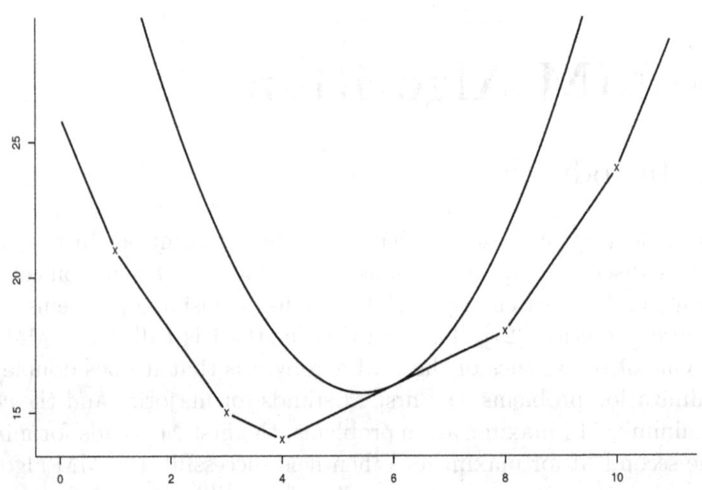

FIGURE 12.1. A Quadratic Majorizing Function for the Piecewise Linear Function $f(x) = |x-1| + |x-3| + |x-4| + |x-8| + |x-10|$ at the Point $x_n = 6$

In the minimization version of the MM algorithm, we minimize the surrogate majorizing function $g(x \mid x_n)$ rather than the actual function $f(x)$. If x_{n+1} denotes the minimum of the surrogate $g(x \mid x_n)$, then we can show that the MM procedure forces $f(x)$ downhill. Indeed, the relations

$$f(x_{n+1}) \leq g(x_{n+1} \mid x_n) \leq g(x_n \mid x_n) = f(x_n) \tag{12.2}$$

follow directly from the definition of x_{n+1} and the majorization conditions (12.1). The descent property (12.2) lends the MM algorithm remarkable numerical stability. Strictly speaking, it depends only on decreasing the surrogate function $g(x \mid x_n)$, not on minimizing it. This fact has practical consequences when the minimum of $g(x \mid x_n)$ cannot be found exactly. When $f(x)$ is strictly convex, one can show with a few additional mild hypotheses that the iterates x_n converge to the global minimum of $f(x)$ regardless of the initial point x_0 [24].

If $g(x \mid x_n)$ majorizes $f(x)$ at an interior point x_n of the domain of $f(x)$, then x_n is a stationary point of the difference $g(x \mid x_n) - f(x)$, and the identity

$$\nabla g(x_n \mid x_n) = \nabla f(x_n) \tag{12.3}$$

holds. Furthermore, the second differential $d^2g(x_n \mid x_n) - d^2f(x_n)$ is positive semidefinite. Problem 2 makes the point that the majorization relation between functions is closed under the formation of sums, nonnegative products, limits, and composition with an increasing function. These rules permit one to work piecemeal in simplifying complicated objective functions. With obvious changes, the MM algorithm also applies to maximization rather than to minimization. To maximize a function $f(x)$, we minorize it by a surrogate function $g(x \mid x_n)$ and maximize $g(x \mid x_n)$ to produce the next iterate x_{n+1}.

The reader might well object that the MM algorithm is not so much an algorithm as a vague philosophy for deriving an algorithm. The same objection applies to the EM algorithm. As we proceed through the current chapter, we hope the various examples will convince the reader of the value of a unifying principle and a framework for attacking concrete problems. The strong connection of the MM algorithm to convexity and inequalities has the natural pedagogical advantage of strengthening skills in these areas.

12.3 Majorization and Minorization

We will feature five methods for constructing majorizing functions. Two of these simply adapt Jensen's inequality

$$f\Big(\sum_i \alpha_i t_i\Big) \leq \sum_i \alpha_i f(t_i)$$

defining a convex function $f(t)$. It is easy to identify convex functions on the real line, so the first method composes such a function with a linear function $c^t x$ to create a new convex function of the vector x. Invoking the definition of convexity with $\alpha_i = c_i y_i / c^t y$ and $t_i = c^t y \, x_i / y_i$ then yields

$$f(c^t x) \leq \sum_i \frac{c_i y_i}{c^t y} f\Big(\frac{c^t y}{y_i} x_i\Big) = g(x \mid y), \tag{12.4}$$

provided all of the components of the vectors c, x, and y are positive. The surrogate function $g(x \mid y)$ equals $f(c^t y)$ when $x = y$. One of the virtues of applying inequality (12.4) in defining a surrogate function is that it separates parameters in the surrogate function. This feature is critically important in high-dimensional problems because it reduces optimization over x to a sequence of one-dimensional optimizations over each component x_i. The argument establishing inequality (12.4) is equally valid if we replace the parameter vector x throughout by a vector-valued function $h(x)$ of x.

To relax the positivity restrictions on the vectors c, x, and y, De Pierro [8] suggested in a medical imaging context the alternative majorization

$$f(c^t x) \leq \sum_i \alpha_i f\left\{\frac{c_i}{\alpha_i}(x_i - y_i) + c^t y\right\} = g(x \mid y) \tag{12.5}$$

for a convex function $f(t)$. Here all $\alpha_i \geq 0$, $\sum_i \alpha_i = 1$, and $\alpha_i > 0$ whenever $c_i \neq 0$. In practice, we must somehow tailor the α_i to the problem at hand. Among the obvious candidates for the α_i are

$$\alpha_i = \frac{|c_i|^p}{\sum_j |c_j|^p}$$

for $p \geq 0$. When $p = 0$, we interpret α_i as 0 if $c_i = 0$ and as $1/q$ if c_i is one among q nonzero coefficients.

Our third method involves the linear majorization

$$f(x) \leq f(y) + df(y)(x - y) = g(x \mid y) \tag{12.6}$$

satisfied by any concave function $f(x)$. Once again we can replace the argument x by a vector-valued function $h(x)$. The most useful concave function in practice is $\ln x$.

Our fourth method applies to functions $f(x)$ with bounded curvature [2, 4]. Assuming that $f(x)$ is twice differentiable, we look for a matrix B satisfying $B \succeq d^2 f(x)$ and $B \succ \mathbf{0}$ in the sense that $B - d^2 f(x)$ is positive semidefinite for all x and B is positive definite. The quadratic bound principle then amounts to the majorization

$$\begin{aligned} f(x) &= f(y) + df(y)(x - y) + \frac{1}{2}(x - y)^t d^2 f(z)(x - y) \\ &\leq f(y) + df(y)(x - y) + \frac{1}{2}(x - y)^t B(x - y) \qquad (12.7) \\ &= g(x \mid y) \end{aligned}$$

of the second-order Taylor expansion of $f(x)$. Here the intermediate point z occurs on the line segment from x to y.

Our fifth and final method exploits the generalized arithmetic-geometric mean inequality

$$\prod_{i=1}^m y_i^{\alpha_i} \leq \sum_{i=1}^m \alpha_i y_i$$

for positive numbers y_i and α_i subject to the constraint $\sum_{i=1}^m \alpha_i = 1$. If we put $y_i = e^{x_i}$, then this inequality is a direct consequence of the strict convexity of the function e^x. Equality holds if and only if all y_i coincide. With this result in mind, Problem 10 asks the reader to prove the majorization

$$\prod_{i=1}^m x_i^{\alpha_i} \leq \left(\prod_{i=1}^m x_{ni}^{\alpha_i}\right) \sum_{i=1}^m \frac{\alpha_i}{\alpha}\left(\frac{x_i}{x_{ni}}\right)^{\alpha} \tag{12.8}$$

for positive numbers x_i, x_{ni}, and α_i and sum $\alpha = \sum_{i=1}^m \alpha_i$. Inequality (12.8) is the key to separating parameters with posynomials. We will use it in sketching an MM algorithm for unconstrained geometric programming.

Any of the first four majorizations can be turned into minorizations by interchanging the adjectives convex and concave and positive definite and negative definite, respectively. Of course, there is an art to applying these methods just as there is an art to applying any mathematical principle. The five highlighted methods hardly exhaust the possibilities for majorization and minorization. Several problems at the end of the chapter sketch other helpful techniques. Readers are also urged to consult the literature on the MM and EM algorithms for a fuller discussion.

12.4 Linear Regression

Because the function $s \mapsto (y-s)^2$ is convex, we can majorize each summand of the least squares criterion $\sum_{i=1}^m (y_i - x_i^t\theta)^2$ using inequality (12.5). It follows that

$$\begin{aligned} \sum_{i=1}^m (y_i - x_i^t\theta)^2 &\le \sum_{i=1}^m \sum_j \alpha_{ij}\left[y_i - \frac{x_{ij}}{\alpha_{ij}}(\theta_j - \theta_{nj}) - x_i^t\theta_n\right]^2 \\ &= g(\theta \mid \theta_n), \end{aligned}$$

with equality when $\theta = \theta_n$. Minimization of $g(\theta \mid \theta_n)$ yields the updates

$$\theta_{n+1,j} = \theta_{nj} + \frac{\sum_{i=1}^m x_{ij}(y_i - x_i^t\theta_n)}{\sum_{i=1}^m \frac{x_{ij}^2}{\alpha_{ij}}} \tag{12.9}$$

and avoids matrix inversion [27]. Although it seems plausible to take $p = 1$ in choosing

$$\alpha_{ij} = \frac{|x_{ij}|^p}{\sum_k |x_{ik}|^p},$$

conceivably other values of p might perform better. In fact, it might accelerate convergence to alternate different values of p as the iterations proceed. For problems involving just a few parameters, this iterative scheme is clearly inferior to the usual single-step solution via matrix inversion. Cyclic coordinate descent also avoids matrix operations, and Problem 12 suggests that it will converge faster than the MM update (12.9).

Least squares estimation suffers from the fact that it is strongly influenced by observations far removed from their predicted values. In least absolute deviation regression, we replace $\sum_{i=1}^m (y_i - x_i^t\theta)^2$ by

$$h(\theta) = \sum_{i=1}^m \left|y_i - x_i^t\theta\right| = \sum_{i=1}^m |r_i(\theta)|, \tag{12.10}$$

where $r_i(\theta) = y_i - x_i^t\theta$ is the ith residual. We are now faced with minimizing a nondifferentiable function. Fortunately, the MM algorithm can be implemented by exploiting the concavity of the function $\sqrt{u}$ in inequality (12.6). Because

$$\sqrt{u} \;\leq\; \sqrt{u_n} + \frac{u - u_n}{2\sqrt{u_n}},$$

we find that

$$\begin{aligned} h(\theta) &= \sum_{i=1}^m \sqrt{r_i^2(\theta)} \\ &\leq h(\theta_n) + \frac{1}{2}\sum_{i=1}^m \frac{r_i^2(\theta) - r_i^2(\theta_n)}{\sqrt{r_i^2(\theta_n)}} \\ &= g(\theta \mid \theta_n). \end{aligned}$$

Minimizing $g(\theta \mid \theta_n)$ is accomplished by minimizing the weighted sum of squares

$$\sum_{i=1}^m w_i(\theta_n) r_i(\theta)^2$$

with ith weight $w_i(\theta_n) = |r_i(\theta_n)|^{-1}$. A slight variation of the usual argument for minimizing a sum of squares leads to the update

$$\theta_{n+1} \;=\; [X^t W(\theta_n) X]^{-1} X^t W(\theta_n) y,$$

where $W(\theta_n)$ is the diagonal matrix with ith diagonal entry $w_i(\theta_n)$. Unfortunately, the possibility that some weight $w_i(\theta_n) = \infty$ cannot be ruled out. The next section, which generalizes the current MM algorithm to multivariate response vectors and least ℓ_p regression, suggests some simple remedies.

12.5 Elliptically Symmetric Densities and ℓ_p Regression

Dutter and Huber [15] introduced an MM algorithm for elliptically symmetric densities

$$f(y) \;=\; \frac{e^{-\frac{1}{2}\kappa[\delta^2(\theta)]}}{(2\pi)^{\frac{k}{2}}[\det \Omega(\theta)]^{\frac{1}{2}}} \tag{12.11}$$

defined for $y \in \mathsf{R}^k$, where $\delta^2 = [y - \mu(\theta)]^t \Omega^{-1}(\theta)[y - \mu(\theta)]$ denotes the Mahalanobis distance between y and $\mu(\theta)$. For our purposes, it is convenient to assume in addition that the function $\kappa(s)$ is strictly increasing and strictly concave. The multivariate t provides a typical example of an elliptically symmetric distribution that can profitably be substituted for the multivariate normal distribution in robust estimation [28, 29].

For a sequence $y_1, \ldots, y_m$ of independent observations from the elliptically symmetric density (12.11) with covariance matrices $\Omega_1(\theta), \ldots, \Omega_m(\theta)$ and means $\mu_1(\theta), \ldots, \mu_m(\theta)$, the multivariate normal loglikelihood

$$g(\theta \mid \theta_n) = -\frac{1}{2}\sum_{i=1}^{m}[w_i\delta_i^2(\theta) + \ln\det\Omega_i(\theta)] + c,$$

with weights $w_i = \kappa'[\delta_i^2(\theta_n)]$ and normalizing constant c minorizes the loglikelihood $L(\theta)$. This follows from inequality (12.6) applied to the concave function $\kappa(s)$. The array of techniques from linear algebra for maximizing the multivariate normal distribution can be brought to bear on maximizing $g(\theta \mid \theta_n)$. See Problems 15 and 16 for how this works for estimation with the multivariate t distribution. For normal/independent distributional families, the Dutter-Huber algorithm usually reduces to an EM algorithm [7, 29].

For $1 \le p \le 2$ and independent univariate observations $y_1, \ldots, y_m$ with unit variances, the choice $\kappa(s) = s^{p/2}$ leads to ℓ_p regression. The Dutter-Huber procedure minimizes

$$h(\theta \mid \theta_n) = \sum_{i=1}^{m} w_i(\theta_n)[y_i - \mu_i(\theta)]^2$$

at each iteration with weights $w_i(\theta_n) = |y_i - \mu_i(\theta_n)|^{p-2}$ as described in Problem 7. Hence, ℓ_p regression can be accomplished by iteratively reweighted least squares. This algorithm, originally proposed by Schlossmacher [36] and Merle and Spath [33], is unfortunately plagued by infinite weights for those observations with zero residuals. To avoid this difficulty, we can redefine the weights to be

$$w_i(\theta_n) = \frac{1}{\epsilon + |y_i - \mu_i(\theta_n)|^{2-p}}$$

for a small $\epsilon > 0$ [29]. This corresponds to the choice

$$\kappa'(s) = \frac{p}{2(\epsilon + s^{1-p/2})}$$

and also leads to an MM algorithm. The revised algorithm for $p = 1$ minimizes the criterion

$$\sum_{i=1}^{m}\{|y_i - \mu_i(\theta)| - \epsilon\ln[\epsilon + |y_i - \mu_i(\theta)|]\}, \tag{12.12}$$

which obviously tends to $\sum_{i=1}^{m}|y_i - \mu_i(\theta)|$ as $\epsilon \to 0$. Problem 17 describes alternative weights for another reasonable criterion.

12.6 Bradley-Terry Model of Ranking

In the sports version of the Bradley and Terry model [3, 16, 19], each team i in a league of teams is assigned a rank parameter $r_i > 0$. Assuming ties are impossible, team i beats team j with probability $r_i/(r_i + r_j)$. If this outcome occurs y_{ij} times during a season of play, then the probability of the whole season is

$$L(r) = \prod_{i,j} \Big(\frac{r_i}{r_i + r_j}\Big)^{y_{ij}},$$

assuming the games are independent. To rank the teams, we find the values $\hat{r}_i$ that maximize $f(r) = \ln L(r)$. The team with largest $\hat{r}_i$ is considered best, the team with smallest $\hat{r}_i$ is considered worst, and so forth. In view of the fact that $\ln u$ is concave, inequality (12.6) implies

$$\begin{aligned} f(r) &= \sum_{i,j} y_{ij}\Big[\ln r_i - \ln(r_i + r_j)\Big] \\ &\geq \sum_{i,j} y_{ij}\Big[\ln r_i - \ln(r_{ni} + r_{nj}) - \frac{r_i + r_j - r_{ni} - r_{nj}}{r_{ni} + r_{nj}}\Big] \\ &= g(r \mid r_n) \end{aligned}$$

with equality when $r = r_n$. Differentiating $g(r \mid r_n)$ with respect to the ith component r_i of r and setting the result equal to 0 produces the next iterate

$$r_{n+1,i} = \frac{\sum_{j \neq i} y_{ij}}{\sum_{j \neq i}(y_{ij} + y_{ji})/(r_{ni} + r_{nj})}.$$

Because $L(r) = L(\beta r)$ for any $\beta > 0$, we constrain $r_1 = 1$ and omit the update $r_{n+1,1}$. In this example, the MM algorithm separates parameters and allows us to maximize $g(r \mid r_n)$ parameter by parameter. Although this model is less sophisticated than Maher's model in Example 11.6.2, it does apply to competitions such as chess that do not involve scoring points.

12.7 A Random Graph Model

Random graphs provide interesting models of connectivity in genetics and internet node ranking. Here we consider a simplification of the random graph model of Chatterjee and Diaconis. In this model we assign a propensity $p_i \geq 0$ to each node i. An edge between nodes i and j then forms independently with probability $p_i p_j/(1 + p_i p_j)$. The most obvious statistical question about the model is how to estimate the p_i from data. Once this is done, we can rank nodes by their estimated propensities.

If E denotes the edge set of the graph, then the loglikelihood can be written as

$$L(p) = \sum_{\{i,j\}\in E} [\ln p_i + \ln p_j] - \sum_{\{i,j\}} \ln(1 + p_i p_j). \qquad (12.13)$$

Here $\{i, j\}$ denotes a generic unordered pair. The logarithms $\ln(1 + p_i p_j)$ are the bothersome terms in the loglikelihood. We will minorize each of these by exploiting the convexity of the function $-\ln(1 + x)$. Application of inequality (12.6) yields

$$-\ln(1 + p_i p_j) \geq -\ln(1 + p_{ni} p_{nj}) - \frac{1}{1 + p_{ni} p_{nj}} (p_i p_j - p_{ni} p_{nj})$$

and eliminates the logarithm. Note that equality holds when $p_i = p_{ni}$ for all i. This minorization is not quite good enough to separate parameters, however. Separation can be achieved by invoking the second minorizing inequality

$$-p_i p_j \geq -\frac{1}{2}\Big(\frac{p_{nj}}{p_{ni}} p_i^2 + \frac{p_{ni}}{p_{nj}} p_j^2\Big).$$

Note again that equality holds when all $p_i = p_{ni}$.

These considerations imply that up to a constant $L(p)$ is minorized by the function

$$g(p \mid p_n) = \sum_{\{i,j\}\in E} [\ln p_i + \ln p_j] - \sum_{\{i,j\}} \frac{1}{1 + p_{ni} p_{nj}} \frac{1}{2}\Big(\frac{p_{nj}}{p_{ni}} p_i^2 + \frac{p_{ni}}{p_{nj}} p_j^2\Big).$$

The fact that $g(p \mid p_n)$ separates parameters allows us to compute $p_{n+1,i}$ by setting the derivative of $g(p \mid p_n)$ with respect to p_i equal to 0. Thus, we must solve

$$0 = \sum_{j \sim i} \frac{1}{p_i} - \sum_{j\sim i} \frac{1}{1 + p_{ni} p_{nj}} \frac{p_{nj}}{p_{ni}} p_i,$$

where $j \sim i$ means $\{i, j\} \in E$. If $d_i = \sum_{j\sim i} 1$ denotes the degree of node i, then the positive square root

$$p_{n+1,i} = \left[\frac{p_{ni} d_i}{\sum_{j\sim i} \frac{p_{nj}}{1+p_{ni} p_{nj}}}\right]^{1/2} \qquad (12.14)$$

is the pertinent solution.

The MM update (12.14) is not particularly intuitive, but it does have the virtue of algebraic simplicity. When $d_i = 0$, it also makes the sensible choice $p_{n+1,i} = 0$. As a check on our derivation, observe that a stationary point of the loglikelihood satisfies

$$0 = \frac{d_i}{p_i} - \sum_{j\sim i} \frac{p_j}{1 + p_i p_j},$$

which is just a rearranged version of the update (12.14) with iteration subscripts suppressed.

The MM algorithm just derived carries with it certain guarantees. It is certain to increase the loglikelihood at every iteration, and if its maximum value is attained at a unique point, then it will also converge to that point. Problem 23 states that the loglikelihood is concave under the reparameterization $p_i = e^{-q_i}$. The requirement of two successive minorizations in our derivation gives us pause because if minorization is not tight, then convergence is slow. On the other hand, if the number of nodes is large, then competing algorithms such as Newton's method entail large matrix inversions and are very expensive.

12.8 Linear Logistic Regression

In linear logistic regression, we observe a sequence of independent Bernoulli trials, each resulting in success or failure. The success probability of the ith trial

$$\pi_i(\theta) = \frac{e^{x_i^t\theta}}{1+e^{x_i^t\theta}}$$

depends on a covariate vector x_i and parameter vector θ by analogy with linear regression. The response y_i at trial i equals 1 for a success and 0 for a failure. In this notation, the likelihood of the data is

$$L(\theta) = \prod_i \pi_i(\theta)^{y_i}[1-\pi_i(\theta)]^{1-y_i}.$$

As usual in maximum likelihood estimation, we pass to the loglikelihood

$$f(\theta) = \sum_i \{y_i \ln \pi_i(\theta) + (1-y_i)\ln[1-\pi_i(\theta)]\}.$$

Straightforward calculations show that

$$\begin{aligned}\nabla f(\theta) &= \sum_i [y_i - \pi_i(\theta)]x_i \\ d^2 f(\theta) &= -\sum_i \pi_i(\theta)[1-\pi_i(\theta)]x_i x_i^t.\end{aligned}$$

The loglikelihood $f(\theta)$ is therefore concave, and we seek to minorize it by a quadratic rather than majorize it by a quadratic as suggested in inequality (12.7). Hence, we must identify a matrix B such that B is negative definite and $B - d^2f(x)$ is negative semidefinite for all x. In view of the scalar inequality $\pi(1-\pi) \leq \frac{1}{4}$, we take $B = -\frac{1}{4}\sum_i x_i x_i^t$. Maximization of the minorizing quadratic

$$f(\theta_n) + df(\theta_n)(\theta - \theta_n) + \frac{1}{2}(\theta - \theta_n)^t B(\theta - \theta_n)$$

is a problem we have met before. It does involve inversion of the matrix B, but once we have computed B^{-1}, we can store and reuse it at every iteration.

12.9 Unconstrained Geometric Programming

The idea behind these algorithms is best understood in a concrete setting. Consider the posynomial

$$f(x) \quad = \quad \frac{1}{x_1^3} + \frac{3}{x_1 x_2^2} + x_1 x_2$$

with the implied constraints $x_1 > 0$ and $x_2 > 0$. (Problem 25 defines a general posynomial.) The majorization (12.8) applied to the third term of $f(x)$ yields

$$\begin{aligned} x_1 x_2 &\leq x_{n1} x_{n2}\left[\frac{1}{2}\left(\frac{x_1}{x_{n1}}\right)^2 + \frac{1}{2}\left(\frac{x_2}{x_{n2}}\right)^2\right] \\ &= \frac{x_{n2}}{2x_{n1}} x_1^2 + \frac{x_{n1}}{2x_{n2}} x_2^2. \end{aligned}$$

When we apply the majorization to the second term of $f(x)$ and replace x_1 by x_1^{-1} and x_2 by x_2^{-1}, we find

$$\begin{aligned} \frac{3}{x_1 x_2^2} &\leq \frac{3}{x_{n1} x_{n2}^2}\left[\frac{1}{3}\left(\frac{x_{n1}}{x_1}\right)^3 + \frac{2}{3}\left(\frac{x_{n2}}{x_2}\right)^3\right] \\ &= \frac{x_{n1}^2}{x_{n2}^2}\frac{1}{x_1^3} + \frac{2x_{n2}}{x_{n1}}\frac{1}{x_2^3}. \end{aligned}$$

The second step of the MM algorithm for minimizing $f(x)$ therefore splits into minimizing the two surrogate functions

$$\begin{aligned} g_1(x_1 \mid x_n) &= \frac{1}{x_1^3} + \frac{x_{n1}^2}{x_{n2}^2}\frac{1}{x_1^3} + \frac{x_{n2}}{2x_{n1}} x_1^2 \\ g_2(x_2 \mid x_n) &= \frac{2x_{n2}}{x_{n1}}\frac{1}{x_2^3} + \frac{x_{n1}}{2x_{n2}} x_2^2. \end{aligned}$$

If we set the derivatives of each of these equal to 0, then we find the solutions

$$x_{n+1,1} \quad = \quad \sqrt[5]{3\left(\frac{x_{n1}^2}{x_{n2}^2} + 1\right)\frac{x_{n1}}{x_{n2}}}$$

$$x_{n+1,2} = \sqrt[5]{6\frac{x_{n2}^2}{x_{n1}^2}}.$$

It is obvious that the point $x = (\sqrt[5]{6}, \sqrt[5]{6})^t$ is a fixed point of these equations and minimizes $f(x)$. Table 12.1 records the iterates of the MM algorithm ignoring this fact. Convergence is slow but sure.

TABLE 12.1. MM Iterates for a Geometric Program

m	x_{n1}	x_{n2}	$f(x_n)$
0	1.00000	2.00000	3.75000
1	1.13397	1.88818	3.56899
2	1.19643	1.75472	3.49766
3	1.24544	1.66786	3.46079
4	1.28395	1.60829	3.44074
5	1.31428	1.56587	3.42942
10	1.39358	1.47003	3.41427
20	1.42699	1.43496	3.41280
30	1.43054	1.43140	3.41279
40	1.43092	1.43101	3.41279
50	1.43096	1.43097	3.41279
51	1.43097	1.43097	3.41279

This analysis carries over to general posynomials except that we cannot expect to derive explicit solutions of the minimization step. (See Problem 25.) Each separated surrogate function is a posynomial in a single variable. If the powers appearing in one of these posynomials are integers, then the derivative of the posynomial is a rational function, and once we equate it to 0, we are faced with solving a polynomial equation. This can be accomplished by bisection or by Newton's method as discussed in Chapter 5. Introducing posynomial constraints is another matter. Box constraints $a_i \leq x_i \leq b_i$ are consistent with parameter separation as developed here, but more complicated posynomial constraints that couple parameters are not.

12.10 Poisson Processes

In preparation for our exposition of transmission tomography in the next section, we now briefly review the theory of Poisson processes, a topic from probability of considerable interest in its own right. A Poisson process involves points randomly scattered in a region S of q-dimensional space R^q [11, 14, 18, 21]. The notion that the points are concentrated on average more in some regions than in others is captured by postulating an intensity function $\lambda(x) \geq 0$ on S. The expected number of points in a subregion T is given by the integral $\omega = \int_T \lambda(x)dx$. If $\omega = \infty$, then an infinite number of random points occur in T. If $\omega < \infty$, then a finite number of random points occur in T, and the probability that this number equals k is given by the Poisson probability

$$p_k(\omega) = \frac{\omega^k}{k!}e^{-\omega}.$$

Derivation of this formula depends critically on the assumption that the numbers N_{T_i} of random points in disjoint regions T_i are independent random variables. This basically means that knowing the values of some of the N_{T_i} tells one nothing about the values of the remaining N_{T_i}. The model also presupposes that random points never coincide.

The Poisson distribution has a peculiar relationship to the multinomial distribution. Suppose a Poisson random variable Z with mean ω represents the number of outcomes from some experiment, say an experiment involving a Poisson process. Let each outcome be independently classified in one of l categories, the kth of which occurs with probability p_k. Then the number of outcomes Z_k falling in category k is Poisson distributed with mean $\omega_k = p_k\omega$. Furthermore, the random variables $Z_1, \ldots, Z_l$ are independent. Conversely, if $Z = \sum_{k=1}^l Z_k$ is a sum of independent Poisson random variables Z_k with means $\omega_k = p_k\omega$, then conditional on $Z = n$, the vector $(Z_1, \ldots, Z_l)^t$ follows a multinomial distribution with n trials and cell probabilities $p_1, \ldots, p_l$. To prove the first two of these assertions, let $n = n_1 + \cdots + n_l$. Then

$$\begin{aligned}
\Pr(Z_1 = n_1, \ldots, Z_l = n_l) &= \frac{\omega^n}{n!}e^{-\omega}\binom{n}{n_1, \ldots, n_l}\prod_{k=1}^l p_k^{n_k} \\
&= \prod_{k=1}^l \frac{\omega_k^{n_k}}{n_k!}e^{-\omega_k} \\
&= \prod_{k=1}^l \Pr(Z_k = n_k).
\end{aligned}$$

To prove the converse, divide the last string of equalities by the probability $\Pr(Z = n) = \omega^n e^{-\omega}/n!$.

The random process of assigning points to categories is termed coloring in the stochastic process literature. When there are just two colors, and only random points of one of the colors are tracked, then the process is termed random thinning. We will see examples of both coloring and thinning in the next section.

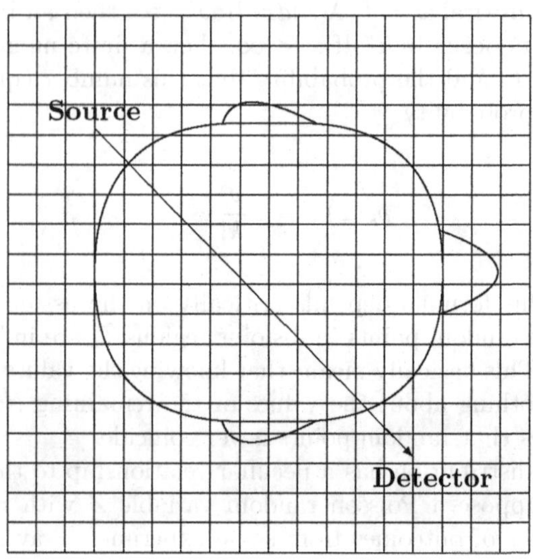

FIGURE 12.2. Cartoon of Transmission Tomography

12.11 Transmission Tomography

Problems in medical imaging often involve thousands of parameters. As an illustration of the MM algorithm, we treat maximum likelihood estimation in transmission tomography. Traditionally, transmission tomography images have been reconstructed by the methods of Fourier analysis. Fourier methods are fast but do not take into account the uncertainties of photon counts. Statistically based methods give better reconstructions with less patient exposure to harmful radiation.

The purpose of transmission tomography is to reconstruct the local attenuation properties of the object being imaged [13]. Attenuation is to be roughly equated with density. In medical applications, material such as bone is dense and stops or deflects X-rays (high-energy photons) better than soft tissue. With enough photons, even small gradations in soft tissue can be detected. A two-dimensional image is constructed from a sequence of photon counts. Each count corresponds to a projection line L drawn from an X-ray source through the imaged object to an X-ray detector. The average number of photons sent from the source along L to the detector is known in advance. The random number of photons actually detected is determined by the probability of a single photon escaping deflection or capture along L. Figure 12.2 shows one such projection line beamed through a cartoon of the human head.

To calculate this probability, let $\mu(x_1, x_2)$ be the intensity (or attenuation coefficient) of photon deflection or capture per unit length at position (x_1, x_2) in the plane. We can imagine that deflection or capture events occur completely randomly along L according to a Poisson process. The first such event effectively prevents the photon from being detected. Thus, the photon is detected with the Poisson probability $p_0(\omega) = e^{-\omega}$ of no such events, where

$$\omega = \int_L \mu(x_1, x_2) ds$$

is the line integral of $\mu(x_1, x_2)$ along L. In actual practice, X-rays are beamed through the object along a large number of different projection lines. We therefore face the inverse problem of reconstructing a function $\mu(x_1, x_2)$ in the plane from a large number of its measured line integrals. Imposing enough smoothness on $\mu(x_1, x_2)$, one can solve this classical deterministic problem by applying Radon transform techniques from Fourier analysis [13].

An alternative to the Fourier method is to pose an explicitly stochastic model and estimate its parameters by maximum likelihood [25, 26]. The MM algorithm suggests itself in this context. The stochastic model depends on dividing the object of interest into small nonoverlapping regions of constant attenuation called pixels. Typically the pixels are squares on a regular grid as depicted in Figure 12.2. The attenuation attributed to pixel j constitutes parameter θ_j of the model. Since there may be thousands of pixels, implementation of maximum likelihood algorithms such as scoring or Newton's method as discussed in Chapter 14 is out of the question.

To summarize our discussion, each observation Y_i is generated by beaming a stream of X-rays or high-energy photons from an X-ray source toward some detector on the opposite side of the object. The observation (or projection) Y_i counts the number of photons detected along the ith line of flight. Naturally, only a fraction of the photons are successfully transmitted from source to detector. If l_{ij} is the length of the segment of projection line i intersecting pixel j, then the probability of a photon escaping attenuation along projection line i is the exponentiated line integral $\exp(-\sum_j l_{ij}\theta_j)$.

In the absence of the intervening object, the number of photons generated and ultimately detected follows a Poisson distribution. We assume that the mean d_i of this distribution for projection line i is known. Ideally, detectors are long tubes aimed at the source. If a photon is deflected, then it is detected neither by the tube toward which it is initially headed nor by any other tube. In practice, many different detectors collect photons simultaneously from a single source. If we imagine coloring the tubes, then each photon is colored by the tube toward which it is directed. Each stream of colored photons is then thinned by capture or deflection. These considerations imply that the counts Y_i are independent and Poisson distributed with means $d_i \exp(-\sum_j l_{ij}\theta_j)$. It follows that we can express the loglikelihood of the observed data $Y_i = y_i$ as the finite sum

$$\sum_i \Big[- d_i e^{-\sum_j l_{ij}\theta_j} - y_i \sum_j l_{ij}\theta_j + y_i \ln d_i - \ln y_i! \Big]. \qquad (12.15)$$

Omitting irrelevant constants, we can rewrite the loglikelihood (12.15) more succinctly as

$$L(\theta) = -\sum_i f_i(l_i^t \theta),$$

where $f_i(s)$ is the convex function $d_i e^{-s} + y_i s$ and $l_i^t \theta = \sum_j l_{ij}\theta_j$ is the inner product of the attenuation parameter vector θ and the vector of intersection lengths l_i for projection i.

To generate a surrogate function, we majorize each $f_i(l_i^t\theta)$ according to the recipe (12.4). This gives the surrogate function

$$g(\theta \mid \theta_n) = -\sum_i \sum_j \frac{l_{ij}\theta_{nj}}{l_i^t \theta_n} f_i\Big(\frac{l_i^t \theta_n}{\theta_{nj}} \theta_j\Big) \qquad (12.16)$$

minorizing $L(\theta)$. Here the inner sum ranges over those pixels j with $l_{ij} > 0$. By construction, maximization of $g(\theta \mid \theta_n)$ separates into a sequence of one-dimensional problems, each of which can be solved approximately by one step of Newton's method. We will take up the details of this in Section 14.8.

The images produced by maximum likelihood estimation in transmission tomography look grainy. The cure is to enforce image smoothness by penalizing large differences between estimated attenuation parameters of neighboring pixels. Geman and McClure [9] recommend multiplying the likelihood of the data by a Gibbs prior $\pi(\theta)$. Equivalently we add the log prior

$$\ln \pi(\theta) = -\gamma \sum_{\{j,k\} \epsilon N} w_{jk} \psi(\theta_j - \theta_k)$$

to the loglikelihood, where γ and the weights w_{jk} are positive constants, N is a set of unordered pairs $\{j, k\}$ defining a neighborhood system, and $\psi(r)$ is called a potential function. This function should be large whenever $|r|$ is large. Neighborhoods have limited extent. For instance, if the pixels are squares, we might define the weights by $w_{jk} = 1$ for orthogonal nearest neighbors sharing a side and $w_{jk} = 1/\sqrt{2}$ for diagonal nearest neighbors sharing only a corner. The constant γ scales the overall strength assigned to the prior. The sum $L(\theta) + \ln \pi(\theta)$ is called the logposterior function; its maximum is the posterior mode.

Choice of the potential function $\psi(r)$ is the most crucial feature of the Gibbs prior. It is convenient to assume that $\psi(r)$ is even and strictly convex.

Strict convexity of $\psi(r)$ leads to the strict concavity of the log posterior function $L(\theta) + \ln \pi(\theta)$ and permits simple modification of the MM algorithm based on the surrogate function $g(\theta \mid \theta^n)$ defined by equation (12.16). Many potential functions exist satisfying these conditions. One natural example is $\psi(r) = r^2$. This choice unfortunately tends to deter the formation of boundaries. The gentler alternatives $\psi(r) = \sqrt{r^2 + \epsilon}$ for a small positive ϵ and $\psi(r) = \ln[\cosh(r)]$ are preferred in practice [10]. Problem 30 asks the reader to verify some of the properties of these two potential functions.

One adverse consequence of introducing a prior is that it couples pairs of parameters in the maximization step of the MM algorithm for finding the posterior mode. One can decouple the parameters by exploiting the convexity and evenness of the potential function $\psi(r)$ through the inequality

$$\begin{aligned}\psi(\theta_j - \theta_k) &= \psi\Big(\frac{1}{2}\Big[2\theta_j - \theta_{nj} - \theta_{nk}\Big] + \frac{1}{2}\Big[-2\theta_k + \theta_{nj} + \theta_{nk}\Big]\Big) \\ &\leq \frac{1}{2}\psi(2\theta_j - \theta_{nj} - \theta_{nk}) + \frac{1}{2}\psi(2\theta_k - \theta_{nj} - \theta_{nk}),\end{aligned}$$

which is strict unless $\theta_j + \theta_k = \theta_{nj} + \theta_{nk}$. This inequality allows us to redefine the surrogate function as

$$\begin{aligned}& g(\theta \mid \theta_n) \\ = \;& -\sum_i \sum_j \frac{l_{ij}\theta_{nj}}{l_i^t \theta_n} f_i\Big(\frac{l_i^t \theta_n}{\theta_{nj}}\theta_j\Big) \\ & -\frac{\gamma}{2} \sum_{\{j,k\}\epsilon N} w_{jk}[\psi(2\theta_j - \theta_{nj} - \theta_{nk}) + \psi(2\theta_k - \theta_{nj} - \theta_{nk})].\end{aligned}$$

Once again the parameters are separated, and the maximization step reduces to a sequence of one-dimensional problems. Maximizing $g(\theta \mid \theta_n)$ drives the logposterior uphill and eventually leads to the posterior mode.

12.12 Problems

1. Code and test any of the algorithms discussed in the text or problems of this chapter.

2. Prove that the majorization relation between functions is closed under the formation of sums, nonnegative products, maxima, limits, and composition with an increasing function. In what sense is the relation also transitive?

3. Demonstrate the majorizing and minorizing inequalities

$$x^q \;\leq\; q x_n^{q-1} x + (1-q) x_n^q$$

$$\begin{aligned}
\ln x &\le \frac{x}{x_n} + \ln x_n - 1 \\
x \ln x &\le \frac{x^2}{x_n} + x \ln x_n - x \\
\|x\|_2 &\ge \frac{x_n^t x}{\|x_n\|_2} \\
xy &\le \frac{y_n}{2x_n} x^2 + \frac{x_n}{2y_n} y^2 \\
-xy &\le -x_n y_n \left[1 + \ln\left(\frac{x}{x_n}\right) + \ln\left(\frac{y}{y_n}\right)\right] \\
\frac{1}{x} &\le \frac{1}{x_n} - \frac{x - x_n}{x_n^2} + \frac{(x - x_n)^2}{c^3}.
\end{aligned}$$

Determine the relevant domains of each variable q, x, x_n, y, y_n, and c, and check that equality occurs in each of the inequalities when $x = x_n$ and $y = y_n$ [12].

4. As alternatives to the fifth and sixth examples of Problem 3, demonstrate the majorizations

$$\begin{aligned}
xy &\le \frac{1}{2}(x^2 + y^2) + \frac{1}{2}(x_n - y_n)^2 - (x_n - y_n)(x - y) \\
-xy &\le \frac{1}{2}(x^2 + y^2) + \frac{1}{2}(x_n + y_n)^2 - (x_n + y_n)(x + y)
\end{aligned}$$

valid for all values of x, y, x_n, and y_n.

5. Based on Problem 4, devise an MM algorithm to minimize Rosenbrock's function

$$f(x) = 100(x_1^2 - x_2)^2 + (x_1 - 1)^2.$$

Show that up to an irrelevant constant $f(x)$ is majorized by the sum of the two functions

$$\begin{aligned}
g_1(x_1 \mid x_{n1}, x_{n2}) &= 200x_1^4 - [200(x_{n1}^2 + x_{n2}) - 1]x_1^2 - 2x_1 \\
g_2(x_2 \mid x_{n1}, x_{n2}) &= 200x_2^2 - 200(x_{n1}^2 + x_{n2})x_2.
\end{aligned}$$

Hence at each iteration one must minimize a quartic in x_1 and a quadratic in x_2. Implement this MM algorithm, and check whether it converges to the global minimum of $f(x)$ at $x = \mathbf{1}$.

6. Prove van Ruitenburg's [39] minorization

$$\ln x \ge -\frac{3x_n}{2x} - \frac{x}{2x_n} + \ln x_n + 2.$$

Deduce the further minorization

$$x \ln x \ge -\frac{3x_n}{2} - \frac{x^2}{2x_n} + x \ln x_n + 2x.$$

7. Suppose $p \in [1, 2]$ and $x_n \neq 0$. Verify the majorizing inequality

$$|x|^p \quad \leq \quad \frac{p}{2}|x_n|^{p-2}x^2 + \left(1 - \frac{p}{2}\right)|x_n|^p.$$

8. Let P be an orthogonal projection onto a subspace of R^m. Demonstrate the majorization

$$\|Px\|_2^2 \quad \leq \quad \|x - x_n\|_2^2 + 2(x - x_n)^t P x_n + \|Px_n\|_2^2.$$

9. The majorization

$$|x| \quad \leq \quad \frac{1}{2|x_n|}\left(x^2 + x_n^2\right) \qquad (12.17)$$

for $x_n \neq 0$ is a special case of Problem 7. Use (12.17) and the identity

$$\max\{x, y\} \quad = \quad \frac{1}{2}|x - y| + \frac{1}{2}x + \frac{1}{2}y$$

to majorize $\max\{x, y\}$ when $x_n \neq y_n$. Note that your majorization contains the product xy up to a negative factor. Describe how one can invoke Problem 3 or Problem 4 to separate x and y.

10. Prove the majorization (12.8) of the text.

11. Consider the function

$$f(x) \quad = \quad \frac{1}{4}x^4 - \frac{1}{2}x^2.$$

This function has global minima at $x = \pm 1$ and a local maximum at $x = 0$. Show that the function

$$g(x \mid x_n) \quad = \quad \frac{1}{4}x^4 + \frac{1}{2}x_n^2 - x x_n$$

majorizes $f(x)$ at x_n and leads to the MM update $x_{n+1} = \sqrt[3]{x_n}$. Prove that the alternative update $x_{n+1} = -\sqrt[3]{x_n}$ leads to the same value of $f(x)$, but the first update always converges while the second oscillates in sign and has two converging subsequences [5].

12. In the regression algorithm (12.9), let p tend to 0. If there are q predictors and all x_{ij} are nonzero, then show that $\alpha_{ij} = 1/q$. This leads to the update

$$\theta_{n+1,j} \quad = \quad \theta_{nj} + \frac{\sum_{i=1}^m x_{ij}(y_i - x_i^t \theta_n)}{q \sum_{i=1}^m x_{ij}^2}.$$

On the other hand, argue that cyclic coordinate descent yields the update

$$\theta_{n+1,j} \;=\; \theta_{nj} + \frac{\sum_{i=1}^m x_{ij}(y_i - x_i^t\theta_n)}{\sum_{i=1}^m x_{ij}^2},$$

which definitely takes larger steps.

13. A number μ is said to be a q quantile of the m numbers $x_1, \ldots, x_m$ if it satisfies

$$\frac{1}{m}\sum_{i:x_i \le \mu} 1 \;\ge\; q, \quad \frac{1}{m}\sum_{i:x_i \ge \mu} 1 \;\ge\; 1-q\,.$$

If we define

$$\rho_q(r) \;=\; \begin{cases} qr & r \ge 0 \\ (1-q)|r| & r < 0\,, \end{cases}$$

then demonstrate that μ is a q quantile if and only if μ minimizes the function $f_q(\omega) = \sum_{i=1}^m \rho_q(x_i - \omega)$. Medians correspond to the case $q = 1/2$.

14. Continuing Problem 13, show that the function $\rho_q(r)$ is majorized by the quadratic

$$\zeta_q(r \mid r_n) \;=\; \frac{1}{4}\left[\frac{r^2}{|r_n|} + (4q-2)r + |r_n|\right].$$

Deduce from this majorization the MM algorithm

$$\mu_{n+1} \;=\; \frac{m(2q-1) + \sum_{i=1}^m w_{ni}x_i}{\sum_{i=1}^m w_{ni}}$$

$$w_{ni} \;=\; \frac{1}{|x_i - \mu_n|}$$

for finding a q quantile. This interesting algorithm involves no sorting, only arithmetic operations.

15. The multivariate t-distribution has density

$$f(x) \;=\; \frac{\Gamma\left[\frac{\nu+p}{2}\right]}{\Gamma\left(\frac{\nu}{2}\right)(\nu\pi)^{p/2}(\det\Omega)^{1/2}\left[1 + \frac{1}{\nu}(x-\mu)^t\Omega^{-1}(x-\mu)\right]^{(\nu+p)/2}}$$

for all $x \in \mathsf{R}^p$. Here μ is the location vector, Ω is the positive definite scale matrix, and $\nu > 0$ is the degrees of freedom. Let $x_1, \ldots, x_m$ be a random sample from this density. Following Section 12.5, derive the MM algorithm

$$\begin{aligned}
\mu_{n+1} &= \frac{\sum_{i=1}^m w_{ni} x_i}{\sum_{i=1}^m w_{ni}} \\
\Omega_{n+1} &= \frac{1}{m} \sum_{i=1}^m w_{ni}(x_i - \mu_{n+1})(x_i - \mu_{n+1})^t
\end{aligned}$$

for estimating μ and Ω when ν is fixed, where

$$w_{ni} = \frac{\nu + p}{\nu + \delta_{ni}^2}, \quad \delta_{ni}^2 = (x_i - \mu_n)^t \Omega_n^{-1}(x_i - \mu_n).$$

(Hint: Extend the logic of Example 11.2.3.)

16. Continuing Problem 15, let $s_n = \sum_{i=1}^m w_{ni}$ be the sum of the case weights. Kent et al. [20] suggest an alternative algorithm that replaces the MM update for Ω by

$$\Omega_{n+1} = \frac{1}{s_n} \sum_{i=1}^m w_{ni}(x_i - \mu_{n+1})(x_i - \mu_{n+1})^t.$$

The update for μ remains the same. To justify this update, take $a = 1/(\nu + p)$ and prove the minorization

$$\begin{aligned}
&-\frac{m}{2} \ln \det \Omega - \frac{\nu + p}{2} \sum_{i=1}^m \ln[\nu + (x_i - \mu)^t \Omega^{-1}(x_i - \mu)] \\
&= -\frac{\nu + p}{2} \sum_{i=1}^m \ln\{(\det \Omega)^a [\nu + (x_i - \mu)^t \Omega^{-1}(x_i - \mu)]\} \\
&\geq -\sum_{i=1}^m \frac{w_i^n}{2(\det \Omega^n)^a} \{(\det \Omega)^a [\nu + (x_i - \mu)^t \Omega^{-1}(x_i - \mu)]\} + c_n
\end{aligned}$$

of the loglikelihood, where c_n is an irrelevant constant depending on neither μ nor Ω. Now argue as in Example 11.2.3. This alternative MM update exhibits much faster convergence than the traditional update of Problem 15.

17. Suppose for a small positive number ϵ we minimize the function

$$h_\epsilon(\theta) = \sum_{i=1}^m \left\{ \left[y_i - \sum_{j=1}^q x_{ij}\theta_j \right]^2 + \epsilon \right\}^{1/2}$$

instead of the function

$$h(\theta) = \sum_{i=1}^m |y_i - \mu_i(\theta)|.$$

Show that the MM algorithm of Section 12.5 applies with revised weights

$$w_i(\theta_n) = \frac{1}{\sqrt{[y_i - \mu_i(\theta_n)]^2 + \epsilon}}.$$

18. In ℓ_1 regression, show that the maximum likelihood estimate satisfies the equality

$$\sum_{i=1}^m \text{sgn}[y_i - \mu_i(\theta)] \nabla \mu_i(\theta) = \mathbf{0},$$

provided no residual $y_i - \mu_i(\theta) = 0$ and the regression functions $\mu_i(\theta)$ are differentiable. What are the corresponding equalities for ℓ_p regression and the modified ℓ_1 criterion (12.12)?

19. Problem 23 of Chapter 11 deals with minimizing the quadratic function $f(x) = \frac{1}{2}x^t Ax + b^t x + c$ subject to the constraints $x_i \geq 0$. If one drops the assumption that $A = (a_{ij})$ is positive definite, it is still possible to devise an MM algorithm. Define matrices A^+ and A^- with entries $\max\{a_{ij}, 0\}$ and $-\min\{a_{ij}, 0\}$, respectively. Based on the fifth and sixth majorizations of Problem 2, derive the MM updates

$$x_{n+1,i} = x_{n,i}\left[\frac{-b_i + \sqrt{b_i^2 + 4(A^+ x_n)_i (A^- x_n)_i}}{2(A^+ x_n)_i}\right]$$

of Sha et al [37]. All entries of the initial point x_0 should be positive.

20. Show that the loglikelihood $f(r) = \ln L(r)$ of the Bradley-Terry model in Section 12.6 is concave under the reparameterization $r_i = e^{\theta_i}$.

21. In the Bradley-Terry model of Section 12.6, suppose we want to include the possibility of ties [16]. One way of doing this is to write the probabilities of the three outcomes of i versus j as

$$\begin{aligned}
\Pr(i \text{ wins}) &= \frac{r_i}{r_i + r_j + \theta\sqrt{r_i r_j}} \\
\Pr(i \text{ ties}) &= \frac{\theta\sqrt{r_i r_j}}{r_i + r_j + \theta\sqrt{r_i r_j}} \\
\Pr(i \text{ loses}) &= \frac{r_j}{r_i + r_j + \theta\sqrt{r_i r_j}},
\end{aligned}$$

where $\theta > 0$ is an additional parameter to be estimated. Let y_{ij} represent the number of times i beats j and t_{ij} the number of times i ties j. Prove that the loglikelihood of the data is

$$L(\theta, r) = \sum_{i,j}\left(y_{ij} \ln \frac{r_i}{r_i + r_j + \theta\sqrt{r_i r_j}} + \frac{t_{ij}}{2} \ln \frac{\theta\sqrt{r_i r_j}}{r_i + r_j + \theta\sqrt{r_i r_j}}\right).$$

One way of maximizing $L(\theta, r)$ is to alternate between updating θ and r. Both of these updates can be derived from the perspective of the MM algorithm. Two minorizations are now involved. The first proceeds using the convexity of $-\ln t$ just as in the text. This produces a function involving $-\sqrt{r_i r_j}$ terms. Use the minorization

$$-\sqrt{r_i r_j} \quad \geq \quad -\frac{r_i}{2}\sqrt{\frac{r_{nj}}{r_{ni}}} - \frac{r_j}{2}\sqrt{\frac{r_{ni}}{r_{nj}}},$$

to minorize $L(\theta, r)$. Finally, determine r_{n+1} for θ fixed at θ_n and θ_{n+1} for r fixed at r_{n+1}. The details are messy, but the overall strategy is straightforward.

22. In the linear logistic model of Section 12.8, it is possible to separate parameters and avoid matrix inversion altogether. In constructing a minorizing function, first prove the inequality

$$\begin{aligned} \ln[1-\pi(\theta)] &= -\ln\left(1+e^{x_i^t\theta}\right) \\ &\geq -\ln\left(1+e^{x_i^t\theta_n}\right) - \frac{e^{x_i^t\theta} - e^{x_i^t\theta_n}}{1+e^{x_i^t\theta_n}}, \end{aligned}$$

with equality when $\theta = \theta_n$. This eliminates the log terms. Now apply the arithmetic-geometric mean inequality to the exponential functions $e^{x_i^t\theta}$ to separate parameters. Assuming that θ has n components and that there are k observations, show that these maneuvers lead to the minorizing function

$$g(\theta \mid \theta_n) \quad = \quad -\frac{1}{n}\sum_{i=1}^{k}\frac{e^{x_i^t\theta_n}}{1+e^{x_i^t\theta_n}}\sum_{j=1}^{n} e^{nx_{ij}(\theta_j-\theta_{nj})} + \sum_{i=1}^{k} y_i x_i^t\theta$$

up to a constant that does not depend on θ. Finally, prove that maximizing $g(\theta \mid \theta_n)$ consists in solving the transcendental equation

$$-\sum_{i=1}^{k}\frac{e^{x_i^t\theta_n}x_{ij}e^{-nx_{ij}\theta_{nj}}}{1+e^{x_i^t\theta_n}}e^{nx_{ij}\theta_j} + \sum_{i=1}^{k} y_i x_{ij} \quad = \quad 0$$

for each j. This can be accomplished numerically.

23. Prove that the loglikelihood (12.13) in the random graph model is concave under the reparameterization $p_i = e^{-q_i}$.

24. In many cases, a random directed graph is more meaningful than a random undirected graph. With a directed graph we deal with ordered pairs (j, k) called arcs rather than unordered pairs $\{j, k\}$ called edges. Let A be the set of arcs generated by a random graph. To each node

j we assign two nonnegative propensities p_j and q_j. The probability that an arc forms from j to k is then $p_j q_k/(1+p_j q_k)$. The object of this exercise is to design an MM algorithm to estimate the propensities. Show that the loglikelihood of A amounts to

$$L(p,q) = \sum_{(j,k)\in A} [\ln p_j + \ln q_k] - \sum_{(j,k)} \ln(1+p_j q_k).$$

Minorize $L(p,q)$ and derive the MM updates

$$\begin{aligned} p_{n+1,j} &= \left[\frac{p_{nj} o_j}{\sum_{k\neq j} \frac{q_{nk}}{1+p_{nj}q_{nk}}}\right]^{1/2} \\ q_{n+1,k} &= \left[\frac{q_{nk} i_k}{\sum_{j\neq k} \frac{p_{nj}}{1+p_{nj}q_{nk}}}\right]^{1/2}, \end{aligned}$$

where o_j is the number of outgoing arcs from j and i_k is the number of incoming arcs to k.

25. Consider the general posynomial of m variables

$$f(x) = \sum_{\alpha} c_\alpha x_1^{\alpha_1} \cdots x_m^{\alpha_m}$$

subject to the constraints $x_i > 0$ for each i. We can assume that at least one $\alpha_i > 0$ and at least one $\alpha_i < 0$ for every i. Otherwise, $f(x)$ can be reduced by sending x_i to ∞ or 0. Demonstrate that $f(x)$ is majorized by the sum

$$\begin{aligned} g(x \mid x_n) &= \sum_{i=1}^m g_i(x_i \mid x_n) \\ g_i(x_i \mid x_n) &= \sum_\alpha c_\alpha \left(\prod_{j=1}^m x_{nj}^{\alpha_j}\right) \frac{|\alpha_i|}{\|\alpha\|_1} \left(\frac{x_i}{x_{ni}}\right)^{\|\alpha\|_1 \operatorname{sgn}(\alpha_i)}, \end{aligned}$$

where $\|\alpha\|_1 = \sum_{j=1}^m |\alpha_j|$ and $\operatorname{sgn}(\alpha_i)$ is the sign function. To prove that the MM algorithm is well defined and produces iterates with positive entries, demonstrate that

$$\lim_{x_i\to\infty} g_i(x_i \mid x_n) = \lim_{x_i \to 0} g_i(x_i \mid x_n) = \infty.$$

Finally change variables by setting

$$\begin{aligned} y_i &= \ln x_i \\ h_i(y_i \mid x_n) &= g_i(x_i \mid x_n) \end{aligned}$$

for each i. Show that $h_i(y_i \mid x_n)$ is strictly convex in y_i and therefore possesses a unique minimum point. The latter property carries over to the surrogate function $g_i(x_i \mid x_n)$.

26. Devise MM algorithms based on Problem 25 to minimize the posynomials

$$\begin{aligned} f_1(x) &= \frac{1}{x_1x_2^2} + x_1x_2^2 \\ f_2(x) &= \frac{1}{x_1x_2^2} + x_1x_2. \end{aligned}$$

In the first case, demonstrate that the MM algorithm iterates according to

$$x_{n+1,1} = \sqrt[3]{\frac{x_{n1}^2}{x_{n2}^2}}, \quad x_{n+1,2} = \sqrt[3]{\frac{x_{n2}}{x_{n1}}}.$$

Furthermore, show that (a) $f_1(x)$ attains its minimum value of 2 whenever $x_1x_2^2 = 1$, (b) the MM algorithm converges after a single iteration to the value 2, and (c) the converged point x_1 depends on the initial point x_0. In the second case, demonstrate that the MM algorithm iterates according to

$$x_{n+1,1} = \sqrt[5]{\frac{x_{n1}^3}{x_{n2}^3}}, \quad x_{n+1,2} = \sqrt[5]{2\frac{x_{n2}^2}{x_{n1}^2}}.$$

Furthermore, show that (a) the infimum of $f_2(x)$ is 0, (b) the MM algorithm satisfies the identities

$$x_{n1}x_{n2}^{3/2} = 2^{3/10}, \quad x_{n+1,2} = 2^{2/25}x_{n2}$$

for all $n \geq 2$, and (c) the minimum value 0 is attained asymptotically with x_{n1} tending to 0 and x_{n2} tending to ∞.

27. A general posynomial of m variables can be represented as

$$h(y) = \sum_{\alpha \in S} c_\alpha e^{\alpha^t y}$$

in the parameterization $y_i = \ln x_i$. Here the index set $S \subset \mathsf{R}^m$ is finite and the coefficients c_α are positive. Show that $h(y)$ is strictly convex if and only if the power vectors $\{\alpha\}_{\alpha \in S}$ span R^m.

28. Show that the loglikelihood (12.15) for the transmission tomography model is concave. State a necessary condition for strict concavity in terms of the number of pixels and the number of projections.

29. In the maximization phase of the MM algorithm for transmission tomography without a smoothing prior, demonstrate that the exact solution of the one-dimensional equation

$$\frac{\partial}{\partial \theta_j} g(\theta \mid \theta_n) = 0$$

exists and is positive when $\sum_i l_{ij} d_i > \sum_i l_{ij} y_i$. Why would this condition typically hold in practice?

30. Prove that the functions $\psi(r) = \sqrt{r^2 + \epsilon}$ and $\psi(r) = \ln[\cosh(r)]$ are even, strictly convex, infinitely differentiable, and asymptotic to $|r|$ as $|r| \to \infty$.

31. In the dictionary model of motif finding [35], a DNA sequence is viewed as a concatenation of words independently drawn from a dictionary having the four letters A, C, G, and T. The words of the dictionary of length k have collective probability q_k. The EM algorithm offers one method of estimating the q_k. Omitting many details, the EM algorithm maximizes the function

$$Q(q \mid q_n) = \sum_{k=1}^{l} c_{nk} \ln q_k - \ln\Big(\sum_{k=1}^{l} k q_k\Big).$$

Here the constants c_{nk} are positive, l is the maximum word length, and maximization is performed subject to the constraints $q_k \geq 0$ for $k = 1, \ldots, l$ and $\sum_{k=1}^{l} q_k = 1$. Because this problem can not be solved in closed form, it is convenient to follow the EM minorization with a second minorization based on the inequality

$$\ln x \leq \ln y + x/y - 1. \tag{12.18}$$

Application of inequality (12.18) produces the minorizing function

$$h(q \mid q_n) = \sum_{k=1}^{l} c_{nk} \ln q_k - \ln\Big(\sum_{k=1}^{l} k q_{nk}\Big) - d_n \sum_{k=1}^{l} k q_k + 1$$

with $d_n = 1/(\sum_{k=1}^{l} k q_{nk})$.

(a) Show that the function $h(q \mid q_n)$ minorizes $Q(q \mid q_n)$.

(b) Maximize $h(q \mid q_n)$ using the method of Lagrange multipliers. At the current iteration, show that the solution has components

$$q_k = \frac{c_{nk}}{d_n k - \lambda}$$

for an unknown Lagrange multiplier λ.

(c) Using the constraints, prove that λ exists and is unique.

(d) Describe a reliable method for computing λ.

(e) As an alternative to the exact method, construct a quadratic approximation to $h(q \mid q_n)$ near q_n of the form

$$\frac{1}{2}(q - q_n)^t A (q - q_n) + b^t (q - q_n) + a.$$

In particular, what are A and b?

(f) Show that the quadratic approximation has maximum

$$q = q_n - A^{-1}\Big(b - \frac{\mathbf{1}^t A^{-1} b}{\mathbf{1}^t A^{-1}\mathbf{1}}\mathbf{1}\Big) \qquad (12.19)$$

subject to the constraint $\sum_{k=1}^{l}(q_k - q_{nk}) = 0$.

(g) In the dictionary model, demonstrate that the solution (12.19) takes the form

$$q_j = q_{nj} + \frac{(q_{nj})^2}{c_{nj}}\left[\frac{c_{nj}}{q_{nj}} - d_{nj} - \frac{1 - \sum_{k=1}^{l} d_{nk}\frac{(q_{nk})^2}{c_{nk}}}{\sum_{k=1}^{l}\frac{(q_{nk})^2}{c_{nk}}}\right]$$

for the jth component of q.

(h) Point out two potential pitfalls of this particular solution in conjunction with maximizing $h(q \mid q_n)$.

32. In the balanced ANOVA model with two factors, we estimate the parameter vector $\theta = (\mu, \alpha^t, \beta^t)^t$ by minimizing the sum of squares

$$f(\theta) = \sum_{i=1}^{I}\sum_{j=1}^{J}\sum_{k=1}^{K} w_{ijk}(y_{ijk} - \mu - \alpha_i - \beta_j)^2$$

with all weights $w_{ijk} = 1$. If some of the observations y_{ijk} are missing, then we take the corresponding weights to be 0. The missing observations are now irrelevant, but it is possible to replace each one by its predicted value

$$\hat{y}_{ijk} = \mu + \alpha_i + \beta_j.$$

If there are missing observations, de Leeuw [4] notes that

$$g(\theta \mid \theta_n) = \sum_{i=1}^{I}\sum_{j=1}^{J}\sum_{k=1}^{K}(z_{ijk} - \mu - \alpha_i - \beta_j)^2$$

majorizes $f(\theta)$ provided we define

$$z_{ijk} = \begin{cases} y_{ijk} & \text{for a regular observation} \\ \hat{y}_{ijk}(\theta_n) & \text{for a missing observation.} \end{cases}$$

Prove this fact and calculate the MM update of θ subject to the constraints $\sum_{i=1}^{I}\alpha_i = 0$ and $\sum_{j=1}^{J}\beta_j = 0$.

33. Inequality (12.5) generates many novel MM algorithms. Consider maximizing a loglikelihood $L(\theta) = -\sum_{i=1}^{p} f_i(c_i^t\theta)$, where $c_i^t\theta$ denotes an inner product and each function $f_i(r)$ is strictly convex. If each

$f_i(r)$ is twice continuously differentiable, then show that $L(\theta)$ has observed information matrix $-d^2L(\theta) = \sum_{i=1}^p f_i''(c_i^t\theta)c_ic_i^t$. Therefore $L(\theta)$ is strictly concave whenever each $f_i''(r)$ is strictly positive and the c_i span the domain of θ.

Now assume nonnegative constants λ_{ij} are given with $\lambda_{ij} > 0$ when $c_{ij} \neq 0$ and with $\sum_{j=1}^q \lambda_{ij} = 1$. If $S_i = \{j : \lambda_{ij} > 0\}$, then demonstrate the inequality

$$\begin{aligned} -\sum_{i=1}^p f_i(c_i^t\theta) &\geq -\sum_{i=1}^p \sum_{j\in S_i} \lambda_{ij} f_i\Big[\frac{c_{ij}}{\lambda_{ij}}(\theta_j - \theta_{nj}) + c_i^t\theta_n\Big] \\ &= g(\theta \mid \theta_n), \end{aligned} \tag{12.20}$$

with equality when $\theta = \theta_n$. Thus, $g(\theta \mid \theta_n)$ minorizes $L(\theta)$ and separates the parameters. Prove that $g(\theta \mid \theta_n)$ attains its maximum when

$$\sum_{i\in T_j} f_i'\Big[\frac{c_{ij}}{\lambda_{ij}}(\theta_j - \theta_{nj}) + c_i^t\theta_n\Big]c_{ij} = 0 \tag{12.21}$$

holds for all j, where $T_j = \{i : \lambda_{ij} > 0\}$. Check that one step of Newton's method provides the approximate maximum

$$\theta_{n+1,j} = \theta_{nj} - \Big[\sum_{i\in T_j} f_i''(c_i^t\theta_n)\frac{c_{ij}^2}{\lambda_{ij}}\Big]^{-1} \sum_{i\in T_j} f_i'(c_i^t\theta_n)c_{ij}. \tag{12.22}$$

The case $f_i(r_i) = (y_i - r_i)^2$ was studied in Section 12.4. The update (12.22) is a special case of the MM gradient algorithm explored in Section 14.8.

34. Continuing Problem 33, show that the functions

$$\begin{aligned} f_i(r) &= -y_ir + m_i\ln(1+e^r) \\ f_i(r) &= -m_ir + y_i\ln(1+e^r) \\ f_i(r) &= -y_ir + e^r \\ f_i(r) &= \nu_ir + y_ie^{-r} \end{aligned}$$

are strictly convex and provide up to sign loglikelihoods for the binomial, negative binomial, Poisson, and gamma densities, respectively. For the binomial, let y_i be the number of successes in m_i trials. For the negative binomial, let y_i be the number of trials until m_i successes. In both cases, $p = e^r/(e^r+1)$ is the success probability. For the Poisson, let e^r be the mean. Finally, for the gamma, let e^r be the scale parameter, assuming the shape parameter ν_i is fixed. For each density, equation (12.21) determining $\theta_{n+1,j}$ appears analytically intractable, but presumably the update (12.22) is viable. This problem has obvious implications for logistic and Poisson regression.

35. Continuing Problem 33, suppose $f_i(r) = |y_i - r|$. These nondifferentiable functions correspond to ℓ_1 regression. Show that the maximum of the surrogate function $Q(\theta \mid \theta_n)$ defined in (12.20) has jth component $\theta_{n+1,j}$ minimizing

$$\begin{aligned} s(\theta_j) &= \sum_{i \in T_j} w_i |d_i - \theta_j| \\ w_i &= |c_{ij}| \\ d_i &= \theta_{nj} + (y_i - c_i^t \theta_n) \frac{\lambda_{ij}}{c_{ij}}, \end{aligned}$$

where $T_j = \{i : \lambda_{ij} > 0\}$. The function $s(\theta_j)$ can be minimized by Edgeworth's algorithm as described in Section 16.5.1.

36. Luce's model [30, 31] is a convenient scheme for ranking items such as candidates in an election, consumer goods in a certain category, or academic departments in a reputational survey. Some people will be too lazy or uncertain to rank each and every item, preferring to focus on just their top choices. How can we use this form of limited voting to rank the entire set of items? A partial ranking by a person is a sequence of random choices $X_1, \ldots, X_m$, with X_1 the highest ranked item, X_2 the second highest ranked item, and so forth. If there are r items, then the index m may be strictly less than r; $m = 1$ is a distinct possibility. The data arrive at our doorstep as a random sample of s independent partial rankings, which we must integrate in some coherent fashion. One possibility is to adopt multinomial sampling without replacement. This differs from ordinary multinomial sampling in that once an item is chosen, it cannot be chosen again. However, remaining items are selected with the conditional probabilities dictated by the original sampling probabilities. Show that the likelihood under Luce's model reduces to

$$\begin{aligned} & \prod_{i=1}^{s} \Pr(X_{i1} = x_{i1}, \ldots, X_{im_i} = x_{im_i}) \\ = & \prod_{i=1}^{s} p_{x_{i1}} \prod_{j=1}^{m_i - 1} \frac{p_{x_{i,j+1}}}{\sum_{k \notin \{x_{i1}, \ldots, x_{ij}\}} p_k}, \end{aligned}$$

where x_{ij} is the jth choice out of m_i choices for person i and p_k is the multinomial probability assigned to item k. If we can estimate the p_k, then we can rank the items accordingly. Thus, the item with largest estimated probability is ranked first and so on.

The model has the added virtue of leading to straightforward estimation by the MM algorithm [16]. Use the supporting hyperplane inequality

$$-\ln t \quad \geq \quad -\ln t_n - \frac{1}{t_n}(t - t_n).$$

to generate the minorization

$$Q(p \mid p^n) \quad = \quad \sum_{i=1}^{s}\sum_{j=1}^{m_i} \ln p_{x_{ij}} - \sum_{i=1}^{s}\sum_{j=1}^{m_i-1} w_{ij}^n \sum_{k \notin \{x_{i1},\ldots,x_{ij}\}} p_k$$

of the loglikelihood up to an irrelevant constant. Specify the positive weights w_{ij}^n and derive the maximization step of the MM algorithm. Show that your update has the intuitive interpretation of equating the expected number of choices of item k to the observed number of choices of item k across all voters. Finally, generalize the model so that person i's choices are limited to a subset S_i of the items. For instance, in rating academic departments, some people may only feel competent to rank those departments in their state or region. What form does the MM algorithm take in this setting?

12.13 References

[1] Becker MP, Yang I, Lange K (1997) EM algorithms without missing data. *Stat Methods Med Res* 6:37–53

[2] Böhning D, Lindsay BG (1988) Monotonicity of quadratic approximation algorithms. *Ann Instit Stat Math* 40:641–663

[3] Bradley RA, Terry ME (1952), Rank analysis of incomplete block designs. *Biometrika*, 39:324–345

[4] De Leeuw J (1994) Block relaxation algorithms in statistics. in *Information Systems and Data Analysis*, Bock HH, Lenski W, Richter MM, Springer, New York, pp 308–325

[5] De Leeuw J (2006) Some majorization techniques. Preprint series, UCLA Department of Statistics.

[6] Dempster AP, Laird NM, Rubin DB (1977) Maximum likelihood from incomplete data via the EM algorithm (with discussion). *J Roy Stat Soc B* 39:1–38

[7] Dempster AP, Laird NM, Rubin DB (1980) Iteratively reweighted least squares for linear regression when the errors are normal/independent distributed. in *Multivariate Analysis V*, Krishnaiah PR, editor, North Holland, Amsterdam, pp 35–57

[8] De Pierro AR (1993) On the relation between the ISRA and EM algorithm for positron emission tomography. *IEEE Trans Med Imaging* 12:328–333

[9] Geman S, McClure D (1985) Bayesian image analysis: An application to single photon emission tomography. *Proc Stat Comput Sec*, Amer Stat Assoc, Washington, DC, pp 12–18

[10] Green P (1990) Bayesian reconstruction for emission tomography data using a modified EM algorithm. *IEEE Trans Med Imaging* 9:84–94

[11] Grimmett GR, Stirzaker DR (1992) *Probability and Random Processes*, 2nd ed. Oxford University Press, Oxford

[12] Heiser WJ (1995) Convergent computing by iterative majorization: theory and applications in multidimensional data analysis. in *Recent Advances in Descriptive Multivariate Analysis*, Krzanowski WJ, Clarendon Press, Oxford pp 157–189

[13] Herman GT (1980) *Image Reconstruction from Projections: The Fundamentals of Computerized Tomography.* Springer, New York

[14] Hoel PG, Port SC, Stone CJ (1971) *Introduction to Probability Theory.* Houghton Mifflin, Boston

[15] Huber PJ (1981) *Robust Statistics*, Wiley, New York

[16] Hunter DR (2004) MM algorithms for generalized Bradley-Terry models. *Annals Stat* 32:386–408

[17] Hunter DR, Lange K (2004) A tutorial on MM algorithms. *Amer Statistician* 58:30–37

[18] Karlin S, Taylor HM (1975) *A First Course in Stochastic Processes*, 2nd ed. Academic Press, New York

[19] Keener JP (1993), The Perron-Frobenius theorem and the ranking of football teams. *SIAM Review*, 35:80–93

[20] Kent JT, Tyler DE, Vardi Y (1994) A curious likelihood identity for the multivariate t-distribution. *Comm Stat Simulation* 23:441–453

[21] Kingman JFC (1993) *Poisson Processes.* Oxford University Press, Oxford

[22] Lange K (1995) A gradient algorithm locally equivalent to the EM algorithm. *J Roy Stat Soc B* 57:425–437

[23] Lange K (2002) *Mathematical and Statistical Methods for Genetic Analysis*, 2nd ed. Springer, New York

[24] Lange K (2004) *Optimization.* Springer, New York

[25] Lange K, Carson R (1984) EM reconstruction algorithms for emission and transmission tomography. *J Computer Assist Tomography* 8:306–316

[26] Lange K, Fessler JA (1995) Globally convergent algorithms for maximum a posteriori transmission tomography. *IEEE Trans Image Processing* 4:1430–1438

[27] Lange K, Hunter D, Yang I (2000) Optimization transfer using surrogate objective functions (with discussion). *J Computational Graphical Stat* 9:1–59

[28] Lange K, Little RJA, Taylor JMG (1989) Robust statistical modeling using the t distribution. *J Amer Stat Assoc* 84:881–896

[29] Lange K, Sinsheimer JS (1993) Normal/independent distributions and their applications in robust regression. *J Comp Graph Stat* 2:175–198

[30] Luce RD (1959) *Individual Choice Behavior: A Theoretical Analysis.* Wiley, New York

[31] Luce RD (1977) The choice axiom after twenty years. *J Math Psychology* 15:215–233

[32] McLachlan GJ, Krishnan T (2008) *The EM Algorithm and Extensions,* 2nd ed. Wiley, New York

[33] Merle G, Spath H (1974) Computational experiences with discrete L_p approximation. *Computing* 12:315–321

[34] Rao CR (1973) *Linear Statistical Inference and its Applications*, 2nd ed. Wiley, New York

[35] Sabatti C, Lange K (2002) Genomewide motif identification using a dictionary model. *Proceedings IEEE* 90:1803–1810

[36] Schlossmacher EJ (1973) An iterative technique for absolute deviations curve fitting. *J Amer Stat Assoc* 68:857–859

[37] Sha F, Saul LK, Lee DD (2003) Multiplicative updates for nonnegative quadratic programming in support vector machines. In *Advances in Neural Information Processing Systems 15*, Becker S, Thrun S, Obermayer K, editors, MIT Press, Cambridge, MA, pp 1065–1073

[38] Steele JM (2004) *The Cauchy-Schwarz Master Class: An Introduction to the Art of Inequalities.* Cambridge University Press and the Mathematical Association of America, Cambridge

[39] van Ruitenburg J (2005) Algorithms for parameter estimation in the Rasch model. Measurement and Research Department Reports 2005–4, CITO, Arnhem, Netherlands

[40] Wu TT, Lange K (2009) The MM alternative to EM. *Stat Sci* (in press)

13

The EM Algorithm

13.1 Introduction

Maximum likelihood is the dominant form of estimation in applied statistics. Because closed-form solutions to likelihood equations are the exception rather than the rule, numerical methods for finding maximum likelihood estimates are of paramount importance. In this chapter we study maximum likelihood estimation by the EM algorithm [2, 8, 9], a special case of the MM algorithm. At the heart of every EM algorithm is some notion of missing data. Data can be missing in the ordinary sense of a failure to record certain observations on certain cases. Data can also be missing in a theoretical sense. We can think of the E (expectation) step of the algorithm as filling in the missing data. This action replaces the loglikelihood of the observed data by a minorizing function, which is then maximized in the M step. Because the surrogate function is usually much simpler than the likelihood, we can often solve the M step analytically. The price we pay for this simplification is iteration. Reconstruction of the missing data is bound to be slightly wrong if the parameters do not already equal their maximum likelihood estimates.

One of the advantages of the EM algorithm is its numerical stability. As an MM algorithm, any EM algorithm leads to a steady increase in the likelihood of the observed data. Thus, the EM algorithm avoids wildly overshooting or undershooting the maximum of the likelihood along its current direction of search. Besides this desirable feature, the EM algorithm handles parameter constraints gracefully. Constraint satisfaction is by definition built into the solution of the M step. In contrast, competing methods of maximization must employ special techniques to cope with parameter constraints. The EM algorithm shares some of the negative features of the more general MM algorithm. For example, the EM algorithm often converges at an excruciatingly slow rate in a neighborhood of the maximum point. This rate directly reflects the amount of missing data in a problem. In the absence of concavity, there is also no guarantee that the EM algorithm will converge to the global maximum. The global maximum can usually be reached by starting the parameters at good but suboptimal estimates such as method-of-moments estimates or by choosing multiple random starting points.

K. Lange, *Numerical Analysis for Statisticians*, Statistics and Computing, 223
DOI 10.1007/978-1-4419-5945-4_13, © Springer Science+Business Media, LLC 2010

13.2 General Definition of the EM Algorithm

A sharp distinction is drawn in the EM algorithm between the observed, incomplete data Y and the unobserved, complete data X of a statistical experiment [2, 8, 13]. Some function $t(X) = Y$ collapses X onto Y. For instance, if we represent X as (Y, Z), with Z as the missing data, then t is simply projection onto the Y component of X. The definition of X is left up to the intuition and cleverness of the statistician. The general idea is to choose X so that maximum likelihood estimation becomes trivial for the complete data.

The complete data are assumed to have a probability density $f(x \mid \theta)$ that is a function of a parameter vector θ as well as of the value x of the complete data X. In the E step of the EM algorithm, we calculate the conditional expectation

$$Q(\theta \mid \theta_n) = \mathrm{E}[\ln f(X \mid \theta) \mid Y = y, \theta_n].$$

Here y is the actual observed data, and θ_n is the current estimated value of θ. In the M step, we maximize $Q(\theta \mid \theta_n)$ with respect to θ. This yields the new parameter estimate θ_{n+1}, and we repeat this two-step process until convergence occurs. Note that θ and θ_n play fundamentally different roles in $Q(\theta \mid \theta_n)$.

If $\ln g(y \mid \theta)$ denotes the loglikelihood of the observed data, then the EM algorithm enjoys the ascent property

$$\ln g(y \mid \theta_{n+1}) \geq \ln g(y \mid \theta_n).$$

The proof of this assertion in the next section unfortunately involves measure theory, so some readers may want to take it on faith and move on to the practical applications of the EM algorithm.

13.3 Ascent Property of the EM Algorithm

The information inequality at the heart of the EM algorithm is a consequence of Jensen's inequality, which relates convex functions to expectations. Section 11.5 reviews the various ways of defining a convex function.

Proposition 13.3.1 (Jensen's Inequality) *Assume that the values of the random variable W are confined to the possibly infinite interval (a, b). If $h(w)$ is convex on (a, b), then $\mathrm{E}[h(W)] \geq h[\mathrm{E}(W)]$, provided both expectations exist. For a strictly convex function $h(w)$, equality holds in Jensen's inequality if and only if $W = \mathrm{E}(W)$ almost surely.*

Proof: For the sake of simplicity, assume that $h(w)$ is differentiable. If we let $u = \mathrm{E}(W)$, then it is clear that u belongs to the interval (a, b). In the

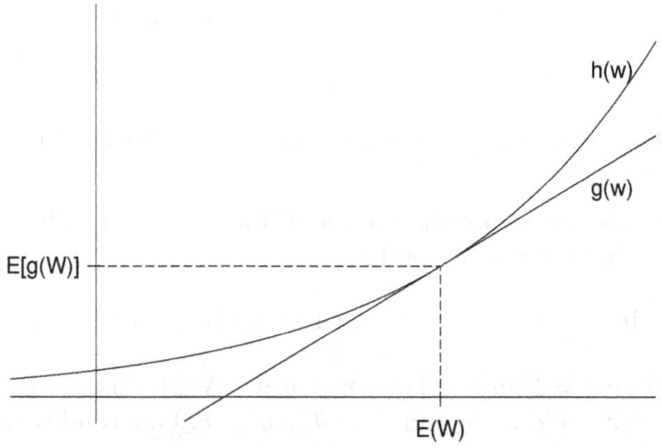

FIGURE 13.1. Geometric Proof of Jensen's Inequality

supporting hyperplane inequality

$$h(w) \geq h(u) + h'(u)(w - u), \tag{13.1}$$

substitute the random variable W for the point w and take expectations. It follows that

$$\mathrm{E}[h(W)] \geq h(u) + h'(u)[\mathrm{E}(W) - u] = h(u).$$

If $h(w)$ is strictly convex, then inequality (13.1) is strict whenever $w \neq u$. Hence, for equality to prevail, $W = \mathrm{E}(W)$ must hold with probability 1. Figure 13.1 visually summarizes the proof. Here the straight line $g(w)$ is tangent to the convex function $h(w)$ at $w = \mathrm{E}(W)$. ■

Proposition 13.3.2 (Information Inequality) *Let f and g be probability densities with respect to a measure μ. Suppose $f > 0$ and $g > 0$ almost everywhere relative to μ. If E_f denotes expectation with respect to the probability measure $f d\mu$, then $\mathrm{E}_f(\ln f) \geq \mathrm{E}_f(\ln g)$, with equality only if $f = g$ almost everywhere relative to μ.*

Proof: Because $-\ln(w)$ is a strictly convex function on $(0, \infty)$, Jensen's inequality applied to the random variable g/f implies

$$\begin{aligned}
\mathrm{E}_f(\ln f) - \mathrm{E}_f(\ln g) &= \mathrm{E}_f\left(-\ln \frac{g}{f}\right) \\
&\geq -\ln \mathrm{E}_f\left(\frac{g}{f}\right)
\end{aligned}$$

$$
\begin{aligned}
&= -\ln \int \frac{g}{f} f d\mu \\
&= -\ln \int g \, d\mu \\
&= 0.
\end{aligned}
$$

Equality holds only if $g/f = \mathrm{E}_f(g/f)$ almost everywhere relative to μ. But $\mathrm{E}_f(g/f) = 1$. ■

To prove the ascent property of the EM algorithm, it suffices to demonstrate the minorization inequality

$$
\ln g(y \mid \theta) \geq Q(\theta \mid \theta_n) + \ln g(y \mid \theta_n) - Q(\theta_n \mid \theta_n), \qquad (13.2)
$$

where $Q(\theta \mid \theta_n) = \mathrm{E}[\ln f(X \mid \theta) \mid Y = y, \theta_n]$. With this end in mind, note that both $f(x \mid \theta)/g(y \mid \theta)$ and $f(x \mid \theta_n)/g(y \mid \theta_n)$ are conditional densities of X on the set $\{x : t(x) = y\}$ with respect to some measure μ_y. The information inequality now indicates that

$$
\begin{aligned}
Q(\theta \mid \theta_n) - \ln g(y \mid \theta) &= \mathrm{E}\left\{\ln\left[\frac{f(X \mid \theta)}{g(Y \mid \theta)}\right] \mid Y = y, \theta_n\right\} \\
&\leq \mathrm{E}\left\{\ln\left[\frac{f(X \mid \theta_n)}{g(Y \mid \theta_n)}\right] \mid Y = y, \theta_n\right\} \\
&= Q(\theta_n \mid \theta_n) - \ln g(y \mid \theta_n).
\end{aligned}
$$

Given the minorization (13.2), the proof of the ascent property parallels the proof of the descent property (12.2). Strict inequality occurs when the conditional density $f(x \mid \theta)/g(y \mid \theta)$ differs at the parameter points θ_n and θ_{n+1} or when $Q(\theta_{n+1} \mid \theta_n) > Q(\theta_n \mid \theta_n)$.

13.3.1 Technical Note

The preceding proof is a little vague as to the meaning of the conditional density $f(x \mid \theta)/g(y \mid \theta)$ and its associated measure μ_y. Commonly the complete data decompose as $X = (Y, Z)$, where Z is considered the missing data and $t(Y, Z) = Y$ is projection onto the observed data. Suppose (Y, Z) has joint density $f(y, z \mid \theta)$ relative to a product measure $\omega \times \mu \, (y, z)$; ω and μ are typically Lebesgue measure or counting measure. In this framework, we define $g(y \mid \theta) = \int f(y, z \mid \theta) d\mu(z)$ and set $\mu_y = \mu$. The function $g(y \mid \theta)$ serves as a density relative to ω. To check that these definitions make sense, it suffices to prove that $\int h(y, z) f(y, z \mid \theta)/g(y \mid \theta) \, d\mu(z)$ is a version of the conditional expectation $\mathrm{E}[h(Y, Z) \mid Y = y]$ for every well-behaved function $h(y, z)$. This assertion can be verified by showing

$$
\mathrm{E}\{1_S(Y) \, \mathrm{E}[h(Y, Z) \mid Y]\} = \mathrm{E}[1_S(Y) h(Y, Z)]
$$

for every measurable set S. With

$$\mathrm{E}[h(Y,Z) \mid Y=y] = \int h(y,z)\frac{f(y,z \mid \theta)}{g(y \mid \theta)}\,d\mu(z),$$

we calculate

$$\begin{aligned}\mathrm{E}\{1_S(Y)\,\mathrm{E}[h(Y,Z) \mid Y]\} &= \int_S \int h(y,z)\frac{f(y,z \mid \theta)}{g(y \mid \theta)}\,d\mu(z)g(y \mid \theta)\,d\omega(y) \\ &= \int_S \int h(y,z)f(y,z \mid \theta)\,d\mu(z)\,d\omega(y) \\ &= \mathrm{E}[1_S(Y)h(Y,Z)].\end{aligned}$$

Hence in this situation, $f(x \mid \theta)/g(y \mid \theta)$ is indeed the conditional density of X given $Y = y$.

13.4 Missing Data in the Ordinary Sense

The most common application of the EM algorithm is to data missing in the ordinary sense. For example, Problem 32 of Chapter 12 considers a balanced ANOVA model with two factors. Missing observations in this setting break the symmetry that permits explicit solution of the likelihood equations. Thus, there is ample incentive for filling in the missing observations. If the observations follow an exponential model, and missing data are missing completely at random, then the EM algorithm replaces the sufficient statistic of each missing observation by its expected value.

The density of a random variable Y from an exponential family can be written as

$$f(y \mid \theta) = g(y)e^{\beta(\theta)+h(y)^t\gamma(\theta)} \tag{13.3}$$

relative to some measure ν [3, 11]. The normal, Poisson, binomial, negative binomial, gamma, beta, and multinomial families are prime examples of exponential families. The function $h(y)$ in equation (13.3) is the sufficient statistic. The maximum likelihood estimate of the parameter vector θ depends on an observation y only through $h(y)$. Predictors of y are incorporated into the functions $\beta(\theta)$ and $\gamma(\theta)$.

To fill in a missing observation y, we take the ordinary expectation

$$\mathrm{E}[\ln f(Y \mid \theta) \mid \theta_n] = \mathrm{E}[\ln g(Y) \mid \theta_n] + \beta(\theta) + \mathrm{E}[h(Y) \mid \theta_n]^t\gamma(\theta)$$

of the complete data loglikelihood. This function is added to the loglikelihood of the regular observations $y_1, \ldots, y_m$ to generate the surrogate function $Q(\theta \mid \theta_n)$. For example, if a typical observation is normally distributed with mean $\mu(\alpha)$ and variance σ^2, then θ is the vector $(\alpha, \sigma^2)^t$, and

$$\begin{aligned} \mathrm{E}[\ln f(Y \mid \theta) \mid \theta_n] &= \ln \frac{1}{\sqrt{2\pi\sigma^2}} - \frac{1}{2\sigma^2} \mathrm{E}\{[Y - \mu(\alpha)]^2 \mid \theta_n\} \\ &= \ln \frac{1}{\sqrt{2\pi\sigma^2}} - \frac{1}{2\sigma^2} \{\sigma_n^2 + [\mu(\alpha_n) - \mu(\alpha)]^2\}. \end{aligned}$$

Once we have filled in the missing data, we can estimate α without reference to σ^2. This is accomplished by adding each square $[\mu_i(\alpha_n) - \mu_i(\alpha)]^2$ corresponding to a missing observation y_i to the sum of squares for the actual observations and then minimizing the entire sum over α. In classical models such as balanced ANOVA, the M step is exact. Once the iterates α_n converge to their limit $\hat{\alpha}$, we can estimate σ^2 in one step by the formula

$$\widehat{\sigma^2} = \frac{1}{m} \sum_{i=1}^{m} [y_i - \mu_i(\hat{\alpha})]^2$$

using only the observed y_i. The reader is urged to work Problem 32 of Chapter 12 to see the whole process in action.

Problem 13 of this chapter takes up the greater challenge of partially missing data for bivariate normal observations. It is important to salvage partial observations rather than discard them and estimate parameters from complete cases alone. Because the missing components are correlated with the observed components, replacing the sufficient statistic of each missing observation by its expected value is no longer valid. The book [8] contains an extended discussion of data imputation techniques that build on Problem 13.

13.5 Bayesian EM

If a prior $\pi(\theta)$ is imposed on the parameter vector θ, then $L(\theta) + \ln \pi(\theta)$ is the logposterior function. Its maximum occurs at the posterior mode. The posterior mode can be found by defining the surrogate function

$$\begin{aligned} Q(\theta \mid \theta_n) &= \mathrm{E}[\ln f(X \mid \theta) + \ln \pi(\theta) \mid Y, \theta_n] \\ &= \mathrm{E}[\ln f(X \mid \theta) \mid Y, \theta_n] + \ln \pi(\theta). \end{aligned}$$

Thus, in the E step of the Bayesian algorithm, one simply adds the logprior to the usual surrogate function. The M-step strategy of maximizing $Q(\theta \mid \theta_n)$ drives the logposterior function uphill. Because the logprior often complicates the M step, it is helpful in some applications to minorize it separately. This is the strategy followed in Example 12.11 on transmission tomography. Alternatively, one can employ the MM gradient algorithm studied in Chapter 14 to maximize the logposterior approximately. We will illustrate the Bayesian version of the EM algorithm in our next example.

13.6 Allele Frequency Estimation

The ABO and Rh genetic loci are usually typed in matching blood donors to blood recipients. The ABO locus exhibits the three alleles A, B, and O and the four observable phenotypes A, B, AB, and O. These phenotypes arise because each person inherits two alleles, one from his mother and one from his father, and the alleles A and B are genetically dominant to allele O. Dominance amounts to a masking of the O allele by the presence of an A or B allele. For instance, a person inheriting an A allele from one parent and an O allele from the other parent is said to have genotype A/O and is phenotypically indistinguishable from a person inheriting an A allele from both parents. This second person is said to have genotype A/A.

The EM algorithm for estimating the population frequencies (proportions) of the three alleles involves an interplay between observed phenotypes and underlying unobserved genotypes. As just noted, both genotypes A/O and A/A correspond to the same phenotype A. Likewise, phenotype B corresponds to either genotype B/O or genotype B/B. Phenotypes AB and O correspond to the single genotypes A/B and O/O, respectively.

As a concrete example, Clarke et al. [1] noted that among their population sample of $n = 521$ duodenal ulcer patients, a total of $n_A = 186$ had phenotype A, $n_B = 38$ had phenotype B, $n_{AB} = 13$ had phenotype AB, and $n_O = 284$ had phenotype O. If we want to estimate the frequencies p_A, p_B, and p_O of the three different alleles from this sample, then we can employ the EM algorithm with the four phenotype counts as the observed data Y and the underlying six genotype counts $n_{A/A}$, $n_{A/O}$, $n_{B/B}$, $n_{B/O}$, $n_{A/B} = n_{AB}$, and $n_{O/O} = n_O$ as the complete data X [12]. Note that the allele frequencies are nonnegative and satisfy the constraint $p_A + p_B + p_O = 1$. Furthermore, the classical Hardy-Weinberg law of population genetics specifies that each genotype frequency equals the product of the corresponding allele frequencies with an extra factor of 2 thrown in to account for ambiguity in parental source when the two alleles differ. For example, genotype A/A has frequency p_A^2, and genotype A/O has frequency $2p_Ap_O$.

With these preliminaries in mind, the complete data loglikelihood becomes

$$\begin{aligned}\ln f(X \mid p) &= n_{A/A} \ln p_A^2 + n_{A/O} \ln(2p_Ap_O) + n_{B/B} \ln p_B^2 \\ &\quad + n_{B/O} \ln(2p_Bp_O) + n_{AB} \ln(2p_Ap_B) + n_O \ln p_O^2 \\ &\quad + \ln \binom{n}{n_{A/A}\ n_{A/O}\ n_{B/B}\ n_{B/O}\ n_{AB}\ n_O}. \qquad (13.4)\end{aligned}$$

In the E step of the EM algorithm we take the expectation of $\ln f(X \mid p)$ conditional on the observed counts n_A, n_B, n_{AB}, and n_O and the current parameter vector $p_m = (p_{mA}, p_{mB}, p_{mO})^t$. It is obvious that

$$\begin{aligned} \mathrm{E}(n_{AB} \mid Y, p_m) &= n_{AB} \\ \mathrm{E}(n_O \mid Y, p_m) &= n_O. \end{aligned}$$

A moment's reflection also yields

$$\begin{aligned} n_{mA/A} &= \mathrm{E}(n_{A/A} \mid Y, p_m) \\ &= n_A \frac{p_{mA}^2}{p_{mA}^2 + 2p_{mA}p_{mO}} \\ n_{mA/O} &= \mathrm{E}(n_{A/O} \mid Y, p_m) \\ &= n_A \frac{2p_{mA}p_{mO}}{p_{mA}^2 + 2p_{mA}p_{mO}}. \end{aligned}$$

The conditional expectations $n_{mB/B}$ and $n_{mB/O}$ are given by similar expressions.

The M step of the EM algorithm maximizes the $Q(p \mid p_m)$ function derived from equation (13.4) by replacing $n_{A/A}$ by $n_{mA/A}$, and so forth. Maximization of $Q(p \mid p_m)$ can be accomplished by finding a stationary point of the Lagrangian as sketched in Example (11.3.1). The solution of the stationarity equations is

$$\begin{aligned} p_{m+1,A} &= \frac{2n_{mA/A} + n_{mA/O} + n_{AB}}{2n} \\ p_{m+1,B} &= \frac{2n_{mB/B} + n_{mB/O} + n_{AB}}{2n} \\ p_{m+1,O} &= \frac{n_{mA/O} + n_{mB/O} + 2n_O}{2n}. \end{aligned}$$

In other words, the EM update is identical to a form of gene counting in which the unknown genotype counts are imputed based on the current allele frequency estimates. Table 13.1 shows the progress of the EM iterates starting from the initial guesses $p_{0A} = 0.3$, $p_{0B} = 0.2$, and $p_{0O} = 0.5$. The EM updates are simple enough to carry out on a pocket calculator. Convergence occurs quickly in this example.

The Dirichlet distribution is the conjugate prior to the multinomial distribution. If we impose the Dirichlet prior

$$\frac{\Gamma(\alpha_A + \alpha_B + \alpha_O)}{\Gamma(\alpha_A)\Gamma(\alpha_B)\Gamma(\alpha_O)} p_A^{\alpha_A - 1} p_B^{\alpha_B - 1} p_O^{\alpha_O - 1}$$

in this example, then it is straightforward to prove that the EM algorithm for finding the posterior mode iterates according to

$$p_{m+1,A} = \frac{2n_{mA/A} + n_{mA/O} + n_{AB} + \alpha_A - 1}{2n + \alpha - 3}$$

TABLE 13.1. Iterations for ABO Duodenal Ulcer Data

Iteration m	p_{mA}	p_{mB}	p_{mO}
0	0.3000	0.2000	0.5000
1	0.2321	0.0550	0.7129
2	0.2160	0.0503	0.7337
3	0.2139	0.0502	0.7359
4	0.2136	0.0501	0.7363
5	0.2136	0.0501	0.7363

$$
\begin{aligned}
p_{m+1,B} &= \frac{2n_{mB/B} + n_{mB/O} + n_{AB} + \alpha_B - 1}{2n + \alpha - 3} \\
p_{m+1,O} &= \frac{n_{mA/O} + n_{mB/O} + 2n_O + \alpha_O - 1}{2n + \alpha - 3},
\end{aligned} \tag{13.5}
$$

where $\alpha = \alpha_A + \alpha_B + \alpha_O$. Thus, imposing the prior can be interpreted as adding $\alpha_A - 1$ pseudo-counts to the A alleles, $\alpha_B - 1$ pseudo-counts to the B alleles, and $\alpha_O - 1$ pseudo-counts to the O alleles.

13.7 Clustering by EM

The k-means clustering algorithm discussed in Example 11.6.3 makes hard choices in cluster assignment. The alternative of soft choices is possible with admixture models [10, 14]. An admixture probability density $h(y)$ can be written as a convex combination

$$
h(y) = \sum_{j=1}^{k} \pi_j h_j(y), \tag{13.6}
$$

where the π_j are nonnegative probabilities that sum to 1 and $h_j(y)$ is the probability density of group j. According to Bayes' rule, the posterior probability that an observation y belongs to group j equals the ratio

$$
\frac{\pi_j h_j(y)}{\sum_{i=1}^{k} \pi_i h_i(y)}. \tag{13.7}
$$

If hard assignment is necessary, then the rational procedure is to assign y to the group with highest posterior probability.

Suppose the observations $y_1, \ldots, y_m$ represent a random sample from the admixture density (13.6). In practice we want to estimate the admixture proportions and whatever further parameters θ characterize the $h_j(y \mid \theta)$. The EM algorithm is natural in this context with group membership as the missing data. If we let z_{ij} be an indicator specifying whether observation

y_i comes from group j, then the complete data loglikelihood amounts to

$$\sum_{i=1}^{m}\sum_{j=1}^{k} z_{ij}\Big[\ln \pi_j + \ln h_j(y_i \mid \theta)\Big].$$

To find the surrogate function, we must find the conditional expectation w_{ij} of z_{ij}. But this reduces to the Bayes' rule (13.7) with θ fixed at θ_n and π_j fixed at π_{nj}, where as usual n indicates iteration number. Note that the property $\sum_{j=1}^k z_{ij} = 1$ entails the property $\sum_{j=1}^k w_{ij} = 1$.

Fortunately, the E step of the EM algorithm separates the π parameters from the θ parameters. The problem of maximizing

$$\sum_{j=1}^{k} c_j \ln \pi_j$$

with $c_j = \sum_{i=1}^m w_{ij}$ should be familiar by now. Since $\sum_{j=1}^k c_j = m$, Example (11.3.1) shows that

$$\pi_{n+1,j} \quad = \quad \frac{c_j}{m}.$$

We now undertake estimation of the remaining parameters assuming the groups are normally distributed with a common variance matrix Ω but different mean vectors $\mu_1, \ldots, \mu_k$. The pertinent part of the surrogate function is

$$\begin{aligned} & \sum_{i=1}^{m}\sum_{j=1}^{k} w_{ij}\Big[-\frac{1}{2}\ln\det\Omega - \frac{1}{2}(y_i-\mu_j)^t\Omega^{-1}(y_i-\mu_j)\Big] \\ = \quad & -\frac{m}{2}\ln\det\Omega - \frac{1}{2}\sum_{j=1}^{k}\sum_{i=1}^{m} w_{ij}(y_i-\mu_j)^t\Omega^{-1}(y_i-\mu_j) \qquad (13.8) \\ = \quad & -\frac{m}{2}\ln\det\Omega - \frac{1}{2}\operatorname{tr}\Big[\Omega^{-1}\sum_{j=1}^{k}\sum_{i=1}^{m} w_{ij}(y_i-\mu_j)(y_i-\mu_j)^t\Big]. \end{aligned}$$

Differentiating the second line of (13.8) with respect to μ_j gives the equation

$$\sum_{i=1}^{m} w_{ij}\Omega^{-1}(y_i-\mu_j) \quad = \quad \mathbf{0}$$

with solution

$$\mu_{n+1,j} \quad = \quad \frac{\sum_{i=1}^m w_{ij} y_i}{\sum_{i=1}^m w_{ij}}.$$

Maximization of the surrogate (13.8) with respect to Ω can be rephrased as maximization of

$$-\frac{m}{2}\ln\det\Omega - \frac{1}{2}\operatorname{tr}(\Omega^{-1}M)$$

for the choice

$$M = \sum_{j=1}^{k}\sum_{i=1}^{m} w_{ij}(y_i - \mu_{n+1,j})(y_i - \mu_{n+1,j})^t.$$

Abstractly this is just the problem we faced in Example 11.2.3. Inspection of the arguments there shows that

$$\Omega_{n+1} = \frac{1}{m}M. \tag{13.9}$$

There is no guarantee of a unique mode in this model. Fortunately, k-means clustering generates good starting values for the parameters. The cluster centers provide the group means. If we set w_{ij} equal to 1 or 0 depending on whether observation i belongs to cluster j or not, then the matrix (13.9) serves as an initial guess of the common variance matrix. The initial admixture proportion π_j can be taken to be the proportion of the observations assigned to cluster j.

13.8 Transmission Tomography

Derivation of the EM algorithm for transmission tomography is more challenging [7]. In this instance, the EM and MM algorithms differ. The MM algorithm is easier to derive and computationally more efficient. In other examples, the opposite is true.

In the transmission tomography example of Section 12.11, it is natural to view the missing data as the number of photons X_{ij} entering each pixel j along each projection line i. These random variables supplemented by the observations Y_i constitute the complete data. If projection line i does not intersect pixel j, then $X_{ij} = 0$. Although X_{ij} and $X_{ij'}$ are not independent, the collection $\{X_{ij}\}_j$ indexed by projection i is independent of the collection $\{X_{i'j}\}_j$ indexed by another projection i'. This allows us to work projection by projection in writing the complete data likelihood. We will therefore temporarily drop the projection subscript i and relabel pixels, starting with pixel 1 adjacent to the source and ending with pixel $m-1$ adjacent to the detector. In this notation X_1 is the number of photons leaving the source, X_j is the number of photons entering pixel j, and $X_m = Y$ is the number of photons detected.

By assumption X_1 follows a Poisson distribution with mean d. Conditional on $X_1, \ldots, X_j$, the random variable X_{j+1} is binomially distributed with X_j trials and success probability $e^{-l_j\theta_j}$. In other words, each of the X_j photons entering pixel j behaves independently and has a chance $e^{-l_j\theta_j}$ of avoiding attenuation in pixel j. It follows that the complete data loglikelihood for the current projection is

$$\begin{aligned}&-d + X_1 \ln d - \ln X_1! \\ &+ \sum_{j=1}^{m-1}\left[\ln\binom{X_j}{X_{j+1}} + X_{j+1}\ln e^{-l_j\theta_j} + (X_j - X_{j+1})\ln(1-e^{-l_j\theta_j})\right].\end{aligned} \tag{13.10}$$

To perform the E step of the EM algorithm, we need only compute the conditional expectations $\mathrm{E}(X_j \mid X_m = y, \theta)$, $j = 1,\ldots,m$. The conditional expectations of other terms such as $\ln\binom{X_j}{X_{j+1}}$ appearing in (13.10) are irrelevant in the subsequent M step.

Reasoning as earlier, we infer that the unconditional mean of X_j is

$$\mu_j \;=\; \mathrm{E}(X_j) = de^{-\sum_{k=1}^{j-1} l_k\theta_k}$$

and that the distribution of X_m conditional on X_j is binomial with X_j trials and success probability

$$\frac{\mu_m}{\mu_j} \;=\; e^{-\sum_{k=j}^{m-1} l_k\theta_k}.$$

In view of our remarks about random thinning in Chapter 12, the joint probability density of X_j and X_m therefore reduces to

$$\Pr(X_j = x_j, X_m = x_m) \;=\; e^{-\mu_j}\frac{\mu_j^{x_j}}{x_j!}\binom{x_j}{x_m}\left(\frac{\mu_m}{\mu_j}\right)^{x_m}\left(1-\frac{\mu_m}{\mu_j}\right)^{x_j-x_m},$$

and the conditional probability density of X_j given X_m becomes

$$\begin{aligned}\Pr(X_j = x_j \mid X_m = x_m) \;&=\; \frac{e^{-\mu_j}\frac{\mu_j^{x_j}}{x_j!}\binom{x_j}{x_m}\left(\frac{\mu_m}{\mu_j}\right)^{x_m}\left(1-\frac{\mu_m}{\mu_j}\right)^{x_j-x_m}}{e^{-\mu_m}\frac{\mu_m^{x_m}}{x_m!}} \\ &=\; e^{-(\mu_j-\mu_m)}\frac{(\mu_j-\mu_m)^{x_j-x_m}}{(x_j-x_m)!}.\end{aligned}$$

In other words, conditional on X_m, the difference $X_j - X_m$ follows a Poisson distribution with mean $\mu_j - \mu_m$. This implies in particular that

$$\begin{aligned}\mathrm{E}(X_j \mid X_m) \;&=\; \mathrm{E}(X_j - X_m \mid X_m) + X_m \\ &=\; \mu_j - \mu_m + X_m.\end{aligned}$$

Reverting to our previous notation, it is now possible to assemble the function $Q(\theta \mid \theta_n)$ of the E step. Define

$$\begin{aligned}M_{ij} \;&=\; d_i\left(e^{-\sum_{k\in S_{ij}} l_{ik}\theta_{nk}} - e^{-\sum_k l_{ik}\theta_{nk}}\right) + y_i \\ N_{ij} \;&=\; d_i\left(e^{-\sum_{k\in S_{ij}\cup\{j\}} l_{ik}\theta_{nk}} - e^{-\sum_k l_{ik}\theta_{nk}}\right) + y_i,\end{aligned}$$

where S_{ij} is the set of pixels between the source and pixel j along projection i. If j' is the next pixel after pixel j along projection i, then

$$\begin{aligned} M_{ij} &= \mathrm{E}(X_{ij} \mid Y_i = y_i, \theta_n) \\ N_{ij} &= \mathrm{E}(X_{ij'} \mid Y_i = y_i, \theta_n). \end{aligned}$$

In view of expression (13.10), we find

$$Q(\theta \mid \theta_n) = \sum_i \sum_j \Big[- N_{ij} l_{ij} \theta_j + (M_{ij} - N_{ij}) \ln(1 - e^{-l_{ij}\theta_j}) \Big]$$

up to an irrelevant constant.

If we try to maximize $Q(\theta \mid \theta_n)$ by setting its partial derivatives equal to 0, we get for pixel j the equation

$$-\sum_i N_{ij} l_{ij} + \sum_i \frac{(M_{ij} - N_{ij}) l_{ij}}{e^{l_{ij}\theta_j} - 1} = 0. \tag{13.11}$$

This is an intractable transcendental equation in the single variable θ_j, and the M step must be solved numerically, say by Newton's method. It is straightforward to check that the left-hand side of equation (13.11) is strictly decreasing in θ_j and has exactly one positive solution. Thus, the EM algorithm like the MM algorithm has the advantages of decoupling the parameters in the likelihood equations and of satisfying the natural boundary constraints $\theta_j \geq 0$. The MM algorithm is preferable to the EM algorithm because the MM algorithm involves far fewer exponentiations in defining its surrogate function.

13.9 Factor Analysis

In some instances, the missing data framework of the EM algorithm offers the easiest way to exploit convexity in deriving an MM algorithm. The complete data for a given problem are often fairly natural, and the difficulty in deriving an EM algorithm shifts toward specifying the E step. Statisticians are particularly adept at calculating complicated conditional expectations connected with sampling distributions. We now illustrate these truths for estimation in factor analysis. Factor analysis explains the covariation among the components of a random vector by approximating the vector by a linear transformation of a small number of uncorrelated factors. Readers with little background in multivariate statistical analysis should review the material in the Appendix at this point.

The classical factor analysis model deals with l independent multivariate observations of the form

$$Y_k = \mu + F X_k + U_k.$$

Here the $p \times q$ factor loading matrix F transforms the unobserved factor score X_k into the observed Y_k. The random vector U_k represents random measurement error. Typically, q is much smaller than p. The random vectors X_k and U_k are independent and normally distributed with means and variances

$$\begin{aligned} \mathrm{E}(X_k) &= \mathbf{0}, & \mathrm{Var}(X_k) &= I \\ \mathrm{E}(U_k) &= \mathbf{0}, & \mathrm{Var}(U_k) &= D, \end{aligned}$$

where I is the $q \times q$ identity matrix and D is a $p \times p$ diagonal matrix with ith diagonal entry d_i. The entries of the mean vector μ, the factor loading matrix F, and the diagonal matrix D constitute the parameters of the model. For a particular random sample $y_1, \ldots, y_l$ from the model, the maximum likelihood estimation of μ is simply the sample mean $\hat{\mu} = \bar{y}$. This fact is a consequence of the reasoning given in Example 11.2.3. Therefore, we replace each y_k by $y_k - \bar{y}$, assume $\mu = \mathbf{0}$, and focus on estimating F and D.

The random vector $\begin{pmatrix} X_k \\ Y_k \end{pmatrix}$ is the obvious choice of the complete data for case k. If $f(x_k)$ is the density of X_k and $g(y_k \mid x_k)$ is the conditional density of Y_k given $X_k = x_k$, then the complete data loglikelihood can be expressed as

$$\begin{aligned} & \sum_{k=1}^{l} \ln f(x_k) + \sum_{k=1}^{l} \ln g(y_k \mid x_k) \\ = & -\frac{l}{2} \ln \det I - \frac{1}{2} \sum_{k=1}^{l} x_k^t x_k - \frac{l}{2} \ln \det D \qquad (13.12) \\ & -\frac{1}{2} \sum_{k=1}^{l} (y_k - F x_k)^t D^{-1} (y_k - F x_k). \end{aligned}$$

We can simplify this by noting that $\ln \det I = 0$ and $\ln \det D = \sum_{i=1}^{p} \ln d_i$.

The key to performing the E step is to note that $\begin{pmatrix} X_k \\ Y_k \end{pmatrix}$ follows a multivariate normal distribution with variance matrix

$$\mathrm{Var} \begin{pmatrix} X_k \\ Y_k \end{pmatrix} = \begin{pmatrix} I & F^t \\ F & FF^t + D \end{pmatrix}.$$

Equation (A.2) of the Appendix therefore permits us to calculate the conditional expectation

$$v_k = \mathrm{E}(X_k \mid Y_k = y_k, F_n, D_n) = F_n^t (F_n F_n^t + D_n)^{-1} y_k$$

and conditional variance

$$A_k = \mathrm{Var}(X_k \mid Y_k = y_k, F_n, D_n) = I - F_n^t (F_n F_n^t + D_n)^{-1} F_n,$$

given the observed data and the current values of the matrices F and D. Combining these results with equation (A.1) of the Appendix yields

$$\begin{aligned}
& \mathrm{E}[(Y_k - FX_k)^t D^{-1}(Y_k - FX_k) \mid Y_k = y_k] \\
= \;& \mathrm{tr}(D^{-1}FA_kF^t) + (y_k - Fv_k)^t D^{-1}(y_k - Fv_k) \\
= \;& \mathrm{tr}\{D^{-1}[FA_kF^t + (y_k - Fv_k)(y_k - Fv_k)^t]\},
\end{aligned}$$

which we use in a moment.

If we define

$$\Lambda = \sum_{k=1}^{l}[A_k + v_k v_k^t], \quad \Gamma = \sum_{k=1}^{l} v_k y_k^t, \quad \Omega = \sum_{k=1}^{l} y_k y_k^t,$$

and take conditional expectations in equation (13.13), then we can write the surrogate function of the E step as

$$\begin{aligned}
& Q(F, D \mid F_n, D_n) \\
= \;& -\frac{l}{2}\sum_{i=1}^{p} \ln d_i - \frac{1}{2}\,\mathrm{tr}[D^{-1}(F\Lambda F^t - F\Gamma - \Gamma^t F^t + \Omega)],
\end{aligned}$$

omitting the additive constant

$$-\frac{1}{2}\sum_{k=1}^{l} \mathrm{E}(X_k^t X_k \mid Y_k = y_k, F_n, D_n),$$

which depends on neither F nor D.

To perform the M step, we first maximize $Q(F, D \mid F_n, D_n)$ with respect to F, holding D fixed. We can do so by permuting factors and completing the square in the trace

$$\begin{aligned}
& \mathrm{tr}[D^{-1}(F\Lambda F^t - F\Gamma - \Gamma^t F^t + \Omega)] \\
= \;& \mathrm{tr}[D^{-1}(F - \Gamma^t\Lambda^{-1})\Lambda(F - \Gamma^t\Lambda^{-1})^t] + \mathrm{tr}[D^{-1}(\Omega - \Gamma^t\Lambda^{-1}\Gamma)] \\
= \;& \mathrm{tr}[D^{-\frac{1}{2}}(F - \Gamma^t\Lambda^{-1})\Lambda(F - \Gamma^t\Lambda^{-1})^t D^{-\frac{1}{2}}] + \mathrm{tr}[D^{-1}(\Omega - \Gamma^t\Lambda^{-1}\Gamma)].
\end{aligned}$$

This calculation depends on the existence of the inverse matrix Λ^{-1}. Now Λ is certainly positive definite if A_k is positive definite, and Problem 23 asserts that A_k is positive definite. It follows that Λ^{-1} not only exists but is positive definite as well. Furthermore, the matrix

$$D^{-\frac{1}{2}}(F - \Gamma^t\Lambda^{-1})\Lambda(F - \Gamma^t\Lambda^{-1})^t D^{-\frac{1}{2}}$$

is positive semidefinite and has a nonnegative trace. Hence, the maximum value of the surrogate function $Q(F, D \mid F_n, D_n)$ with respect to F is attained at the point $F = \Gamma^t\Lambda^{-1}$, regardless of the value of D. In other words, the EM update of F is $F_{n+1} = \Gamma^t\Lambda^{-1}$. It should be stressed that Γ

and Λ implicitly depend on the previous values F_n and D_n. Once F_{n+1} is determined, the equation

$$\begin{aligned} 0 &= \frac{\partial}{\partial d_i} Q(F, D \mid F_n, D_n) \\ &= -\frac{l}{2d_i} + \frac{1}{2d_i^2}(F\Lambda F^t - F\Gamma - \Gamma^t F^t + \Omega)_{ii} \end{aligned}$$

provides the update

$$d_{n+1,i} = \frac{1}{l}(F_{n+1}\Lambda F_{n+1}^t - F_{n+1}\Gamma - \Gamma^t F_{n+1}^t + \Omega)_{ii}.$$

One of the frustrating features of factor analysis is that the factor loading matrix F is not uniquely determined. To understand the source of the ambiguity, consider replacing F by FO, where O is a $q \times q$ orthogonal matrix. The distribution of each random vector Y_k is normal with mean μ and variance matrix $FF^t + D$. If we substitute FO for F, then the variance $FOO^tF^t + D = FF^t + D$ remains the same. Another problem in factor analysis is the existence of more than one local maximum. Which one of these points the EM algorithm converges to depends on its starting value [4]. For a suggestion of how to improve the chances of converging to the dominant mode, see the article [17].

13.10 Problems

1. Code and test any of the algorithms discussed in the text or problems of this chapter.

2. The entropy of a probability density $p(x)$ on R^m is defined by
$$-\int p(x) \ln p(x) dx. \tag{13.13}$$
Among all densities with a fixed mean vector $\mu = \int x p(x) dx$ and variance matrix $\Omega = \int (x - \mu)(x - \mu)^t p(x) dx$, prove that the multivariate normal has maximum entropy. (Hint: Apply Proposition 13.3.2.)

3. In statistical mechanics, entropy is employed to characterize the equilibrium distribution of many independently behaving particles. Let $p(x)$ be the probability density that a particle is found at position x in phase space R^m, and suppose that each position x is assigned an energy $u(x)$. If the average energy $U = \int u(x)p(x)dx$ per particle is fixed, then Nature chooses $p(x)$ to maximize entropy as defined in

equation (13.13). Show that if constants α and β exist satisfying

$$\begin{aligned} \int \alpha e^{\beta u(x)} dx &= 1 \\ \int u(x) \alpha e^{\beta u(x)} dx &= U, \end{aligned}$$

then $p(x) = \alpha e^{\beta u(x)}$ does indeed maximize entropy subject to the average energy constraint. The density $p(x)$ is the celebrated Maxwell-Boltzmann density.

4. Show that the normal, Poisson, binomial, negative binomial, gamma, beta, and multinomial families are exponential by writing their densities in the form (13.3). What are the corresponding measure and sufficient statistic in each case?

5. In the EM algorithm suppose that the complete data X possess a regular exponential density

$$f(x \mid \theta) = g(x) e^{\beta(\theta) + h(x)^t \theta}$$

relative to some measure ν. Prove that the unconditional mean of the sufficient statistic $h(X)$ is given by the negative gradient $-\nabla\beta(\theta)$ and that the EM update is characterized by the condition

$$\mathrm{E}[h(X) \mid Y, \theta_n] = -\nabla\beta(\theta_{n+1}).$$

6. Without mentioning missing data, derive the ABO allele frequency estimation algorithm as an MM algorithm. (Hint: Apply the majorization (12.4) to the loglikelihood.)

7. Suppose the phenotypic counts in the ABO allele frequency estimation example satisfy $n_A + n_{AB} > 0$, $n_B + n_{AB} > 0$, and $n_O > 0$. Show that the loglikelihood is strictly concave and possesses a single global maximum on the interior of the feasible region.

8. Program the EM updates (13.5) for the ABO allele frequency data with prior counts $\alpha_A - 1 = \alpha_B - 1 = \alpha_O - 1 = k$. Plot the posterior mode as a function of k, and comment on its behavior as k tends to ∞.

9. In a genetic linkage experiment, 197 animals are randomly assigned to four categories according to the multinomial distribution with cell probabilities $\pi_1 = \frac{1}{2} + \frac{\theta}{4}$, $\pi_2 = \frac{1-\theta}{4}$, $\pi_3 = \frac{1-\theta}{4}$, and $\pi_4 = \frac{\theta}{4}$. If the corresponding observations are

$$y = (y_1, y_2, y_3, y_4)^t = (125, 18, 20, 34)^t,$$

then devise an EM algorithm and use it to estimate $\hat{\theta} = .6268$ [11]. (Hint: Split the first category into two so that there are five categories for the complete data.)

10. Suppose light bulbs have an exponential lifetime with mean θ. Two experiments are conducted. In the first, the lifetimes $y_1, \ldots, y_m$ of m independent bulbs are observed. In the second, p independent bulbs are observed to burn out before time t, and q independent bulbs are observed to burn out after time t. In other words, the lifetimes in the second experiment are both left and right censored. Construct an EM algorithm for finding the maximum likelihood estimate of θ [6].

TABLE 13.2. Death Notices from *The London Times*

Deaths i	**Frequency** n_i	**Deaths** i	**Frequency** n_i
0	162	5	61
1	267	6	27
2	271	7	8
3	185	8	3
4	111	9	1

11. Consider the data from *The London Times* [14] during the years 1910-1912 given in Table 13.2. The two columns labeled "Deaths i" refer to the number of deaths of women 80 years and older reported by day. The columns labeled "Frequency n_i" refer to the number of days with i deaths. A Poisson distribution gives a poor fit to these data, possibly because of different patterns of deaths in winter and summer. A mixture of two Poissons provides a much better fit. Under the Poisson admixture model, the likelihood of the observed data is

$$\prod_{i=0}^{9} \left[\alpha e^{-\mu_1} \frac{\mu_1^i}{i!} + (1-\alpha) e^{-\mu_2} \frac{\mu_2^i}{i!} \right]^{n_i},$$

where α is the admixture parameter and μ_1 and μ_2 are the means of the two Poisson distributions.

Formulate an EM algorithm for this model. Let $\theta = (\alpha, \mu_1, \mu_2)^t$ and

$$z_i(\theta) = \frac{\alpha e^{-\mu_1} \mu_1^i}{\alpha e^{-\mu_1} \mu_1^i + (1-\alpha) e^{-\mu_2} \mu_2^i}$$

be the posterior probability that a day with i deaths belongs to Poisson population 1. Show that the EM algorithm is given by

$$\alpha_{m+1} = \frac{\sum_i n_i z_i(\theta_m)}{\sum_i n_i}$$

$$\begin{aligned}\mu_{m+1,1} &= \frac{\sum_i n_i i z_i(\theta_m)}{\sum_i n_i z_i(\theta_m)}\\ \mu_{m+1,2} &= \frac{\sum_i n_i i[1 - z_i(\theta_m)]}{\sum_i n_i[1 - z_i(\theta_m)]}.\end{aligned}$$

From the initial estimates $\alpha_0 = 0.3$, $\mu_{0,1} = 1.0$ and $\mu_{0,2} = 2.5$, compute via the EM algorithm the maximum likelihood estimates $\hat{\alpha} = 0.3599$, $\hat{\mu}_1 = 1.2561$, and $\hat{\mu}_2 = 2.6634$. Note how slowly the EM algorithm converges in this example.

12. Let $x_1, \ldots, x_m$ be an i.i.d. sample from a normal density with mean μ and variance σ^2. Suppose for each x_i we observe $y_i = |x_i|$ rather than x_i. Formulate an EM algorithm for estimating μ and σ^2, and show that its updates are

$$\begin{aligned}\mu_{n+1} &= \frac{1}{m}\sum_{i=1}^{m}(w_{ni1}y_i - w_{ni2}y_i)\\ \sigma^2_{n+1} &= \frac{1}{m}\sum_{i=1}^{m}[w_{ni1}(y_i - \mu_{n+1})^2 + w_{ni2}(-y_i - \mu_{n+1})^2]\end{aligned}$$

with weights

$$\begin{aligned}w_{ni1} &= \frac{f(y_i \mid \theta_n)}{f(y_i \mid \theta_n) + f(-y_i \mid \theta_n)}\\ w_{ni2} &= \frac{f(-y_i \mid \theta_n)}{f(y_i \mid \theta_n) + f(-y_i \mid \theta_n)},\end{aligned}$$

where $f(x \mid \theta)$ is the normal density with $\theta = (\mu, \sigma^2)^t$. Demonstrate that the modes of the likelihood of the observed data come in symmetric pairs differing only in the sign of μ. This fact does not prevent accurate estimation of $|\mu|$ and σ^2.

13. Consider an i.i.d. sample drawn from a bivariate normal distribution with mean vector and covariance matrix

$$\mu = \begin{pmatrix}\mu_1\\ \mu_2\end{pmatrix}, \qquad \Omega = \begin{pmatrix}\sigma_1^2 & \sigma_{12}\\ \sigma_{12} & \sigma_2^2\end{pmatrix}.$$

Suppose through some random accident that the first p observations are missing their first component, the next q observations are missing their second component, and the last r observations are complete. Design an EM algorithm to estimate the five mean and variance parameters, taking as complete data the original data before the accidental loss.

TABLE 13.3. Bivariate Normal Data for the EM Algorithm

Obs	Obs	Obs	Obs	Obs	Obs
(1,1)	(1,-1)	(-1,1)	(-1,-1)	(2,*)	(2,*)
(-2,*)	(-2,*)	(*,2)	(*,2)	(*,-2)	(*,-2)

14. Suppose the data displayed in Table 13.3 constitute a random sample from a bivariate normal distribution with both means 0, variances σ_1^2 and σ_2^2, and correlation coefficient ρ. The asterisks indicate missing values. Show that the observed loglikelihood has symmetric global maxima when $\sigma_1^2 = \sigma_2^2 = \frac{8}{3}$ and $\rho = \pm\frac{1}{2}$ and a saddle point when $\sigma_1^2 = \sigma_2^2 = \frac{5}{2}$ and $\rho = 0$. If the EM algorithm designed in Problem 13 starts with $\rho = 0$, prove that it converges to the saddle point [16].

15. The standard linear regression model can be written in matrix notation as $X = A\beta + U$. Here X is the $r \times 1$ vector of responses, A is the $r \times s$ design matrix, β is the $s \times 1$ vector of regression coefficients, and U is the $r \times 1$ normally distributed error vector with mean $\mathbf{0}$ and covariance $\sigma^2 I$. The responses are right censored if for each i there is a constant c_i such that only $Y_i = \min\{c_i, X_i\}$ is observed. The EM algorithm offers a vehicle for estimating the parameter vector $\theta = (\beta^t, \sigma^2)^t$ in the presence of censoring [2, 13]. Show that

$$\begin{aligned}\beta_{n+1} &= (A^tA)^{-1}A^t\,\mathrm{E}(X \mid Y, \theta_n) \\ \sigma_{n+1}^2 &= \frac{1}{r}\,\mathrm{E}[(X - A\beta_{n+1})^t(X - A\beta_{n+1}) \mid Y, \theta_n].\end{aligned}$$

To compute the conditional expectations appearing in these formulas, let a_i be the ith row of A and define

$$H(v) = \frac{\frac{1}{\sqrt{2\pi}}e^{-\frac{v^2}{2}}}{\frac{1}{\sqrt{2\pi}}\int_v^\infty e^{-\frac{w^2}{2}}\,dw}.$$

For a censored observation $y_i = c_i < \infty$, prove that

$$\mathrm{E}(X_i \mid Y_i = c_i, \theta_n) = a_i\beta_n + \sigma_n H\Big(\frac{c_i - a_i\beta_n}{\sigma_n}\Big)$$

and that

$$\mathrm{E}(X_i^2 \mid Y_i = c_i, \theta_n) = (a_i\beta_n)^2 + \sigma_n^2 + \sigma_n(c_i + a_i\beta_n)H\Big(\frac{c_i - a_i\beta_n}{\sigma_n}\Big).$$

Use these formulas to complete the specification of the EM algorithm.

16. In the transmission tomography model it is possible to approximate the solution of equation (13.11) to good accuracy in certain situations. Verify the expansion

$$\frac{1}{e^s - 1} = \frac{1}{s} - \frac{1}{2} + \frac{s}{12} + O(s^2).$$

Using the approximation $1/(e^s - 1) \approx 1/s - 1/2$ for $s = l_{ij}\theta_j$, show that

$$\theta_{n+1,j} = \frac{\sum_i (M_{ij} - N_{ij})}{\frac{1}{2}\sum_i (M_{ij} + N_{ij}) l_{ij}}$$

results. Can you motivate this result heuristically?

17. Suppose that the complete data in the EM algorithm involve N binomial trials with success probability θ per trial. Here N can be random or fixed. If M trials result in success, then the complete data likelihood can be written as $\theta^M (1-\theta)^{N-M} c$, where c is an irrelevant constant. The E step of the EM algorithm reduces to forming

$$Q(\theta \mid \theta_n) = \mathrm{E}(M \mid Y, \theta_n) \ln \theta + \mathrm{E}(N - M \mid Y, \theta_n) \ln(1-\theta) + \ln c.$$

The binomial trials are hidden because only a function Y of them is directly observed. Show that the EM update amounts to

$$\theta_{n+1} = \frac{\mathrm{E}(M \mid Y, \theta_n)}{\mathrm{E}(N \mid Y, \theta_n)}.$$

Prove that this is equivalent to the update

$$\theta_{n+1} = \theta_n + \frac{\theta_n(1-\theta_n)}{\mathrm{E}(N \mid Y, \theta_n)} \frac{d}{d\theta} L(\theta_n),$$

where $L(\theta)$ is the loglikelihood of the observed data Y [15]. (Hint: Apply identity (12.3) of Chapter 12.)

18. As an example of hidden binomial trials as discussed in Problem 17, consider a random sample of twin pairs. Let u of these pairs consist of male pairs, v consist of female pairs, and w consist of opposite sex pairs. A simple model to explain these data involves a random Bernoulli choice for each pair dictating whether it consists of identical or nonidentical twins. Suppose that identical twins occur with probability p and nonidentical twins with probability $1-p$. Once the decision is made as to whether the twins are identical, then sexes are assigned to the twins. If the twins are identical, one assignment of sex is made. If the twins are nonidentical, then two independent assignments of sex are made. Suppose boys are chosen with probability q and girls with probability $1-q$. Model these data as hidden binomial trials. Derive the EM algorithm for estimating p and q.

19. Chun Li has derived an EM update for hidden multinomial trials. Let N denote the number of hidden trials, θ_i the probability of outcome i of k possible outcomes, and $L(\theta)$ the loglikelihood of the observed data Y. Derive the EM update

$$\theta_{n+1,i} = \theta_{ni} + \frac{\theta_{ni}}{\mathrm{E}(N \mid Y, \theta_n)} \left[\frac{\partial}{\partial \theta_i} L(\theta_n) - \sum_{j=1}^{k} \theta_{nj} \frac{\partial}{\partial \theta_j} L(\theta_n) \right]$$

following the reasoning of Problem 17.

20. In this problem you are asked to formulate models for hidden Poisson and exponential trials [15]. If the number of trials is N and the mean per trial is θ, then show that the EM update in the Poisson case is

$$\theta_{n+1} = \theta_n + \frac{\theta_n}{\mathrm{E}(N \mid Y, \theta_n)} \frac{d}{d\theta} L(\theta_n)$$

and in the exponential case is

$$\theta_{n+1} = \theta_n + \frac{\theta_n^2}{\mathrm{E}(N \mid Y, \theta_n)} \frac{d}{d\theta} L(\theta_n),$$

where $L(\theta)$ is the loglikelihood of the observed data Y.

21. In many discrete probability models, only data with positive counts are observed. Counts that are 0 are missing. Show that the likelihoods for the binomial, Poisson, and negative binomial models truncated at 0 amount to

$$\begin{aligned} L_1(p) &= \prod_i \frac{\binom{m_i}{x_i} p^{x_i} (1-p)^{m_i - x_i}}{1 - (1-p)^{m_i}} \\ L_2(\lambda) &= \prod_i \frac{\lambda^{x_i} e^{-\lambda}}{x_i!(1 - e^{-\lambda})} \\ L_3(p) &= \prod_i \frac{\binom{m_i + x_i - 1}{x_i} (1-p)^{x_i} p^{m_i}}{1 - p^{m_i}}. \end{aligned}$$

For observation i of the binomial model, there are x_i successes out of m_i trials with success probability p per trial. For observation i of the negative binomial model, there are x_i failures before m_i required successes. For each model, devise an EM algorithm that fills in the missing observations by imputing a geometrically distributed number of truncated observations for every real observation. Show that the EM updates reduce to

$$p_{n+1} = \frac{\sum_i x_i}{\sum_i \frac{m_i}{1-(1-p_n)^{m_i}}}$$

$$\begin{aligned}
\lambda_{n+1} &= \frac{\sum_i x_i}{\sum_i \frac{1}{1-e^{-\lambda_n}}} \\
p_{n+1} &= \frac{\sum_i \frac{m_i}{1-p_n^{m_i}}}{\sum_i (x_i + \frac{m_i}{1-p_n^{m_i}})}
\end{aligned}$$

for the three models.

22. Demonstrate that the EM updates of the previous problem can be derived as MM updates based on the minorization

$$-\ln(1-u) \;\; \geq \;\; -\ln(1-u_n) + \frac{u_n}{1-u_n} \ln \frac{u}{u_n}$$

for u and u_n in the interval $(0,1)$. Prove this minorization first. (Hint: If you rearrange the minorization, then Proposition 13.3.2 applies.)

23. Suppose that Σ is a positive definite matrix. Prove that the matrix $I - F^t(FF^t + \Sigma)^{-1}F$ is also positive definite. This result is used in the derivation of the EM algorithm in Section 13.9. (Hint: Consider theoretical properties of the sweep operator in Chapter 7.)

24. A certain company asks consumers to rate movies on an integer scale from 1 to 5. Let M_i be the set of movies rated by person i. Denote the cardinality of M_i by $|M_i|$. Each rater does so in one of two modes that we will call "quirky" and "consensus". In quirky mode, i has a private rating distribution $(q_{i1}, q_{i2}, q_{i3}, q_{i4}, q_{i5})$ that applies to every movie regardless of its intrinsic merit. In consensus mode, rater i rates movie j according to the distribution $(c_{j1}, c_{j2}, c_{j3}, c_{j4}, c_{j5})$ shared with all other raters in consensus mode. For every movie i rates, he or she makes a quirky decision with probability π_i and a consensus decision with probability $1-\pi_i$. These decisions are made independently across raters and movies. If x_{ij} is the rating given to movie j by rater i, then prove that the likelihood of the data is

$$L \;\; = \;\; \prod_i \prod_{j \in M_i} [\pi_i q_{ix_{ij}} + (1-\pi_i)c_{jx_{ij}}].$$

Once we estimate the parameters, we can rank the reliability of rater i by the estimate $\hat{\pi}_i$ and the popularity of movie j by its estimated average rating $\sum_k k\hat{c}_{jk}$.

If we choose the natural course of estimating the parameters by maximum likelihood, then it is possible to derive an EM or MM algorithm. From the right perspectives, these two algorithms coincide. Let n denote iteration number and w_{nij} the weight

$$w_{nij} \;\; = \;\; \frac{\pi_{ni} q_{nix_{ij}}}{\pi_{ni} q_{nix_{ij}} + (1-\pi_{ni})c_{njx_{ij}}}.$$

Derive either algorithm and show that it updates the parameters by

$$\begin{aligned}
\pi_{n+1,i} &= \frac{1}{|M_i|}\sum_{j\in M_i} w_{nij} \\
q_{n+1,ix} &= \frac{\sum_{j\in M_i} 1_{\{x_{ij}=x\}} w_{nij}}{\sum_{j\in M_i} w_{nij}} \\
c_{n+1,jx} &= \frac{\sum_i 1_{\{x_{ij}=x\}}(1-w_{nij})}{\sum_i (1-w_{nij})}.
\end{aligned}$$

These updates are easy to implement. Can you motivate them as ratios of expected counts?

13.11 References

[1] Clarke CA, Price Evans DA, McConnell RB, Sheppard PM (1959) Secretion of blood group antigens and peptic ulcers. *Brit Med J* 1:603–607

[2] Dempster AP, Laird NM, Rubin DB (1977) Maximum likelihood from incomplete data via the EM algorithm (with discussion). *J Roy Stat Soc B* 39:1–38

[3] Dobson AJ (1990) *An Introduction to Generalized Linear Models.* Chapman & Hall, London

[4] Duan J-C, Simonato J-G (1993) Multiplicity of solutions in maximum likelihood factor analysis. *J Stat Computation Simulation* 47:37–47

[5] Fessler JA, Clinthorne NH, Rogers WL (1993) On complete-data spaces for PET reconstruction algorithms. *IEEE Trans Nuclear Sci* 40:1055–1061

[6] Flury B, Zoppè A (2000) Exercises in EM. *Amer Statistician* 54:207–209

[7] Lange K, Carson R (1984) EM reconstruction algorithms for emission and transmission tomography. *J Computer Assist Tomography* 8:306–316

[8] Little RJA, Rubin DB (2002) *Statistical Analysis with Missing Data,* 2nd ed. Wiley, New York

[9] McLachlan GJ, Krishnan T (2008) *The EM Algorithm and Extensions,* 2nd ed. Wiley, New York

[10] McLachlan GJ, Peel D (2000) *Finite Mixture Models.* Wiley, New York

[11] Rao CR (1973) *Linear Statistical Inference and Its Applications*, 2nd ed. Wiley, New York

[12] Smith CAB (1957) Counting methods in genetical statistics. *Ann Hum Genet* 21:254–276

[13] Tanner MA (1993) *Tools for Statistical Inference: Methods for the Exploration of Posterior Distributions and Likelihood Functions*, 2nd ed. Springer, New York

[14] Titterington DM, Smith AFM, Makov UE (1985) *Statistical Analysis of Finite Mixture Distributions.* Wiley, New York

[15] Weeks DE, Lange K (1989) Trials, tribulations, and triumphs of the EM algorithm in pedigree analysis. *IMA J Math Appl Med Biol* 6:209–232

[16] Wu CF (1983) On the convergence properties of the EM algorithm. *Ann Stat* 11:95–103

[17] Zhou H, Lange K (2009) On the bumpy road to the dominant mode. *Scandinavian J Stat* (in press)

14

Newton's Method and Scoring

14.1 Introduction

The MM and EM algorithms are hardly the only methods of optimization. Newton's method is better known and more widely applied. We encountered Newton's method in Section 5.4 of Chapter 5. Here we focus on the multidimensional version. Despite its defects, Newton's method is the gold standard for speed of convergence and forms the basis of many modern optimization algorithms. Its variants seek to retain its fast convergence while taming its defects. The variants all revolve around the core idea of locally approximating the objective function by a strictly convex quadratic function. At each iteration the quadratic approximation is optimized. Safeguards are introduced to keep the iterates from veering toward irrelevant stationary points.

Statisticians are among the most avid consumers of optimization techniques. Statistics, like other scientific disciplines, has a special vocabulary. We will meet some of that vocabulary in this chapter as we discuss optimization methods important in computational statistics. Thus, we will take up Fisher's scoring algorithm and the Gauss-Newton method of nonlinear least squares. We have already encountered likelihood functions and the device of passing to loglikelihoods in estimation. In statistics, the gradient of the loglikelihood is called the score, and the negative of the second differential is called the observed information. One major advantage of maximizing the loglikelihood rather than the likelihood is that the loglikelihood, score, and observed information are all additive functions of independent observations.

14.2 Newton's Method and Root Finding

One of the virtues of Newton's method of root finding is that it readily extends to multiple dimensions. Consider a function $f(x)$ mapping R^p into R^p, and suppose a root of $f(x) = \mathbf{0}$ occurs at y. If the differential in the approximate identity

$$\mathbf{0} - f(x) \quad = \quad f(y) - f(x) \quad \approx \quad df(x)(y - x)$$

is invertible, then we can solve for y as

$$y \quad = \quad x - df(x)^{-1} f(x).$$

K. Lange, *Numerical Analysis for Statisticians*, Statistics and Computing,
DOI 10.1007/978-1-4419-5945-4_14,

Thus, Newton's method iterates according to

$$x_{n+1} = x_n - df(x_n)^{-1}f(x_n). \tag{14.1}$$

This generally works well if the initial point x_1 is close to y.

Example 14.2.1 *Newton's Method of Matrix Inversion*

Newton's method for finding the reciprocal of a number can be generalized to compute the inverse of a matrix A [12]. The differentials in question are messy to calculate, but guided by the one-dimensional case featured in Section 5.4.1, we are led to consider the iteration scheme

$$B_{n+1} = 2B_n - B_nAB_n.$$

Rearranging this equation yields

$$A^{-1} - B_{n+1} = (A^{-1} - B_n)A(A^{-1} - B_n),$$

which implies that

$$\|A^{-1} - B_{n+1}\| \leq \|A\| \cdot \|A^{-1} - B_n\|^2$$

for every matrix norm. It follows that the sequence B_n converges at a quadratic rate to A^{-1} if B_1 is sufficiently close to A^{-1}. ■

14.3 Newton's Method and Optimization

We now take up the topic of maximizing a loglikelihood $L(\theta)$. A second-order Taylor expansion around the current point θ_n gives

$$L(\theta) \approx L(\theta_n) + dL(\theta_n)(\theta - \theta_n) + \frac{1}{2}(\theta - \theta_n)^t d^2L(\theta_n)(\theta - \theta_n). \tag{14.2}$$

In Newton's method one maximizes the right-hand side of (14.2) by equating its gradient

$$\nabla L(\theta_n) + d^2L(\theta_n)(\theta - \theta_n) = \mathbf{0}$$

and solving for the next iterate

$$\theta_{n+1} = \theta_n - d^2L(\theta_n)^{-1}\nabla L(\theta_n).$$

Obviously, any stationary point of $L(\theta)$ is a fixed point of Newton's method.

There are two potential problems with Newton's method. First, it can be expensive computationally to evaluate and invert the observed information. Second, far from $\hat{\theta}$, Newton's method is equally happy to head uphill or down. In other words, Newton's method is not an ascent algorithm in the sense that $L(\theta_{n+1}) > L(\theta_n)$. To generate an ascent algorithm, we can replace the observed information $-d^2L(\theta_n)$ by a positive definite approximating matrix A_n. With this substitution, the proposed increment $\Delta\theta_n = A_n^{-1}\nabla L(\theta_n)$, if sufficiently contracted, forces an increase in $L(\theta)$. For a nonstationary point, this assertion follows from the first-order Taylor expansion

$$\begin{aligned} L(\theta_n + s\Delta\theta_n) - L(\theta_n) &= dL(\theta_n)s\Delta\theta_n + o(s) \\ &= sdL(\theta_n)A_n^{-1}\nabla L(\theta_n) + o(s), \end{aligned}$$

where the error ratio $o(s)/s$ tends to 0 as the positive contraction constant s tends to 0. Thus, a positive definite modification of the observed information combined with some form of backtracking leads to an ascent algorithm.

Backtracking is a crude form of one-dimensional optimization. The methods of golden section search and cubic interpolation introduced in Sections 5.5 and 5.6 are applicable. A simpler and less expensive form of backtracking is step-halving. If the initial increment $\Delta\theta_n$ does not produce an increase in $L(\theta)$, then try $\Delta\theta_n/2$. If $\Delta\theta_n/2$ fails, then try $\Delta\theta_n/4$, and so forth. Note that we have said nothing about how well A_n approximates $d^2L(\theta_n)$. The quality of this approximation obviously affects the rate of convergence toward a local maximum.

If we maximize $L(\theta)$ subject to the linear equality constraints $V\theta = d$, then maximization of the approximating quadratic can be accomplished as indicated in Example 11.3.3 of Chapter 11. Because

$$V(\theta_{n+1} - \theta_n) = \mathbf{0},$$

the revised increment $\Delta\theta_n = \theta_{n+1} - \theta_n$ is

$$\Delta\theta_n = -\left[A_n^{-1} - A_n^{-1}V^t(VA_n^{-1}V^t)^{-1}VA_n^{-1}\right]\nabla L(\theta_n). \tag{14.3}$$

This can be viewed as the projection of the unconstrained increment onto the null space of V. Problem 5 shows that backtracking also works for the projected increment.

Finally, the reader may want to review Section 5.7 on stopping rules for iterative processes. The question of when to stop arises for all optimization algorithms, not just for Newton's method and scoring. Keep in mind that the stopping criterion (5.12) should be satisfied across all coordinates.

14.4 Ad Hoc Approximations of Hessians

In minimization problems, we have emphasized the importance of approximating $d^2f(\theta)$ by a positive definite matrix. Three key ideas drive the process of approximation. One is the recognition that symmetric outer product matrices are positive semidefinite. Another is a feel for when terms are small on average. Usually this involves comparing random variables and their means. Finally, it is almost always advantageous to avoid the explicit calculation of complicated second derivatives.

For example, consider the problem of least squares estimation with nonlinear regression functions. Let us formulate the problem slightly more generally as one of minimizing the weighted sum of squares

$$f(\theta) \;=\; \frac{1}{2}\sum_{i=1}^{m} w_i[y_i - \mu_i(\theta)]^2$$

involving a weight $w_i > 0$ and response y_i for case i. Here y_i is a realization of a random variable Y_i with mean $\mu_i(\theta)$. In linear regression, the mean $\mu_i(\theta) = \sum_j x_{ij}\theta_j$. To implement Newton's method, we need

$$\begin{aligned}\nabla f(\theta) &= -\sum_{i=1}^{m} w_i[y_i - \mu_i(\theta)]\nabla\mu_i(\theta) \qquad (14.4)\\ d^2 f(\theta) &= \sum_{i=1}^{m} w_i \nabla\mu_i(\theta) d\mu_i(\theta) - \sum_{i=1}^{m} w_i[y_i - \mu_i(\theta)]d^2\mu_i(\theta).\end{aligned}$$

In the Gauss-Newton algorithm, we approximate

$$d^2 f(\theta) \;\approx\; \sum_{i=1}^{m} w_i \nabla\mu_i(\theta) d\mu_i(\theta) \qquad (14.5)$$

on the rationale that either the weighted residuals $w_i[y_i - \mu_i(\theta)]$ are small or the regression functions $\mu_i(\theta)$ are nearly linear. In both instances, the Gauss-Newton algorithm shares the fast convergence of Newton's method.

Maximum likelihood estimation with the Poisson distribution furnishes another example. Here the count data $y_1, \ldots, y_m$ have loglikelihood, score, and negative observed information

$$\begin{aligned}L(\theta) &= \sum_{i=1}^{m} [y_i \ln \lambda_i(\theta) - \lambda_i(\theta) - \ln y_i!]\\ \nabla L(\theta) &= \sum_{i=1}^{m} \Big[\frac{y_i}{\lambda_i(\theta)}\nabla\lambda_i(\theta) - \nabla\lambda_i(\theta)\Big]\\ d^2 L(\theta) &= \sum_{i=1}^{m} \Big[-\frac{y_i}{\lambda_i(\theta)^2}\nabla\lambda_i(\theta)d\lambda_i(\theta) + \frac{y_i}{\lambda_i(\theta)}d^2\lambda_i(\theta) - d^2\lambda_i(\theta)\Big],\end{aligned}$$

where $\mathrm{E}(y_i) = \lambda_i(\theta)$. Given that the ratio $y_i/\lambda_i(\theta)$ has average value 1, the negative semidefinite approximations

$$\begin{aligned}d^2 L(\theta) &\approx -\sum_{i=1}^{m} \frac{y_i}{\lambda_i(\theta)^2}\nabla\lambda_i(\theta)d\lambda_i(\theta)\\ &\approx -\sum_{i=1}^{m} \frac{1}{\lambda_i(\theta)}\nabla\lambda_i(\theta)d\lambda_i(\theta)\end{aligned}$$

are reasonable. The second of these leads to the scoring algorithm discussed in the next section.

The exponential distribution offers a third illustration. Now the data have means $\mathrm{E}(y_i) = \lambda_i(\theta)^{-1}$. The loglikelihood

$$L(\theta) = \sum_{i=1}^{m} [\ln \lambda_i(\theta) - y_i \lambda_i(\theta)]$$

yields the score and negative observed information

$$\begin{aligned}
\nabla L(\theta) &= \sum_{i=1}^{m} \Big[\frac{1}{\lambda_i(\theta)} \nabla \lambda_i(\theta) - y_i \nabla \lambda_i(\theta)\Big] \\
d^2 L(\theta) &= \sum_{i=1}^{m} \Big[-\frac{1}{\lambda_i(\theta)^2} \nabla \lambda_i(\theta) d\lambda_i(\theta) + \frac{1}{\lambda_i(\theta)} d^2 \lambda_i(\theta) - y_i d^2 \lambda_i(\theta)\Big].
\end{aligned}$$

Replacing observations by their means suggests the approximation

$$d^2 L(\theta) \approx -\sum_{i=1}^{m} \frac{1}{\lambda_i(\theta)^2} \nabla \lambda_i(\theta) d\lambda_i(\theta)$$

made in the scoring algorithm. Table 14.1 summarizes the scoring algorithm with means $\mu_i(\theta)$ replacing intensities $\lambda_i(\theta)$.

Our final example involves maximum likelihood estimation with the multinomial distribution. The observations $y_1, \ldots, y_m$ are now cell counts over n independent trials. Cell i is assigned probability $p_i(\theta)$ and averages a total of $np_i(\theta)$ counts. The loglikelihood, score, and negative observed information amount to

$$\begin{aligned}
L(\theta) &= \sum_{i=1}^{m} y_i \ln p_i(\theta) \\
\nabla L(\theta) &= \sum_{i=1}^{m} \frac{y_i}{p_i(\theta)} \nabla p_i(\theta) \\
d^2 L(\theta) &= \sum_{i=1}^{m} \Big[-\frac{y_i}{p_i(\theta)^2} \nabla p_i(\theta) dp_i(\theta) + \frac{y_i}{p_i(\theta)} d^2 p_i(\theta)\Big].
\end{aligned}$$

In light of the identity $\mathrm{E}(y_i) = np_i(\theta)$, the approximation

$$\sum_{i=1}^{m} \frac{y_i}{p_i(\theta)} d^2 p_i(\theta) \approx n \sum_{i=1}^{m} d^2 p_i(\theta) = n\, d^2 1 = 0$$

is reasonable. This suggests the further negative semidefinite approximations

$$d^2 L(\theta) \approx -\sum_{i=1}^{m} \frac{y_i}{p_i(\theta)^2} \nabla p_i(\theta) dp_i(\theta)$$

$$\approx \quad -n \sum_{i=1}^{m} \frac{1}{p_i(\theta)} \nabla p_i(\theta) dp_i(\theta),$$

the second of which coincides with the scoring algorithm.

14.5 Scoring and Exponential Families

One can approximate the observed information in a variety of ways. The method of steepest ascent replaces the observed information by the identity matrix I. The usually more efficient scoring algorithm replaces the observed information by the expected information $J(\theta) = \mathrm{E}[-d^2L(\theta)]$. The alternative representation $J(\theta) = \mathrm{Var}[\nabla L(\theta)]$ of $J(\theta)$ as a covariance matrix shows that it is positive semidefinite [25]. An extra dividend of scoring is that the inverse matrix $J(\hat{\theta})^{-1}$ immediately supplies the asymptotic variances and covariances of the maximum likelihood estimate $\hat{\theta}$ [25]. Scoring shares this benefit with Newton's method since the observed information is under natural assumptions asymptotically equivalent to the expected information.

It is possible to compute $J(\theta)$ explicitly for exponential families of densities following the approach of Jennrich and Moore [15]. (See also the references [1, 3, 11, 24], where the connections between scoring and iteratively reweighted least squares are emphasized.) The general form of an exponential family is displayed in equation (13.3). Most of the distributional families commonly encountered in statistics are exponential families. The score and expected information can be expressed in terms of the mean vector $\mu(\theta) = \mathrm{E}[h(X)]$ and covariance matrix $\Sigma(\theta) = \mathrm{Var}[h(X)]$ of the sufficient statistic $h(X)$. If $d\gamma(\theta)$ is the matrix of partial derivatives of the column vector $\gamma(\theta)$, then the first differential amounts to

$$dL(\theta) \quad = \quad d\ln f(x \mid \theta) \quad = \quad d\beta(\theta) + h(x)^t d\gamma(\theta). \tag{14.6}$$

If $\gamma(\theta)$ is linear in θ, then $J(\theta) = -d^2L(\theta) = -d^2\beta(\theta)$, and scoring coincides with Newton's method. If, in addition, $J(\theta)$ is positive definite, then $L(\theta)$ is strictly concave and possesses at most one local maximum.

The score conveniently has vanishing expectation because

$$\mathrm{E}[dL(\theta)] \quad = \quad \int \frac{df(x \mid \theta)}{f(x \mid \theta)} f(x \mid \theta)\, d\nu(x) \quad = \quad d\int f(x \mid \theta)\, d\nu(x)$$

and $\int f(x \mid \theta)\, d\nu(x) = 1$. (Differentiation under the expectation sign is incidentally permitted for exponential families [21].) For an exponential family, this fact can be restated as

$$d\beta(\theta) + \mu(\theta)^t d\gamma(\theta) \quad = \quad \mathbf{0}^t. \tag{14.7}$$

Subtracting equation (14.7) from equation (14.6) yields the alternative representation

$$dL(\theta) \quad = \quad [h(x) - \mu(\theta)]^t d\gamma(\theta) \tag{14.8}$$

of the first differential. From this it follows directly that the expected information is given by

$$J(\theta) \quad = \quad \mathrm{Var}[\nabla L(\theta)] \quad = \quad d\gamma(\theta)^t \Sigma(\theta) d\gamma(\theta). \tag{14.9}$$

To eliminate $d\gamma(\theta)$ in equations (14.8) and (14.9), note that

$$\begin{aligned} d\mu(\theta) &= \int h(x) df(x \mid \theta)\, d\nu(x) \\ &= \int h(x) dL(\theta) f(x \mid \theta)\, d\nu(x) \\ &= \int h(x)[h(x) - \mu(\theta)]^t d\gamma(\theta) f(x \mid \theta)\, d\nu(x) \\ &= \Sigma(\theta) d\gamma(\theta). \end{aligned}$$

When $\Sigma(\theta)$ is invertible, this calculation implies $d\gamma(\theta) = \Sigma(\theta)^{-1} d\mu(\theta)$, which in view of equations (14.8) and (14.9) yields

$$dL(\theta) \quad = \quad [h(x) - \mu(\theta)]^t \Sigma(\theta)^{-1} d\mu(\theta) \tag{14.10}$$

and

$$J(\theta) \quad = \quad d\mu(\theta)^t \Sigma(\theta)^{-1} d\mu(\theta). \tag{14.11}$$

In fact, the representations in (14.10) and (14.11) hold even when $\Sigma(\theta)$ is not invertible, provided the generalized inverse $\Sigma(\theta)^-$ is substituted for $\Sigma(\theta)^{-1}$ [15].

Based on equations (13.3), (14.10), and (14.11), Table 14.1 displays the loglikelihood, score vector, and expected information matrix for some commonly applied exponential families. In this table, x represents a single observation from the binomial, Poisson, and exponential families. For the multinomial family with m categories, $x = (x_1, \ldots, x_m)$ gives the category-by-category counts. The quantity μ denotes the mean of x for the Poisson and exponential families. For the binomial family, we express the mean np as the product of the number of trials n and the success probability p per trial. A similar convention holds for the multinomial family.

The multinomial family deserves further comment. Straightforward calculation shows that the covariance matrix $\Sigma(\theta)$ has entries

$$n[1_{\{i=j\}} p_i(\theta) - p_i(\theta) p_j(\theta)].$$

Here the matrix $\Sigma(\theta)$ is singular, so the generalized inverse applies in formula (14.11). In this case it is easier to derive the expected information by taking the expectation of the observed information given in Section 14.4.

TABLE 14.1. Score and Information for Some Exponential Families

Family	$L(\theta)$	$\nabla L(\theta)$	$J(\theta)$
Binomial	$x\ln\frac{p}{1-p}+n\ln(1-p)$	$\frac{x-np}{p(1-p)}\nabla p$	$\frac{n}{p(1-p)}\nabla p dp$
Multinomial	$\sum_i x_i \ln p_i$	$\sum_i \frac{x_i}{p_i}\nabla p_i$	$\sum_i \frac{n}{p_i}\nabla p_i dp_i$
Poisson	$-\mu + x\ln\mu$	$-\nabla\mu + \frac{x}{\mu}\nabla\mu$	$\frac{1}{\mu}\nabla\mu d\mu$
Exponential	$-\ln\mu - \frac{x}{\mu}$	$-\frac{1}{\mu}\nabla\mu + \frac{x}{\mu^2}\nabla\mu$	$\frac{1}{\mu^2}\nabla\mu d\mu$

In the ABO allele frequency estimation problem studied in Chapter 13, scoring can be implemented by taking as basic parameters p_A and p_B and expressing $p_O = 1 - p_A - p_B$. Scoring then leads to the same maximum likelihood point $(\hat{p}_A, \hat{p}_B, \hat{p}_O) = (.2136, .0501, .7363)$ as the EM algorithm. The quicker convergence of scoring here – four iterations as opposed to five starting from (.3, .2, .5) – is often more dramatic in other problems. Scoring also has the advantage over EM of immediately providing asymptotic standard deviations of the parameter estimates. These are (.0135, .0068, .0145) for the estimates $(\hat{p}_A, \hat{p}_B, \hat{p}_O)$.

14.6 The Gauss-Newton Algorithm

In nonlinear regression with normally distributed errors, the scoring algorithm metamorphoses into the Gauss-Newton algorithm. Suppose that the m independent responses $x_1, \ldots, x_m$ are normally distributed with means $\mu_i(\theta)$ and variances σ^2/w_i, where the w_i are known constants. To estimate the mean parameter vector θ and the variance parameter σ^2 by scoring, we first write the loglikelihood up to a constant as the function

$$L(\phi) = -\frac{m}{2}\ln\sigma^2 - \frac{1}{2\sigma^2}\sum_{i=1}^m w_i[x_i - \mu_i(\theta)]^2$$

of the parameters $\phi^t = (\theta^t, \sigma^2)$. Straightforward differentiations and integrations yield the score

$$\nabla L(\phi) = \begin{pmatrix} \frac{1}{\sigma^2}\sum_{i=1}^m w_i[x_i - \mu_i(\theta)]\nabla\mu_i(\theta) \\ -\frac{m}{2\sigma^2} + \frac{1}{2\sigma^4}\sum_{i=1}^m w_i[x_i - \mu_i(\theta)]^2 \end{pmatrix}$$

and the expected information

$$J(\phi) = \begin{pmatrix} \frac{1}{\sigma^2}\sum_{i=1}^m w_i \nabla\mu_i(\theta) d\mu_i(\theta) & \mathbf{0} \\ \mathbf{0}^t & \frac{m}{2\sigma^4} \end{pmatrix}.$$

The upper block of the score is a special case of formula (14.10) and coincides with formula (14.4) except for a factor of $-\sigma^{-2}$. The upper-left block of the expected information is a special case of formula (14.11) and coincides with formula (14.5) except for a factor of σ^{-2}. Scoring updates θ by

$$\theta_{n+1} = \theta_n+[\sum_{i=1}^{m} w_i\nabla\mu_i(\theta_n)d\mu(\theta_n)]^{-1}\sum_{i=1}^{m} w_i[x_i - \mu_i(\theta_n)]\nabla\mu_i(\theta_n) \quad (14.12)$$

and σ^2 by

$$\sigma^2_{n+1} = \frac{1}{m}\sum_{i=1}^{m} w_i[x_i - \mu_i(\theta_n)]^2.$$

The iterations (14.12) on θ can be carried out blithely neglecting those on σ^2.

TABLE 14.2. AIDS Data from Australia during 1983-1986

Quarter	Deaths	Quarter	Deaths	Quarter	Deaths
1	0	6	4	11	20
2	1	7	9	12	25
3	2	8	18	13	37
4	3	9	23	14	45
5	1	10	31		

14.7 Generalized Linear Models

The generalized linear model [24] deals with exponential families (13.3) in which the sufficient statistic $h(X)$ is X and the mean μ of X completely determines the distribution of X. In many applications it is natural to postulate that $\mu(\theta) = q(z^t\theta)$ is a monotonic function q of some linear combination of known covariates z. The inverse of q is called the link function.In this setting, $\nabla\mu(\theta) = q'(z^t\theta)z$. It follows from equations (14.10) and (14.11) that if $x_1, \ldots, x_m$ are independent observations with corresponding variances $\sigma_1^2, \ldots, \sigma_m^2$ and covariate vectors $z_1, \ldots, z_m$, then the score and expected information can be written as

$$\begin{aligned}\nabla L(\theta) &= \sum_{i=1}^{m} \frac{x_i - \mu_i(\theta)}{\sigma_i^2} q'(z_i^t\theta)z_i \\ J(\theta) &= \sum_{i=1}^{m} \frac{1}{\sigma_i^2} q'(z_i^t\theta)^2 z_i z_i^t.\end{aligned}$$

Table 14.2 contains quarterly data on AIDS deaths in Australia that illustrate the application of a generalized linear model [10, 27]. A simple plot of the data suggests exponential growth. A plausible model therefore involves Poisson distributed observations x_i with means $\mu_i(\theta) = e^{\theta_1 + i\theta_2}$. Because this parameterization renders scoring equivalent to Newton's method, scoring gives the quick convergence noted in Table 14.3.

TABLE 14.3. Scoring Iterates for the AIDS Model

Iteration	**Step-halves**	θ_1	θ_2
1	0	0.0000	0.0000
2	3	-1.3077	0.4184
3	0	0.6456	0.2401
4	0	0.3744	0.2542
5	0	0.3400	0.2565
6	0	0.3396	0.2565

14.8 MM Gradient Algorithm

Often it is impossible to solve the optimization step of the MM algorithm exactly. If $f(\theta)$ is the objective function and $g(\theta \mid \theta_n)$ minorizes or majorizes $f(\theta)$ at θ_n, then one step of Newton's method can be applied to approximately optimize $g(\theta \mid \theta_n)$. Thus, the MM gradient algorithm iterates according to

$$\begin{aligned} \theta_{n+1} &= \theta_n - d^2 g(\theta_n \mid \theta_n)^{-1} \nabla g(\theta_n \mid \theta_n) \\ &= \theta_n - d^2 g(\theta_n \mid \theta_n)^{-1} \nabla f(\theta_n). \end{aligned}$$

Here derivatives are taken with respect to the left argument of $g(\theta \mid \theta_n)$. Substitution of $\nabla f(\theta_n)$ for $\nabla g(\theta_n \mid \theta_n)$ can be justified by appealing to equation (12.3). In most practical examples, the surrogate function $g(\theta \mid \theta_n)$ is either convex or concave, and its second differential $d^2 g(\theta_n \mid \theta_n)$ gives a descent or ascent algorithm with backtracking. The MM gradient algorithm and the MM algorithm enjoy the same rate of convergence approaching the optimal point $\hat{\theta}$. Furthermore, in the vicinity of $\hat{\theta}$, the MM gradient algorithm also satisfies the appropriate ascent or descent condition $f(\theta_{n+1}) > f(\theta_n)$ or $f(\theta_{n+1}) < f(\theta_n)$ without backtracking [18].

Example 14.8.1 *Newton's Method in Transmission Tomography*

In the transmission tomography model of Chapters 12 and 13, the surrogate function $g(\theta \mid \theta_n)$ of equation (12.16) minorizes the loglikelihood $L(\theta)$ in the absence of a smoothing prior. Differentiating $g(\theta \mid \theta_n)$ with respect to θ_j gives the transcendental equation

$$0 = \sum_i l_{ij}\Big[d_i e^{-l_i^t \theta_n \theta_j/\theta_{nj}} - y_i\Big].$$

One step of Newton's method starting at $\theta_j = \theta_{nj}$ produces the next iterate

$$\begin{aligned} \theta_{n+1,j} &= \theta_{nj} + \frac{\theta_{nj}\sum_i l_{ij}(d_i e^{-l_i^t\theta_n} - y_i)}{\sum_i l_{ij} l_i^t \theta_n d_i e^{-l_i^t\theta_n}} \\ &= \theta_{nj}\frac{\sum_i l_{ij}[d_i e^{-l_i^t\theta_n}(1 + l_i^t\theta_n) - y_i]}{\sum_i l_{ij} l_i^t \theta_n d_i e^{-l_i^t\theta_n}}. \end{aligned}$$

This step typically increases $L(\theta)$. ■

Example 14.8.2 *Estimation with the Dirichlet Distribution*

As another example, consider parameter estimation for the Dirichlet distribution [17]. This distribution has probability density

$$\frac{\Gamma(\sum_{i=1}^m \theta_i)}{\prod_{i=1}^m \Gamma(\theta_i)} \prod_{i=1}^m y_i^{\theta_i - 1} \tag{14.13}$$

on the simplex $\{y = (y_1, \ldots, y_m)^t : y_1 > 0, \ldots, y_m > 0, \sum_{i=1}^m y_i = 1\}$ endowed with the uniform measure. The Dirichlet distribution is used to represent random proportions. All components θ_i of its parameter vector θ are positive.

If $y_1, \ldots, y_l$ are randomly sampled vectors from the Dirichlet distribution, then their loglikelihood is

$$L(\theta) = l \ln\Gamma\Big(\sum_{i=1}^m \theta_i\Big) - l\sum_{i=1}^m \ln\Gamma(\theta_i) + \sum_{j=1}^l\sum_{i=1}^m (\theta_i - 1)\ln y_{ji}.$$

Except for the first term on the right, the parameters are separated. Fortunately the function $\ln\Gamma(t)$ is convex [20]. Denoting its derivative by $\psi(t)$, we exploit the minorization

$$\ln\Gamma\Big(\sum_{i=1}^m \theta_i\Big) \ge \ln\Gamma\Big(\sum_{i=1}^m \theta_{ni}\Big) + \psi\Big(\sum_{i=1}^m \theta_{ni}\Big)\sum_{i=1}^m (\theta_i - \theta_{ni})$$

and create the surrogate function

$$\begin{aligned} g(\theta \mid \theta_n) &= l\ln\Gamma\Big(\sum_{i=1}^m \theta_{ni}\Big) + l\psi\Big(\sum_{i=1}^m \theta_{ni}\Big)\sum_{i=1}^m(\theta_i - \theta_{ni}) \\ &\quad - l\sum_{i=1}^m \ln\Gamma(\theta_i) + \sum_{j=1}^l\sum_{i=1}^m(\theta_i - 1)\ln y_{ji}. \end{aligned}$$

Owing to the presence of the terms $\ln \Gamma(\theta_i)$, the maximization step is intractable. However, the MM gradient algorithm can be readily implemented because the parameters are now separated and the functions $\psi(t)$ and $\psi'(t)$ can be computed as suggested in Problem 18. The whole process is carried out on actual data in the references [18, 23]. ■

14.9 Quasi-Newton Methods

Quasi-Newton methods of maximum likelihood update the current approximation A_n to the observed information $-d^2L(\theta_n)$ by a low-rank perturbation satisfying a secant condition. The secant condition originates from the first-order Taylor approximation

$$\nabla L(\theta_n) - \nabla L(\theta_{n+1}) \quad \approx \quad d^2L(\theta_{n+1})(\theta_n - \theta_{n+1}).$$

If we set

$$\begin{aligned} g_n &= \nabla L(\theta_n) - \nabla L(\theta_{n+1}) \\ s_n &= \theta_n - \theta_{n+1}, \end{aligned} \tag{14.14}$$

then the secant condition is $-A_{n+1}s_n = g_n$. The unique symmetric rank-one update to A_n satisfying the secant condition is furnished by Davidon's formula [7]

$$A_{n+1} = A_n - c_n v_n v_n^t \tag{14.15}$$

with constant c_n and vector v_n specified by

$$\begin{aligned} c_n &= \left[(g_n + A_n s_n)^t s_n\right]^{-1} \\ v_n &= g_n + A_n s_n. \end{aligned} \tag{14.16}$$

Historically, symmetric rank-two updates such as those associated with Davidon, Fletcher, and Powell (DFP) or with Broyden, Fletcher, Goldfarb, and Shanno (BFGS) were considered superior to the more parsimonious update (14.15). However, numerical analysts [5, 16] now better appreciate the virtues of Davidon's formula. To put it into successful practice, one must usually monitor A_n for positive definiteness. An immediate concern is that the constant c_n is undefined when the inner product $(g_n + A_n s_n)^t s_n = 0$. In such situations or when $|(g_n + A_n s_n)^t s_n|$ is small compared to $\|g_n + A_n s_n\|_2 \|s_n\|_2$, one can ignore the secant requirement and simply take $A_{n+1} = A_n$.

If A_n is positive definite and $c_n \leq 0$, then A_{n+1} is certainly positive definite. If $c_n > 0$, then it may be necessary to shrink c_n to maintain positive definiteness. In order for A_{n+1} to be positive definite, it is necessary that $\det A_{n+1} > 0$. In view of formula (7.10) of Chapter 7,

$$\det A_{n+1} = (1 - c_n v_n^t A_n^{-1} v_n) \det A_n,$$

and $\det A_{n+1} > 0$ requires $1 - c_n v_n^t A_n^{-1} v_n > 0$. Conversely, the condition

$$1 - c_n v_n^t A_n^{-1} v_n \quad > \quad 0 \tag{14.17}$$

is also sufficient to ensure positive definiteness of A_{n+1}. This fact can be most easily demonstrated by invoking the Sherman-Morrison formula

$$[A_n - c_n v_n v_n^t]^{-1} \quad = \quad A_n^{-1} + \frac{c_n}{1 - c_n v_n^t A_n^{-1} v_n} A_n^{-1} v_n [A_n^{-1} v_n]^t. \tag{14.18}$$

Formula (14.18) shows that $[A_n - c_n v_n v_n^t]^{-1}$ exists and is positive definite under condition (14.17). Since the inverse of a positive definite matrix is positive definite, it follows that $A_n - c_n v_n v_n^t$ is positive definite as well.

The above analysis suggests the possibility of choosing c_n so that not only does A_{n+1} remain positive definite, but $\det A_{n+1}$ always exceeds a small constant $\epsilon > 0$. This strategy can be realized by replacing c_n by

$$\min\left\{c_n, \left(1 - \frac{\epsilon}{\det A_n}\right)\frac{1}{v_n^t A_n^{-1} v_n}\right\}$$

in updating A_n. An even better strategy that monitors the condition number of A_n is sketched in Problem 22.

In successful applications of quasi-Newton methods, choice of the initial matrix A_1 is critical. Setting $A_1 = I$ is convenient, but often poorly scaled for a particular problem. A better choice is $A_1 = J(\theta_1)$ when the expected information matrix $J(\theta_1)$ is available. In some problems, $J(\theta)$ is expensive to compute and manipulate for general θ but cheap to compute and manipulate for special θ. The special θ can furnish good starting points for a quasi-Newton search. For instance, $J(\theta)$ can be diagonal in certain circumstances.

TABLE 14.4. Quasi-Newton Iterates for the AIDS Model

Iteration	**Step-halves**	θ_1	θ_2
1	0	0.0000	0.0000
2	2	0.0222	0.2490
3	2	0.2501	0.2624
4	3	0.3747	0.2517
5	0	0.3404	0.2568
6	0	0.3395	0.2565
7	0	0.3396	0.2565

Table 14.4 displays the performance of the quasi-Newton method on the AIDS data of this chapter. Comparison of Tables 14.3 and 14.4 demonstrates that the quasi-Newton method is nearly as fast as Newton's method. Reliability is another matter. The identity matrix approximation to the

Hessian is poor in this example, and the first step of the algorithm takes θ into a region of parameter space where the exponential function overflows for the argument $\theta_1 + i\theta_2$. This disaster is avoided by shrinking all parameter increments to have Euclidean length at most 1. Comparison of the two tables also shows that the quasi-Newton method experiences more step halving, a clear indication that it is struggling in the early iterations.

It is noteworthy that the strategy of updating the approximation A_n to $-d^2L(\theta_n)$ can be reformulated to update the approximation $H_n = A_n^{-1}$ to $-d^2L(\theta_n)^{-1}$ instead. Restating the secant condition $-A_{n+1}s_n = g_n$ as the inverse secant condition $-H_{n+1}g_n = s_n$ leads to the symmetric rank-one update

$$\begin{aligned} H_{n+1} &= H_n - b_n w_n w_n^t \\ b_n &= \left[(s_n + H_n g_n)^t g_n\right]^{-1} \\ w_n &= s_n + H_n g_n. \end{aligned} \tag{14.19}$$

This strategy has the advantage of avoiding explicit inversion of A_n in calculating the quasi-Newton direction $\Delta\theta_n = H_n \nabla L(\theta_n)$. However, monitoring positive definiteness of H_n forces us to invert it. Whichever strategy one adopts, monitoring positive definiteness is most readily accomplished by carrying forward simultaneously A_n and $H_n = A_n^{-1}$ and applying the Sherman-Morrison formula to update either H_n or A_n.

14.10 Accelerated MM

We now consider the question of how to accelerate the often excruciatingly slow convergence of the MM algorithm. The simplest device is to just double each MM step [8, 18]. Thus, if $F(x_n)$ is the MM algorithm map from R^p to R^p, then we move to $x_n + 2[F(x_n) - x_n]$ rather than to $F(x_n)$. Step doubling is a standard tactic that usually halves the number of iterations until convergence. However, in many problems something more radical is necessary. Because Newton's method enjoys exceptionally quick convergence in a neighborhood of the optimal point, an attractive strategy is to amend the MM algorithm so that it resembles Newton's method. The papers [13, 14, 19] take up this theme from the perspective of optimizing the objective function by Newton's method. It is also possible to apply Newton's method to find a root of the equation $\mathbf{0} = x - F(x)$. This alternative perspective has the advantage of dealing directly with the iterates of the MM algorithm. Let $G(x)$ denote the difference $G(x) = x - F(x)$. Because $G(x)$ has differential $dG(x) = I - dF(x)$, Newton's method iterates according to

$$x_{n+1} = x_n - dG(x_n)^{-1}G(x_n) = x_n - [I - dF(x_n)]^{-1}G(x_n). \tag{14.20}$$

If we can approximate $dF(x_n)$ by a low-rank matrix M, then we can replace $I - dF(x_n)$ by $I - M$ and explicitly form the inverse $(I - M)^{-1}$. Let us see where this strategy leads.

Quasi-Newton methods operate by secant approximations. It is easy to generate a secant condition by taking two MM iterates starting from the current point x_n. Close to the optimal point x_∞, the linear approximation

$$F \circ F(x_n) - F(x_n) \quad \approx \quad M[F(x_n) - x_n]$$

holds, where $M = dF(x_\infty)$. If v is the vector $F \circ F(x_n) - F(x_n)$ and u is the vector $F(x_n) - x_n$, then the secant condition is $Mu = v$. In fact, the best results may require several secant conditions $Mu_i = v_i$ for $i = 1, \ldots, q$, where $q \leq p$. These can be generated at the current iterate x_n and the previous $q - 1$ iterates. For convenience represent the secant conditions in the matrix form $MU = V$ for $U = (u_1, \ldots, u_q)$ and $V = (v_1, \ldots, v_q)$. Example 11.3.4 shows that the choice $M = V(U^tU)^{-1}U^t$ minimizes the Frobenius norm of M subject to the secant constraint $MU = V$. In practice, it is better to make a controlled approximation to $dF(x_\infty)$ than a wild guess.

To apply the approximation, we must invert the matrix $I - V(U^tU)^{-1}U^t$. Fortunately, we have the explicit inverse

$$[I - V(U^tU)^{-1}U^t]^{-1} \quad = \quad I + V[U^tU - U^tV]^{-1}U^t. \tag{14.21}$$

The reader can readily check this variant of the Sherman-Morrison formula. It is noteworthy that the $q \times q$ matrix $U^tU - U^tV$ is trivial to invert for q small even when p is large. With these results in hand, the Newton update (14.20) can be replaced by the quasi-Newton update

$$\begin{aligned} x_{n+1} &= x_n - [I - V(U^tU)^{-1}U^t]^{-1}[x_n - F(x_n)] \\ &= x_n - [I + V(U^tU - U^tV)^{-1}U^t][x_n - F(x_n)] \\ &= F(x_n) - V(U^tU - U^tV)^{-1}U^t[x_n - F(x_n)]. \end{aligned}$$

The special case $q = 1$ is interesting in its own right. A brief calculation shows that the quasi-Newton update for $q = 1$ is

$$\begin{aligned} x_{n+1} &= (1 - c_n)F(x_n) + c_nF \circ F(x_n) \qquad (14.22) \\ c_n &= -\frac{\|F(x_n) - x_n\|_2^2}{[F \circ F(x_n) - 2F(x_n) + x_n]^t[F(x_n) - x_n]}. \end{aligned}$$

Our quasi-Newton acceleration enjoys several desirable properties in high-dimensional problems. First, the computational effort per iteration is relatively light: two MM updates and a few matrix times vector multiplications. Second, memory demands are also light. If we fix q in advance, the most onerous requirement is storage of the secant matrices U and V. These two matrices can be updated by replacing the earliest retained secant pair by the latest secant pair generated. Third, the whole scheme is

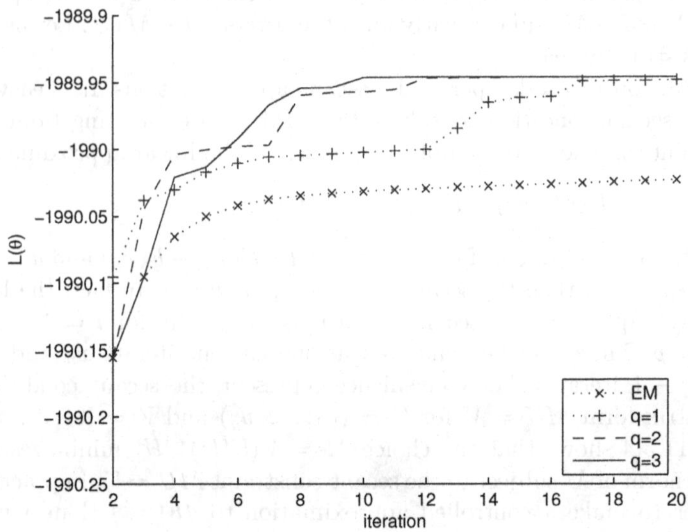

FIGURE 14.1. MM Acceleration for the Mixture of Poissons Example

consistent with linear constraints. Thus, if the parameter space satisfies a linear constraint $w^t x = a$ for all feasible x, then the quasi-Newton iterates also satisfy $w^t x_n = a$ for all n. This claim follows from the equalities $w^t F(x) = a$ and $w^t V = \mathbf{0}$. Finally, if the quasi-Newton update at x_n fails the ascent or descent test, then one can always revert to the second MM update $F \circ F(x_n)$. Balanced against these advantages is the failure of the quasi-Newton acceleration to respect parameter lower and upper bounds.

Example 14.10.1 *A Mixture of Poissons*

Problem 11 of Chapter 13 describes a Poisson mixture model for mortality data from *The London Times.* Starting from the method of moments estimates $(\mu_{01}, \mu_{02}, \pi_0) = (1.101, 2.582, .2870)$, the EM algorithm takes an excruciating 535 iterations for the loglikelihood $L(\theta)$ to attain its maximum of -1989.946. Even worse, it takes 1749 iterations for the parameters to reach the maximum likelihood estimates $(\hat{\mu_1}, \hat{\mu_2}, \hat{\pi}) = (1.256, 2.663, .3599)$. The sizable difference in convergence rates to the maximum loglikelihood and the maximum likelihood estimates indicates that the likelihood surface is quite flat. In contrast, the accelerated EM algorithm converges to the maximum loglikelihood in about 10 to 150 iterations, depending on the value of q. Figure 14.1 plots the progress of the EM algorithm and the different versions of the quasi-Newton acceleration. Titterington et al. [26] report that Newton's method typically takes 8 to 11 iterations to converge when it converges for these data. For about a third of their initial points, Newton's method fails. ■

14.11 Problems

1. Consider the map

$$f(x) = \begin{pmatrix} x_1^2 + x_2^2 - 2 \\ x_1 - x_2 \end{pmatrix}$$

of the plane into itself [9]. Show that $f(x) = \mathbf{0}$ has the roots $-\mathbf{1}$ and $\mathbf{1}$ and no other roots. Prove that Newton's method iterates according to

$$x_{n+1,1} = x_{n+1,2} = \frac{x_{n1}^2 + x_{n2}^2 + 2}{2(x_{n1} + x_{n2})}$$

and that these iterates converge to the root $-\mathbf{1}$ if $x_{01}+x_{02}$ is negative and to the root $\mathbf{1}$ if $x_{01} + x_{02}$ is positive. If $x_{01} + x_{02} = 0$, then the first iterate is undefined. Finally, prove that

$$\lim_{n\to\infty} \frac{|x_{n+1,1} - y_1|}{|x_{n1} - y_1|^2} = \lim_{n\to\infty} \frac{|x_{n+1,2} - y_2|}{|x_{n2} - y_2|^2} = \frac{1}{2},$$

where y is the root relevant to the initial point x_0.

2. Continuing Example 14.2.1, consider iterating according to

$$B_{n+1} = B_n \sum_{i=0}^{j} (I - AB_n)^i \tag{14.23}$$

to find A^{-1} [12]. Example 14.2.1 is the special case $j = 1$. Verify the alternative representation

$$B_{n+1} = \sum_{i=0}^{j} (I - B_n A)^i B_n,$$

and use it to prove that B_{n+1} is symmetric whenever A and B_n are. Also show that

$$A^{-1} - B_{n+1} = (A^{-1} - B_n)[A(A^{-1} - B_n)]^j.$$

From this last identity deduce the norm inequality

$$\|A^{-1} - B_{n+1}\| \le \|A\|^j \|A^{-1} - B_n\|^{j+1}.$$

Thus, the algorithm converges at a cubic rate when $j = 2$, at a quartic rate when $j = 3$, and so forth.

3. Problem 14 of Chapter 5 can be adapted to extract the mth root of a positive semidefinite matrix A [5]. Consider the iteration scheme

$$B_{n+1} = \frac{m-1}{m}B_n + \frac{1}{m}B_n^{-m+1}A$$

starting with $B_0 = cI$ for some positive constant c. Show by induction that (a) B_n commutes with A, (b) B_n is symmetric, and (c) B_n is positive definite. To prove that B_n converges to $A^{1/m}$, consider the spectral decomposition $A = UDU^t$ of A with D diagonal and U orthogonal. Show that B_n has a similar spectral decomposition $B_n = UD_nU^t$ and that the ith diagonal entries of D_n and D satisfy

$$d_{n+1,i} = \frac{m-1}{m}d_{ni} + \frac{1}{m}d_{ni}^{-m+1}d_i.$$

Problem 14 of Chapter 5 implies that d_{ni} converges to $\sqrt[m]{d_i}$ when $d_i > 0$. This convergence occurs at a fast quadratic rate. If $d_i = 0$, then d_{ni} converges to 0 at the slower linear rate $\frac{m-1}{m}$.

4. Program the algorithm of Problem 3 and extract the square roots of the two matrices

$$\begin{pmatrix} 1 & 1 \\ 1 & 1 \end{pmatrix}, \quad \begin{pmatrix} 2 & 1 \\ 1 & 2 \end{pmatrix}.$$

Describe the apparent rate of convergence in each case and any difficulties you encounter with roundoff error.

5. Prove that the increment (14.3) can be expressed as

$$\begin{aligned} \Delta\theta_n &= -A_n^{-1/2}\left[I - A_n^{-1/2}V^t(VA_n^{-1}V^t)^{-1}VA_n^{-1/2}\right]A_n^{-1/2}\nabla L(\theta_n) \\ &= -A_n^{-1/2}(I-P_n)A_n^{-1/2}\nabla L(\theta_n) \end{aligned}$$

using the symmetric square root $A_n^{-1/2}$ of A_n^{-1}. Check that the matrix P_n is a projection in the sense that $P_n^t = P_n$ and $P_n^2 = P_n$ and that these properties carry over to $I - P_n$. Now argue that

$$-dL(\theta_n)\Delta\theta_n = \|(I-P_n)A_n^{-1/2}\nabla L(\theta_n)\|_2^2$$

and consequently that backtracking is bound to produce a decrease in $L(\theta)$ if

$$(I-P_n)A_n^{-1/2}\nabla L(\theta_n) \neq \mathbf{0}.$$

6. Show that Newton's method converges in one iteration to the minimum of

$$f(\theta) = d + e^t\theta + \frac{1}{2}\theta^t F\theta$$

when the symmetric matrix F is positive definite. Note that this implies that the Gauss-Newton algorithm (14.12) converges in a single step when the regression functions $\mu_i(\theta)$ are linear.

7. Formally prove that the two expressions $\mathrm{E}[-d^2L(\theta)]$ and $\mathrm{Var}[\nabla L(\theta)]$ for the expected information $J(\theta)$ coincide. You may assume that integration and differentiation can be interchanged as needed.

8. Verify the score and information entries in Table 14.1.

9. Prove that the inverse matrix $\Sigma(\theta)^{-1}$ appearing in equation (14.10) can be replaced by the generalized inverse $\Sigma(\theta)^{-}$ [15]. (Hints: Show that the difference $h(X) - \mu(\theta)$ is almost surely in the range of $\Sigma(\theta)$ and hence that

$$\Sigma(\theta)\Sigma(\theta)^{-}[h(X) - \mu(\theta)] \quad = \quad h(X) - \mu(\theta)$$

almost surely. To validate the claim about the range of $h(X) - \mu(\theta)$, let P denote perpendicular projection onto the range of $\Sigma(\theta)$. Then show that $\mathrm{E}(\|(I - P)[h(X) - \mu(\theta)]\|_2^2) = 0$.)

10. A quantal response model involves independent binomial observations $x_1, \ldots, x_m$ with n_i trials and success probability $\pi_i(\theta)$ per trial for the ith observation. If z_i is a covariate vector and θ a parameter vector, then the specification

$$\pi_i(\theta) \quad = \quad \frac{e^{z_i^t\theta}}{1 + e^{z_i^t\theta}}$$

gives a generalized linear model. Estimate $\hat{\theta} = (-5.132, 0.0677)^t$ for the ingot data of Cox [6] displayed in Table 14.5.

TABLE 14.5. Ingot Data for a Quantal Response Model

Trials n_i	**Observation x_i**	**Covariate z_{i1}**	**Covariate z_{i2}**
55	0	1	7
157	2	1	14
159	7	1	27
16	3	1	57

11. Let $g(x)$ and $h(x)$ be probability densities defined on the real line. Show that the admixture density $f(x) = \theta g(x) + (1 - \theta)h(x)$ for $\theta \in [0, 1]$ has score and expected information

$$L'(\theta) \quad = \quad \frac{g(x) - h(x)}{\theta g(x) + (1 - \theta)h(x)}$$

$$\begin{aligned} J(\theta) &= \int \frac{[g(x)-h(x)]^2}{\theta g(x)+(1-\theta)h(x)}dx \\ &= \frac{1}{\theta(1-\theta)}\Big[1-\int \frac{g(x)h(x)}{\theta g(x)+(1-\theta)h(x)}dx\Big]. \end{aligned}$$

What happens to $J(\theta)$ when $g(x)$ and $h(x)$ coincide? What does $J(\theta)$ equal when $g(x)$ and $h(x)$ have nonoverlapping domains? (Hint: The identities

$$h-g = \frac{\theta g+(1-\theta)h-g}{1-\theta}, \quad g-h = \frac{\theta g+(1-\theta)h-h}{\theta}$$

will help.)

12. In robust regression it is useful to consider location-scale families with densities of the form

$$\frac{c}{\sigma}e^{-\rho(\frac{x-\mu}{\sigma})}, \qquad x \in (-\infty,\infty). \tag{14.24}$$

Here $\rho(r)$ is a strictly convex even function, decreasing to the left of 0 and symmetrically increasing to the right of 0. Without loss of generality, one can take $\rho(0)=0$. The normalizing constant c is determined by $c\int_{-\infty}^{\infty} e^{-\rho(r)}dr = 1$. Show that a random variable X with density (14.24) has mean μ and variance

$$\operatorname{Var}(X) = c\sigma^2\int_{-\infty}^{\infty} r^2 e^{-\rho(r)}\,dr.$$

If μ depends on a parameter vector θ, demonstrate that the score corresponding to a single observation $X=x$ amounts to

$$\nabla L(\phi) = \begin{pmatrix} \frac{1}{\sigma}\rho'(\frac{x-\mu}{\sigma})\nabla\mu(\theta) \\ -\frac{1}{\sigma}+\rho'(\frac{x-\mu}{\sigma})\frac{x-\mu}{\sigma^2} \end{pmatrix}$$

for $\phi=(\theta^t,\sigma)^t$. Finally, prove that the expected information $J(\phi)$ is block diagonal with upper-left block

$$\frac{c}{\sigma^2}\int_{-\infty}^{\infty}\rho''(r)e^{-\rho(r)}\,dr\nabla\mu(\theta)d\mu(\theta)$$

and lower-right block

$$\frac{c}{\sigma^2}\int_{-\infty}^{\infty}\rho''(r)r^2e^{-\rho(r)}\,dr+\frac{1}{\sigma^2}.$$

13. In the context of Problem 12, take $\rho(r)=\ln\cosh^2(\frac{r}{2})$. Show that this corresponds to the logistic distribution with density

$$f(x) = \frac{e^{-x}}{(1+e^{-x})^2}.$$

Compute the integrals

$$\begin{aligned}
\frac{\pi^2}{3} &= c\int_{-\infty}^{\infty} r^2 e^{-\rho(r)}\,dr \\
\frac{1}{3} &= c\int_{-\infty}^{\infty} \rho''(r) e^{-\rho(r)}\,dr \\
\frac{1}{3}+\frac{\pi^2}{9} &= c\int_{-\infty}^{\infty} \rho''(r) r^2 e^{-\rho(r)}\,dr + 1
\end{aligned}$$

determining the variance and expected information of the density (14.24) for this choice of $\rho(r)$.

14. Continuing Problems 12 and 13, compute the normalizing constant c and the three integrals determining the variance and expected information for Huber's function

$$\rho(r) = \begin{cases} \frac{r^2}{2} & |r| \le k \\ k|r| - \frac{k^2}{2} & |r| > k. \end{cases}$$

15. A family of discrete density functions $p_j(\theta)$ defined on $\{0, 1, \ldots\}$ and indexed by a parameter $\theta > 0$ is said to be a power series family if for all j

$$p_j(\theta) = \frac{c_j \theta^j}{g(\theta)}, \tag{14.25}$$

where $c_j \ge 0$ and $g(\theta) = \sum_{k=0}^{\infty} c_k \theta^k$ is the appropriate normalizing constant. If $x_1, \ldots, x_m$ are independent observations from the discrete density (14.25), then show that the maximum likelihood estimate of θ is a root of the equation

$$\frac{1}{m}\sum_{i=1}^{m} x_i = \frac{\theta g'(\theta)}{g(\theta)}. \tag{14.26}$$

Prove that the expected information in a single observation is

$$J(\theta) = \frac{\sigma^2(\theta)}{\theta^2},$$

where $\sigma^2(\theta)$ is the variance of the density (14.25).

16. Continuing problem 15, equation (14.26) suggests that one can find the maximum likelihood estimate $\hat{\theta}$ by iterating via

$$\theta_{n+1} = \frac{\bar{x} g(\theta_n)}{g'(\theta_n)} = f(\theta_n),$$

where $\bar{x}$ is the sample mean. The question now arises whether this iteration scheme is likely to converge to $\hat{\theta}$. Local convergence hinges on the condition $|f'(\hat{\theta})| < 1$. When this condition is true, the map $\theta_{n+1} = f(\theta_n)$ is locally contractive near the fixed point $\hat{\theta}$. Prove that

$$f'(\hat{\theta}) \quad = \quad 1 - \frac{\sigma^2(\hat{\theta})}{\mu(\hat{\theta})},$$

where

$$\mu(\theta) \quad = \quad \frac{\theta g'(\theta)}{g(\theta)}$$

is the mean of a single realization. Thus, convergence depends on the ratio of the variance to the mean. (Hints: By differentiating $g(\theta)$ it is easy to compute the mean and the second factorial moment

$$\mathrm{E}[X(X-1)] \quad = \quad \frac{\theta^2 g''(\theta)}{g(\theta)}.$$

Substitute this in $f'(\hat{\theta})$, recall $\mathrm{Var}(X) = \mathrm{E}[X(X-1)] + \mathrm{E}(X) - \mathrm{E}(X)^2$, and invoke equality (14.26).)

17. In the Gauss-Newton algorithm (14.12), the matrix

$$\sum_{i=1}^m w_i \nabla \mu_i(\theta_n) d\mu(\theta_n)$$

can be singular or nearly so. To cure this ill, Marquardt suggested substituting

$$A_n \quad = \quad \sum_{i=1}^m w_i \nabla \mu_i(\theta_n) d\mu(\theta_n) + \lambda I$$

for it and iterating according to

$$\theta_{n+1} \quad = \quad \theta_n + A_n^{-1} \sum_{i=1}^m w_i [x_i - \mu_i(\theta_n)] \nabla \mu_i(\theta_n). \qquad (14.27)$$

Prove that the increment $\Delta\theta_n = \theta_{n+1} - \theta_n$ proposed in equation (14.27) minimizes the criterion

$$\frac{1}{2} \sum_{i=1}^m w_i [x_i - \mu_i(\theta_n) - d\mu_i(\theta_n) \Delta\theta_n]^2 + \frac{\lambda}{2} \|\Delta\theta_n\|_2^2.$$

18. In Example 14.8.2, digamma and trigamma functions must be evaluated. Show that these functions satisfy the recurrence relations

$$\begin{aligned} \psi(t) &= -t^{-1} + \psi(t+1) \\ \psi'(t) &= t^{-2} + \psi'(t+1). \end{aligned}$$

Thus, if $\psi(t)$ and $\psi'(t)$ can be accurately evaluated via asymptotic expansions for large t, then they can be accurately evaluated for small t. For example, Stirling's formula and its extension give

$$\begin{aligned} \psi(t) &= \ln t - (2t)^{-1} + O(t^{-2}) \\ \psi'(t) &= t^{-1} + (\sqrt{2}t)^{-2} + O(t^{-3}) \end{aligned}$$

as $t \to \infty$.

19. Compute the score vector and the observed and expected information matrices for the Dirichlet distribution (14.13). Explicitly invert the expected information using the Sherman-Morrison formula.

20. Consider the quadratic function

$$L(\theta) = -(1,1)\theta - \frac{1}{2}\theta^t \begin{pmatrix} 2 & 1 \\ 1 & 1 \end{pmatrix} \theta$$

defined on R^2. Compute the iterates of the quasi-Newton method for maximizing $L(\theta)$ using the inverse update (14.19) for H_n and starting from $\theta_1 = \mathbf{0}$ and $H_1 = \begin{pmatrix} 1 & 0 \\ 0 & 1 \end{pmatrix}$.

21. Let A be a positive definite matrix. Prove [2] that

$$\operatorname{tr}(A) - \ln \det(A) \geq \ln[\operatorname{cond}_2(A)], \tag{14.28}$$

where $\operatorname{cond}_2(A) = \|A\|_2 \|A^{-1}\|_2$. (Hint: Express $\operatorname{tr}(A) - \ln\det(A)$ in terms of the eigenvalues of A. Then use the inequalities $\lambda - \ln \lambda \geq 1$ and $\lambda > 2 \ln \lambda$ for all $\lambda > 0$.)

22. In Davidon's symmetric rank-one update (14.15), it is possible to control the condition number of A_{n+1} by shrinking the constant c_n. Suppose a moderately sized number d is chosen. Due to inequality (14.28), one can avoid ill-conditioning in the matrices A_n by imposing the constraint $\operatorname{tr}(A_n) - \ln\det(A_n) \leq d$. To see how this fits into the updating scheme (14.15), verify that

$$\begin{aligned} \ln\det(A_{n+1}) &= \ln\det(A_n) + \ln(1 - c_n v_n^t A_n^{-1} v_n) \\ \operatorname{tr}(A_{n+1}) &= \operatorname{tr}(A_n) - c_n \|v_n\|_2^2. \end{aligned}$$

Employing these results, deduce that $\operatorname{tr}(A_{n+1}) - \ln\det(A_{n+1}) \leq d$ provided c_n satisfies

$$-c_n\|v_n\|_2^2 - \ln(1 - c_n v_n^t A_n^{-1} v_n) \leq d - \operatorname{tr}(A_n) + \ln\det(A_n).$$

23. Survival analysis deals with nonnegative random variables T modeling random lifetimes. Let such a random variable $T \geq 0$ have density function $f(t)$ and distribution function $F(t)$. The hazard function

$$\begin{aligned} h(t) &= \lim_{s \downarrow 0} \frac{\Pr(t < T \leq t + s \mid T > t)}{s} \\ &= \frac{f(t)}{1 - F(t)} \end{aligned}$$

represents the instantaneous rate of death under lifetime T. Statisticians call the right-tail probability $1 - F(t) = S(t)$ the survival function and view $h(t)$ as the derivative

$$h(t) = -\frac{d}{dt} \ln S(t).$$

The cumulative hazard function $H(t) = \int_0^t h(s)ds$ obviously satisfies the identity

$$S(t) = e^{-H(t)}.$$

In Cox's proportional hazards model, longevity depends not only on time but also covariates. This is formalized by taking

$$h(t) = \lambda(t)e^{x^t \alpha},$$

where x and α are column vectors of predictors and regression coefficients, respectively. For instance, x might be $(1, d)^t$, where d indicates dosage of a life-prolonging drug.

Many clinical trials involve right censoring. In other words, instead of observing a lifetime $T = t$, we observe $T > t$. Censored and ordinary data can be mixed in the same study. Generally, each observation T comes with a censoring indicator W. If T is censored, then $W = 1$; otherwise, $W = 0$.

(a) Show that

$$H(t) = \Lambda(t)e^{x^t \alpha},$$

where

$$\Lambda(t) = \int_0^t \lambda(s)ds.$$

In the Weibull proportional hazards model, $\lambda(t) = \beta t^{\beta - 1}$. Show that this translates into the survival and density functions

$$\begin{aligned} S(t) &= e^{-t^\beta e^{x^t \alpha}} \\ f(t) &= \beta t^{\beta - 1} e^{x^t \alpha - t^\beta e^{x^t \alpha}}. \end{aligned}$$

(b) Consider independent possibly censored observations $t_1, \ldots, t_m$ with corresponding predictor vectors $x_1 \ldots, x_m$ and censoring indicators $w_1, \ldots, w_m$. Prove that the loglikelihood of the data is

$$L(\alpha, \beta) = \sum_{i=1}^{m} w_i \ln S_i(t_i) + \sum_{i=1}^{m} (1 - w_i) \ln f_i(t_i),$$

where $S_i(t)$ and $f_i(t)$ are the survival and density functions of the ith case.

(c) Calculate the score and observed information for the Weibull model as posed. The observed information is

$$\begin{aligned} -d^2 L(\alpha, \beta) &= \sum_{i=1}^{m} t_i^{\beta} e^{x_i^t \alpha} \begin{pmatrix} x_i \\ \ln t_i \end{pmatrix} \begin{pmatrix} x_i \\ \ln t_i \end{pmatrix}^t \\ &\quad + \sum_{i=1}^{m} (1 - w_i) \begin{pmatrix} 0 & 0 \\ 0 & \beta^{-2} \end{pmatrix}. \end{aligned}$$

(d) Show that the loglikelihood $L(\alpha, \beta)$ for the Weibull model is concave. Demonstrate that it is strictly concave if and only if the m vectors $x_1, \ldots, x_m$ span R^p, where α has p components.

24. In the survival model of Problem 23, implement Newton's method for finding the maximum likelihood estimate of the parameter vector (α, β). What difficulties do you encounter? Why is concavity of the loglikelihood helpful?

25. Write a computer program and reproduce the iterates displayed in Table 14.4.

26. Let $x_1, \ldots, x_m$ be a random sample from the gamma density

$$f(x) = \Gamma(\alpha)^{-1} \beta^{\alpha} x^{\alpha - 1} e^{-\beta x}$$

on $(0, \infty)$. Find the score, observed information, and expected information for the parameters α and β, and demonstrate that Newton's method and scoring coincide.

27. Continuing Problem 26, derive the method of moments estimators

$$\hat{\alpha} = \frac{\overline{x}^2}{s^2}, \qquad \hat{\beta} = \frac{\overline{x}}{s^2},$$

where $\overline{x} = \frac{1}{m} \sum_{i=1}^m x_i$ and $s^2 = \frac{1}{m} \sum_{i=1}^m (x_i - \overline{x})^2$ are the sample mean and variance, respectively. These are not necessarily the best explicit estimators of the two parameters. Show that setting the score function equal to $\mathbf{0}$ implies that $\beta = \alpha/\bar{x}$ is a stationary point of the loglikelihood $L(\alpha, \beta)$ of the sample $x_1, \ldots, x_m$ for α fixed. Why does $\beta = \alpha/\bar{x}$ furnish the maximum? Now argue that substituting it in the loglikelihood reduces maximum likelihood estimation to optimization of the profile loglikelihood

$$L(\alpha) = m\alpha \ln \alpha - m\alpha \ln \bar{x} - m \ln \Gamma(\alpha) + m(\alpha - 1)\overline{\ln x} - m\alpha.$$

Here $\overline{\ln x} = \frac{1}{m}\sum_{i=1}^m \ln x_i$. There are two nasty terms in $L(\alpha)$. One is $\alpha \ln \alpha$, and the other is $\ln \Gamma(\alpha)$. We can eliminate both by appealing to a version of Stirling's formula. Ordinarily Stirling's formula is only applied for large factorials. This limitation is inconsistent with small α. However, Gosper's version of Stirling's formula is accurate for all arguments. This little-known version of Stirling's formula says that

$$\Gamma(\alpha + 1) \approx \sqrt{(\alpha + 1/6)2\pi}\,\alpha^\alpha e^{-\alpha}.$$

Given that $\Gamma(\alpha) = \Gamma(\alpha + 1)/\alpha$, show that the application of Gosper's formula leads to the approximate maximum likelihood estimate

$$\hat{\alpha} = \frac{3 - d + \sqrt{(3-d)^2 + 24d}}{12d},$$

where $d = \ln \bar{x} - \overline{\ln x}$ [4]. Why is this estimate of α positive? Why does one take the larger root of the defining quadratic?

28. In the multilogit model, items are draw from m categories. Let y_i denote the ith of l independent draws and x_i a corresponding predictor vector. The probability π_{ij} that $y_i = j$ is given by

$$\pi_{ij}(\theta) = \begin{cases} \frac{e^{x_i^t \theta_j}}{1+\sum_{k=1}^{m-1} e^{x_i^t \theta_k}} & 1 \le j < m \\ \frac{1}{1+\sum_{k=1}^{m-1} e^{x_i^t \theta_k}} & j = m\,. \end{cases}$$

Find the loglikelihood, score, observed information, and expected information. Demonstrate that Newton's method and scoring coincide. (Hint: You can achieve compact expressions by stacking vectors and using matrix Kronecker products.)

29. Derive formulas (14.21) and (14.22).

14.12 References

[1] Bradley EL (1973) The equivalence of maximum likelihood and weighted least squares estimates in the exponential family. *J Amer Stat Assoc* 68:199–200

[2] Byrd RH, Nocedal J (1989) A tool for the analysis of quasi-Newton methods with application to unconstrained minimization. *SIAM J Numer Anal* 26:727–739

[3] Charnes A, Frome EL, Yu PL (1976) The equivalence of generalized least squares and maximum likelihood in the exponential family. *J Amer Stat Assoc* 71:169–171

[4] Choi SC, Wette R (1969) Maximum likelihood estimation of the parameters of the gamma distribution and their bias. *Technometrics* 11:683–690

[5] Conn AR, Gould NIM, Toint PL (1991) Convergence of quasi-Newton matrices generated by the symmetric rank-one update. *Math Prog* 50:177–195

[6] Cox DR (1970) *Analysis of Binary Data.* Methuen, London

[7] Davidon WC (1959) Variable metric methods for minimization. *AEC Research and Development Report ANL–5990*, Argonne National Laboratory, USA

[8] De Leeuw J, Heiser WJ (1980) Multidimensional scaling with restrictions on the configuration. In *Multivariate Analysis, Volume V*, Krishnaiah PR, North-Holland, Amsterdam, pp 501-522

[9] de Souza PN, Silva J-N (2001) *Berkeley Problems in Mathematics*, 2nd ed. Springer, New York

[10] Dobson AJ (1990) *An Introduction to Generalized Linear Models.* Chapman & Hall, London

[11] Green PJ (1984) Iteratively reweighted least squares for maximum likelihood estimation and some robust and resistant alternatives (with discussion). *J Roy Stat Soc B* 46:149–192

[12] Householder AS (1975) *The Theory of Matrices in Numerical Analysis.* Dover, New York

[13] Jamshidian M, Jennrich RI (1995) Acceleration of the EM algorithm by using quasi-Newton methods. *J Roy Stat Soc B* 59:569–587

[14] Jamshidian M, Jennrich RI (1997) Quasi-Newton acceleration of the EM algorithm. *J Roy Stat Soc B* 59:569–587

[15] Jennrich RI, Moore RH (1975) Maximum likelihood estimation by means of nonlinear least squares. *Proceedings of the Statistical Computing Section: Amer Stat Assoc* 57–65

[16] Khalfan HF, Byrd RH, Schnabel RB (1993) A theoretical and experimental study of the symmetric rank-one update. *SIAM J Optimization* 3:1–24

[17] Kingman JFC (1993) *Poisson Processes.* Oxford University Press, Oxford

[18] Lange K (1995) A gradient algorithm locally equivalent to the EM algorithm. *J Roy Stat Soc B* 57:425–437

[19] Lange K (1995) A quasi-Newton acceleration of the EM algorithm. *Statistica Sinica* 5:1–18

[20] Lange K (2004) *Optimization.* Springer, New York

[21] Lehmann EL (1986) *Testing Statistical Hypotheses*, 2nd ed. Wiley, New York

[22] Magnus JR, Neudecker H (1988) *Matrix Differential Calculus with Applications in Statistics and Econometrics.* Wiley, New York

[23] Narayanan A (1991) Algorithm AS 266: maximum likelihood estimation of the parameters of the Dirichlet distribution. *Appl Stat* 40:365–374

[24] Nelder JA, Wedderburn RWM (1972) Generalized linear models. *J Roy Stat Soc A* 135:370–384

[25] Rao CR (1973) *Linear Statistical Inference and its Applications*, 2nd ed. Wiley, New York

[26] Titterington DM, Smith AFM, Makov UE (1985) *Statistical Analysis of Finite Mixture Distributions.* Wiley, New York

[27] Whyte BM, Gold J, Dobson AJ, Cooper DA (1987) Epidemiology of acquired immunodeficiency syndrome in Australia. *Med J Aust* 147:65–69

15

Local and Global Convergence

15.1 Introduction

Proving convergence of the various optimization algorithms is a delicate exercise. In general, it is helpful to consider local and global convergence patterns separately. The local convergence rate of an algorithm provides a useful benchmark for comparing it to other algorithms. On this basis, Newton's method wins hands down. However, the tradeoffs are subtle. Besides the sheer number of iterations until convergence, the computational complexity and numerical stability of an algorithm are critically important. The MM algorithm is often the epitome of numerical stability and computational simplicity. Scoring lies somewhere between these two extremes. It tends to converge more quickly than the MM algorithm and to behave more stably than Newton's method. Quasi-Newton methods also occupy this intermediate zone. Because the issues are complex, all of these algorithms survive and prosper in certain computational niches.

The following overview of convergence manages to cover only some highlights. The books [2, 9, 14] provide a fuller survey. Quasi-Newton methods are given especially short shrift here. The efforts of a generation of numerical analysts in understanding quasi-Newton methods defy easy summary or digestion. Interested readers can consult one of the helpful references [2, 4, 9, 12]. We emphasize the MM and related algorithms, partially because a fairly coherent theory for them can be reviewed in a few pages.

15.2 Calculus Preliminaries

As a prelude to our study of convergence, let us review some ideas from advanced calculus. A function $h : \mathsf{R}^m \to \mathsf{R}^n$ is differentiable at a point $x \in \mathsf{R}^m$ if and only if an $n \times m$ matrix A exists such that

$$\|h(x+w) - h(x) - Aw\| \quad = \quad o(\|w\|)$$

as $\|w\| \to 0$. Because of the equivalence of vector norms, any pair of norms on R^m and R^n will do in this definition. The matrix A is typically written $dh(x)$. Its ith row consists of the partial derivatives of the ith component $h_i(x)$ of $h(x)$. To avoid certain pathologies, we usually make the simplifying assumption that all first partial derivatives of $h(x)$ exist and are continuous. This continuity assumption guarantees that the differential of $h(x)$ exists

K. Lange, *Numerical Analysis for Statisticians*, Statistics and Computing,
DOI 10.1007/978-1-4419-5945-4_15, © Springer Science+Business Media, LLC 2010

for all x and can be identified with the Jacobi matrix $dh(x)$. Differentiability of $h(x)$ obviously entails continuity of $h(x)$.

We will need an inequality substitute for the mean value theorem. The desired bound can be best developed by introducing the vector-valued integral $\int_a^b g(t)dt$ of a continuous vector-valued function $g : [a, b] \to \mathsf{R}^n$. The components of $\int_a^b g(t)dt$ are just the integrals $\int_a^b g_i(t)dt$ of the components $g_i(t)$ of $g(t)$. If $a = t_0 < t_1 < \cdots < t_{k-1} < t_k = b$ is a partition of $[a, b]$, the Riemann sum $\sum_{i=1}^k g(t_i)(t_i - t_{i-1})$ approximates the integral $\int_a^b g(t)dt$ and satisfies the norm inequality

$$\Big\|\sum_{i=1}^k g(t_i)(t_i - t_{i-1})\Big\| \;\le\; \sum_{i=1}^k \|g(t_i)\|(t_i - t_{i-1}).$$

Passing to the limit as the mesh size $\max_i(t_i - t_{i-1}) \to 0$, one can readily verify that $\|\int_a^b g(t)dt\| \le \int_a^b \|g(t)\|dt$. Applying this inequality, the fundamental theorem of calculus, and the chain rule leads to the bound

$$\begin{aligned}
\|h(y) - h(x)\| &= \Big\|\int_0^1 dh[x + t(y-x)](y-x)dt\Big\| \\
&\le \int_0^1 \|dh[x + t(y-x)](y-x)\|dt \\
&\le \int_0^1 \|dh[x + t(y-x)]\| \cdot \|y - x\|dt \qquad (15.1)\\
&\le \sup_{t\in[0,1]} \|dh[x + t(y-x)]\| \cdot \|y - x\|.
\end{aligned}$$

The mean value inequality (15.1) can be improved. Suppose that along the line segment $\{z = x + t(y-x) : t \in [0,1]\}$ the differential $dh(z)$ satisfies the Lipschitz inequality

$$\|dh(u) - dh(v)\| \;\le\; \lambda\|u - v\| \qquad (15.2)$$

for some constant $\lambda > 0$. This is the case if the second differential $d^2h(z)$ exists and is continuous in z, for then inequality (15.2) follows from an analog of inequality (15.1). Assuming the truth of inequality (15.2), we find that

$$\begin{aligned}
&\|h(y) - h(x) - dh(x)(y-x)\| \\
&= \Big\|\int_0^1 \{dh[x + t(y-x)] - dh(x)\}(y-x)dt\Big\| \\
&\le \int_0^1 \|dh[x + t(y-x)] - dh(x)\| \cdot \|y - x\|dt \qquad (15.3)\\
&\le \lambda\|y - x\|^2 \int_0^1 t\,dt \\
&= \frac{\lambda}{2}\|y - x\|^2.
\end{aligned}$$

15.3 Local Rates of Convergence

Local convergence of many optimization algorithms hinges on the following result of Ostrowski [13, 15].

Proposition 15.3.1 *Let the map $M : \mathsf{R}^m \to \mathsf{R}^m$ have fixed point x_∞. If $M(x)$ is differentiable at x_∞, and the spectral radius $\rho[dM(x_\infty)]$ of its differential satisfies $\rho[dM(x_\infty)] < 1$, then the iterates $x_{n+1} = M(x_n)$ are locally attracted to x_∞ at a linear rate or better.*

Proof: As mentioned in Proposition 6.3.2, there exists a vector norm $\|x\|$ such that the induced matrix norm $\|dM(x_\infty)\|$ comes arbitrarily close to the spectral radius $\rho[dM(x_\infty)]$. Accordingly, choose an appropriate norm with $\|dM(x_\infty)\| = \sigma < 1$ and then a constant $\epsilon > 0$ with $\epsilon + \sigma < 1$. Because $M(x)$ is differentiable, there is a ball $B = \{x : \|x - x_\infty\| < \delta\}$ such that

$$\|M(x) - M(x_\infty) - dM(x_\infty)(x - x_\infty)\| \quad \leq \quad \epsilon\|x - x_\infty\|$$

for $x \in B$. It follows that $x \in B$ implies

$$\begin{aligned} \|M(x) - x_\infty\| &= \|M(x) - M(x_\infty)\| \\ &\leq \|M(x) - M(x_\infty) - dM(x_\infty)(x - x_\infty)\| \\ &\quad + \|dM(x_\infty)(x - x_\infty)\| \\ &\leq (\epsilon + \sigma)\|x - x_\infty\|. \end{aligned}$$

One can now argue inductively that if the initial iterate x_1 belongs to B, then all subsequent iterates $x_{n+1} = M(x_n)$ belong to B as well and that

$$\|x_{n+1} - x_\infty\| \quad \leq \quad (\sigma + \epsilon)^n \|x_1 - x_\infty\|.$$

In other words, x_n converges to x_∞ at least as fast as $(\sigma + \epsilon)^n \to 0$. ■

Our intention is to apply Ostrowski's result to iteration maps of the type

$$M(x) \quad = \quad x - A(x)^{-1}\nabla f(x), \tag{15.4}$$

where $f(x)$ is an objective function such as a sum of squares or a loglikelihood and $A(x)$ equals $d^2 f(x)$ or an approximation to it. For instance, when $f(x)$ is a loglikelihood, $-A(x)$ is the corresponding observed information in Newton's method and the expected information in scoring. In an MM gradient algorithm for minimizing $f(x)$, we take $A(x)$ to be the Hessian $d^{20}g(x \mid x)$ of the surrogate function $g(x \mid x)$. For the sake of convenience in the remainder of this chapter, we limit ourselves to minimization.

Our first order of business is to compute the differential $dM(x_\infty)$ at a local optimum x_∞ of $f(x)$. If x_∞ is a strict local minimum, then ordinarily $d^2 f(x_\infty)$ is positive definite. We make this assumption as well as the assumption that $d^2 f(x)$ is continuous in a neighborhood of x_∞. Thus, the iteration map is certainly well defined in some neighborhood of x_∞ for Newton's method. For scoring and the MM gradient algorithm, we likewise assume that $A(x)$ is continuous and positive definite in a neighborhood of x_∞.

Because $\nabla f(x_\infty) = \mathbf{0}$, the differential of $A(x)^{-1}$ at $x = x_\infty$ is irrelevant in determining $dM(x_\infty)$. Thus, it is plausible to conjecture that

$$dM(x_\infty) = I - A(x_\infty)^{-1} d^2 f(x_\infty).$$

This claim can be verified by noting that

$$\begin{aligned}
& M(x) - M(x_\infty) - dM(x_\infty)(x - x_\infty) \\
= \;& x - A(x)^{-1}\nabla f(x) - x_\infty - dM(x_\infty)(x - x_\infty) \\
= \;& -A(x)^{-1}\nabla f(x) + [I - dM(x_\infty)](x - x_\infty) \\
= \;& -A(x)^{-1}\nabla f(x) + A(x_\infty)^{-1} d^2 f(x_\infty)(x - x_\infty).
\end{aligned}$$

As a consequence of this representation and the identity $\nabla f(x_\infty) = \mathbf{0}$, we deduce the inequality

$$\begin{aligned}
& \|M(x) - M(x_\infty) - dM(x_\infty)(x - x_\infty)M(x)\| \\
= \;& \| - A(x)^{-1}\nabla f(x) + A(x_\infty)^{-1} d^2 f(x_\infty)(x - x_\infty)\| \\
\le \;& \|A(x)^{-1}[A(x_\infty) - A(x)]A(x_\infty)^{-1} d^2 f(x_\infty)(x - x_\infty)\| \\
& + \|A(x)^{-1}[\nabla f(x) - \nabla f(x_\infty) - d^2 f(x_\infty)(x - x_\infty)]\| \\
\le \;& \|A(x)^{-1}\| \cdot \|A(x_\infty) - A(x)\| \cdot \|A(x_\infty)^{-1}\| \cdot \|d^2 f(x_\infty)\| \cdot \|x - x_\infty\| \\
& + \|A(x)^{-1}\| o(\|x - x_\infty\|).
\end{aligned}$$

Because $\|A(x_\infty) - A(x)\| \to 0$ as $\|x - x_\infty\| \to 0$ and $\|A(x)^{-1}\|$ is bounded in a neighborhood of x_∞, the overall error in the linear approximation of $M(x)$ around x_∞ is consequently $o(\|x - x_\infty\|)$.

Calculation of the differential $dM(x_\infty)$ of an MM algorithm map at a local minimum x_∞ is equally interesting. In Section 16.6.1 we prove that

$$\begin{aligned}
dM(x_\infty) &= d^{20}g(x_\infty \mid x_\infty)^{-1}[d^{20}g(x_\infty \mid x_\infty) - d^2 f(x_\infty)] \\
&= I - d^{20}g(x_\infty \mid x_\infty)^{-1} d^2 f(x_\infty), \qquad (15.5)
\end{aligned}$$

which is precisely the differential just computed for the MM gradient algorithm. Thus, substituting a single Newton step for the full solution of the optimization step of the MM algorithm does not slow convergence locally.

Proposition 15.3.2 *Both the MM algorithm and the MM gradient algorithm are locally attracted to a local optimum x_∞ at a linear rate equal to the spectral radius of $I - d^{20}g(x_\infty \mid x_\infty)^{-1} d^2 f(x_\infty)$.*

Proof: Let $M(x)$ be the iteration map. According to Proposition 15.3.1, it suffices to show that all eigenvalues of the differential $dM(x_\infty)$ lie on the half-open interval $[0, 1)$. In view of the discussion at the end of Section 8.3, $dM(x_\infty)$ has eigenvalues determined by the stationary values of the Rayleigh quotient

$$\begin{aligned} R(v) &= \frac{v^t[d^{20}g(x_\infty \mid x_\infty) - d^2f(x_\infty)]v}{v^t d^{20}g(x_\infty \mid x_\infty)v} \\ &= 1 - \frac{v^t d^2 f(x_\infty)v}{v^t d^{20}g(x_\infty \mid x_\infty)v}. \end{aligned} \tag{15.6}$$

Because both $d^2f(x_\infty)$ and $d^{20}g(x_\infty \mid x_\infty)$ are positive definite, $R(v) < 1$ for all vectors v of unit length. It follows that the maximum of $R(v)$ is strictly less than 1. On the other hand, $R(v) \geq 0$ since the difference $d^{20}g(x_\infty \mid x_\infty) - d^2f(x_\infty)$ is positive semidefinite. ■

The next proposition validates local convergence of Newton's method.

Proposition 15.3.3 *Newton's method is locally attracted to a local optimum x_∞ at a rate faster than linear. If the second differential $d^2f(x)$ satisfies*

$$\|d^2f(y) - d^2f(x)\| \leq \lambda\|y - x\| \tag{15.7}$$

in some neighborhood of x_∞, then the Newton iterates x_n satisfy

$$\|x_{n+1} - x_\infty\| \leq 2\lambda\|d^2f(x_\infty)^{-1}\| \cdot \|x_n - x_\infty\|^2 \tag{15.8}$$

close to x_∞.

Proof: If $M(x)$ represents the Newton iteration map, then

$$dM(x_\infty) = I - d^2f(x_\infty)^{-1}d^2f(x_\infty) = \mathbf{0}.$$

Hence, Proposition 15.3.1 implies local attraction to x_∞ at a rate faster than linear. If, in addition, inequality (15.7) holds, then inequality (15.3) is true for $h(x) = \nabla f(x)$. Inequalities (15.3) and (15.7) together imply

$$\begin{aligned} & \|x_{n+1} - x_\infty\| \\ = \; & \|x_n - d^2f(x_n)^{-1}\nabla f(x_n) - x_\infty\| \\ \leq \; & \|-d^2f(x_n)^{-1}[\nabla f(x_n) - \nabla f(x_\infty) - d^2f(x_\infty)(x_n - x_\infty)]\| \\ & + \|d^2f(x_n)^{-1}[d^2f(x_n) - d^2f(x_\infty)](x_n - x_\infty)\| \\ \leq \; & \Big(\frac{\lambda}{2} + \lambda\Big)\|d^2f(x_n)^{-1}\| \cdot \|x_n - x_\infty\|^2, \end{aligned}$$

which yields inequality (15.8) for x_n sufficiently close to x_∞ by virtue of the assumed continuity and invertibility of $d^2f(x)$. ■

Local convergence of the scoring algorithm is not guaranteed by Proposition 15.3.1 because nothing prevents an eigenvalue of

$$dM(x_\infty) = I + J(x_\infty)^{-1}d^2L(x_\infty)$$

from falling below -1. Scoring with a fixed partial step as specified by

$$x_{n+1} = x_n + \alpha J(x_n)^{-1} \nabla L(x_n)$$

will converge locally for $\alpha > 0$ sufficiently small. In practice, no adjustment is usually necessary. For reasonably large sample sizes, the expected information matrix $J(x_\infty)$ approximates the observed information matrix $-d^2 L(x_\infty)$ well, and the spectral radius of $dM(x_\infty)$ is nearly 0.

Finally, let us consider local convergence of block relaxation. The argument $x = (x_{[1]}, x_{[2]}, \ldots, x_{[b]})$ of the objective function $f(x)$ now splits into disjoint blocks, and $f(x)$ is minimized along each block of components $x_{[i]}$ in turn. Let $M_i(x)$ denote the update to block i. To compute the differential of the full update $M(x)$ at a local optimum x_∞, we need compact notation. Set $y = x_\infty$ and let $d_i f(x)$ denote the partial differential of $f(x)$ with respect to block i; the transpose of $d_i f(x)$ is the partial gradient $\nabla_i f(x)$. The updates satisfy the partial gradient equations

$$\mathbf{0} = \nabla_i f[M_1(x), \ldots, M_i(x), x_{[i+1]}, \ldots, x_{[b]}]. \tag{15.9}$$

Now let $d_j \nabla_i f(x)$ denote the partial differential of the partial gradient $\nabla_i f(x)$ with respect to block j. Taking the partial differential of equation (15.9) with respect to block j, applying the chain rule, and substituting the optimal point $y = M(y)$ for x yield

$$\begin{aligned} \mathbf{0} &= \sum_{k=1}^{i} d_k \nabla_i f(y) d_j M_k(y), \quad j \le i \\ \mathbf{0} &= \sum_{k=1}^{i} d_k \nabla_i f(y) d_j M_k(y) + d_j \nabla_i f(y), \quad j > i. \end{aligned} \tag{15.10}$$

It is helpful to express these equations in block matrix form.

For example in the case of $b = 3$ blocks, the linear system of equations (15.10) can be represented as $L\, dM(y) = D - U$, where $U = L^t$ and

$$\begin{aligned} dM(y) &= \begin{pmatrix} d_1 M_1(y) & d_2 M_1(y) & d_3 M_1(y) \\ d_1 M_2(y) & d_2 M_2(y) & d_3 M_2(y) \\ d_1 M_3(y) & d_2 M_3(y) & d_3 M_3(y) \end{pmatrix} \\ L &= \begin{pmatrix} d_1 \nabla_1 f(y) & \mathbf{0} & \mathbf{0} \\ d_1 \nabla_2 f(y) & d_2 \nabla_2 f(y) & \mathbf{0} \\ d_1 \nabla_3 f(y) & d_2 \nabla_3 f(y) & d_3 \nabla_3 f(y) \end{pmatrix} \\ D &= \begin{pmatrix} d_1 \nabla_1 f(y) & \mathbf{0} & \mathbf{0} \\ \mathbf{0} & d_2 \nabla_2 f(y) & \mathbf{0} \\ \mathbf{0} & \mathbf{0} & d_3 \nabla_3 f(y) \end{pmatrix}. \end{aligned}$$

The identity $d_j \nabla_i f(y)^t = d_i \nabla_j f(y)$ between two nontrivial blocks of U and L is a consequence of the equality of mixed partials. The matrix equation

$L\,dM(y) = D - U$ can be explicitly solved in the form $dM(y) = L^{-1}(D-U)$. Here L is invertible provided its diagonal blocks $d_i\nabla_i f(y)$ are invertible. At an optimal point y, the partial Hessian matrix $d_i\nabla_i f(y)$ is always positive semidefinite and usually positive definite as well.

Local convergence of block relaxation hinges on whether the spectral radius ρ of the matrix $L^{-1}(U - D)$ satisfies $\rho < 1$. Suppose that λ is an eigenvalue of $L^{-1}(D - U)$ with eigenvector v. These can be complex. The equality $L^{-1}(D - U)v = \lambda v$ implies $(1 - \lambda)Lv = (L + U - D)v$. Premultiplying this by the conjugate transpose v^* gives

$$\frac{1}{1-\lambda} \;=\; \frac{v^* L v}{v^*(L+U-D)v}.$$

Hence, the real part of $1/(1 - \lambda)$ satisfies

$$\begin{aligned}\mathsf{Re}\Big(\frac{1}{1-\lambda}\Big) &= \frac{v^*(L+U)v}{2v^*(L+U-D)v} \\ &= \frac{1}{2}\left[1 + \frac{v^* D v}{v^* d^2 f(y) v}\right] \\ &> \frac{1}{2}\end{aligned}$$

for $d^2 f(y)$ positive definite. If $\lambda = \alpha + \beta i$, then the last inequality entails

$$\frac{1-\alpha}{(1-\alpha)^2 + \beta^2} > \frac{1}{2},$$

which is equivalent to $|\lambda|^2 = \alpha^2 + \beta^2 < 1$. Hence, the spectral radius $\rho < 1$.

15.4 Global Convergence of the MM Algorithm

In this section and the next, we tackle global convergence. We begin with the MM algorithm and consider without loss of generality minimization of the objective function $f(x)$ via the majorizing surrogate $g(x \mid x_n)$. In studying global convergence, we must carefully specify the parameter domain U. Let us take U to be any open convex subset of R^m. To avoid colliding with the boundary of U, we assume that $f(x)$ is coercive as defined in Section 11.2. Whenever necessary we also assume that $f(x)$ and $g(x \mid x_n)$ and their various first and second differentials are jointly continuous in x and x_n.

Finally, we demand that the second differential $d^2 g(x \mid x_n)$ be positive definite. This implies that $g(x \mid x_n)$ is strictly convex. Note that the objective function $f(x)$ is not required to be convex. Strict convexity of $g(x \mid x_n)$ in turn implies that the solution x_{n+1} of the minimization step is unique. Existence of a solution fortunately is guaranteed by coerciveness. Indeed, the closed set

$$\{x \in U : g(x \mid x_n) \le g(x_n \mid x_n) = f(x_n)\}$$

is compact because it is contained within the compact set

$$\{x \in U : f(x) \leq f(x_n)\}.$$

Finally, the implicit function theorem [7, 14] shows that the iteration map $x_{n+1} = M(x_n)$ is continuously differentiable in a neighborhood of every point x_n. Local differentiability of $M(x)$ clearly extends to global differentiability.

The gradient algorithm (15.4) has the property that stationary points of the objective function and fixed points of the iteration map coincide. This property also applies to the MM algorithm. Here we recall the two identities $\nabla g(x_{n+1} \mid x_n) = \mathbf{0}$ and $\nabla g(x_n \mid x_n) = \nabla f(x_n)$ and the strict convexity of $g(x \mid x_n)$. By the same token, stationary points and only stationary points give equality in the descent inequality $f[M(x)] \leq f(x)$.

At this juncture, we remind the reader that a point y is a cluster point of a sequence x_n provided there is a subsequence x_{n_k} that tends to y. One can easily verify that any limit of a sequence of cluster points is also a cluster point and that a bounded sequence has a limit if and only if it has at most one cluster point. With these facts in mind, we now state and prove a version of Liapunov's theorem for discrete dynamical systems [9].

Proposition 15.4.1 (Liapunov) *Let Γ be the set of cluster points generated by the sequence $x_{n+1} = M(x_n)$ starting from some initial x_1. Then Γ is contained in the set S of stationary points of $f(x)$.*

Proof: The sequence x_n stays within the compact set

$$\{x \in U : f(x) \leq f(x_1)\}.$$

Consider a cluster point $z = \lim_{k\to\infty} x_{n_k}$. Since the sequence $f(x_n)$ is monotonically decreasing and bounded below, $\lim_{m\to\infty} f(x_n)$ exists. Hence, taking limits in the inequality $f[M(x_{n_k})] \leq f(x_{n_k})$ and using the continuity of $M(x)$ and $f(x)$, we infer that $f[M(z)] = f(z)$. Thus, z is a stationary point of $f(x)$. ■

The next two propositions are adapted from reference [11]. In the second of these, recall that a point x in a set S is isolated if and only if there exists a radius $r > 0$ such that $S \cap \{y : \|y - x\| < r\} = \{x\}$.

Proposition 15.4.2 *The set of cluster points Γ of $x_{n+1} = M(x_n)$ is compact and connected.*

Proof: Γ is a closed subset of the compact set $\{x \in U : f(x) \leq f(x_1)\}$ and is therefore itself compact. According to Proposition 8.2.1, Γ is connected provided $\lim_{m\to\infty} \|x_{n+1} - x_n\| = 0$. If this sufficient condition fails, then the compactness of $\{x \in U : f(x) \leq f(x_1)\}$ makes it possible to extract a subsequence x_{n_k} such that $\lim_{k\to\infty} x_{n_k} = u$ and $\lim_{k\to\infty} x_{n_k+1} = v$ both exist, but $v \neq u$. However, the continuity of $M(x)$ requires $v = M(u)$ while the descent condition implies

$$f(v) \;=\; f(u) \;=\; \lim_{n\to\infty} f(x_n).$$

The equality $f(v) = f(u)$ forces the contradiction that u is a fixed point of $M(x)$. Hence, the stated sufficient condition for connectivity holds. ■

Proposition 15.4.3 *Suppose that all stationary points of $f(x)$ are isolated and that the differentiability, coerciveness, and convexity assumptions are true. Then any sequence of iterates $x_{n+1} = M(x_n)$ generated by the iteration map $M(x)$ of the MM algorithm possesses a limit, and that limit is a stationary point of $f(x)$. If $f(x)$ is strictly convex, then $\lim_{m\to\infty} x_n$ is the minimum point.*

Proof: In the compact set $\{x \in U : f(x) \leq f(x_1)\}$ there can only be a finite number of stationary points. An infinite number of stationary points would admit a convergent sequence whose limit would not be isolated. Since the set of cluster points Γ is a connected subset of this finite set of stationary points, Γ reduces to a single point. ■

Two remarks on Proposition 15.4.3 are in order. First, except when strict convexity prevails for $f(x)$, the proposition offers no guarantee that the limit x_∞ of the sequence x_n furnishes a global minimum. Problem 14 of Chapter 13 contains a counterexample of Wu [16] exhibiting convergence to a saddle point in the EM algorithm. Fortunately in practice, descent algorithms almost always converge to at least a local minimum of the objective function. Second, if the set S of stationary points is not discrete, then there exists a sequence $z_n \in S$ converging to $z \in S$ with $z_n \neq z$ for all m. Because the surface of the unit sphere in R^n is compact, we can extract a subsequence such that

$$\lim_{k\to\infty} \frac{z_{n_k} - z}{\|z_{n_k} - z\|} \;=\; v$$

exists and is nontrivial. Taking limits in

$$\begin{aligned} \mathbf{0} &= \frac{1}{\|z_{n_k} - z\|}[\nabla f(z_{n_k}) - \nabla f(z)] \\ &= \int_0^1 d^2 f[z + t(z_{n_k} - z)] \frac{z_{n_k} - z}{\|z_{n_k} - z\|} dt \end{aligned}$$

then produces $\mathbf{0} = d^2 f(z)v$. In other words, the second differential at z is singular. If one can rule out such degeneracies, then all stationary points are isolated [15]. Interested readers can consult the literature on Morse functions for further commentary on this subject [5].

15.5 Global Convergence of Block Relaxation

Verification of global convergence of block relaxation parallels the MM algorithm case. Careful scrutiny of the proof of Proposition 15.4.3 shows that it relies on five properties of the objective function $f(x)$ and the iteration map $M(x)$:

(a) $f(x)$ is coercive on its convex open domain U,

(b) $f(x)$ has only isolated stationary points,

(c) $M(x)$ is continuous,

(d) y is a fixed point of $M(x)$ if and only if it is a stationary point of $f(x)$,

(e) $f[M(y)] \leq f(y)$, with equality if and only if y is a fixed point of $M(x)$.

Let us suppose for notational simplicity that the argument $x = (v, w)$ breaks into just two blocks. Criteria (a) and (b) can be demonstrated for many objective functions and are independent of the algorithm chosen to minimize $f(x)$. In block relaxation we ordinarily take U to be the Cartesian product $V \times W$ of two convex open sets. If we assume that $f(v, w)$ is strictly convex in v for fixed w and vice versa, then the block relaxation updates are well defined. If $f(v, w)$ is twice continuously differentiable, and $d_{11}f(v, w)$ and $d_{22}f(v, w)$ are invertible matrices, then application of the implicit function theorem demonstrates that the iteration map $M(x)$ is a composition of two differentiable maps. Criterion (c) is therefore valid. A fixed point $x = (v, w)$ satisfies the two equations $\nabla_1 f(v, w) = \mathbf{0}$ and $\nabla_2 f(v, w) = \mathbf{0}$, and criterion (d) follows. Finally, both block updates decrease $f(x)$. They give a strict decrease if and only if they actually change either argument v or w. Hence, criterion (e) is true. We emphasize that collectively these are sufficient but not necessary conditions. Observe that we have not assumed that $f(v, w)$ is convex in both variables simultaneously.

15.6 Global Convergence of Gradient Algorithms

We now turn to the question of global convergence for gradient algorithms of the sort (15.4). The assumptions concerning $f(x)$ made in the previous sections remain in force. A major impediment to establishing the global convergence of any minimization algorithm is the possible failure of the descent property

$$f(x_{n+1}) \leq f(x_n)$$

enjoyed by the MM and block relaxation algorithms. Provided the matrix $A(x_n)$ is positive definite, the direction $v_n = -A(x_n)^{-1}\nabla f(x_n)$ is guaranteed to point locally downhill. Hence, if we elect the natural strategy of instituting a limited line search along the direction v_n emanating from x_n, then we can certainly find an x_{n+1} that decreases $f(x)$.

Although an exact line search is tempting, we may pay too great a price for precision when we need mere progress. The step-halving tactic mentioned in Chapter 14 is better than a full line search but not quite adequate for theoretical purposes. Instead, we require a sufficient decrease along a descent direction v. This is summarized by the Armijo rule of considering only steps tv satisfying the inequality

$$f(x+tv) \quad \leq \quad f(x) + \alpha t df(x)v \tag{15.11}$$

for t and some fixed α in $(0,1)$. To avoid too stringent a test, we take a low value of α such as 0.01. In combining Armijo's rule with regular step decrementing, we first test the step v. If it satisfies Armijo's rule we are done. If it fails, we choose $\sigma \in (0,1)$ and test σv. If this fails, we test $\sigma^2 v$ and so forth until we encounter and take the first partial step $\sigma^k v$ that works. In step halving, obviously $\sigma = 1/2$.

Step halving can be combined with a partial line search. For instance, suppose the line search has been confined to the interval $t \in [0,s]$. If the point $x + sv$ passes Armijo's test, then we accept it. Otherwise, we fit a cubic to the function $t \mapsto f(x+tv)$ on the interval $[0,s]$ as described in Section 5.6. If the minimum point t of the cubic approximation satisfies $t \geq \sigma s$ and passes Armijo's test, then we accept $x + tv$. Otherwise, we replace the interval $[0,s]$ by the interval $[0,\sigma s]$ and proceed inductively. For the sake of simplicity in the sequel, we will ignore this elaboration of step halving and concentrate on the unadorned version.

We would like some guarantee that the exponent k of the step decrementing power σ^k does not grow too large. Mindful of this criterion, we suppose that the positive definite matrix $A(x)$ depends continuously on x. This is not much of a restriction for Newton's method, the Gauss-Newton algorithm, the MM gradient algorithm, or scoring. If we combine continuity with coerciveness, then we can conclude that there exist positive constants β, γ, and ϵ such that $\|A(x)\|_2 \leq \beta$, $\|A(x)^{-1}\|_2 \leq \gamma$, and $\|\nabla f(x)\|_2 \leq \epsilon$ on the compact set $D = \{x \in U : f(x) \leq f(x_1)\}$ where any descent algorithm acts.

Before we tackle Armijo's rule, let us consider the more pressing question of whether the proposed points $x+v$ lie in the domain U of $f(x)$. This is too much to hope for, but it is worth considering whether $x + \sigma^d v$ always lies in U for some fixed power σ^d. Fortunately, $v(x) = -A(x)^{-1}\nabla f(x)$ satisfies the bound

$$\|v(x)\|_2 \quad \leq \quad \gamma\epsilon$$

on D. Now suppose no single power σ^k is adequate for all $x \in D$. Then there exists a sequence of points $x_k \in D$ with $y_k = x_k + \sigma^k v(x_k) \notin U$. Passing to a subsequence if necessary, we can assume that x_k converges to

$x \in D$. Because σ^k is tending to 0, and $v(x)$ is bounded on D, the sequence y_k likewise converges to x. Since the complement of U is closed, x must lie in the complement of U as well as in D. This contradiction proves our contention.

In dealing with Armijo's rule, we will majorize $f(x+tv)$ by a quadratic. The standard second-order Taylor expansion

$$f(x+u) = f(x) + df(x)u + \frac{1}{2}u^t d^2 f(z)u$$

is valid for z on the line segment connecting x and $x+u$. The remainder is bounded above by

$$\frac{1}{2}u^t d^2 f(z)u \leq \frac{\|u\|_2^2}{2} \sup_{0 \leq s \leq 1} \|d^2 f(x+su)\|_2.$$

Now the function

$$h(x,u) = \sup_{0 \leq s \leq 1} \|d^2 f(x+su)\|_2$$

is jointly continuous in (x,u) and attains its maximum δ on the compact set $\{(x,u) : x \in D, x+u \in D\}$. It follows that

$$f(x+tv) \leq f(x) + tdf(x)v + \frac{\delta}{2}t^2\|v\|_2^2 \tag{15.12}$$

for all triples (x,v,t) with x in D, $x+tv$ in D, and t in $[0,1]$.

Finally, we are ready to consider Armijo's rule. Taking into account the upper bound β on $\|A(x)\|_2$ and the identity

$$\|A(x)^{1/2}\|_2 = \|A(x)\|_2^{1/2}$$

entailed by Proposition 6.3.1, we have

$$\begin{aligned} \|\nabla f(x)\|_2^2 &= \|A(x)^{1/2}A(x)^{-1/2}\nabla f(x)\|_2^2 \\ &\leq \|A(x)^{1/2}\|_2^2\|A(x)^{-1/2}\nabla f(x)\|_2^2 \\ &\leq \beta df(x)A(x)^{-1}\nabla f(x). \end{aligned} \tag{15.13}$$

It follows that

$$\begin{aligned} \|v\|_2^2 &= \|A(x)^{-1}\nabla f(x)\|_2^2 \\ &\leq \gamma^2\|\nabla f(x)\|_2^2 \\ &\leq -\beta\gamma^2 df(x)v. \end{aligned}$$

Combining this last inequality with the majorization (15.12) yields

$$f(x+tv) \leq f(x) + t\left(1 - \frac{\beta\gamma^2\delta}{2}t\right) df(x)v.$$

Hence, as soon as σ^k satisfies

$$1 - \frac{\beta\gamma^2\delta}{2}\sigma^k \geq \alpha,$$

Armijo's rule (15.11) holds. In terms of k, backtracking is guaranteed to succeed in at most

$$k_{\max} = \max\left\{d, \left\lceil \frac{1}{\ln\sigma} \ln \frac{2(1-\alpha)}{\beta\gamma^2\delta} \right\rceil\right\}$$

decrements. Of course, a lower value of k may suffice.

Proposition 15.6.1 *Suppose that all stationary points of $f(x)$ are isolated and that the continuity, differentiability, positive definiteness, and coerciveness assumptions are true. Then any sequence of iterates generated by the iteration map $M(x) = x - tA(x)^{-1}\nabla f(x)$ with t chosen by step decrementing possesses a limit, and that limit is a stationary point of $f(x)$. If $f(x)$ is strictly convex, then $\lim_{n\to\infty} x_n$ is the minimum point.*

Proof: Let $v_n = -A(x_n)^{-1}\nabla f(x_n)$ and $x_{n+1} = x_n + \sigma^{k_n}v_n$. The sequence $f(x_n)$ is decreasing by construction. Because the function $f(x)$ is bounded below on the compact set $D = \{x \in U : f(x) \leq f(x_1)\}$, $f(x_n)$ is bounded below as well and possesses a limit. Based on Armijo's rule (15.11) and inequality (15.13), we calculate

$$\begin{aligned} f(x_n) - f(x_{n+1}) &\geq -\alpha\sigma^{k_n} df(x_n)v_n \\ &= \alpha\sigma^{k_n} df(x_n)A(x_n)^{-1}\nabla f(x_n) \\ &\geq \frac{\alpha\sigma^{k_n}}{\beta}\|\nabla f(x_n)\|_2^2. \end{aligned}$$

Since $\sigma^{k_n} \geq \sigma^{k_{\max}}$, and the difference $f(x_n) - f(x_{n+1})$ tends to 0, we deduce that $\|\nabla f(x_n)\|_2$ tends to 0. This conclusion and the inequality

$$\begin{aligned} \|x_{n+1} - x_n\|_2 &= \sigma^{k_n}\|A(x_n)^{-1}\nabla f(x_n)\|_2 \\ &\leq \sigma^{k_n}\gamma\|\nabla f(x_n)\|_2, \end{aligned}$$

demonstrate that $\|x_{n+1} - x_n\|_2$ tends to 0 as well.

Given these results, the conclusions of Propositions 15.4.1 and 15.4.2 are true. All of the claims of the current proposition now follow as in the proof of Proposition 15.4.3. ∎

15.7 Problems

1. Define $f : \mathsf{R}^2 \to \mathsf{R}$ by $f(\mathbf{0}) = 0$ and $f(x) = x_1^3/(x_1^2 + x_2^2)$ for $x \neq \mathbf{0}$. Show that $f(x)$ is differentiable along every straight line in R^2 but lacks a differential at $\mathbf{0}$.

2. For $f : \mathsf{R}^2 \to \mathsf{R}^2$ given by $f_1(x) = x_1^3$ and $f_2(x) = x_1^2$, show that no $\bar{x}$ exists on the line segment from $\mathbf{0} = (0,0)^t$ to $\mathbf{1} = (1,1)^t$ such that

$$f(\mathbf{1}) - f(\mathbf{0}) = df(\bar{x})(\mathbf{1} - \mathbf{0}).$$

3. Let $f : U \to \mathsf{R}$ be a continuously differentiable function defined on an open connected set $U \subset \mathsf{R}^m$. Suppose that $\nabla f(x) = \mathbf{0}$ for all $x \in U$. Show that $f(x)$ is constant on U. (Hint: There is a polygonal path between any two points along which one can integrate $\nabla f(x)$.)

4. Suppose a continuously differentiable function $f : U \to \mathsf{R}^n$ satisfies the Lipschitz bound (15.2) on the convex open set $U \subset \mathsf{R}^m$. Prove that

$$\|f(u) - f(v) - df(x)(u - v)\| \leq \frac{\lambda}{2}(\|u - x\| + \|v - x\|)\|u - v\|$$

for any triple of points u, v, and x contained in U. (Hint: Mimic the derivation of the bound (15.3).)

5. In the context of Problem 4, suppose $m = n$ and the matrix $df(x)$ is invertible. Show that there exist positive constants α, β, and ϵ such that

$$\alpha\|u - v\| \leq \|f(u) - f(v)\| \leq \beta\|u - v\|$$

for all u and v with $\max\{\|u - x\|, \|v - x\|\} \leq \epsilon$. (Hints: Write

$$f(u) - f(v) = f(u) - f(v) - df(x)(u - v) + df(x)(u - v),$$

and apply the bound $\|u - v\| \leq \|df(x)^{-1}\| \cdot \|df(x)(u - v)\|$ and the result of Problem 4.)

6. Demonstrate that cyclic coordinate descent either diverges or converges to a saddle point of the function $f : \mathsf{R}^2 \to \mathsf{R}$ defined by

$$f(x_1, x_2) = (x_1 - x_2)^2 - 2x_1x_2.$$

This function of de Leeuw [1] has no minimum.

7. Consider the function $f(x) = (x_1^2 + x_2^2)^{-1} + \ln(x_1^2 + x_2^2)$ for $x \neq \mathbf{0}$. Explicitly find the minimum value of $f(x)$. Specify the coordinate descent algorithm for finding the minimum. Note any ambiguities in the implementation of coordinate descent, and describe the possible cluster points of the algorithm as a function of the initial point. (Hint: Coordinate descent, properly defined, converges in a finite number of iterations.)

8. In block relaxation with b blocks, let $B_i(x)$ be the map that updates block i and leaves the other blocks fixed. Show that the overall iteration map $M(x) = B_b \circ \cdots \circ B_1(x)$ has differential $dB_b(y) \cdots dB_1(y)$ at a fixed point y. Write $dB_i(y)$ as a block matrix and identify the blocks by applying the implicit function theorem as needed. Do not confuse $B_i(x)$ with the update $M_i(x)$ of the text. In fact, $M_i(x)$ only summarizes the update of block i, and its argument is the value of x at the start of the current round of updates.

9. Consider a Poisson-distributed random variable Y with mean $a\theta + b$, where a and b are known positive constants and $\theta \geq 0$ is a parameter to be estimated. An EM algorithm for estimating θ can be concocted that takes as complete data independent Poisson random variables U and V with means $a\theta$ and b and sum $U + V = Y$. If $Y = y$ is observed, then show that the EM iterates are defined by

$$\theta_{n+1} = \frac{y\theta_n}{a\theta_n + b}.$$

Show that these iterates converge monotonically to the maximum likelihood estimate $\max\{0, (y-b)/a\}$. When $y = b$, verify that convergence to the boundary value 0 occurs at a rate slower than linear [5]. (Hint: When $y = b$, check that $\theta_{n+1} = b\theta_1/(na\theta_1 + b)$.)

10. The sublinear convergence of the EM algorithm exhibited in the previous problem occurs in other problems. Here is a conceptually harder example by Robert Jennrich. Suppose that $W_1, \ldots, W_m$ and B are independent normally distributed random variables with 0 means. Let σ_w^2 be the common variance of the W's and σ_b^2 be the variance of B. If the values y_i of the linear combinations $Y_i = B + W_i$ are observed, then show that the EM algorithm amounts to

$$\begin{aligned}
\sigma_{n+1,b}^2 &= \left(\frac{m\sigma_{nb}^2 \bar{y}}{m\sigma_{nb}^2 + \sigma_{nw}^2}\right)^2 + \frac{\sigma_{nb}^2 \sigma_{nw}^2}{m\sigma_{nb}^2 + \sigma_{nw}^2} \\
\sigma_{n+1,w}^2 &= \frac{m-1}{m} s_y^2 + \left(\frac{\sigma_{nw}^2 \bar{y}}{m\sigma_{nb}^2 + \sigma_{nw}^2}\right)^2 + \frac{\sigma_{nb}^2 \sigma_{nw}^2}{m\sigma_{nb}^2 + \sigma_{nw}^2},
\end{aligned}$$

where $\bar{y} = \frac{1}{m}\sum_{i=1}^m y_i$ and $s_y^2 = \frac{1}{m-1}\sum_{i=1}^m (y_i - \bar{y})^2$ are the sample mean and variance. Although one can formally calculate the maximum likelihood estimates $\hat{\sigma}_w^2 = s_y^2$ and $\hat{\sigma}_b^2 = \bar{y}^2 - s_y^2/m$, these are only valid provided $\hat{\sigma}_b^2 \geq 0$. If for instance $\bar{y} = 0$, then the EM iterates will converge to $\sigma_w^2 = (m-1)s_y^2/m$ and $\sigma_b^2 = 0$. Show that convergence is sublinear when $\bar{y} = 0$.

11. Suppose the MM gradient iterates θ_n converge to a local maximum θ_∞ of the loglikelihood $L(\theta)$. Under the hypotheses of the text, prove

that for all sufficiently large n, either $\theta_n = \theta_\infty$ or $L(\theta_{n+1}) > L(\theta_n)$ [6]. (Hints: Let $g(\theta \mid \theta_n)$ be the surrogate function. Show that

$$\begin{aligned} & L(\theta_{n+1}) - L(\theta_n) \\ = \ & \frac{1}{2}(\theta_{n+1} - \theta_n)^t[d^2L(\phi_n) - 2d^{20}g(\theta_n \mid \theta_n)](\theta_{n+1} - \theta_n), \end{aligned}$$

where ϕ_n lies on the line segment between θ_n and θ_{n+1}. Then use a continuity argument, noting that $d^2L(\theta_\infty) - d^{20}g(\theta_\infty \mid \theta_\infty)$ is positive semidefinite and $d^{20}g(\theta_\infty \mid \theta_\infty)$ is negative definite.)

12. Let $M(\theta)$ be the MM algorithm or MM gradient algorithm map. Consider the modified algorithm $M_t(\theta) = \theta + t[M(\theta) - \theta]$ for $t > 0$. At a local maximum θ_∞, show that the spectral radius ρ_t of the differential $dM_t(\theta_\infty) = (1-t)I + tdM(\theta_\infty)$ satisfies $\rho_t < 1$ whenever $0 < t < 2$. Hence, Ostrowski's theorem implies local attraction of $M_t(\theta)$ to θ_∞. If the largest and smallest eigenvalues of $dM(\theta_\infty)$ are $\omega_{\max}$ and $\omega_{\min}$, then prove that ρ_t is minimized by taking $t = [1-(\omega_{\min}+\omega_{\max})/2]^{-1}$. In practice, the eigenvalues of $dM(\theta_\infty)$ are impossible to predict in advance of knowing θ_∞, but for many problems, the value $t = 2$ works well [6]. (Hint: To every eigenvalue ω of $dM(\theta_\infty)$, there corresponds an eigenvalue $\omega_t = 1 - t + t\omega$ of $dM_t(\theta_\infty)$ and vice versa.)

13. In the notation of Chapter 13, prove the EM algorithm formula

$$d^2L(\theta) \quad = \quad d^{20}Q(\theta \mid \theta) + \operatorname{Var}[d \ln f(X \mid \theta) \mid Y, \theta]$$

of Louis [8].

14. Consider independent observations $y_1, \ldots, y_m$ from the univariate t-distribution. These data have loglikelihood

$$\begin{aligned} L &= -\frac{m}{2}\ln\sigma^2 - \frac{\nu+1}{2}\sum_{i=1}^m \ln(\nu + \delta_i^2) \\ \delta_i^2 &= \frac{(y_i - \mu)^2}{\sigma^2}. \end{aligned}$$

To illustrate the occasionally bizarre behavior of the MM algorithm, take $\nu = 0.05$, $m = 4$, and the data vector $y = (-20, 1, 2, 3)^t$. According to Problem 15 of Chapter 12, the MM algorithm for estimating μ with σ^2 fixed at 1 has iterates

$$\mu_{n+1} \quad = \quad \frac{\sum_{i=1}^m w_{ni} y_i}{\sum_{i=1}^m w_{ni}}, \qquad w_{ni} \quad = \quad \frac{\nu + 1}{\nu + (y_i - \mu_n)^2}.$$

Plot the likelihood curve and show that it has the four local maxima -19.993, 1.086, 1.997, and 2.906 and the three local minima -14.516,

1.373, and 2.647. Demonstrate numerically convergence to a local maximum that is not the global maximum. Show that the algorithm converges to a local minimum in one step starting from −1.874 or −0.330 [10].

15. In our exposition of least ℓ_1 regression in Chapter 12, we considered a modified iteration scheme that minimizes the criterion

$$\sum_{i=1}^{m}\{|y_i - \mu_i(\theta)| - \epsilon \ln[\epsilon + |y_i - \mu(\theta)|]\}. \tag{15.14}$$

For a sequence of constants ϵ_n tending to 0, let θ_n be a corresponding sequence minimizing (15.14). If θ_∞ is a cluster point of this sequence and the regression functions $\mu_i(\theta)$ are continuous, then show that θ_∞ minimizes $\sum_{i=1}^m |y_i - \mu_i(\theta)|$. If in addition the minimum point θ_∞ of $\sum_{i=1}^m |y_i - \mu_i(\theta)|$ is unique and $\lim_{\|\theta\|\to\infty} \sum_{i=1}^m |\mu_i(\theta)| = \infty$, then prove that $\lim_{n\to\infty} \theta_n = \theta_\infty$. (Hints: For the first assertion, take limits in

$$\sum_{i=1}^{m} h_\epsilon[s_i(\theta_n)] \;\le\; \sum_{i=1}^{m} h_\epsilon[s_i(\theta)],$$

where $s_i(\theta) = y_i - \mu_i(\theta)$ and $h_\epsilon(s) = |s| - \epsilon\ln(\epsilon + |s|)$. Note that $h_\epsilon(s)$ is jointly continuous in ϵ and s. For the second assertion, it suffices that the sequence θ_n be confined to a bounded set. To prove this fact, demonstrate and use the inequalities

$$\begin{aligned}\sum_{i=1}^{m} h_\epsilon(s_i) &\ge \frac{1}{2}\sum_{i=1}^{m} 1_{\{|s_i|\ge 1\}}|s_i| \\ &\ge \frac{1}{2}\sum_{i=1}^{m}|\mu_i(\theta)| - \frac{1}{2}\sum_{i=1}^{m}|y_i| - \frac{m}{2} \\ \sum_{i=1}^{m} h_\epsilon(s_i) &\le \sum_{i=1}^{m}[|s_i| - \epsilon\ln\epsilon] \\ &\le \sum_{i=1}^{m}|s_i| + \frac{m}{e}\end{aligned}$$

for $0 \le \epsilon < \frac{1}{2}$.)

16. Example 14.2.1 and Problem 2 of Chapter 14 suggest a method of accelerating the MM gradient algorithm. Denote the loglikelihood of the observed data by $L(\theta)$ and the surrogate function by $g(\theta \mid \theta_n)$. To accelerate the MM gradient algorithm, we can replace the positive definite matrix $B(\theta)^{-1} = -d^{20}g(\theta \mid \theta)$ by a matrix that better approximates the observed information $A(\theta) = -d^2L(\theta)$. Note that

often $d^{20}g(\theta \mid \theta)$ is diagonal and therefore trivial to invert. Now consider the formal expansion

$$\begin{aligned} A^{-1} &= (B^{-1} + A - B^{-1})^{-1} \\ &= \{B^{-\frac{1}{2}}[I - B^{\frac{1}{2}}(B^{-1} - A)B^{\frac{1}{2}}]B^{-\frac{1}{2}}\}^{-1} \\ &= B^{\frac{1}{2}}\sum_{i=0}^{\infty}[B^{\frac{1}{2}}(B^{-1} - A)B^{\frac{1}{2}}]^i B^{\frac{1}{2}}. \end{aligned}$$

If we truncate this series after a finite number of terms, then we recover the first iterate of equation (14.23) in the disguised form

$$S_j = B^{\frac{1}{2}}\sum_{i=0}^{j}[B^{\frac{1}{2}}(B^{-1} - A)B^{\frac{1}{2}}]^i B^{\frac{1}{2}}.$$

The accelerated algorithm

$$\theta_{n+1} = \theta_n + S_j(\theta_n)\nabla L(\theta_n) \tag{15.15}$$

has several desirable properties.

(a) Show that S_j is positive definite and hence that the update (15.15) is an ascent algorithm. (Hint: Use the fact that $B^{-1} - A$ is positive semidefinite.)

(b) Algorithm (15.15) has differential

$$I + S_j(\theta_\infty)d^2L(\theta_\infty) = I - S_j(\theta_\infty)A(\theta_\infty)$$

at a local maximum θ_∞. If $d^2L(\theta_\infty)$ is negative definite, then prove that all eigenvalues of this differential lie on $[0, 1)$. (Hint: The eigenvalues are determined by the stationary points of the Rayleigh quotient $v^t[A^{-1}(\theta_\infty) - S_j(\theta_\infty)]v/v^tA^{-1}(\theta_\infty)v$.)

(c) If ρ_j is the spectral radius of the differential, then demonstrate that $\rho_j \le \rho_{j-1}$, with strict inequality when $B^{-1}(\theta_\infty) - A(\theta_\infty)$ is positive definite.

In other words, the accelerated algorithm (15.15) is guaranteed to converge faster than the MM gradient algorithm. It will be particularly useful for maximum likelihood problems with many parameters because it entails no matrix inversion or multiplication, just matrix times vector multiplication. When $j = 1$, it takes the simple form

$$\theta_{n+1} = \theta_n + [2B(\theta_n) - B(\theta_n)A(\theta_n)B(\theta_n)]\nabla L(\theta_n).$$

17. In Problem 33 of Chapter 12, we considered maximizing functions of the form $L(\theta) = -\sum_{i=1}^{p} f_i(c_i^t\theta)$, with each $f_i(s)$ strictly convex. If we

choose nonnegative constants λ_{ij} such that $\sum_j \lambda_{ij} = 1$ and $\lambda_{ij} > 0$ when $c_{ij} \neq 0$, then the function

$$g(\theta \mid \theta_n) = -\sum_{i=1}^{p} \sum_{j \in S_i} \lambda_{ij} f_i\Big[\frac{c_{ij}}{\lambda_{ij}}(\theta_j - \theta_{nj}) + c_i^t \theta_n\Big]$$

with $S_i = \{j : \lambda_{ij} > 0\}$ serves as a surrogate function minorizing $L(\theta)$. In the resulting MM algorithm, we suggested that a reasonable choice for λ_{ij} might be $\lambda_{ij} = |c_{ij}|^\alpha / \|c_i\|_\alpha^\alpha$, where $\|c_i\|_\alpha^\alpha = \sum_j |c_{ij}|^\alpha$ and $\alpha > 0$. It would be helpful to determine the α yielding the fastest rate of convergence. As pointed out in Proposition 15.3.2, the rate of convergence is given by the maximum of the Rayleigh quotient (15.6). This fact suggests that we should choose α to minimize $-v^t d^{20} g(\theta \mid \theta) v$ over all unit vectors v. This appears to be a difficult problem. A simpler problem is to minimize $\mathrm{tr}[-d^{20} g(\theta \mid \theta)]$. Show that this substitute problem has solution $\alpha = 1$ regardless of the point θ selected. (Hint: Multiply the inequality

$$\Big(\sum_{j \in S_i} |c_{ij}|\Big)^2 \leq \sum_{j \in S_i} \frac{c_{ij}^2}{\lambda_{ij}}$$

by $f_i''(c_i^t \theta)$ and sum on i.)

18. Consider a sequence x_n in R^m. Verify that the set of cluster points of x_n is closed. If x_n is bounded, then show that it has a limit if and only if it has at most one cluster point.

15.8 References

[1] de Leeuw J (1994) Block relaxation algorithms in statistics. *Information Systems and Data Analysis,* Bock HH, Lenski W, Richter MM, Springer, Berlin, pp 308–325

[2] Dennis JE Jr, Schnabel RB (1983) *Numerical Methods for Unconstrained Optimization and Nonlinear Equations.* Prentice-Hall, Englewood Cliffs, NJ

[3] Fessler JA, Clinthorne NH, Rogers WL (1993) On complete-data spaces for PET reconstruction algorithms. *IEEE Trans Nuclear Sci* 40:1055–1061

[4] Gill PE, Murray W, Wright MH (1981) *Practical Optimization.* Academic Press, New York

[5] Guillemin V, Pollack A (1974) *Differential Topology.* Prentice-Hall, Englewood Cliffs, NJ

[6] Lange K (1995) A gradient algorithm locally equivalent to the EM algorithm. *J Roy Stat Soc B* 57:425–437

[7] Lange, K (2004) *Optimization.* Springer, New York

[8] Louis TA (1982) Finding the observed information matrix when using the EM algorithm. *J Roy Stat Soc B* 44:226–233

[9] Luenberger DG (1984) *Linear and Nonlinear Programming*, 2nd ed. Addison-Wesley, Reading, MA

[10] McLachlan GJ, Krishnan T (2008) *The EM Algorithm and Extensions,* 2nd ed. Wiley, New York

[11] Meyer RR (1976) Sufficient conditions for the convergence of monotonic mathematical programming algorithms. *J Computer System Sci* 12:108–121

[12] Nocedal J (1991) Theory of algorithms for unconstrained optimization. *Acta Numerica 1991*:199–242.

[13] Ortega JM (1990) *Numerical Analysis: A Second Course.* SIAM, Philadelphia

[14] Ortega JM, Rheinboldt WC (1970) *Iterative Solution of Nonlinear Equations in Several Variables.* Academic Press, San Diego

[15] Ostrowski AM (1960) *Solution of Equations and Systems of Equations.* Academic Press, New York

[16] Wu CF (1983) On the convergence properties of the EM algorithm. *Ann Stat* 11:95–103

16

Advanced Optimization Topics

16.1 Introduction

Our final chapter on optimization provides a concrete introduction to several advanced topics. The first vignette describes classical penalty and barrier methods for constrained optimization [22, 37, 45]. Penalty methods operate on the exterior and barrier methods on the interior of the feasible region. Fortunately, it is fairly easy to prove global convergence for both methods under reasonable hypotheses.

In convex programming, adaptive barrier methods are attractive alternatives to standard barrier methods. Adaptive algorithms encourage rapid convergence to a boundary by gradually diminishing the strength of the corresponding barrier. The MM algorithm perspective suggests a way of accomplishing this feat while steadily decreasing the objective function [7, 30, 48]. As our examples demonstrate, adaptive barrier methods have novel applications to linear and geometric programming. For a proof that the adaptive barrier algorithm converges, see reference [31].

Specifying an interior feasible point is the first issue that must be faced in using a barrier method. Dykstra's algorithm [2, 12, 14] finds the closest point to the intersection $\cap_{i=0}^{r-1} C_i$ of a finite number of closed convex sets. If C_i is defined by the convex constraint $h_i(x) \leq 0$, then one obvious tactic is to replace C_i by the set $C_i(\epsilon)$ defined by the constraint $h_j(x) \leq -\epsilon$ for some small number $\epsilon > 0$. Projecting onto the intersection of the $C_i(\epsilon)$ then produces an interior point.

Our fourth topic is lasso penalized estimation. The lasso is an ℓ_1 penalty that shrinks parameter estimates toward zero and performs a kind of continuous model selection [13, 49]. The predictors whose estimated regression coefficients are exactly zero are candidates for elimination from the model. With the enormous data sets now confronting statisticians, considerations of model parsimony have taken on greater urgency. In addition to this philosophical justification, imposition of lasso penalties also has an enormous impact on computational speed. Standard methods of regression require matrix diagonalization, matrix inversion, or, at the very least, the solution of large systems of linear equations. Because the number of arithmetic operations for these processes scales as the cube of the number of predictors, problems with tens of thousands of predictors appear intractable. Recent research has shown this assessment to be too pessimistic [4, 17, 29, 40, 51]. Coordinate descent methods mesh well with the lasso and are simple, fast, and stable. We will see how their potential to transform data mining plays

K. Lange, *Numerical Analysis for Statisticians*, Statistics and Computing,
DOI 10.1007/978-1-4419-5945-4_16, © Springer Science+Business Media, LLC 2010

out in both ℓ_1 and ℓ_2 regression.

Our final topic touches on the estimation of parameter asymptotic standard errors. The MM and EM algorithms do not automatically deliver asymptotic standard errors, and it is important to see how any MM algorithm can be adapted to produce them [28, 38]. The chapter ends with a brief discussion of the computation of asymptotic standard errors in the presence of linear equality constraints.

16.2 Barrier and Penalty Methods

In general, unconstrained optimization problems are easier to solve than constrained optimization problems, and equality constrained problems are easier to solve than inequality constrained problems. To simplify analysis, mathematical scientists rely on several devices. For instance, one can replace the inequality constraint $g(x) \leq 0$ by the equality constraint $g_+(x) = 0$, where $g_+(x) = \max\{g(x), 0\}$. This tactic is not entirely satisfactory because $g_+(x)$ has kinks along the boundary $g(x) = 0$. The smoother substitute $g_+(x)^2$ avoids the kinks in first derivatives. Alternatively, one can introduce an extra parameter y and require $g(x) + y = 0$ and $y \geq 0$. This tactic substitutes a simple inequality constraint for a complex inequality constraint.

The addition of barrier and penalty terms to the objective function $f(x)$ is a more systematic approach. Later in the chapter we will discuss the role of penalties in producing sparse solutions. In the current section, penalties are introduced to steer the optimization process toward the feasible region. In the penalty method we construct a continuous nonnegative penalty $p(x)$ that is 0 on the feasible region and positive outside it. We then optimize the functions $f(x) + \lambda_k p(x)$ for an increasing sequence of tuning constants λ_k that tend to ∞. The penalty method works from the outside of the feasible region inward. Under the right hypotheses, the sequence of unconstrained solutions x_k tends to a solution of the constrained optimization problem.

In contrast, the barrier method works from the inside of the feasible region outward. We now construct a continuous barrier function $b(x)$ that is finite on the interior of the feasible region and infinite on its boundary. We then optimize the sequence of functions $f(x) + \mu_k b(x)$ as the decreasing sequence of tuning constants μ_k tends to 0. Again under the right hypotheses, the sequence of unconstrained solutions x_k tends to the solution of the constrained optimization problem.

Example 16.2.1 *Linear Regression with Linear Constraints*

Consider the regression problem of minimizing $\|Y - X\beta\|_2^2$ subject to the linear constraints $V\beta = d$. If we take the penalty function $p(\beta) = \|V\beta - d\|_2^2$, then we must minimize at each stage the function

$$h_k(\beta) = \|Y - X\beta\|_2^2 + \lambda_k \|V\beta - d\|_2^2.$$

Setting the gradient

$$\nabla h_k(\beta) = -2X^t(Y - X\beta) + 2\lambda_k V^t(V\beta - d)$$

equal to **0** yields the sequence of solutions

$$\beta_k = (X^t X + \lambda_k V^t V)^{-1}(X^t Y + \lambda_k V^t d). \tag{16.1}$$

In a moment we will demonstrate that the β_k tend to the constrained solution as λ_k tends to ∞. ■

Example 16.2.2 *Estimation of Multinomial Proportions*

In estimating multinomial proportions, we minimize the negative loglikelihood $-\sum_{i=1}^m n_i \ln p_i$ subject to the constraints $\sum_{i=1}^m p_i = 1$ and $p_i \geq 0$ for all i. An appropriate barrier function is $-\sum_{i=1}^m \ln p_i$. The minimum of the function

$$h_k(p) = -\sum_{i=1}^m n_i \ln p_i - \mu_k \sum_{i=1}^m \ln p_i$$

subject to the constraint $\sum_{i=1}^m p_i = 1$ occurs at the point with coordinates

$$p_{ki} = \frac{n_i + \mu_k}{n + m\mu_k},$$

where $n = \sum_{i=1}^m n_i$. In this example, it is clear that the solution vector p_k occurs on the interior of the parameter space and tends to the maximum likelihood estimate as μ_k tends to 0. ■

The next proposition derives the ascent and descent properties of the penalty and barrier methods.

Proposition 16.2.1 *Consider two real-valued functions $f(x)$ and $g(x)$ on a common domain and two positive constants $\alpha < \omega$. Suppose the linear combination $f(x) + \alpha g(x)$ attains its minimum value at y and the linear combination $f(x)+\omega g(x)$ attains its minimum value at z. Then $f(y) \leq f(z)$ and $g(y) \geq g(z)$.*

Proof: Adding the two inequalities

$$\begin{aligned} f(z) + \omega g(z) &\leq f(y) + \omega g(y) \\ -f(z) - \alpha g(z) &\leq -f(y) - \alpha g(y) \end{aligned}$$

and dividing by the constant $\omega - \alpha$ validates the claim $g(y) \geq g(z)$. The claim $f(y) \leq f(z)$ is proved by interchanging the roles of $f(x)$ and $g(x)$ and considering the functions $g(x) + \alpha^{-1} f(x)$ and $g(x) + \omega^{-1} f(x)$. ■

It is fairly easy to prove a version of global convergence for the penalty method.

Proposition 16.2.2 *Suppose that both the objective function $f(x)$ and the penalty function $p(x)$ are continuous on R^m and that the penalized functions $h_k(x) = f(x) + \lambda_k p(x)$ are coercive on R^m. Then one can extract a corresponding sequence of minimum points x_k such that $f(x_k) \le f(x_{k+1})$. Furthermore, any cluster point of this sequence resides in the feasible region $C = \{x : p(x) = 0\}$ and attains the minimum value of $f(x)$ there. Finally, if $f(x)$ is coercive and possesses a unique minimum point in C, then the sequence x_k converges to that point.*

Proof: By virtue of the coerciveness assumption (c), the minimum points x_k exist. Proposition 16.2.1 confirms the ascent property. Now suppose that z is a cluster point of the sequence x_k and y is any point in C. If we take limits in the inequalities

$$f(y) = f(y) + \lambda_k p(y) \;\ge\; f(x_k) + \lambda_k p(x_k) \;\ge\; f(x_k)$$

along the corresponding subsequence x_{k_l}, then it is clear that $f(y) \ge f(z)$. Furthermore, because the λ_k tend to infinity, the inequality

$$\limsup_{l\to\infty} \lambda_{k_l} p(x_{k_l}) \;\le\; f(y) - \lim_{l\to\infty} f(x_{k_l}) \;=\; f(y) - f(z)$$

can only hold if $p(z) = \lim_{l\to\infty} p(x_{k_l}) = 0$.

If $f(x)$ possesses a unique minimum point y in C, then to prove that x_k converges to y, it suffices to prove that x_k is bounded. If $f(x)$ is coercive, then it is possible to choose r so that $f(x) > f(y)$ for all x with $\|x\|_2 \ge r$. The assumption $\|x_k\|_2 \ge r$ consequently implies

$$h_k(x_k) \;\ge\; f(x_k) \;>\; f(y) \;=\; f(y) + \lambda_k p(y),$$

which contradicts the assumption that x_k minimizes $h_k(x)$. Hence, all x_k satisfy $\|x_k\|_2 < r$. ■

Here is the corresponding result for the barrier method.

Proposition 16.2.3 *Suppose the real-valued function $f(x)$ is continuous on the bounded open set U and its closure V. Also suppose the barrier function $b(x)$ is continuous and coercive on U. If the tuning constants μ_k decrease to 0, then the linear combinations $h_k(x) = f(x) + \mu_k b(x)$ attain their minima at a sequence of points x_k in U satisfying the descent property $f(x_{k+1}) \le f(x_k)$. Furthermore, any cluster point of the sequence furnishes the minimum value of $f(x)$ on V. If the minimum point of $f(x)$ in V is unique, then the sequence x_k converges to this point.*

Proof: Each of the continuous functions $h_k(x)$ is coercive on U, being the sum of a coercive function and a function bounded below. Therefore, the sequence x_k exists. An appeal to Proposition 16.2.1 establishes the descent property. If z is a cluster point of x_k and x is any point of U, then taking limits in the inequality

$$f(x_k) + \mu_k b(x_k) \;\le\; f(x) + \mu_k b(x)$$

along the relevant subsequence x_{k_l} produces

$$f(z) \le \lim_{l\to\infty} f(x_{k_l}) + \limsup_{l\to\infty} \mu_{k_l} b(x_{k_l}) \le f(x).$$

It follows that $f(z) \le f(x)$ for every x in V as well. If the minimum point of $f(x)$ on V is unique, then every cluster point of the bounded sequence x_k coincides with this point. Hence, the sequence itself converges to the point. ■

Despite the elegance of the penalty and barrier methods, they suffer from three possible defects. First, they are predicated on finding the minimum point of the surrogate function for each value of the tuning constant. This entails iterations within iterations. Second, there is no obvious prescription for deciding how fast to send the tuning constants to their limits. Third, too large a value of λ_k in the penalty method or too small a value μ_k in the barrier method can lead to numerical instability.

16.3 Adaptive Barrier Methods

The standard convex programming problem involves minimizing a convex function $f(x)$ subject to affine equality constraints $a_i^t x - b_i = 0$ for $1 \le i \le p$ and convex inequality constraints $h_j(x) \le 0$ for $1 \le j \le q$. This formulation renders the feasible region convex. To avoid distracting negative signs in this section, we will replace the constraint $h_j(x) \le 0$ by the constraint $v_j(x) \ge 0$ for $v_j(x) = -h_j(x)$. In the logarithmic barrier method, we define the barrier function

$$b(x) = \sum_{j=1}^{q} \ln v_j(x) \tag{16.2}$$

and optimize $h_k(x) = f(x) + \mu_k b(x)$ subject to the equality constraints. The presence of the barrier term $\ln v_j(x)$ keeps an initially inactive constraint $v_j(x)$ inactive throughout the search. Proposition 16.2.3 demonstrates convergence under specific hypotheses.

One way of improving the barrier method is to change the barrier constant as the iterations proceed [7, 30, 48]. This sounds vague, but matters simplify enormously if we view the construction of an adaptive barrier method from the perspective of the MM algorithm. Consider the following inequalities

$$\begin{aligned}
& -v_j(x_k)\ln v_j(x) + v_j(x_k)\ln v_j(x_k) + dv_j(x_k)(x - x_k) \\
\ge\; & -\frac{v_j(x_k)}{v_j(x_k)}[v_j(x) - v_j(x_k)] + dv_j(x_k)(x - x_k) \qquad (16.3) \\
=\; & -v_j(x) + v_j(x_k) + dv_j(x_k)(x - x_k) \\
\ge\; & 0
\end{aligned}$$

based on the concavity of the functions $\ln y$ and $v_j(x)$. Because equality holds throughout when $x = x_k$, we have identified a novel function majorizing 0 and incorporating a barrier for $v_j(x)$. (Such functions are known as Bregman distances in the literature [3].) The import of this discovery is that the surrogate function

$$\begin{aligned} g(x \mid x_k) &= f(x) - \gamma \sum_{j=1}^{q} v_j(x_k) \ln v_j(x) \\ & \quad + \gamma \sum_{j=1}^{q} dv_j(x_k)(x - x_k) \end{aligned} \tag{16.4}$$

majorizes $f(x)$ up to an irrelevant additive constant. Here γ is a fixed positive constant. Minimization of the surrogate function drives $f(x)$ downhill while keeping the inequality constraints inactive. In the limit, one or more of the inequality constraints may become active.

Because minimization of the surrogate function $g(x \mid x_k)$ cannot be accomplished in closed form, we must revert to the MM gradient algorithm. In performing one step of Newton's method, we need the first and second differentials

$$\begin{aligned} dg(x_k \mid x_k) &= df(x_k) \\ d^2 g(x_k \mid x_k) &= d^2 f(x_k) - \gamma \sum_{j=1}^{q} d^2 v_j(x_k) \\ & \quad + \gamma \sum_{j=1}^{q} \frac{1}{v_j(x_k)} \nabla v_j(x_k) dv_j(x_k). \end{aligned}$$

In view of the convexity of $f(x)$ and the concavity of the $v_j(x)$, it is obvious that $d^2 g(x_k \mid x_k)$ is positive semidefinite. If either $f(x)$ is strictly convex or the sum $\sum_{j=1}^{q} v_j(x)$ is strictly concave, then $d^2 g(x_k \mid x_k)$ is positive definite. As a safeguard in Newton's method, it is always a good idea to contract any proposed step so that $f(x_{k+1}) < f(x_k)$ and $v_j(x_{k+1}) > \epsilon v_j(x_k)$ for all j and a small ϵ such as 0.1.

The surrogate function (16.4) does not exhaust the possibilities for majorizing the objective function. If we replace the concave function $\ln y$ by the concave function $-y^{-\alpha}$ in our derivation (16.3), then we can construct for each $\alpha > 0$ and β the alternative surrogate

$$\begin{aligned} g(x \mid x_k) &= f(x) + \gamma \sum_{j=1}^{q} v_j(x_k)^{\alpha+\beta} v_j(x)^{-\alpha} \\ & \quad + \gamma\alpha \sum_{j=1}^{q} v_j(x_k)^{\beta-1} dv_j(x_k)(x - x_k) \end{aligned} \tag{16.5}$$

majorizing $f(x)$ up to an irrelevant additive constant. This surrogate also exhibits an adaptive barrier that prevents the constraint $v_j(x)$ from becoming prematurely active. Imposing the condition $\alpha + \beta > 0$ is desirable because we want a barrier to relax as it is approached. For this particular surrogate, straightforward differentiation yields

$$\begin{aligned} dg(x_k \mid x_k) &= df(x_k) && (16.6) \\ d^2g(x_k \mid x_k) &= d^2f(x_k) - \gamma\alpha \sum_{j=1}^{q} v_j(x_k)^{\beta-1} d^2 v_j(x_k) && (16.7) \\ &\quad + \gamma\alpha(\alpha+1) \sum_{j=1}^{q} v_j(x_k)^{\beta-2} \nabla v_j(x_k) dv_j(x_k). \end{aligned}$$

Example 16.3.1 *A Geometric Programming Example*

Consider the typical geometric programming problem of minimizing

$$f(x) = \frac{1}{x_1 x_2 x_3} + x_2 x_3$$

subject to

$$v(x) = 4 - 2x_1x_3 - x_1x_2 \geq 0$$

and positive values for the x_i. Making the change of variables $x_i = e^{y_i}$ transforms the problem into a convex program. With the choice $\gamma = 1$, the MM gradient algorithm with the exponential parameterization and the log surrogate (16.4) produces the iterates displayed in the top half of Table 16.1. In this case Newton's method performs well, and none of the safeguards is needed. The MM gradient algorithm with the power surrogate (16.5) does somewhat better. The results shown in the bottom half of Table 16.1 reflect the choices $\gamma = 1$, $\alpha = 1/2$, and $\beta = 1$. ■

In the presence of linear constraints, both updates for the adaptive barrier method rely on the quadratic approximation of the surrogate function $g(x \mid x_k)$ using the calculated first and second differentials. This quadratic approximation is then minimized subject to the equality constraints as prescribed in Example 11.3.3.

Example 16.3.2 *Linear Programming*

Consider the standard linear programming problem of minimizing $c^t x$ subject to $Vx = b$ and $x \geq \mathbf{0}$ [19]. At iteration $k+1$ of the adaptive barrier method with the power surrogate (16.5), we minimize the quadratic approximation

$$c^t x_k + c^t(x - x_k) + \tfrac{1}{2}\gamma\alpha(\alpha+1)\sum_{j=1}^{n} x_{kj}^{\beta-2}(x_j - x_{kj})^2$$

TABLE 16.1. Solution of a Geometric Programming Problem

Iterates for the Log Surrogate

Iteration m	$f(x_k)$	x_{k1}	x_{k2}	x_{k3}
1	2.0000	1.0000	1.0000	1.0000
2	1.7299	1.4386	0.9131	0.6951
3	1.6455	1.6562	0.9149	0.6038
4	1.5993	1.7591	0.9380	0.5685
5	1.5700	1.8256	0.9554	0.5478
10	1.5147	1.9614	0.9903	0.5098
15	1.5034	1.9910	0.9977	0.5023
20	1.5008	1.9979	0.9995	0.5005
25	1.5002	1.9995	0.9999	0.5001
30	1.5000	1.9999	1.0000	0.5000
35	1.5000	2.0000	1.0000	0.5000

Iterates for the Power Surrogate

Iteration m	$f(x_k)$	x_{k1}	x_{k2}	x_{k3}
1	2.0000	1.0000	1.0000	1.0000
2	1.6478	1.5732	1.0157	0.6065
3	1.5817	1.7916	0.9952	0.5340
4	1.5506	1.8713	1.0011	0.5164
5	1.5324	1.9163	1.0035	0.5090
10	1.5040	1.9894	1.0011	0.5008
15	1.5005	1.9986	1.0002	0.5001
20	1.5001	1.9998	1.0000	0.5000
25	1.5000	2.0000	1.0000	0.5000

subject to $V(x - x_k) = \mathbf{0}$. Note here the application of the two identities (16.6) and (16.7). According to Example 11.3.3 and equation (14.3), this minimization problem has solution

$$x_{k+1} = x_k - [D_k^{-1} - D_k^{-1}V^t(VD_k^{-1}V^t)^{-1}VD_k^{-1}]c,$$

where D_k is a diagonal matrix with jth diagonal entry $\gamma\alpha(\alpha+1)x_{kj}^{\beta-2}$. It is convenient here to take $\gamma\alpha(\alpha+1) = 1$ and to step halve along the search direction $x_{k+1} - x_k$ whenever necessary. The case $\beta = 0$ bears a strong resemblance to Karmarkar's celebrated method of linear programming. ■

We will not undertake a systematic convergence analysis of the adaptive barrier algorithms. The next example illustrates that the local rate of convergence can be linear even when one of the constraints $v_i(x) \geq 0$ is active at the minimum. Further partial results appear in [30].

Example 16.3.3 *Convergence for the Multinomial Distribution*

As pointed out in Example 11.3.1, the loglikelihood for a multinomial distribution reduces to $\sum_{i=1}^m n_i \ln p_i$, where n_i is the observed number of counts for category i and p_i is the probability attached to category i. Maximizing the loglikelihood subject to the constraints $p_i \geq 0$ and $\sum_{i=1}^m p_i = 1$ gives the explicit maximum likelihood estimates $p_i = n_i/n$ for n trials. To compute the maximum likelihood estimates iteratively using the surrogate function (16.4), we find a stationary point of the Lagrangian

$$-\sum_{i=1}^m n_i \ln p_i - \gamma \sum_{i=1}^m p_{ki} \ln p_i + \gamma \sum_{i=1}^m (p_i - p_{ki}) + \delta\Big(\sum_{i=1}^m p_i - 1\Big).$$

Setting the ith partial derivative of the Lagrangian equal to 0 gives

$$-\frac{n_i}{p_i} - \frac{\gamma p_{ki}}{p_i} + \gamma + \delta \quad = \quad 0. \tag{16.8}$$

Multiplying equation (16.8) by p_i, summing on i, and solving for δ yield $\delta = n$. Substituting this value back in equation (16.8) produces

$$p_{k+1,i} \quad = \quad \frac{n_i + \gamma p_{ki}}{n + \gamma}.$$

At first glance it is not obvious that p_{ki} tends to n_i/n, but the algebraic rearrangement

$$\begin{aligned} p_{k+1,i} - \frac{n_i}{n} &= \frac{n_i + \gamma p_{ki}}{n+\gamma} - \frac{n_i}{n} \\ &= \frac{\gamma}{n+\gamma}\Big(p_{ki} - \frac{n_i}{n}\Big) \end{aligned}$$

shows that p_{ki} approaches n_i/n at the linear rate $\gamma/(n+\gamma)$. This is true regardless of whether $n_i/n = 0$ or $n_i/n > 0$. ■

16.4 Dykstra's Algorithm

Dykstra's algorithm deals with the projection of points onto a nonempty closed convex set. Before we discuss the algorithm, it is important to check that the notion of projection is well defined. The pertinent facts are summarized in the next proposition. Note that the projection operator discussed is usually nonlinear.

Proposition 16.4.1 *If C is a closed convex set in R^n, then there is a closest point $P_C(x)$ to x in C. The projection $P_C(x)$ can be characterized as the unique point satisfying the inequality*

$$[y - P_C(x)]^t[x - P_C(x)] \quad \leq \quad 0 \tag{16.9}$$

for every $y \in C$. Furthermore, $P_C[P_C(x)] = P_C(x)$ and

$$\|P_C(x) - P_C(y)\|_2 \quad \leq \quad \|x - y\|_2.$$

Proof: Obviously, $P_C(x) = x$ for $x \in C$. This makes the idempotent identity $P_C[P_C(x)] = P_C(x)$ obvious. For $x \notin C$, we will show that $P_C(x)$ exists and is unique by employing the easily verified parallelogram law

$$\|u+v\|_2^2 + \|u-v\|_2^2 = 2\Big(\|u\|_2^2 + \|v\|_2^2\Big).$$

If $d = \inf\{\|x-y\|_2 : y \in C\}$, then there exists a sequence $y_k \in C$ such that $\lim_{k\to\infty} \|x - y_k\|_2 = d$. By virtue of the parallelogram law, we have

$$\|y_j - y_k\|_2^2 + \|y_j + y_k - 2x\|_2^2 = 2\Big(\|y_j - x\|_2^2 + \|x - y_k\|_2^2\Big),$$

or equivalently

$$\frac{1}{4}\|y_j - y_k\|_2^2 = \frac{1}{2}\Big(\|x - y_j\|_2^2 + \|x - y_k\|_2^2\Big) - \|x - \frac{1}{2}\Big(y_j + y_k\Big)\|_2^2.$$

Because C is convex, $\frac{1}{2}(y_j + y_k) \in C$, and it follows from the definition of d that

$$\frac{1}{4}\|y_j - y_k\|_2^2 \le \frac{1}{2}\Big(\|x - y_j\|_2^2 + \|x - y_k\|_2^2\Big) - d^2. \tag{16.10}$$

Letting j and k tend to ∞ confirms that y_m is a Cauchy sequence whose limit y_∞ must lie in the closed set C. In view of the continuity of the norm, we have $\|x - y_\infty\|_2 = d$. To prove uniqueness, suppose there is a second point $z_\infty \in C$ satisfying $\|x - z_\infty\|_2 = d$. Then substituting y_∞ for y_j and z_∞ for y_k in inequality (16.10) shows that $z_\infty = y_\infty$.

To prove the inequality $[x - P_C(x)]^t[y - P_C(x)] \le 0$ for $y \in C$, note that the definition of $P_C(x)$ and the convexity of C imply

$$\begin{aligned}
& \|P_C(x) - x\|_2^2 \\
\le\ & \|(1-s)P_C(x) + sy - x\|_2^2 \\
=\ & \|P_C(x) - x\|_2^2 + 2s[P_C(x) - x]^t[y - P_C(x)] + s^2\|y - P_C(x)\|_2^2
\end{aligned}$$

for any $s \in (0,1]$. Canceling $\|P_C(x) - x\|_2^2$ from both sides and dividing by s yields

$$0 \le 2[P_C(x) - x]^t[y - P_C(x)] + s\|y - P_C(x)\|_2^2.$$

Sending s to 0 now gives the result. Conversely, if $z \in C$ satisfies the inequality $(y - z)^t(x - z) \le 0$ for every $y \in C$, then we have

$$\begin{aligned}
[P_C(x) - z]^t[x - z] &\le 0 \\
[z - P_C(x)]^t\,[x - P_C(x)] &\le 0.
\end{aligned}$$

Adding these two inequalities produces $\|z - P_C(x)\|_2^2 \le 0$. This can only be true if $z = P_C(x)$.

Finally, to prove the inequality $\|P_C(x) - P_C(z)\|_2 \leq \|x - z\|_2$, we add the inequalities

$$\begin{aligned}
[x - P_C(x)]^t[P_C(z) - P_C(x)] &\leq 0 \\
[z - P_C(z)]^t[P_C(x) - P_C(z)] &\leq 0
\end{aligned}$$

and rearrange. This gives

$$\begin{aligned}
[P_C(x) - P_C(z)]^t[P_C(x) - P_C(z)] &\leq (x - z)^t[P_C(x) - P_C(z)] \\
&\leq \|x - z\|_2\|P_C(x) - P_C(z)\|_2
\end{aligned}$$

by virtue of the Cauchy-Schwarz inequality. Dividing by $\|P_C(x) - P_C(z)\|_2$ now completes the proof. ■

For most closed convex sets C, it is impossible to give an explicit formula for the projection operator P_C. A notable exception is projection onto the range C of a matrix A of full column rank. This puts us back in the familiar terrain of least squares estimation, where $P_C(x) = A(A^tA)^{-1}A^tx$. One can easily check that the projection matrix $P_C = A(A^tA)^{-1}A^t$ satisfies $P_C^t = P_C$ and $P_C^2 = P_C$. Furthermore, since $A^t(I - P_C) = \mathbf{0}$, any $y = Av$ in the range of A gives equality in the necessary inequality (16.9).

Projection onto a closed rectangle $C = \prod_{i=1}^n [a_i, b_i]$ yields the simple formula

$$P_C(x)_i = \begin{cases} a_i & x_i < a_i \\ x_i & x_i \in [a_i, b_i] \\ b_i & x_i > b_i \end{cases}$$

for the ith coordinate of $P_C(x)$. This formula extends to the cases $a_i = -\infty$, $a_i = b_i$, or $b_i = \infty$. Verification of inequality (16.9) reduces to showing that

$$[y_i - P_C(x)_i][x_i - P_C(x)_i] \leq 0$$

for all $y_i \in [a_i, b_i]$.

Subspaces and closed rectangles appear in the following brief list of the explicit projections that are most useful in practice.

Example 16.4.1 *Examples of Projection Operators*

Closed Ball: If $C = \{y \in \mathsf{R}^n : \|y - z\|_2 \leq r\}$, then

$$P_C(x) = \begin{cases} z + r\frac{(x-z)}{\|x-z\|_2} & x \notin C \\ x & x \in C. \end{cases}$$

Closed Rectangle: If $C = [a, b]$ is a closed rectangle in R^n, then

$$P_C(x)_i = \begin{cases} a_i & x_i < a_i \\ x_i & x_i \in [a_i, b_i] \\ b_i & x_i > b_i. \end{cases}$$

Hyperplane: If $C = \{y \in \mathsf{R}^n : a^t y = b\}$ for $a \neq \mathbf{0}$, then

$$P_C(x) = x - \frac{a^t x - b}{\|a\|_2^2} a.$$

Closed Halfspace: If $C = \{y \in \mathsf{R}^n : a^t y \leq b\}$ for $a \neq \mathbf{0}$, then

$$P_C(x) = \begin{cases} x - \frac{a^t x - b}{\|a\|_2^2} a & a^t x > b \\ x & a^t x \leq b\,. \end{cases}$$

Subspace: If C is the range of a matrix A with full column rank, then

$$P_C(x) = A(A^t A)^{-1} A^t x.$$

Positive Semidefinite Matrices: Let M be an $n \times n$ symmetric matrix with spectral decomposition $M = UDU^t$, where U is an orthogonal matrix and D is a diagonal matrix with ith diagonal entry d_i. The projection of M onto the set S of positive semidefinite matrices is given by $P_S(M) = UD_+U^t$, where D_+ is diagonal with ith diagonal entry $\max\{d_i, 0\}$. Problem 8 asks the reader to check this fact.

Dykstra's algorithm [2, 12, 14] is designed to find the projection of a point onto a finite intersection of closed convex sets. The intersection inherits the properties of closedness and convexity. Here are some possible situations where Dykstra's algorithm applies.

Example 16.4.2 *Applications of Dykstra's Algorithm*

Linear Equalities: Any solution of the system of linear equations $Ax = b$ belongs to the intersection of the hyperplanes $a_i^t x = b_i$, where a_i^t is the ith row of A.

Linear Inequalities: Any solution of the system of linear inequalities $Ax \leq b$ belongs to the intersection of the halfspaces $a_i^t x \leq b_i$, where a_i^t is the ith row of A.

Isotone Regression: The least squares problem of minimizing the sum $\sum_{i=1}^n (x_i - w_i)^2$ subject to the constraints $w_i \leq w_{i+1}$ corresponds to projection of x onto the intersection of the halfspaces

$$H_i = \{w \in \mathsf{R}^n : w_i - w_{i+1} \leq 0\}, \quad 1 \leq i \leq n-1.$$

Convex Regression: The least squares problem of minimizing the sum $\sum_{i=1}^n (x_i - w_i)^2$ subject to the constraints $w_i \leq \frac{1}{2}(w_{i-1} + w_{i+1})$ corresponds to projection of x onto the intersection of the halfspaces

$$H_i = \left\{w \in \mathsf{R}^n : w_i - \frac{1}{2}(w_{i-1} + w_{i+1}) \leq 0\right\}, \quad 2 \leq i \leq n-1.$$

Quadratic Programming: To minimize the strictly convex quadratic form $\frac{1}{2}x^tAx + b^tx + c$ subject to $Dx = e$ and $Fx \leq g$, we make the change of variables $y = Ux$, where $L = U^t$ is the Cholesky decomposition of A. This transforms the problem to one of minimizing

$$\begin{aligned} \frac{1}{2}x^tAx + b^tx + c &= \frac{1}{2}\|y\|_2^2 + b^tU^{-1}y + c \\ &= \frac{1}{2}\|y + L^{-1}b\|_2^2 - \frac{1}{2}b^tA^{-1}b + c \end{aligned}$$

subject to $DU^{-1}y = e$ and $FU^{-1}y \leq g$. The solution in the y coordinates is determined by projecting $-L^{-1}b$ onto the convex feasible region determined by $DU^{-1}y = e$ and $FU^{-1}y \leq g$.

To state Dykstra's algorithm, it is helpful to label the closed convex sets $C_0, \ldots, C_{r-1}$ and denote their nonempty intersection by $C = \cap_{i=0}^{r-1} C_i$. The algorithm keeps track of a primary sequence x_n and a companion sequence e_n. In the limit, x_n tends to $P_C(x)$. To initiate the process, we set $x_{-1} = x$ and $e_{-r} = \cdots = e_{-1} = \mathbf{0}$. For $n \geq 0$ we then iterate via

$$\begin{aligned} x_n &= P_{C_{n \bmod r}}(x_{n-1} + e_{n-r}) \\ e_n &= x_{n-1} + e_{n-r} - x_n. \end{aligned}$$

Here $n \bmod r$ is the nonnegative remainder after dividing n by r. In essence, the algorithm cycles among the convex sets and projects the sum of the current vector and the relevant previous companion vector onto the current convex set. The proof that Dykstra's algorithm converges to $P_C(x)$ is not beyond us conceptually, but we omit it in the interests of brevity.

As an example, suppose $r = 2$, C_0 is the closed unit ball in R^2, and C_1 is the closed halfspace with $x_1 \geq 0$. The intersection C is the right half ball centered at the origin. Table 16.2 records the iterates of Dykstra's algorithm starting from the point $x = (-1, 2)$ and their eventual convergence to the geometrically obvious solution $(0, 1)$.

When C_i is a subspace, Dykstra's algorithm can dispense with the corresponding companion subsequence of e_n. In this case, e_n is perpendicular to C_i whenever $n \bmod r = i$. Indeed, since $P_{C_i}(y)$ is a projection matrix, we have

$$\begin{aligned} x_n &= P_{C_i}(x_{n-1} + e_{n-r}) \\ &= P_{C_i}x_{n-1} + P_{C_i}e_{n-r} \\ &= P_{C_i}x_{n-1} \end{aligned}$$

under the perpendicularity assumption. The initial condition $e_{i-r} = \mathbf{0}$, the identity

$$\begin{aligned} e_n &= x_{n-1} - x_n + e_{n-r} \\ &= [I - P_{C_i}]x_{n-1} + e_{n-r} \\ &= P_{C_i^\perp}x_{n-1} + e_{n-r}, \end{aligned}$$

TABLE 16.2. Iterates of Dykstra's Algorithm

Iteration m	x_{m1}	x_{m2}
0	-1.00000	2.00000
1	-0.44721	0.89443
2	0.00000	0.89443
3	-0.26640	0.96386
4	0.00000	0.96386
5	-0.14175	0.98990
10	-0.01814	0.99984
15	0.00000	0.99999
20	-0.00057	1.00000
25	0.00000	1.00000
30	-0.00002	1.00000
35	0.00000	1.00000

and induction show that e_n belongs to the perpendicular complement $C_i^{\perp}$ if $n \bmod r = i$. When all of the C_i are subspaces, Dykstra's algorithm reduces to the method of alternating projections first studied by von Neumann.

If one merely wants to find any element of the closed convex intersection C, it is possible to dispense with the companion sequence of Dykstra's algorithm. The Elsner-Koltracht-Neumann theorem says that the sequence $x_n = P_{C_{n \bmod r}}(x_{n-1})$ converges to some element of C regardless of its starting point [6, 18]. For instance, in our toy example of the half ball, a single round $P_{C_1} \circ P_{C_0}(x)$ of projection lands in the half ball. Except for special x, the converged point is not the closest point in C to x.

16.5 Model Selection and the Lasso

We now turn to penalized regression and continuous model selection. Our focus will be on the lasso penalty and its application in regression problems where the number of predictors p exceeds the number of cases n [8, 9, 43, 47, 49]. The lasso also finds applications in generalized linear models. In each of these contexts, we let y_i be the response for case i, x_{ij} be the value of predictor j for case i, and β_j be the regression coefficient corresponding to predictor j. In practice one should standardize each predictor to have mean 0 and variance 1. Standardization puts all regression coefficients on a common scale as implicitly demanded by the lasso penalty.

The intercept α is ignored in the lasso penalty, whose strength is determined by the positive tuning constant λ. If $\theta = (\alpha, \beta_1, \ldots, \beta_p)$ is the parameter vector and $g(\theta)$ is the loss function ignoring the penalty, then the lasso minimizes the criterion

$$f(\theta) \quad = \quad g(\theta) + \lambda \sum_{j=1}^{p} |\beta_j|,$$

where $g(\theta) = \frac{1}{2}\sum_{i=1}^{n}(y_i - x_i^t\theta)^2$ in ℓ_2 regression and $g(\theta) = \sum_{i=1}^{n} |y_i - x_i^t\theta|$ in ℓ_1 regression. The penalty $\lambda \sum_j |\beta_j|$ shrinks each β_j toward the origin and tends to discourage models with large numbers of irrelevant predictors. The lasso penalty is more effective in this regard than the ridge penalty $\lambda \sum_j \beta_j^2$ because $|b|$ is much bigger than b^2 for small b.

Lasso penalized estimation raises two issues. First, what is the most effective method of minimizing the objective function $f(\theta)$? In the current section we highlight the method of coordinate descent [10, 23, 25, 54]. Second, how does one choose the tuning constant λ? The standard answer is cross-validation. Although this is a good reply, it does not resolve the problem of how to minimize average cross-validation error as measured by the loss function. Recall that in k-fold cross-validation, one divides the data into k equal batches (subsamples) and estimates parameters k times, leaving one batch out per time. The testing error (total loss) for each omitted batch is computed using the estimates derived from the remaining batches, and the cross-validation error $c(\lambda)$ is computed by averaging testing error across the k batches.

Unless carefully planned, evaluation of $c(\lambda)$ on a grid of points may be computationally costly, particularly if grid points occur near $\lambda = 0$. Because coordinate descent is fastest when λ is large and the vast majority of β_j are estimated as 0, it makes sense to start with a very large value and work downward. One advantage of this tactic is that parameter estimates for a given λ can be used as parameter starting values for the next lower λ. For the initial value of λ, the starting value $\theta = \mathbf{0}$ is recommended. It is also helpful to set an upper bound on the number of active parameters allowed and abort downward sampling of λ when this bound is exceeded. Once a fine enough grid is available, visual inspection usually suggests a small interval flanking the minimum. Application of golden section search over the flanking interval will then quickly lead to the minimum.

Coordinate descent comes in several varieties. The standard version cycles through the parameters and updates each in turn. An alternative version is greedy and updates the parameter giving the largest decrease in the objective function. Because it is impossible to tell in advance the extent of each decrease, the greedy version uses the surrogate criterion of steepest descent. In other words, for each parameter we compute forward and backward directional derivatives and update the parameter with the most negative directional derivative, either forward or backward. The overhead of keeping track of the directional derivative works to the detriment of the greedy method. For ℓ_1 regression, the overhead is relatively light, and greedy coordinate descent converges faster than cyclic coordinate descent.

Although the lasso penalty is nondifferentiable, it does possess directional derivatives along each forward or backward coordinate direction. For instance, if e_j is the coordinate direction along which β_j varies, then

$$d_{e_j} f(\theta) = \lim_{t \downarrow 0} \frac{f(\theta + te_j) - f(\theta)}{t} = d_{e_j} g(\theta) + \begin{cases} \lambda & \beta_j \geq 0 \\ -\lambda & \beta_j < 0, \end{cases}$$

and

$$d_{-e_j} f(\theta) = \lim_{t \downarrow 0} \frac{f(\theta - te_j) - f(\theta)}{t} = d_{-e_j} g(\theta) + \begin{cases} -\lambda & \beta_j > 0 \\ \lambda & \beta_j \leq 0. \end{cases}$$

In ℓ_1 regression, the loss function is also nondifferentiable, and a brief calculation shows that the coordinate directional derivatives are

$$d_{e_j} \sum_{i=1}^n |y_i - x_i^t \theta| = \sum_{i=1}^n \begin{cases} -x_{ij} & y_i - x_i^t \theta > 0 \\ x_{ij} & y_i - x_i^t \theta < 0 \\ |x_{ij}| & y_i - x_i^t \theta = 0 \end{cases}$$

and

$$d_{-e_j} \sum_{i=1}^n |y_i - x_i^t \theta| = \sum_{i=1}^n \begin{cases} x_{ij} & y_i - x_i^t \theta > 0 \\ -x_{ij} & y_i - x_i^t \theta < 0 \\ |x_{ij}| & y_i - x_i^t \theta = 0 \end{cases}$$

with predictor vector $x_i^t = (1, z_i^t)$ for case i. Fortunately, when a function is differentiable, its directional derivative along e_j coincides with its ordinary partial derivative, and its directional derivative along $-e_j$ coincides with the negative of its ordinary partial derivative.

When we visit parameter β_j in cyclic coordinate descent, we evaluate $d_{e_j} f(\theta)$ and $d_{-e_j} f(\theta)$. If both are nonnegative, then we skip the update for β_j. This decision is defensible when $g(\theta)$ is convex because the sign of a directional derivative fully determines whether improvement can be made in that direction. If either directional derivative is negative, then we must solve for the minimum in that direction. If the current slope parameter β_j is parked at zero and the partial derivative $\frac{\partial}{\partial \beta_j} g(\theta)$ exists, then

$$d_{e_j} f(\theta) = \frac{\partial}{\partial \beta_j} g(\theta) + \lambda, \quad d_{-e_j} f(\theta) = -\frac{\partial}{\partial \beta_j} g(\theta) + \lambda.$$

Hence, β_j moves to the right if $\frac{\partial}{\partial \beta_j} g(\theta) < -\lambda$, to the left if $\frac{\partial}{\partial \beta_j} g(\theta) > \lambda$, and stays fixed otherwise. In underdetermined problems with just a few relevant predictors, most updates are skipped, and the parameters never budge from their starting values of 0. This simple fact plus the complete absence of matrix operations explains the speed of coordinate descent. It inherits its numerical stability from the descent property of each update.

16.5.1 Application to ℓ_1 Regression

In lasso constrained ℓ_1 regression, greedy coordinate descent is quick because directional derivatives are trivial to update. Indeed, if updating β_j does not alter the sign of the residual $y_i - x_i^t\theta$ for case i, then the contributions of case i to the various directional derivatives do not change. When the residual $y_i - x_i^t\theta$ changes sign, these contributions change by $\pm 2x_{ij}$. When a residual changes from 0 to nonzero or vice versa, the increment depends on the sign of the nonzero residual and the sign of x_{ij}.

Updating the value of the chosen parameter can be achieved by the nearly forgotten algorithm of Edgeworth [15, 16], which for a long time was considered a competitor of least squares. Portnoy and Koenker [42] trace the history of the algorithm from Boscovich to Laplace to Edgeworth. It is fair to say that the algorithm has managed to cling to life despite decades of obscurity both before and after its rediscovery by Edgeworth.

To illustrate Edgeworth's algorithm in operation, consider minimizing the two-parameter model

$$g(\theta) \quad = \quad \sum_{i=1}^{n} |y_i - \alpha - z_i\beta|$$

with a single slope β. To update α, we recall the well-known connection between ℓ_1 regression and medians and replace α for fixed β by the sample median of the numbers $v_i = y_i - z_i\beta$. This action drives $g(\theta)$ downhill. Updating β for α fixed depends on writing

$$g(\theta) \quad = \quad \sum_{i=1}^{n} |z_i| \left| \frac{y_i - \alpha}{z_i} - \beta \right|,$$

sorting the numbers $v_i = (y_i - \alpha)/z_i$, and finding the weighted median with weight $w_i = |z_i|$ assigned to v_i. We replace β by the order statistic $v_{[i]}$ whose index i satisfies

$$\sum_{j=1}^{i-1} w_{[j]} \;<\; \frac{1}{2}\sum_{j=1}^{n} w_{[j]}, \qquad \sum_{j=1}^{i} w_{[j]} \;\geq\; \frac{1}{2}\sum_{j=1}^{n} w_{[j]}.$$

Problem 3 demonstrates that this choice is valid. Edgeworth's algorithm easily generalizes to multiple linear regression. Implementing the algorithm with a lasso penalty requires viewing the penalty terms as the absolute values of pseudo-residuals. Thus, we write

$$\lambda|\beta_j| \quad = \quad |y - x^t\theta|$$

by taking $y = 0$ and $x_k = \lambda 1_{\{k=j\}}$.

Two criticisms have been leveled at Edgeworth's algorithm. First, although it drives the objective function steadily downhill, it sometimes converges to an inferior point. See Problem 20 for an example. The second criticism is that convergence often occurs in a slow seesaw pattern. These defects are not completely fatal. As late as 1978, Armstrong and Kung published a computer implementation of Edgeworth's algorithm in *Applied Statistics* [1].

16.5.2 Application to ℓ_2 Regression

In ℓ_2 regression with a lasso penalty, we minimize the objective function

$$f(\theta) \quad = \quad \frac{1}{2}\sum_{i=1}^{n}(y_i - \alpha - z_i^t\beta)^2 + \lambda\sum_{j=1}^{p}|\beta_j| \quad = \quad g(\theta) + \lambda\sum_{j=1}^{p}|\beta_j|.$$

The update of the intercept parameter can be written as

$$\hat{\alpha} \quad = \quad \frac{1}{n}\sum_{i=1}^{n}(y_i - z_i^t\beta) \quad = \quad \alpha - \frac{1}{n}\frac{\partial}{\partial\alpha}g(\theta).$$

For the parameter β_k, there are separate solutions to the left and right of 0. These boil down to

$$\begin{aligned} \hat{\beta}_{k,-} &= \min\left\{0, \beta_k - \frac{\frac{\partial}{\partial\beta_k}g(\theta) - \lambda}{\sum_{i=1}^{n} z_{ik}^2}\right\} \\ \hat{\beta}_{k,+} &= \max\left\{0, \beta_k - \frac{\frac{\partial}{\partial\beta_k}g(\theta) + \lambda}{\sum_{i=1}^{n} z_{ik}^2}\right\}. \end{aligned}$$

The reader can check that only one of these two solutions can be nonzero. The partial derivatives

$$\frac{\partial}{\partial\alpha}g(\theta) \quad = \quad -\sum_{i=1}^{n} r_i, \qquad \frac{\partial}{\partial\beta_k}g(\theta) \quad = \quad -\sum_{i=1}^{n} r_i z_{ik}$$

of $g(\theta)$ are easy to compute if we keep track of the residuals $r_i = y_i - \alpha - z_i^t\beta$. The residual r_i starts with the value y_i and is reset to $r_i + \alpha - \hat{\alpha}$ when α is updated and to $r_i + z_{ij}(\beta_j - \hat{\beta}_j)$ when β_j is updated. Organizing all updates around residuals promotes fast evaluation of $g(\theta)$. At the expense of somewhat more complex code [24], a better tactic is to exploit the identity

$$\sum_{i=1}^{n} r_i z_{ik} \quad = \quad \sum_{i=1}^{n} y_i z_{ik} - \alpha\sum_{i=1}^{n} z_{ik} - \sum_{j:|\beta_j|>0}\left(\sum_{i=1}^{n} z_{ij}z_{ik}\right)\beta_j.$$

This representation suggests storing and reusing the inner products

$$\sum_{i=1}^{n} y_i z_{ik}, \qquad \sum_{i=1}^{n} z_{ik}, \qquad \sum_{i=1}^{n} z_{ij}z_{ik}$$

for the active predictors.

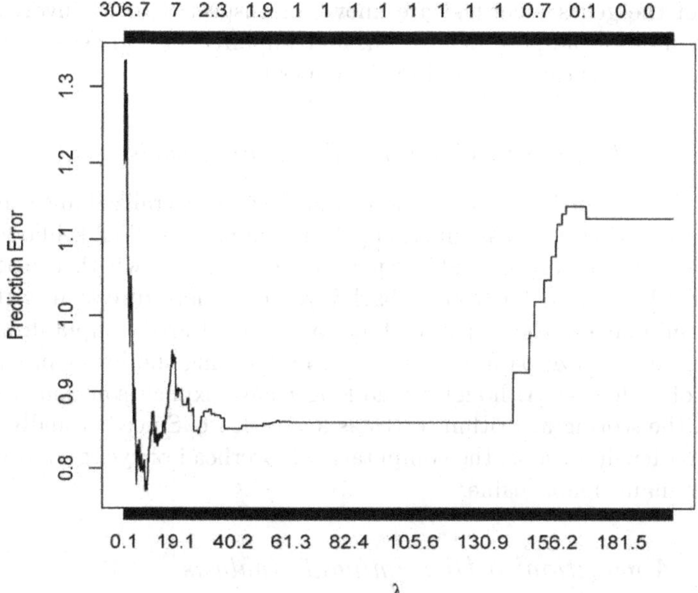

FIGURE 16.1. The Cross-Validation Curve $c(\lambda)$ for Obesity in Mice

Example 16.5.1 *Obesity and Gene Expression in Mice*

We now consider a genetics example involving gene expression levels and obesity in mice. Wang et al. [52] measured abdominal fat mass on $n = 311$ F2 mice (155 males and 156 females). The F2 mice were created by mating two inbred strains and then mating brother-sister pairs from the resulting offspring. Wang et al. also recorded the expression levels in liver of $p =$ 23,388 genes in each mouse. A reasonable model postulates

$$y_i \quad = \quad 1_{\{i \,\text{male}\}}\alpha_1 + 1_{\{i \,\text{female}\}}\alpha_2 + \sum_{j=1}^{p} x_{ij}\beta_j + \epsilon_i,$$

where y_i measures fat mass on mouse i, x_{ij} is the expression level of gene j in mouse i, and ϵ_i is random error. Since male and female mice exhibit across the board differences in size and physiology, it is prudent to estimate a different intercept for each sex. Figure 16.1 plots average prediction error as a function of λ (lower horizontal axis) and the average number of nonzero predictors (upper horizontal axis). Here we use ℓ_2 penalized regression and 10-fold cross-validation. Examination of the cross-validation curve $c(\lambda)$ over a fairly dense grid shows an optimal λ of 7.8 with 41 nonzero predictors. For ℓ_1 penalized regression, the optimal λ is around 3.5 with 77 nonzero predictors. The preferred ℓ_1 and ℓ_2 models share 27 predictors in common.

Several of the genes identified are known or suspected to be involved in lipid metabolism, adipose deposition, and impaired insulin sensitivity in mice. More details can be found in the paper [54].

16.5.3 Application to Generalized Linear Models

The tactics described for ℓ_2 regression carry over to generalized linear models. In this setting, the loss function $g(\theta)$ is the negative loglikelihood. In many cases, $g(\theta)$ is convex, and it is possible to determine whether progress can be made along a forward or backward coordinate direction without actually minimizing the objective function. It is clearly computationally beneficial to organize parameter updates by tracking the linear predictor $\alpha + z_i^t\beta$ of each case. Although we no longer have explicit solutions to fall back on, the scoring algorithm serves as a substitute. Since it usually converges in a few iterations, the computational overhead of cyclic coordinate descent remains manageable.

16.5.4 Application to Discriminant Analysis

Discriminant analysis is another attractive candidate for penalized estimation. In discriminant analysis with two categories, each case i is characterized by a feature vector z_i and a category membership indicator y_i taking the values -1 or 1. In the machine learning approach to discriminant analysis [46, 50], the hinge loss function $[1 - y_i(\alpha + z_i^t\beta)]_+$ plays a prominent role. Here $(u)_+$ is shorthand for the convex function $\max\{u, 0\}$. Just as in ordinary regression, we can penalize the overall loss

$$g(\theta) = \sum_{i=1}^{m}[1 - y_i(\alpha + z_i^t\beta)]_+$$

by imposing a lasso or ridge penalty. Note that the linear regression function $h_i(\theta) = \alpha + z_i^t\beta$ predicts either -1 or 1. If $y_i = 1$ and $h_i(\theta)$ over-predicts in the sense that $h_i(\theta) > 1$, then there is no loss. Similarly, if $y_i = -1$ and $h_i(\theta)$ under-predicts in the sense that $h_i(\theta) < -1$, then there is no loss.

Most strategies for estimating θ pass to the dual of the original minimization problem. A simpler strategy is to majorize each contribution to the loss by a quadratic and minimize the surrogate loss plus penalty [26]. A little calculus shows that $(u)_+$ is majorized at $u_n \neq 0$ by the quadratic

$$q(u \mid u_n) = \frac{1}{4|u_n|}(u + |u_n|)^2. \tag{16.11}$$

See Problem 24. In fact, this is the best quadratic majorizer [11]. To avoid the singularity at 0, we recommend replacing $q(u \mid u_n)$ by

$$r(u \mid u_n) = \frac{1}{4|u_n| + \epsilon}(u + |u_n|)^2.$$

In double precision, a good choice of ϵ is 10^{-5}. Of course, the dummy variable u is identified in case i with $1 - y_i(\alpha + z_i^t\beta)$.

If we impose a ridge penalty, then the majorization (16.11) leads to a pure MM algorithm exploiting weighted least squares. Coordinate descent algorithms with a lasso or ridge penalty are also enabled by majorization, but each ℓ_2 coordinate update merely decreases the objective function along the given coordinate direction. Fortunately, this drawback is outweighed by the gain in numerical simplicity in circumventing hinge loss. The decisions to use a lasso or ridge penalty and apply pure MM or coordinate descent with majorization will be dictated in practical problems by the number of potential predictors. If a lasso penalty is imposed, then cyclic coordinate descent can be applied, with the surrogate function substituting for the objective function in each parameter update.

In discriminant analysis with more than two categories, it is convenient to pass to ϵ-insensitive loss and multiple linear regression. The story is too long to tell here, but it is worth mentioning that the conjunction of a parsimonious loss function and an efficient MM algorithm produces one of the most effective discriminant analysis methods tested [32].

16.6 Standard Errors

In this section we tackle two issues that arise in computing asymptotic standard errors of parameter estimates.

16.6.1 Standard Errors and the MM Algorithm

Under appropriate large sample assumptions [21], it is possible to demonstrate that in unconstrained estimation a maximum likelihood estimator has asymptotic variance matrix equal to the inverse of the expected information matrix. In practice, the expected information matrix is often well approximated by the observed information matrix $-d^2L(\theta)$. Thus, once the maximum likelihood estimate $\hat{\theta}$ has been found, its standard errors can be obtained by taking square roots of the diagonal entries of the inverse of $-d^2L(\hat{\theta})$. In MM and EM algorithm problems, however, direct calculation of $-d^2L(\hat{\theta})$ can be difficult. Here we discuss a numerical approximation to this matrix that exploits quantities readily obtained by running an MM algorithm [28]. Let $g(\theta \mid \theta_n)$ denote a minorizing function of the loglikelihood $L(\theta)$ at the point θ_n, and define $M(\theta)$ to be the MM algorithm map taking θ_n to θ_{n+1}.

Our numerical approximation of $-d^2L(\hat{\theta})$ is based on the differential $dM(\hat{\theta})$. Assuming that $L(\theta)$ and $g(\theta \mid \phi)$ are sufficiently smooth and that $d^2g(\theta \mid \theta)$ is invertible, the implicit function theorem [31] applies to the equation

$$\nabla g[M(\theta) \mid \theta] = \mathbf{0}$$

and shows that $M(\theta)$ is continuously differentiable with differential

$$dM(\theta) = -d^2 g[M(\theta) \mid \theta]^{-1} d^{11} g[M(\theta) \mid \theta]. \quad (16.12)$$

Here $d^{11}g(\theta \mid \phi)$ denotes the differential of $\nabla g(\theta \mid \phi)$ with respect to ϕ. Further simplification can be achieved by taking the differential of

$$\nabla L(\theta) - \nabla g(\theta \mid \theta) = \mathbf{0}$$

and setting $\theta = \hat{\theta}$. These actions give

$$d^2 L(\hat{\theta}) - d^2 g(\hat{\theta} \mid \hat{\theta}) - d^{11} g(\hat{\theta} \mid \hat{\theta}) = \mathbf{0}.$$

This last equation can be solved for $d^{11}g(\hat{\theta} \mid \hat{\theta})$, and the result substituted in equation (16.12). It follows that

$$\begin{aligned} dM(\hat{\theta}) &= -d^2 g(\hat{\theta} \mid \hat{\theta})^{-1} [d^2 L(\hat{\theta}) - d^2 g(\hat{\theta} \mid \hat{\theta})] \\ &= I - d^2 g(\hat{\theta} \mid \hat{\theta})^{-1} d^2 L(\hat{\theta}) \end{aligned}$$

and therefore that

$$d^2 L(\hat{\theta}) = d^2 g(\hat{\theta} \mid \hat{\theta}) \left[I - dM(\hat{\theta}) \right]. \quad (16.13)$$

This formula is the basis of the SEM algorithm of Meng and Rubin [38].

Approximation of $d^2L(\hat{\theta})$ based on equation (16.13) requires numerical approximation of the Jacobi matrix $dM(\theta)$, whose i, j entry equals

$$\frac{\partial}{\partial \theta_j} M_i(\theta) = \lim_{s \to 0} \frac{M_i(\theta + s e_j) - M_i(\theta)}{s}.$$

Since $M(\hat{\theta}) = \hat{\theta}$, the jth column of $dM(\hat{\theta})$ may be approximated using output from the corresponding MM algorithm by (a) iterating until $\hat{\theta}$ is found, (b) altering the jth component of $\hat{\theta}$ by a small amount s, (c) applying the MM algorithm to this altered θ, (d) subtracting $\hat{\theta}$ from the result, and (e) dividing by s.

16.6.2 Standard Errors under Linear Constraints

To calculate the asymptotic covariance matrix of an estimated parameter vector $\hat{\theta}$ subject to linear equality constraints $V\theta = d$, we reparameterize. Suppose that the $m \times n$ matrix V has full row rank $m < n$ and that α is a particular solution of $V\theta = d$. The Gram-Schmidt process allows us to construct an $n \times (n-m)$ matrix W with $n-m$ linearly independent columns $w_1, \ldots, w_{n-m}$ orthogonal to the rows $v_1^t, \ldots, v_m^t$ of V. Now consider the reparameterization $\theta = \alpha + W\beta$. By virtue of our choice of α and the identity $VW = \mathbf{0}$, it is clear that $V(\alpha + W\beta) = d$. Since the range of W and the null space of V both have dimension $n - m$, it is also clear that all solutions of $V\theta = d$ are generated as image vectors $\alpha + W\beta$. Thus, β can be viewed as a replacement for θ.

Under the preceding reparameterization, the loglikelihood $L(\alpha + W\beta)$ has score $W^t \nabla L(\alpha + W\beta)$, observed information $-W^t d^2 L(\alpha + W\beta)W$, and expected information $-W^t \,\mathrm{E}[d^2 L(\alpha + W\beta)]W$. If we let $J(\theta)$ represent either the observed information $-d^2 L(\theta)$ or the expected information $\mathrm{E}[-d^2 L(\theta)]$ of the original parameters, then an asymptotic variance matrix of the estimated parameter vector $\hat{\beta}$ is $[W^t J(\hat{\theta}) W]^{-1}$. The brief calculation

$$\mathrm{Var}(\hat{\theta}) \;=\; \mathrm{Var}(\alpha + W\hat{\beta}) \;=\; \mathrm{Var}(W\hat{\beta}) \;=\; W\,\mathrm{Var}(\hat{\beta})W^t$$

shows that $\hat{\theta}$ has asymptotic covariance matrix $W[W^t J(\hat{\theta})W]^{-1}W^t$, which unfortunately appears to depend on the particular reparameterization chosen. This is an illusion because if we replace W by WT, where T is any invertible matrix, then

$$\begin{aligned} WT[T^t W^t J(\hat{\theta}) WT]^{-1} T^t W^t &= WTT^{-1}[W^t J(\hat{\theta})W]^{-1}(T^t)^{-1}T^t W^t \\ &= W[W^t J(\hat{\theta})W]^{-1}W^t. \end{aligned}$$

Problems 31 and 32 sketch Silvey [44] and Jennrich's method of computing $W[W^t J(\hat{\theta})W]^{-1}W^t$. This method relies on the sweep operator rather than Gram-Schmidt orthogonalization.

In calculating the asymptotic covariance matrix of $\hat{\theta}$ in the presence of linear inequality constraints, the traditional procedure is to ignore the inactive constraints and to append the active constraints to the existing linear equality constraints. This creates a larger constraint matrix V and a corresponding smaller matrix W orthogonal to V.

16.7 Problems

1. In Example 16.2.1 prove directly that the solution displayed in equation (16.1) converges to the minimum point of $\|Y - X\beta\|_2^2$ subject to the linear constraints $V\beta = d$. (Hints: Assume that the matrix V has full column rank and apply Example 11.3.3, Proposition 11.3.2, and Woodbury's formula (7.9).)

2. Consider the convex programming problem of minimizing the convex function $f(x)$ subject to the affine equality constraints $g_i(x) = b_i$ for $1 \le i \le p$ and the convex inequality constraints $h_j(x) \le c_j$ for

$1 \leq j \leq q$. Let $R(b, c)$ denote the feasible region for a particular choice of the vectors $b = (b_1, \ldots, b_p)^t$ and $c = (c_1, \ldots, c_q)^t$. Show that

$$s(b, c) \quad = \quad \inf\{f(x) : x \in R(b, c)\}$$

is a convex function of the vector (b, c).

3. The convex function $f(x)$ is defined on a convex set C. Suppose that $y \in C$ has nonnegative directional derivative $d_v f(y)$ for every direction $v = x - y$ defined by an $x \in C$. Demonstrate that y minimizes $f(x)$ on C. Conversely, show that if any such directional derivative $d_v f(y)$ is negative, then y cannot minimize $f(x)$ on C. (Hint: The difference quotient

$$\frac{f(y + s[x - y]) - f(y)}{s}$$

is increasing in $s > 0$.)

4. Prove the surrogate function (16.5) majorizes $f(x)$ up to an irrelevant additive constant.

5. Suppose C is a closed convex set. If y is on the line segment between $x \notin C$ and $P_C(x)$, then prove that $P_C(y) = P_C(x)$.

6. Suppose C is a closed convex set and $P_C(y) = y$. Demonstrate that equality occurs in the projection inequality

$$\|P_C(x) - P_C(y)\|_2 \quad \leq \quad \|x - y\|_2$$

if and only if $P_C(x) = x$. An operator $P_C(x)$ having this property is said to be paracontractive relative to the given norm.

7. Let C be a closed convex set in R^n. Show that

   (a) $P_{C+y}(x + y) = P_C(x) + y$ for all x and y.

   (b) $P_{aC}(ax) = aP_C(x)$ for all x and real a.

   Let S be a subspace of R^n. Show that

   (a) $P_S(x + y) = P_S(x) + y$ for all $x \in \mathsf{R}^n$ and $y \in S$.

   (b) $P_S(ax) = aP_S(x)$ for all $x \in \mathsf{R}^n$ and real a.

8. Let M be an $n \times n$ symmetric matrix with spectral decomposition $M = UDU^t$, where U is an orthogonal matrix and D is a diagonal matrix with ith diagonal entry d_i. Prove that the Frobenius norm $\|M - P_S(M)\|_F$ is minimized over the set S of positive semidefinite matrices by taking $P_S(M) = UD_+U^t$, where D_+ is diagonal with ith diagonal entry $\max\{d_i, 0\}$.

9. Suppose A is an $n \times n$ matrix and Sym_n is the set of $n \times n$ symmetric matrices. Find the matrix M in Sym_n that minimizes the Frobenius norm $\|A - M\|_F$.

10. Let A be a full rank $m \times n$ matrix and b be an $m \times 1$ vector with $m < n$. The set $H = \{x \in \mathsf{R}^n : Ax = b\}$ defines a plane in R^n. If $m = 1$, H is a hyperplane. Given $y \in \mathsf{R}^n$, prove that the closest point to y in H is

$$P(y) \quad = \quad y - A^t(AA^t)^{-1}Ay + A^t(AA^t)^{-1}b.$$

(Hint: Verify the criterion (16.9).)

11. Suppose C is a closed convex set wholly contained within an affine subspace $V = \{y \in \mathsf{R}^n : Ay = b\}$. For $x \notin V$ demonstrate the projection identity $P_C(x) = P_C \circ P_V(x)$ [39]. (Hint: Consider the equality

$$\begin{aligned}[x - P_C(x)]^t[y - P_C(x)] \quad &= \quad [x - P_V(x)]^t[y - P_C(x)] \\ &\quad + [P_V(x) - P_C(x)]^t[y - P_C(x)]\end{aligned}$$

with the criterion (16.9) in mind.)

12. For positive numbers $c_1, \ldots, c_n$ and nonnegative numbers $b_1, \ldots, b_n$ satisfying $\sum_{i=1}^n c_i b_i \le 1$, define the truncated simplex

$$S \quad = \quad \Big\{ y \in \mathsf{R}^n : \sum_{i=1}^n c_i y_i = 1,\ y_i \ge b_i,\ 1 \le i \le n \Big\}.$$

If $x \in \mathsf{R}^n$ has coordinate sum $\sum_{i=1}^n c_i x_i = 1$, then prove that the closest point y in S to x satisfies the Lagrange multiplier conditions

$$y_i - x_i + \lambda c_i - \mu_i \quad = \quad 0$$

for appropriate multipliers λ and $\mu_i \ge 0$. Further show that

$$\lambda \quad = \quad \frac{c^t \mu}{\|c\|_2^2} \quad \ge \quad 0.$$

Why does it follow that $y_i = b_i$ whenever $x_i < b_i$? Prove that the Lagrange multiplier conditions continue to hold when $x_i < b_i$ if we replace x_i by b_i and μ_i by λc_i. Since the Lagrange multiplier conditions are sufficient as well as necessary in convex programming, this demonstrates that (a) we can replace each coordinate x_i by $\max\{x_i, b_i\}$ without changing the projection y of x onto S, and (b) y can be viewed as a point in a similar simplex in a reduced number of dimensions when one or more $x_i \le b_i$ [39].

13. Michelot's [39] algorithm for projecting a point x onto the simplex S defined in Problem 12 cycles through the following steps:

    (a) Projection onto the affine subspace $V_n = \{y \in \mathsf{R}^n : \sum_i c_i y_i = 1\}$,
    (b) Replacement of every coordinate x_i by $\max\{x_i, b_i\}$,
    (c) Reduction of the dimension n whenever some $x_i = b_i$.

    In view of Problems 11 and 12, demonstrate that Michelot's algorithm converges to the correct solution in at most n steps. Explicitly solve the Lagrange multiplier problem corresponding to step (a). Program and test the algorithm.

14. A polyhedron is the intersection of a finite number of halfspaces. Program Dykstra's algorithm for projection onto an arbitrary polyhedron.

15. In weighted isotone regression one minimizes the weighted sum of squares $\sum_{i=1}^n w_i(y_i - x_i)^2$ subject to the constraints $x_1 \leq \cdots \leq x_n$ and positive weights w_i. Show how Dykstra's algorithm can be adapted to handle this problem. As an alternative, the pool adjacent violators algorithm starts with $x_i = y_i$ for all i. It then cycles through adjacent pairs (x_i, x_{i+1}). When $x_i \leq x_{i+1}$, it leaves the pair untouched. When $x_i > x_{i+1}$, it replaces the pair by their weighted average

$$\bar{x} = \frac{w_i x_i + w_{i+1} x_{i+1}}{w_i + w_{i+1}}$$

    with weight $w_i + w_{i+1}$ attached to $\bar{x}$. This procedure reduces the number of pairs by 1. It is repeated until no new violators are found. At that moment, one recovers x_i as the value assigned to the pool containing index i. Program Dykstra's algorithm and this alternative. Compare them in terms of speed and reliability.

16. The power plant production problem [45] involves minimizing

$$f(x) = \sum_{i=1}^n f_i(x_i), \quad f_i(x_i) = a_i x_i + \frac{1}{2} b_i x_i^2$$

    subject to the constraints $0 \leq x_i \leq u_i$ for each i and $\sum_{i=1}^n x_i \geq c$. For plant i, x_i is the power output, u_i is the capacity, and $f_i(x_i)$ is the cost. The total demand is c, and the cost constants a_i and b_i are positive. This problem can be solved by the adaptive barrier method or Dykstra's algorithm. Program either method and test it on a simple example with at least two power plants. Argue that the minimum is unique.

17. In Problem 16 investigate the performance of cyclic coordinate descent. Explain why it fails.

18. Show that $\hat{\mu}$ minimizes $f(\mu) = \sum_{i=1}^n w_i|x_i - \mu|$ if and only if

$$\sum_{x_i < \hat{\mu}} w_i \leq \frac{1}{2}\sum_{i=1}^n w_i, \quad \sum_{x_i \leq \hat{\mu}} w_i \geq \frac{1}{2}\sum_{i=1}^n w_i.$$

Assume that the weights w_i are positive. (Hint: Apply Problem 3.)

19. Consider the piecewise linear function

$$f(\mu) = c\mu + \sum_{i=1}^n w_i|x_i - \mu|,$$

where the positive weights satisfy $\sum_{i=1}^n w_i = 1$ and the points satisfy $x_1 < x_2 < \cdots < x_n$. Show that $f(\mu)$ has no minimum when $|c| > 1$. What happens when $c = 1$ or $c = -1$? This leaves the case $|c| < 1$. Show that a minimum occurs when

$$\sum_{x_i > \mu} w_i - \sum_{x_i \leq \mu} w_i \leq c \text{ and } \sum_{x_i \geq \mu} w_i - \sum_{x_i < \mu} w_i \geq c.$$

(Hints: A crude plot of $f(\mu)$ might help. What conditions on the right-hand and left-hand derivatives of $f(\mu)$ characterize a minimum?)

20. Show that Edgeworth's algorithm [36] for ℓ_1 regression converges to an inferior point for the data values (0.3,-1.0), (-0.4,-0.1), (-2.0,-2.9), (-0.9,-2.4), and (-1.1,2.2) for the pairs (x_i, y_i) and parameter starting values $(\alpha, \beta) = (-1.0, 3.5)$.

21. Implement and test greedy coordinate descent for lasso penalized ℓ_1 regression or cyclic coordinate descent for lasso penalized ℓ_2 regression.

22. In lasso penalized regression, suppose the convex loss function $g(\theta)$ is differentiable. A stationary point θ of coordinate descent satisfies the conditions $d_{e_j} f(\theta) \geq 0$ and $d_{-e_j} f(\theta) \geq 0$ for all j. Here α varies along the coordinate direction e_0. Calculate the general directional derivative

$$d_v f(\theta) = \sum_j \frac{\partial}{\partial \theta_j} g(\theta) v_j + \lambda \sum_{j>0} \begin{cases} v_j & \theta_j > 0 \\ -v_j & \theta_j < 0 \\ |v_j| & \theta_j = 0 \end{cases}$$

and show that

$$d_v f(\theta) = \sum_{v_j > 0} d_{e_j} f(\theta) v_j + \sum_{v_j < 0} d_{-e_j} f(\theta)|v_j|.$$

Conclude that every directional derivative is nonnegative at a stationary point. In view of Problem 3, stationary points therefore coincide with minimum points. This result does not hold for lasso penalized ℓ_1 regression.

23. Consider the function $\|x\|_0 = \sum_{i=1}^m 1_{\{x_i \neq 0\}}$ defined on R^m. What properties of a norm does it enjoy? Demonstrate that

$$\|x\|_0 = \lim_{\epsilon \downarrow 0} \sum_{i=1}^m \frac{\ln\left(1 + \frac{|x_i|}{\epsilon}\right)}{\ln\left(1 + \frac{1}{\epsilon}\right)}.$$

Note that the same limit applies if one substitutes x_i^2 for $|x_i|$. Now prove the majorization

$$\ln(\epsilon + y) \leq \ln(\epsilon + y_n) + \frac{1}{\epsilon + y_n}(y - y_n)$$

for nonnegative scalars y and y_n, and show how it can be employed to majorize approximations to $\|x\|_0$ based on the choices $|x_i|$ and x_i^2. See the references [5, 20, 53] for applications to sparse estimation and machine learning.

24. Show that the function $(u)_+ = \max\{u, 0\}$ is majorized by the quadratic function (16.11) at a point $u_n \neq 0$. Why does it suffice to prove that $(u)_+$ and $q(u \mid u_n)$ have the same value and same derivative at u_n and $-u_n$? (Hint: Draw rough graphs of $(u)_+$ and $q(u \mid u_n)$.)

25. Implement and test one of the discriminant analysis algorithms that depend on quadratic majorization of hinge loss.

26. Nonnegative matrix factorization was introduced by Lee and Seung [33, 34] as an analog of principal components and vector quantization with applications in data compression and clustering. In mathematical terms, one approximates a matrix U with nonnegative entries u_{ij} by a product VW of two low-rank matrices with nonnegative entries v_{ij} and w_{ij}. If the entries u_{ij} are integers, then they can be viewed as realizations of independent Poisson random variables with means $\sum_k v_{ik}w_{kj}$. In this setting the loglikelihood is

$$L(V, W) = \sum_i \sum_j \Big[u_{ij} \ln\Big(\sum_k v_{ik}w_{kj}\Big) - \sum_k v_{ik}w_{kj}\Big].$$

Maximization with respect to V and W should lead to a good factorization. Lee and Seung construct a block relaxation algorithm that hinges on the minorization

$$\ln\Big(\sum_k v_{ik}w_{kj}\Big) \geq \sum_k \frac{a_{ikj}^n}{b_{ij}^n} \ln\Big(\frac{b_{ij}^n}{a_{ikj}^n} v_{ik}w_{kj}\Big),$$

where

$$a_{ikj}^n = v_{ik}^n w_{kj}^n, \quad b_{ij}^n = \sum_k v_{ik}^n w_{kj}^n,$$

and n indicates the current iteration. Prove this minorization and derive the Lee-Seung algorithm with alternating multiplicative updates

$$v_{ik}^{n+1} = v_{ik}^{n} \frac{\sum_j u_{ij} \frac{w_{kj}^n}{b_{ij}^n}}{\sum_j w_{kj}^n}$$

and

$$w_{kj}^{n+1} = w_{kj}^{n} \frac{\sum_i u_{ij} \frac{v_{ik}^n}{b_{ij}^n}}{\sum_i v_{ik}^n}.$$

27. Continuing Problem 26, consider minimizing the squared Frobenius norm

$$\|U - VW\|_F^2 = \sum_i \sum_j \Big(u_{ij} - \sum_k v_{ik} w_{kj}\Big)^2.$$

Demonstrate the majorization

$$\Big(u_{ij} - \sum_k v_{ik} w_{kj}\Big)^2 \le \sum_k \frac{a_{ikj}^n}{b_{ij}^n} \Big(u_{ij} - \frac{b_{ij}^n}{a_{ikj}^n} v_{ik} w_{kj}\Big)^2$$

based on the notation of Problem 26. Now derive the block relaxation algorithm with multiplicative updates

$$v_{ik}^{n+1} = v_{ik}^{n} \frac{\sum_j u_{ij} w_{kj}^n}{\sum_j b_{ij}^n w_{kj}^n}$$

and

$$w_{kj}^{n+1} = w_{kj}^{n} \frac{\sum_i u_{ij} v_{ik}^n}{\sum_i b_{ij}^n v_{ik}^n}.$$

28. In the matrix factorizations described in Problems 26 and 27, it may be worthwhile shrinking the estimates of the entries of V and W toward 0 [41]. Let λ and μ be positive constants, and consider the revised objective functions

$$\begin{aligned} l(V,W) &= L(V,W) - \lambda \sum_i \sum_k v_{ik} - \mu \sum_k \sum_j w_{kj} \\ r(V,W) &= \|U - VW\|_F^2 + \lambda \sum_i \sum_k v_{ik}^2 + \mu \sum_k \sum_j w_{kj}^2 \end{aligned}$$

with lasso and ridge penalties, respectively. Derive the block relaxation updates

$$v_{ik}^{n+1} = v_{ik}^{n} \frac{\sum_j u_{ij} \frac{w_{kj}^n}{b_{ij}^n}}{\sum_j w_{kj}^n + \lambda}, \quad w_{kj}^{n+1} = w_{kj}^{n} \frac{\sum_i u_{ij} \frac{v_{ik}^n}{b_{ij}^n}}{\sum_i v_{ik}^n + \mu}$$

for $l(V, W)$ and the block relaxation updates

$$v_{ik}^{n+1} = v_{ik}^n \frac{\sum_j u_{ij} w_{kj}^n}{\sum_j b_{ij}^n w_{kj}^n + \lambda v_{ik}^n}, \quad w_{kj}^{n+1} = w_{kj}^n \frac{\sum_i u_{ij} v_{ik}^n}{\sum_i b_{ij}^n v_{ik}^n + \mu w_{kj}^n}$$

for $r(V, W)$. These updates maintain positivity. Shrinkage is obvious, with stronger shrinkage for the lasso penalty with small parameters.

29. Let $y_1, \ldots, y_m$ be a random sample from a multivariate normal distribution on R^p. Example 11.2.3 demonstrates that the sample mean $\bar{y}$ and sample variance matrix S are the maximum likelihood estimates of the theoretical mean μ and variance Ω. The implicit assumption here is that $m \geq p$ and S is invertible. Unfortunately, S is singular whenever $m < p$. Furthermore, the entries of S typically have high variance in this situation. To avoid these problems, Levina et al. [35] pursue lasso penalized estimation of Ω^{-1}. If we assume that Ω is invertible and let $\Omega = LL^t$ be its Cholesky decomposition, then $\Omega^{-1} = (L^t)^{-1}L^{-1} = RR^t$ for the upper triangular matrix $R = (r_{ij}) = (L^t)^{-1}$. With the understanding $\hat{\mu} = \bar{y}$, show that the loglikelihood of the sample is

$$m \ln \det R - \frac{m}{2} \operatorname{tr}(R^t S R) = m \sum_i \ln r_{ii} - \frac{m}{2} \sum_j r_j^t S r_j,$$

where r_j is column j of R. In lasso penalized estimation of R, we minimize the objective function

$$f(R) = -m \sum_i \ln r_{ii} + \frac{m}{2} \sum_j r_j^t S r_j + \lambda \sum_{j>i} |r_{ij}|.$$

The diagonal entries of R are not penalized because we want R to be invertible. Why is $f(R)$ a convex function? For $r_{ij} \neq 0$, show that

$$\begin{aligned} \frac{\partial}{\partial r_{ij}} f(R) &= -1_{\{j=i\}} \frac{m}{r_{ii}} + m s_{ii} r_{ij} + m \sum_{k \neq i} s_{ik} r_{kj} \\ & \quad + 1_{\{j \neq i\}} \begin{cases} \lambda & r_{ij} > 0 \\ -\lambda & r_{ij} < 0. \end{cases} \end{aligned}$$

Demonstrate that this leads to the cyclic coordinate descent update

$$\hat{r}_{ii} = \frac{-\sum_{k \neq i} s_{ik} r_{ki} + \sqrt{(\sum_{k \neq i} s_{ik} r_{ki})^2 + 4 s_{ii}}}{2 s_{ii}}.$$

Finally for $j \neq i$, demonstrate that the cyclic coordinate descent update chooses

$$\hat{r}_{ij} = -\frac{m \sum_{k \neq i} s_{ik} r_{kj} + \lambda}{m s_{ii}}$$

when this quantity is positive, it chooses

$$\hat{r}_{ij} = -\frac{m\sum_{k\neq i} s_{ik}r_{kj} - \lambda}{ms_{ii}}$$

when this second quantity is negative, and it chooses 0 otherwise. In organizing cyclic coordinate descent, it is helpful to retain and periodically update the sums $\sum_{k\neq i} s_{ik}r_{kj}$. The matrix R can be traversed column by column.

30. Based on the theory of Section 16.6, show that the asymptotic covariance matrix of the maximum likelihood estimates in Example 11.3.1 has entries

$$\widehat{\mathrm{Cov}}(\hat{p}_i, \hat{p}_j) = \begin{cases} \frac{1}{n}\hat{p}_i(1-\hat{p}_i) & i = j \\ -\frac{1}{n}\hat{p}_i\hat{p}_j & i \neq j. \end{cases}$$

For n large, these are close to the true values

$$\mathrm{Cov}(\hat{p}_i, \hat{p}_j) = \begin{cases} \frac{1}{n}p_i(1-p_i) & i = j \\ -\frac{1}{n}p_ip_j & i \neq j. \end{cases}$$

(Hints: You will need to use the Sherman-Morrison formula. For the sake of simplicity, assume that all $n_i > 0$.)

31. In the notation of Section 16.6, show that the asymptotic covariance matrix $W[W^tJ(\hat{\theta})W]^{-1}W^t$ appears as the upper-left block of the matrix inverse

$$\begin{pmatrix} J & V^t \\ V & \mathbf{0} \end{pmatrix}^{-1} = \begin{pmatrix} W[W^tJW]^{-1}W^t & J^{-1}V^t[VJ^{-1}V^t]^{-1} \\ [VJ^{-1}V^t]^{-1}VJ^{-1} & -[VJ^{-1}V^t]^{-1} \end{pmatrix}.$$

(Hint: Show that the matrices

$$\begin{aligned} P_1 &= JW[W^tJW]^{-1}W^t \\ P_2 &= V^t[VJ^{-1}V^t]^{-1}VJ^{-1} \end{aligned}$$

satisfy $P_1^2 = P_1$, $P_2^2 = P_2$, $P_1P_2 = P_2P_1 = \mathbf{0}$, and $P_1 + P_2 = I_n$.)

32. Continuing Problem 31, it may be impossible to invert the matrix

$$M = \begin{pmatrix} J & V^t \\ V & \mathbf{0} \end{pmatrix}$$

by sweeping on its diagonal entries. However, suppose that $\theta^tJ\theta > 0$ whenever $\theta \neq \mathbf{0}$ and $V\theta = \mathbf{0}$. Then the matrix

$$M(s) = \begin{pmatrix} J + sV^tV & V^t \\ V & \mathbf{0} \end{pmatrix}$$

for sufficiently large $s > 0$ serves as a substitute for M. Show that (a) if $M(s)$ is invertible for one s, then it is invertible for all s, (b) the upper-left block of $M(s)^{-1}$ is $W[W^tJW]^{-1}W^t$, and (c) $M(s)$ can be inverted by sweeping on its diagonal entries if s is sufficiently large. (Hint: Write

$$M(s) = \begin{pmatrix} I_n & sV^t \\ \mathbf{0} & I_m \end{pmatrix} \begin{pmatrix} J & V^t \\ V & \mathbf{0} \end{pmatrix}$$

and invert. Part (c) is a direct consequence of Theorem 6.1 of [27].)

33. When the number of components n of the parameter vector θ of a model is large, inverting the observed or expected information matrix $J(\theta)$ can be very expensive. If we want to compute the asymptotic variance of a linear combination $u^t\theta$ of the parameters, then there is an alternative way of bounding $\mathrm{Var}(u^t\hat{\theta}) = u^tJ(\hat{\theta})^{-1}u$ that is cheaper. Let $\sum_{j=1}^n \sigma_j v_j v_j^t$ be the singular value decomposition of $J(\hat{\theta})$. Suppose that we can peel off the singular values $\sigma_n \le \cdots \le \sigma_k$ and corresponding singular vectors $v_n, \ldots, v_k$ of $J(\hat{\theta})$. Demonstrate that

$$\begin{aligned} \sum_{j=k}^n \frac{1}{\sigma_j}(u^tv_j)^2 &\le \mathrm{Var}(u^t\hat{\theta}) \\ &\le \sum_{j=k}^n \frac{1}{\sigma_j}(u^tv_j)^2 + \frac{1}{\sigma_k}\Big[\|u\|_2^2 - \sum_{j=k}^n (u^tv_j)^2\Big]. \end{aligned}$$

Describe how one could adapt the algorithm developed in Problem 8 of Chapter 8 to peel off the singular values and singular vectors one by one. Discuss potential pitfalls of the method.

16.8 References

[1] Armstrong RD, Kung MT (1978) Algorithm AS 132: least absolute value estimates for a simple linear regression problem. *Appl Stat* 27:363–366

[2] Boyle JP, Dykstra RL (1985) A method for finding projections onto the intersection of convex sets in Hilbert space. In *Advances in Order Restricted Statistical Inference*, Lecture Notes in Statistics, Springer, New York, 28–47

[3] Bregman LM (1965) The method of successive projection for finding a common point of convex sets. *Soviet Math Doklady* 6:688–692

[4] Candes EJ, Tao T (2007) The Danzig selector: statistical estimation when p is much larger than n. *Annals Stat* 35:2313–2351

[5] Candes EJ, Wakin M, Boyd S (2007) Enhancing sparsity by reweighted ℓ_1 minimization. *J Fourier Anal Appl* 14:877–905

[6] Censor Y, Reich S (1996) Iterations of paracontractions and firmly nonexpansive operators with applications to feasibility and optimization. *Optimization* 37:323–339

[7] Censor Y, Zenios SA (1992) Proximal minimization with D-functions. *J Optimization Theory Appl* 73:451–464

[8] Chen SS, Donoho DL, Saunders MA (1998) Atomic decomposition by basis pursuit. *SIAM J Sci Comput* 20:33–61

[9] Claerbout J, Muir F (1973) Robust modeling with erratic data. *Geophysics* 38:826–844

[10] Daubechies I, Defrise M, De Mol C (2004) An iterative thresholding algorithm for linear inverse problems with a sparsity constraint. *Comm Pure Appl Math* 57:1413–1457

[11] de Leeuw J, Lange K (2007) Sharp quadratic majorization in one dimension.

[12] Deutsch F (2001) *Best Approximation in Inner Product Spaces.* Springer, New York

[13] Donoho D, Johnstone I (1994) Ideal spatial adaptation by wavelet shrinkage. *Biometrika* 81:425–455

[14] Dykstra RL (1983) An algorithm for restricted least squares estimation. *J Amer Stat Assoc* 78:837–842

[15] Edgeworth FY (1887) On observations relating to several quantities. *Hermathena* 6:279–285

[16] Edgeworth FY (1888) On a new method of reducing observations relating to several quantities. *Philosophical Magazine* 25:184–191

[17] Efron B, Hastie T, Johnstone I, Tibshirani R (2004) Least angle regression. *Annals Stat* 32:407–499

[18] Elsner L, Koltracht L, Neumann M (1992) Convergence of sequential and asynchronous nonlinear paracontractions. *Numerische Mathematik* 62:305–319

[19] Fang S-C, Puthenpura S (1993) *Linear Optimization and Extensions: Theory and Algorithms.* Prentice-Hall, Englewood Cliffs, NJ

[20] Fazel M, Hindi M, Boyd S (2003) Log-det heuristic for matrix rank minimization with applications to Hankel and Euclidean distance matrices. *Proceedings American Control Conference* 3:2156–2162

[21] Ferguson TS (1996) *A Course in Large Sample Theory.* Chapman & Hall, London

[22] Forsgren A, Gill PE, Wright MH (2002) Interior point methods for nonlinear optimization. *SIAM Review* 44:523–597

[23] Friedman J, Hastie T, Tibshirani R (2007) Pathwise coordinate optimization. *Ann Appl Stat* 1:302–332

[24] Friedman J, Hastie T, Tibshirani R (2009) Regularized paths for generalized linear models via coordinate descent. Technical Report, Stanford University Department of Statistics

[25] Fu WJ (1998) Penalized regressions: the bridge versus the lasso. *J Comp Graph Stat* 7:397–416

[26] Groenen PJF, Nalbantov G, Bioch JC (2007) Nonlinear support vector machines through iterative majorization and I-splines. *Studies in Classification, Data Analysis, and Knowledge Organization*, Lenz HJ, Decker R, Springer, Heidelberg-Berlin, pp 149–161

[27] Hestenes MR (1981) *Optimization Theory: The Finite Dimensional Case.* Robert E Krieger Publishing, Huntington, NY

[28] Hunter DR, Lange K (2004) A tutorial on MM algorithms. *Amer Statistician* 58:30–37

[29] Hunter DR, Li R (2005) Variable selection using MM algorithms. *Annals Stat* 33:1617–1642

[30] Lange K (1994) An adaptive barrier method for convex programming. *Methods Applications Analysis* 1:392–402

[31] Lange, K (2004) *Optimization.* Springer, New York

[32] Lange K, Wu T (2007) An MM algorithm for multicategory vertex discriminant analysis. *J Computational Graphical Stat* 17:527–544

[33] Lee DD, Seung HS (1999) Learning the parts of objects by non-negative matrix factorization. *Nature* 401:788–791

[34] Lee DD, Seung HS (2001) Algorithms for non-negative matrix factorization. *Advances in Neural Information Processing Systems* 13:556–562

[35] Levina E, Rothman A, Zhu J (2008) Sparse estimation of large covariance matrices via a nested lasso penalty. *Ann Appl Stat* 2:245–263

[36] Li Y, Arce GR (2004) A maximum likelihood approach to least absolute deviation regression. *EURASIP J Applied Signal Proc* 2004:1762–1769

[37] Luenberger DG (1984) *Linear and Nonlinear Programming,* 2nd ed. Addison-Wesley, Reading, MA

[38] Meng X-L, Rubin DB (1991) Using EM to obtain asymptotic variance-covariance matrices: the SEM algorithm, *J Amer Stat Assoc* 86: 899–909

[39] Michelot C (1986) A finite algorithm for finding the projection of a point onto the canonical simplex in R^n. *J Optimization Theory Applications* 50:195–200

[40] Park MY, Hastie T (2008) Penalized logistic regression for detecting gene interactions. *Biostatistics* 9:30–50

[41] Pauca VP, Piper J, Plemmons RJ (2006) Nonnegative matrix factorization for spectral data analysis. *Linear Algebra Applications* 416:29–47

[42] Portnoy S, Koenker R (1997) The Gaussian hare and the Laplacian tortoise: computability of squared-error versus absolute-error estimators *Stat Sci* 12:279-300

[43] Santosa F, Symes WW (1986) Linear inversion of band-limited reflection seimograms. *SIAM J Sci Stat Comput* 7:1307–1330

[44] Silvey SD (1975) *Statistical Inference.* Chapman & Hall, London

[45] Ruszczynski A (2006) *Nonlinear Optimization.* Princeton University Press, Princeton, NJ

[46] Schölkopf B, Smola AJ (2002) *Learning with Kernels: Support Vector Machines, Regularization, Optimization, and Beyond.* MIT Press, Cambridge, MA

[47] Taylor H, Banks SC, McCoy JF (1979) Deconvolution with the ℓ_1 norm. *Geophysics* 44:39–52

[48] Teboulle M (1992) Entropic proximal mappings with applications to nonlinear programming. *Math Operations Research* 17:670–690

[49] Tibshirani R (1996) Regression shrinkage and selection via the lasso. *J Roy Stat Soc, Series B* 58:267–288

[50] Vapnik V (1995) *The Nature of Statistical Learning Theory.* Springer, New York

[51] Wang L, Gordon MD, Zhu J (2006) Regularized least absolute deviations regression and an efficient algorithm for parameter tuning. *Proceedings of the Sixth International Conference on Data Mining (ICDM'06).* IEEE Computer Society, pp 690–700

[52] Wang S, Yehya N, Schadt EE, Wang H, Drake TA, Lusis AJ (2006) Genetic and genomic analysis of a fat mass trait with complex inheritance reveals marked sex specificity. *PLoS Genet* 2:148–159

[53] Weston J, Elisseeff A, Schölkopf B, Tipping M (2003) Use of the zero-norm with linear models and kernel methods. *J Machine Learning Research* 3:1439–1461

[54] Wu TT, Lange K (2008) Coordinate descent algorithms for lasso penalized regression. *Ann Appl Stat* 2:224–244

17

Concrete Hilbert Spaces

17.1 Introduction

In this chapter we consider an infinite-dimensional generalization of Euclidean space introduced by the mathematician David Hilbert. This generalization preserves two fundamental geometric notions of Euclidean space—namely, distance and perpendicularity. Both of these geometric properties depend on the existence of an inner product. In the infinite-dimensional case, however, we take the inner product of functions rather than of vectors. Our emphasis here will be on concrete examples of Hilbert spaces relevant to statistics. To keep our discussion within bounds, some theoretical facts are stated without proof. Relevant proofs can be found in almost any book on real or functional analysis [6, 12]. Applications of our examples to numerical integration, wavelets, and other topics appear in later chapters.

The chapter ends with a brief introduction to reproducing kernel Hilbert spaces. This vast topic is not as well documented in textbooks as it should be. It provides a bridge to important research done by computer scientists and applied mathematicians on inverse problems. Our treatment stresses the traditional application to spline regression. Readers interested in more modern applications and a comprehensive theoretical overview should consult the books [2, 5, 14, 15]. Also worth reading are the tutorials [3, 11], the early mathematical article of Aronszajn [1], and the conversation with Emanuel Parzen [9] on the discovery of the statistical importance of reproducing kernel methods.

17.2 Definitions and Basic Properties

An inner product space H is a vector space over the real or complex numbers equipped with an inner product $\langle f, g\rangle$ on pairs of vectors f and g from H. If the underlying field is the real numbers, then $\langle f, g\rangle$ is always real. If the field is the complex numbers, then, in general, $\langle f, g\rangle$ is complex. An inner product satisfies the following postulates:

(a) $\langle f, g\rangle$ is linear in f for g fixed,

(b) $\langle f, g\rangle = \langle g, f\rangle^*$, where * denotes complex conjugate,

(c) $\langle f, f\rangle \geq 0$, with equality if and only $f = \mathbf{0}$.

K. Lange, *Numerical Analysis for Statisticians*, Statistics and Computing,
DOI 10.1007/978-1-4419-5945-4_17, © Springer Science+Business Media, LLC 2010

The inner product allows one to define a vector norm $\|f\| = \langle f, f\rangle^{1/2}$ on H, just as in linear algebra. Furthermore, the Cauchy-Schwarz inequality immediately generalizes. This says that any two vectors f and g in H satisfy

$$|\langle f, g\rangle| \le \|f\| \cdot \|g\|,$$

with equality only when f and g are linearly dependent. An inner product space is said to be complete if every Cauchy sequence converges. In other words, if for some sequence $\{f_n\}_{n=1}^{\infty}$ the norm $\|f_m - f_n\|$ can be made arbitrarily small by taking m and n large enough, then the limit of f_n exists as $n \to \infty$. A complete inner product space is called a Hilbert space.

Example 17.2.1 *Finite-Dimensional Vector Spaces*

The Euclidean space $\mathsf{R}^m$of m-dimensional vectors with real components is a Hilbert space over the real numbers with the usual inner product $\langle f, g\rangle = \sum_{i=1}^m f_i g_i$. If we consider the space C^m of m-dimensional vectors with complex components, then we get a Hilbert space over the complex numbers with inner product $\langle f, g\rangle = \sum_{i=1}^m f_i g_i^*$. Note that the first of these spaces is embedded in the second space in a way that preserves inner products and distances. Since the other Hilbert spaces met in this chapter also exist in compatible real and complex versions, we ordinarily omit specifying the number field. ■

Example 17.2.2 *Space of Square-Integrable Functions*

This is the canonical example of a Hilbert space. Let μ be a measure on some Euclidean space R^m. The vector space $\mathrm{L}^2(\mu)$ of real (or complex) square-integrable functions with respect to μ is a Hilbert space over the real (or complex) numbers with inner product

$$\langle f, g\rangle = \int f(x)g(x)^* d\mu(x).$$

If μ is the uniform measure on an interval (a, b), then we denote the corresponding space of square-integrable functions by $\mathrm{L}^2(a, b)$.

It is a fairly deep fact that the set of continuous functions with compact support is dense in $\mathrm{L}^2(\mu)$ [6]. This means that every square-integrable function f can be approximated to within an arbitrarily small $\epsilon > 0$ by a continuous function g vanishing outside some bounded interval; in symbols, $\|f - g\| < \epsilon$. On the real line the step functions with compact support also constitute a dense set of $\mathrm{L}^2(\mu)$. Both of these dense sets contain countable, dense subsets. In general, a Hilbert space H with a countable, dense set is called separable. Most concrete Hilbert spaces possess this property, so we append it as an additional postulate. ■

A finite or infinite sequence $\{\psi_n\}_{n\ge1}$ of nonzero vectors in a Hilbert space H is said to be orthogonal if $\langle \psi_m, \psi_n\rangle = 0$ for $m \ne n$. An orthogonal sequence $\{\psi_n\}_{n\ge1}$ is orthonormal if $\|\psi_n\| = 1$ for every n. Given a function $f \in H$, one can compute its Fourier coefficients $\langle f, \psi_n \rangle$ relative to an orthonormal sequence $\{\psi_n\}_{n\geq 1}$. The finite expansion $\sum_{n=1}^{m} \langle f, \psi_n \rangle \psi_n$ provides the best approximation to f in the sense that

$$\begin{aligned} \Big\| f - \sum_{n=1}^{m} \langle f, \psi_n \rangle \psi_n \Big\|^2 &= \|f\|^2 - \sum_{n=1}^{m} |\langle f, \psi_n \rangle|^2 \\ &\leq \Big\| f - \sum_{n=1}^{m} c_n \psi_n \Big\|^2 \end{aligned} \tag{17.1}$$

for any other finite sequence of coefficients $\{c_n\}_{n=1}^{m}$. Inequality (17.1) incidentally entails Bessel's inequality

$$\sum_{n=1}^{m} |\langle f, \psi_n \rangle|^2 \leq \|f\|^2. \tag{17.2}$$

An orthonormal sequence $\{\psi_n\}_{n\geq 1}$ is said to be complete (or constitute a basis for H) if

$$f = \sum_{n\geq 1} \langle f, \psi_n \rangle \psi_n$$

for every $f \in H$. (This usage of the word "complete" conflicts with the topological notion of completeness involving Cauchy sequences.) The next proposition summarizes and extends our discussion thus far.

Proposition 17.2.1 *The following statements about an orthonormal sequence $\{\psi_n\}_{n\geq 1}$ are equivalent:*

(a) The sequence is a basis for H.

(b) For each $f \in H$ and $\epsilon > 0$, there is a corresponding $m(f, \epsilon)$ such that

$$\Big\| f - \sum_{n=1}^{m} \langle f, \psi_n \rangle \psi_n \Big\| \leq \epsilon$$

for all $m \geq m(f, \epsilon)$.

(c) If a vector $f \in H$ satisfies $\langle f, \psi_n \rangle = 0$ for every n, then $f = \mathbf{0}$.

(d) For every $f \in H$,

$$f = \sum_{n\geq 1} \langle f, \psi_n \rangle \psi_n.$$

(e) For every $f \in H$,

$$\|f\|^2 = \sum_{n\geq 1} |\langle f, \psi_n \rangle|^2. \tag{17.3}$$

Proof: This basic characterization is proved in standard mathematical texts such as [6, 12]. ■

A linear functional ℓ maps a Hilbert space H into its scalar field of real or complex numbers in a linear fashion. For instance, the map

$$\ell(f) = \int_0^1 f(x)\,dx$$

is a linear functional on $\mathrm{L}^2(0,1)$. The most important linear functionals are continuous. Continuity is intimately tied to boundedness. A linear functional ℓ is said to be bounded if there exists a constant m such that $|\ell(f)| \le m\|f\|$ for all $f \in H$. The next proposition clarifies the relationship between boundedness and continuity.

Proposition 17.2.2 *The following three assertions concerning a linear functional ℓ are equivalent:*

(a) ℓ is continuous,

(b) ℓ is continuous at the origin,

(c) ℓ is bounded.

Proof: Assertion (a) clearly implies assertion (b). Assertion (c) implies assertion (a) because of the inequality

$$|\ell(f) - \ell(g)| = |\ell(f-g)| \le m\|f-g\|.$$

To complete the proof, we must show that assertion (b) implies assertion (c). If the linear functional ℓ is unbounded, then there exists a sequence $f_n \neq \mathbf{0}$ with $|\ell(f_n)| \ge n\|f_n\|$. If we set

$$g_n = \frac{1}{n\|f_n\|} f_n,$$

then g_n converges to $\mathbf{0}$, but $|\ell(g_n)| \ge 1$ does not converge to 0. ■

One can generate a continuous linear functional ℓ_g from a vector g by defining $\ell_g(f) = \langle f, g\rangle$. In fact, this procedure generates all continuous linear functionals.

Proposition 17.2.3 (Reisz) *Every continuous linear functional ℓ on a Hilbert space H can be represented as ℓ_g for some $g \in H$.*

Proof: Define $\mathrm{Ker}(\ell)$ to be the closed subspace of vectors f with $\ell(f) = 0$. If $\mathrm{Ker}(\ell) = H$, then take $g = \mathbf{0}$. Otherwise, there exists a nontrivial vector h that is perpendicular to $\mathrm{Ker}(\ell)$. Without loss of generality, we assume $\ell(h) = 1$ and define $g = h/\|h\|^2$. It follows that

$$\langle f, g\rangle = \langle f - \ell(f)h, g\rangle + \ell(f)\langle h, g\rangle = \ell(f)\frac{\langle h, h\rangle}{\|h\|^2} = \ell(f)$$

because $f - \ell(f)h$ is in $\mathrm{Ker}(\ell)$. ■

17.3 Fourier Series

The complex exponentials $\{e^{2\pi i n x}\}_{n=-\infty}^{\infty}$ provide an orthonormal basis for the space of square-integrable functions with respect to the uniform distribution on [0,1]. Indeed, the calculation

$$\begin{aligned} \int_0^1 e^{2\pi i m x} e^{-2\pi i n x} dx &= \begin{cases} 1 & m = n \\ \frac{e^{2\pi i (m-n)x}}{2\pi i (m-n)}\Big|_0^1 & m \neq n \end{cases} \\ &= \begin{cases} 1 & m = n \\ 0 & m \neq n \end{cases} \end{aligned}$$

shows that the sequence is orthonormal. Completeness is essentially a consequence of Fejér's theorem [4], which says that any periodic, continuous function can be uniformly approximated by a linear combination of sines and cosines. (Fejér's theorem is a special case of the more general Stone-Weierstrass theorem [6].) In dealing with a square-integrable function $f(x)$ on [0,1], it is convenient to extend it periodically to the whole real line via the equation $f(x+1) = f(x)$. (We consider only functions with period 1 in this chapter.) The Fourier coefficients of $f(x)$ are computed according to the standard recipe

$$c_n = \int_0^1 f(x) e^{-2\pi i n x} dx.$$

The Fourier series $\sum_{n=-\infty}^{\infty} c_n e^{2\pi i n x}$ is guaranteed to converge to $f(x)$ in mean square. The more delicate issue of pointwise convergence is partially covered by the next proposition.

Proposition 17.3.1 *Assume that the square-integrable function $f(x)$ on [0,1] is continuous at x_0 and possesses both one-sided derivatives there. Then*

$$\lim_{m\to\infty} \sum_{n=-m}^{m} c_n e^{2\pi i n x_0} = f(x_0).$$

Proof: Extend $f(x)$ to be periodic, and consider the associated periodic function

$$g(x) = \frac{f(x+x_0) - f(x_0)}{e^{-2\pi i x} - 1}.$$

Applying l'Hôpital's rule yields

$$\lim_{x\to 0+} g(x) = \frac{\frac{d}{dx} f(x_0^+)}{-2\pi i},$$

where $\frac{d}{dx} f(x_0^+)$ denotes the one-sided derivative from the right. A similar expression holds for the limit from the left. Since these two limits are finite and $\int_0^1 |f(x)|^2 dx < \infty$, we have $\int_0^1 |g(x)|^2 dx < \infty$ as well.

Now let d_n be the nth Fourier coefficient of $g(x)$. Because

$$f(x + x_0) = f(x_0) + (e^{-2\pi i x} - 1)g(x),$$

it follows that

$$\begin{aligned} c_n e^{2\pi i n x_0} &= \int_0^1 f(x)e^{-2\pi i n(x-x_0)}dx \\ &= \int_0^1 f(x+x_0)e^{-2\pi i n x}dx \\ &= f(x_0)1_{\{n=0\}} + d_{n+1} - d_n. \end{aligned}$$

Therefore,

$$\begin{aligned} \sum_{n=-m}^{m} c_n e^{2\pi i n x_0} &= f(x_0) + \sum_{n=-m}^{m}(d_{n+1} - d_n) \\ &= f(x_0) + d_{m+1} - d_{-m}. \end{aligned}$$

To complete the proof, observe that

$$\begin{aligned} \lim_{|m|\to\infty} d_m &= \lim_{|m|\to\infty} \int_0^1 g(x)e^{-2\pi i m x}dx \\ &= 0 \end{aligned}$$

by the Riemann-Lebesgue lemma to be proved in Proposition 19.4.1 of Chapter 19. ■

Example 17.3.1 *Bernoulli Functions*

There are Bernoulli polynomials $B_n(x)$ and periodic Bernoulli functions $b_n(x)$. Let us start with the Bernoulli polynomials. These are defined by the three conditions

$$\begin{aligned} B_0(x) &= 1 \\ \frac{d}{dx}B_n(x) &= nB_{n-1}(x), \quad n > 0 \\ \int_0^1 B_n(x)dx &= 0, \quad n > 0. \end{aligned} \tag{17.4}$$

For example, we calculate recursively

$$\begin{aligned} B_1(x) &= x - \frac{1}{2} \\ B_2(x) &= 2\Big(\frac{x^2}{2} - \frac{x}{2} + \frac{1}{12}\Big). \end{aligned}$$

The Bernoulli function $b_n(x)$ coincides with $B_n(x)$ on $[0,1)$. Outside $[0,1)$, $b_n(x)$ is extended periodically. In particular, $b_0(x) = B_0(x) = 1$ for all x. Note that $b_1(x)$ is discontinuous at $x = 1$ while $b_2(x)$ is continuous. All subsequent $b_n(x)$ are continuous at $x = 1$ because

$$\begin{aligned} B_n(1) - B_n(0) &= \int_0^1 \frac{d}{dx} B_n(x)dx \\ &= n\int_0^1 B_{n-1}(x)dx \\ &= 0 \end{aligned}$$

by assumption.

To compute the Fourier series expansion $\sum_k c_{nk} e^{2\pi ikx}$ of $b_n(x)$ for $n > 0$, note that $c_{n0} = \int_0^1 B_n(x)dx = 0$. For $k \neq 0$, we have

$$\begin{aligned} c_{nk} &= \int_0^1 b_n(x)e^{-2\pi ikx}dx \\ &= b_n(x)\frac{e^{-2\pi ikx}}{-2\pi ik}\Big|_0^1 + \frac{1}{2\pi ik}\int_0^1 \frac{d}{dx}b_n(x)e^{-2\pi ikx}dx \qquad (17.5) \\ &= b_n(x)\frac{e^{-2\pi ikx}}{-2\pi ik}\Big|_0^1 + \frac{n}{2\pi ik}\int_0^1 b_{n-1}(x)e^{-2\pi ikx}dx. \end{aligned}$$

From the integration-by-parts formula (17.5), we deduce that $b_1(x)$ has Fourier series expansion

$$-\frac{1}{2\pi i}\sum_{k\neq 0}\frac{e^{2\pi ikx}}{k}.$$

This series converges pointwise to $b_1(x)$ except at $x = 0$ and $x = 1$. For $n > 1$, the boundary terms in (17.5) vanish, and

$$c_{nk} = \frac{nc_{n-1,k}}{2\pi ik}. \qquad (17.6)$$

Formula (17.6) and Proposition 17.3.1 together imply that

$$b_n(x) = -\frac{n!}{(2\pi i)^n}\sum_{k\neq 0}\frac{e^{2\pi ikx}}{k^n} \qquad (17.7)$$

for all $n > 1$ and all x.

The constant term $B_n = B_n(0)$ is known as a Bernoulli number. One can compute B_{n-1} recursively by expanding $B_n(x)$ in a Taylor series around $x = 0$. In view of the defining properties (17.4),

$$\begin{aligned} B_n(x) &= \sum_{k=0}^n \frac{1}{k!}\frac{d^k}{dx^k}B_n(0)x^k \\ &= \sum_{k=0}^n \frac{1}{k!}n^{\underline{k}}B_{n-k}x^k, \end{aligned}$$

where

$$n^{\underline{k}} = n(n-1)\cdots(n-k+1)$$

denotes a falling power. The continuity and periodicity of $b_n(x)$ for $n \geq 2$ therefore imply that

$$\begin{aligned} B_n &= B_n(1) \\ &= \sum_{k=0}^{n} \binom{n}{k} B_{n-k}. \end{aligned}$$

Subtracting B_n from both sides of this equality gives the recurrence relation

$$0 = \sum_{k=1}^{n} \binom{n}{k} B_{n-k}$$

for computing B_{n-1} from $B_0, \ldots, B_{n-2}$. For instance, starting from $B_0 = 1$, we calculate $B_1 = -1/2$, $B_2 = 1/6$, $B_3 = 0$, and $B_4 = -1/30$. From the expansion (17.7), evidently $B_n = 0$ for all odd integers $n > 1$. ■

17.4 Orthogonal Polynomials

The subject of orthogonal polynomials has a distinguished history and many applications in physics and engineering [4, 7, 8]. Although it is a little under-appreciated in statistics, subsequent chapters will illustrate that it is well worth learning. Our goal here is simply to provide some concrete examples of orthogonal polynomials. The next proposition is useful in checking that an orthonormal sequence of polynomials is complete. In applying it, note that condition (17.8) holds whenever the probability measure μ possesses a moment generating function. In particular, if μ is concentrated on a finite interval, then its moment generating function exists.

Proposition 17.4.1 *Let μ be a probability measure on the line R such that for some $\alpha > 0$*

$$\int e^{\alpha|x|} d\mu(x) < \infty. \tag{17.8}$$

Then the polynomials $1, x, x^2, \ldots$ generate an orthonormal sequence of polynomials $\{p_n(x)\}_{n\geq 0}$ that is complete in the Hilbert space $L^2(\mu)$ of square-integrable functions.

Proof: This is proved as Proposition 43.1 of [10]. ■

We now discuss some concrete examples of orthogonal polynomial sequences. Because no universally accepted conventions exist for most of the classical sequences, we adopt conventions that appear best suited to the purposes of probability and statistics.

Example 17.4.1 *Poisson–Charlier Polynomials*

Let μ be the Poisson probability measure with mean λ. This probability measure attributes mass $\mu(\{x\}) = e^{-\lambda}\lambda^x/x!$ to the nonnegative integer x. Consider the exponential generating function

$$\begin{aligned} p(x,t) &= e^{-t}\Big(1+\frac{t}{\lambda}\Big)^x \\ &= \sum_{n=0}^{\infty} \frac{p_n^{(\lambda)}(x)}{n!} t^n, \end{aligned}$$

where t is a real parameter. Expanding e^{-t} and $(1+t/\lambda)^x$ in power series and equating coefficients of t^n give

$$\begin{aligned} p_n^{(\lambda)}(x) &= n!\sum_{k=0}^{n}\binom{x}{k}\lambda^{-k}\frac{(-1)^{n-k}}{(n-k)!} \\ &= \sum_{k=0}^{n}\binom{n}{k}(-1)^{n-k}\lambda^{-k}x^{\underline{k}}. \end{aligned}$$

The polynomial $p_n^{(\lambda)}(x)$ is the nth-degree Poisson–Charlier polynomial.

These polynomials form an orthonormal sequence if properly normalized. Indeed, on the one hand,

$$\begin{aligned} \int p(x,s)p(x,t)d\mu(x) &= \sum_{x=0}^{\infty} e^{-s}\Big(1+\frac{s}{\lambda}\Big)^x e^{-t}\Big(1+\frac{t}{\lambda}\Big)^x e^{-\lambda}\frac{\lambda^x}{x!} \\ &= e^{-(s+t+\lambda)}\sum_{x=0}^{\infty}\frac{1}{x!}\Big[\Big(1+\frac{s}{\lambda}\Big)\Big(1+\frac{t}{\lambda}\Big)\lambda\Big]^x \\ &= e^{-(s+t+\lambda)}e^{(1+\frac{s}{\lambda})(1+\frac{t}{\lambda})\lambda} \\ &= e^{\frac{st}{\lambda}} \\ &= \sum_{n=0}^{\infty}\frac{1}{\lambda^n n!}s^n t^n. \end{aligned}$$

On the other hand,

$$\int p(x,s)p(x,t)d\mu(x) = \sum_{m=0}^{\infty}\sum_{n=0}^{\infty}\frac{s^m t^n}{m!n!}\int p_m^{(\lambda)}(x)p_n^{(\lambda)}(x)d\mu(x).$$

Equating coefficients of $s^m t^n$ in these two expressions shows that

$$\int p_m^{(\lambda)}(x)p_n^{(\lambda)}(x)d\mu(x) = \begin{cases} 0 & m \neq n \\ \frac{n!}{\lambda^n} & m = n. \end{cases}$$

Proposition 17.4.1 implies that the sequence $\{p_n^{(\lambda)}(x)\sqrt{\lambda^n/n!}\}_{n\geq 0}$ is a complete orthonormal sequence for the Poisson distribution. ■

Example 17.4.2 *Hermite Polynomials*

If μ is the probability measure associated with the standard normal distribution, then

$$\mu(S) = \frac{1}{\sqrt{2\pi}} \int_S e^{-\frac{1}{2}x^2} dx$$

for any measurable set S. The Hermite polynomials have exponential generating function

$$\begin{aligned} p(x,t) &= e^{xt-\frac{1}{2}t^2} \\ &= \sum_{n=0}^{\infty} \frac{H_n(x)}{n!} t^n \end{aligned}$$

for t real. The fact that $H_n(x)$ is a polynomial of degree n follows from evaluating the nth partial derivative of $e^{xt-\frac{1}{2}t^2}$ with respect to t at $t = 0$. To prove that the Hermite polynomials yield an orthogonal sequence, equate coefficients of $s^m t^n$ in the formal expansion

$$\begin{aligned} \sum_{m=0}^{\infty}\sum_{n=0}^{\infty} s^m t^n \int \frac{H_m(x)}{m!}\frac{H_n(x)}{n!} d\mu(x) &= \int p(x,s)p(x,t)d\mu(x) \\ &= \frac{1}{\sqrt{2\pi}} \int_{-\infty}^{\infty} e^{st} e^{-\frac{1}{2}(x-s-t)^2} dx \\ &= e^{st}. \end{aligned}$$

This gives

$$\int \frac{H_m(x)}{m!}\frac{H_n(x)}{n!} d\mu(x) = \begin{cases} 0 & m \neq n \\ \frac{1}{n!} & m = n, \end{cases}$$

and Proposition 17.4.1 implies that $\{H_n(x)/\sqrt{n!}\}_{n\geq 0}$ is a complete orthonormal sequence for the standard normal distribution.

An explicit expression for $H_n(x)$ can be derived by writing

$$\begin{aligned} e^{xt-\frac{1}{2}t^2} &= e^{xt}e^{-\frac{1}{2}t^2} \\ &= \Big(\sum_{i=0}^{\infty} \frac{x^i t^i}{i!}\Big)\Big(\sum_{j=0}^{\infty} \frac{(-1)^j t^{2j}}{2^j j!}\Big). \end{aligned}$$

This shows that

$$H_n(x) = \sum_{j=0}^{\lfloor \frac{n}{2} \rfloor} \frac{n!(-1)^j x^{n-2j}}{2^j j!(n-2j)!}.$$

In practice, the recurrence relation for $H_n(x)$ given in Section 2.3.3 of Chapter 2 and repeated in equation (17.11) ahead is more useful. ■

Example 17.4.3 *Laguerre Polynomials*

If we let μ be the probability measure associated with the gamma distribution with scale parameter 1 and shape parameter α, then

$$\mu(S) = \frac{1}{\Gamma(\alpha)} \int_S x^{\alpha-1} e^{-x} dx$$

for any measurable set $S \subset (0, \infty)$. The sequence of Laguerre polynomials $\{L_n^{(\alpha)}(x)\}_{n=0}^{\infty}$ has exponential generating function

$$\begin{aligned} p(x,t) &= \frac{1}{(1-t)^{\alpha}} e^{-\frac{tx}{1-t}} \\ &= \sum_{n=0}^{\infty} \frac{L_n^{(\alpha)}(x)}{n!} t^n \end{aligned}$$

for $t \in (-1, 1)$. If we let $\alpha^{\overline{n}} = \alpha(\alpha+1)\cdots(\alpha+n-1)$ denote a rising power, then equating coefficients of $s^m t^n$ in

$$\begin{aligned} & \sum_{m=0}^{\infty} \sum_{n=0}^{\infty} s^m t^n \int \frac{L_m^{(\alpha)}(x)}{m!} \frac{L_n^{(\alpha)}(x)}{n!} d\mu(x) \\ &= \int p(x,s) p(x,t) d\mu(x) \\ &= \frac{1}{(1-s)^{\alpha}(1-t)^{\alpha}\Gamma(\alpha)} \int_0^{\infty} x^{\alpha-1} e^{-\frac{sx}{1-s} - \frac{tx}{1-t} - x} dx \\ &= \frac{1}{(1-s)^{\alpha}(1-t)^{\alpha}\Gamma(\alpha)} \int_0^{\infty} x^{\alpha-1} e^{-\frac{x(1-st)}{(1-s)(1-t)}} dx \\ &= \frac{1}{(1-st)^{\alpha}} \\ &= \sum_{n=0}^{\infty} \frac{\alpha^{\overline{n}}}{n!} s^n t^n \end{aligned}$$

shows that the sequence of polynomials

$$\left\{ \left[\frac{1}{n! \alpha^{\overline{n}}} \right]^{\frac{1}{2}} L_n^{(\alpha)}(x) \right\}_{n=0}^{\infty}$$

is orthonormal for the gamma distribution. In view of Proposition 17.4.1, it is also complete.

It is possible to find explicit expressions for the Laguerre polynomials by expanding

$$\frac{1}{(1-t)^{\alpha}} e^{-\frac{tx}{1-t}} = \sum_{m=0}^{\infty} \frac{(-1)^m x^m}{m!} \frac{t^m}{(1-t)^{m+\alpha}}$$

$$\begin{aligned}
&= \sum_{m=0}^{\infty} \frac{(-1)^m x^m}{m!} \sum_{k=0}^{\infty} \binom{-m-\alpha}{k} (-1)^k t^{k+m} \\
&= \sum_{m=0}^{\infty} \frac{(-1)^m x^m}{m!} \sum_{k=0}^{\infty} \frac{\Gamma(k+m+\alpha)}{\Gamma(m+\alpha)} \frac{t^{m+k}}{k!} \\
&= \sum_{m=0}^{\infty} (-1)^m x^m \sum_{n=m}^{\infty} \frac{\Gamma(n+\alpha)}{\Gamma(m+\alpha)} \binom{n}{m} \frac{t^n}{n!} \\
&= \sum_{n=0}^{\infty} \frac{t^n}{n!} \Gamma(n+\alpha) \sum_{m=0}^{n} \binom{n}{m} \frac{(-1)^m x^m}{\Gamma(m+\alpha)}.
\end{aligned}$$

Thus,

$$L_n^{(\alpha)}(x) = \Gamma(n+\alpha) \sum_{m=0}^{n} \binom{n}{m} \frac{(-1)^m x^m}{\Gamma(m+\alpha)}.$$

Again, this is not the most convenient form for computing. ■

Example 17.4.4 *Beta Distribution Polynomials*

The measure μ associated with the beta distribution assigns probability

$$\mu(S) = \frac{1}{B(\alpha, \beta)} \int_S x^{\alpha-1}(1-x)^{\beta-1} dx$$

to any measurable set $S \subset (0,1)$, where $B(\alpha,\beta) = \Gamma(\alpha)\Gamma(\beta)/\Gamma(\alpha+\beta)$ is the usual normalizing constant. No miraculous generating function exists in this case, but if we abbreviate the beta density by

$$w(x) = \frac{1}{B(\alpha, \beta)} x^{\alpha-1}(1-x)^{\beta-1},$$

then the functions

$$\begin{aligned}
\psi_n^{(\alpha,\beta)}(x) &= \frac{1}{w(x)} \frac{d^n}{dx^n} \Big[w(x) x^n (1-x)^n \Big] \\
&= x^{-\alpha+1}(1-x)^{-\beta+1} \frac{d^n}{dx^n} \Big[x^{n+\alpha-1}(1-x)^{n+\beta-1} \Big]
\end{aligned}$$

will prove to be orthogonal polynomials.

We can demonstrate inductively that $\psi_n^{(\alpha,\beta)}(x)$ is a polynomial of degree n by noting first that $\psi_0^{(\alpha,\beta)}(x) = 1$ and second that

$$\begin{aligned}
&\psi_n^{(\alpha,\beta)}(x) \\
&= (n+\alpha-1)(1-x)\psi_{n-1}^{(\alpha,\beta+1)}(x) - (n+\beta-1)x\psi_{n-1}^{(\alpha+1,\beta)}(x). \quad (17.9)
\end{aligned}$$

Equality (17.9) follows from the definition of $\psi_n^{(\alpha,\beta)}(x)$ and the identity

$$\begin{aligned}
&\frac{d^n}{dx^n}\Big[x^{n+\alpha-1}(1-x)^{n+\beta-1}\Big] \\
&= (n+\alpha-1)\frac{d^{n-1}}{dx^{n-1}}\Big[x^{n+\alpha-2}(1-x)^{n+\beta-1}\Big] \\
&\quad -(n+\beta-1)\frac{d^{n-1}}{dx^{n-1}}\Big[x^{n+\alpha-1}(1-x)^{n+\beta-2}\Big].
\end{aligned}$$

To show that the polynomials $\psi_n^{(\alpha,\beta)}(x)$ are orthogonal, note that for $m \le n$ repeated integration by parts leads to

$$\begin{aligned}
&\int_0^1 \psi_m^{(\alpha,\beta)}(x)\psi_n^{(\alpha,\beta)}(x)w(x)dx \\
&= \int_0^1 \psi_m^{(\alpha,\beta)}(x)\frac{d^n}{dx^n}\Big[w(x)x^n(1-x)^n\Big]dx \\
&= (-1)^n\int_0^1 \frac{d^n}{dx^n}\Big[\psi_m^{(\alpha,\beta)}(x)\Big]w(x)x^n(1-x)^n dx
\end{aligned}$$

since all boundary contributions vanish. If $m < n$, then

$$\frac{d^n}{dx^n}\psi_m^{(\alpha,\beta)}(x) = 0,$$

and this proves orthogonality. When $m = n$,

$$\frac{d^n}{dx^n}\Big[\psi_n^{(\alpha,\beta)}(x)\Big] = n!c_n^{(\alpha,\beta)},$$

where $c_n^{(\alpha,\beta)}$ is the coefficient of x^n in $\psi_n^{(\alpha,\beta)}(x)$. It follows that

$$\begin{aligned}
\int_0^1 \psi_n^{(\alpha,\beta)}(x)\psi_n^{(\alpha,\beta)}(x)w(x)dx &= (-1)^n n!c_n^{(\alpha,\beta)}\int_0^1 w(x)x^n(1-x)^n dx \\
&= (-1)^n n!c_n^{(\alpha,\beta)}\frac{B(\alpha+n,\beta+n)}{B(\alpha,\beta)}.
\end{aligned}$$

Because the beta distribution is concentrated on a finite interval, the polynomial sequence

$$\left\{\sqrt{\frac{B(\alpha,\beta)}{(-1)^n n!c_n^{(\alpha,\beta)}B(\alpha+n,\beta+n)}}\psi_n^{(\alpha,\beta)}(x)\right\}_{n=0}^{\infty}$$

provides an orthonormal basis.

Finally, we claim that $c_n^{(\alpha,\beta)} = (-1)^n(2n+\alpha+\beta-2)^{\underline{n}}$. This assertion is certainly true when $n = 0$ because $c_0^{(\alpha,\beta)} = 1$. In general, the recurrence relation (17.9) and induction imply

$$\begin{aligned} c_n^{(\alpha,\beta)} &= -(n+\alpha-1)c_{n-1}^{(\alpha,\beta+1)} - (n+\beta-1)c_{n-1}^{(\alpha+1,\beta)} \\ &= -(n+\alpha-1)(-1)^{n-1}(2[n-1]+\alpha+\beta+1-2)^{\underline{n-1}} \\ &\quad -(n+\beta-1)(-1)^{n-1}(2[n-1]+\alpha+1+\beta-2)^{\underline{n-1}} \\ &= (-1)^n(2n+\alpha+\beta-2)(2n+\alpha+\beta-3)^{\underline{n-1}} \\ &= (-1)^n(2n+\alpha+\beta-2)^{\underline{n}}. \end{aligned}$$

This proves the asserted formula. ■

The next proposition permits straightforward recursive computation of orthogonal polynomials. Always use high-precision arithmetic when applying the proposition for a particular value of x.

Proposition 17.4.2 *Let a_n and b_n be the coefficients of x^n and x^{n-1} in the nth term $p_n(x)$ of an orthogonal polynomial sequence with respect to a probability measure μ. Then*

$$p_{n+1}(x) = (A_n x + B_n)p_n(x) - C_n p_{n-1}(x), \tag{17.10}$$

where

$$A_n = \frac{a_{n+1}}{a_n}, \quad B_n = \frac{a_{n+1}}{a_n}\Big(\frac{b_{n+1}}{a_{n+1}} - \frac{b_n}{a_n}\Big), \quad C_n = \frac{a_{n+1}a_{n-1}}{a_n^2}\frac{\|p_n\|^2}{\|p_{n-1}\|^2},$$

and $a_{-1} = 0$.

Proof: We will repeatedly use the fact that $\int p_n(x)q(x)d\mu(x) = 0$ for any polynomial $q(x)$ of degree $n-1$ or lower. This follows because $q(x)$ must be a linear combination of $p_0(x), \ldots, p_{n-1}(x)$. Now given the definition of A_n, it is clear that

$$p_{n+1}(x) - A_n x p_n(x) = B_n p_n(x) - C_n p_{n-1}(x) + r_{n-2}(x)$$

for as yet undetermined constants B_n and C_n and a polynomial $r_{n-2}(x)$ of degree $n-2$. If $0 \le k \le n-2$, then

$$\begin{aligned} 0 &= \int p_{n+1}(x)p_k(x)d\mu(x) \\ &= \int p_n(x)[A_n x + B_n]p_k(x)d\mu(x) - C_n\int p_{n-1}(x)p_k(x)d\mu(x) \\ &\quad + \int r_{n-2}(x)p_k(x)d\mu(x), \end{aligned}$$

and consequently $\int r_{n-2}(x)p_k(x)d\mu(x) = 0$. This forces $r_{n-2}(x) = 0$ because $r_{n-2}(x)$ is a linear combination of $p_0(x), \ldots, p_{n-2}(x)$.

If we write

$$xp_{n-1}(x) = \frac{a_{n-1}}{a_n}p_n(x) + q_{n-1}(x),$$

where $q_{n-1}(x)$ is a polynomial of degree $n-1$, then

$$\begin{aligned}
0 &= \int p_{n+1}(x)p_{n-1}(x)d\mu(x) \\
&= A_n\int p_n(x)xp_{n-1}(x)d\mu(x) + B_n\int p_n(x)p_{n-1}(x)d\mu(x) \\
&\quad - C_n\int p_{n-1}^2(x)d\mu(x) \\
&= A_n\frac{a_{n-1}}{a_n}\int p_n^2(x)d\mu(x) + A_n\int p_n(x)q_{n-1}(x)d\mu(x) - C_n\|p_{n-1}\|^2 \\
&= A_n\frac{a_{n-1}}{a_n}\|p_n\|^2 - C_n\|p_{n-1}\|^2.
\end{aligned}$$

This gives C_n. Finally, equating coefficients of x^n in equation (17.10) yields $b_{n+1} = A_n b_n + B_n a_n$, and this determines B_n. ■

After tedious calculations, Proposition 17.4.2 translates into the following recurrence relations for the orthogonal polynomials considered in Examples 17.4.1 through 17.4.4:

$$\begin{aligned}
p_{n+1}^{(\lambda)}(x) &= \Big(\frac{x}{\lambda} - \frac{n}{\lambda} - 1\Big)p_n^{(\lambda)}(x) - \frac{n}{\lambda}p_{n-1}^{(\lambda)}(x) \\
H_{n+1}(x) &= xH_n(x) - nH_{n-1}(x) \\
L_{n+1}^{(\alpha)}(x) &= (2n+\alpha-x)L_n^{(\alpha)}(x) - n(n+\alpha-1)L_{n-1}^{(\alpha)}(x) \qquad (17.11) \\
\psi_{n+1}^{(\alpha,\beta)}(x) &= \frac{(2n+\alpha+\beta)(2n+\alpha+\beta-1)}{n+\alpha+\beta-1} \\
&\quad \times\Big[\frac{(n+1)(n+\alpha)}{2n+\alpha+\beta} - \frac{n(n+\alpha-1)}{2n+\alpha+\beta-2} - x\Big]\psi_n^{(\alpha,\beta)}(x) \\
&\quad - \frac{n(2n+\alpha+\beta)(n+\alpha-1)(n+\beta-1)}{(n+\alpha+\beta-1)(2n+\alpha+\beta-2)}\psi_{n-1}^{(\alpha,\beta)}(x).
\end{aligned}$$

17.5 Reproducing Kernel Hilbert Spaces

One of the most frustrating features of $\mathrm{L}^2(0,1)$ and similar Hilbert spaces is that the vectors are not functions in the ordinary sense. They are only defined almost everywhere. As a consequence, the linear evaluation functionals $\mathrm{ev}_q(f) = f(q)$ are not well defined. If we limit the class of functions and revise the definition of the inner product, then this annoyance fades away. In many novel Hilbert spaces, the evaluation functionals are well defined and continuous. Before we construct some examples, we would like to focus on the axiomatic development of reproducing kernel Hilbert spaces. For such a Hilbert space H of ordinary functions, one postulates the continuity of every evaluation functional. According to Proposition 17.2.3, we can then identify the evaluation functional ev_q with a function $K_q \in H$ in the sense that $f(q) = \langle f, K_q \rangle$ for every f. Taking $f = K_p$ defines a function $K(p,q) = \langle K_p, K_q \rangle$ called the reproducing kernel of H. This function summarizes the structure of H.

The reproducing kernel $K(p,q)$ inherits some useful properties from the inner product. First, we have conjugate symmetry

$$K(p,q) \quad = \quad \langle K_p, K_q \rangle \quad = \quad \langle K_q, K_p \rangle^* \quad = \quad K(q,p)^*.$$

If $p_1, \ldots, p_n$ is a finite sequence of vectors and $c_1, \ldots, c_n$ a corresponding sequence of scalars, then we also have

$$\begin{aligned} \sum_i \sum_j c_i K(p_i, p_j) c_j^* &= \Big\langle \sum_i c_i K_{p_i}, \sum_i c_i K_{p_i} \Big\rangle \\ &= \Big\| \sum_i c_i K_{p_i} \Big\|^2 \\ &\geq 0. \end{aligned}$$

In other words, for all possible sequences $p_1, \ldots, p_n$, the matrix $K(p_i, p_j)$ is positive semidefinite. This compels $K(p,p) \geq 0$ and leads to the Cauchy-Schwarz inequality

$$K(p,p)K(q,q) - |K(p,q)|^2 \quad = \quad \det \begin{pmatrix} K(p,p) & K(p,q) \\ K(q,p) & K(q,q) \end{pmatrix} \quad \geq \quad 0.$$

The Cauchy-Schwarz inequality follows from the fact that the determinant is the product of the eigenvalues of the displayed 2×2 positive semidefinite matrix.

Here are some examples of kernels with domain R^m:

$$\begin{aligned} K(p,q) &= p^t q \\ K(p,q) &= (1 + p^t q)^d, \quad d \text{ a positive integer,} \\ K(p,q) &= \hat{f}(p-q) \\ K(p,q) &= (p-c)_+(q-c)_+, \quad m = 1. \end{aligned}$$

It is easy to verify that the first example generates positive semidefinite matrices. The second example also qualifies because the set of positive semidefinite matrices is closed under entry-wise addition and entry-wise multiplication; see Problems 20 and 21 of Chapter 8. The function $\hat{f}(y)$ in the third example is the Fourier transform

$$\hat{f}(y) \quad = \quad \int e^{iy^t x} f(x)\, dx, \qquad i = \sqrt{-1}$$

of a nonnegative integrable function $f(x)$. The one-dimensional Fourier transform is covered in Chapter 19. We demonstrate there that $\hat{f}(y)$ is real if and only if $f(x)$ is even. The Gaussian kernel

$$\hat{f}(y) = e^{-\sigma^2 y^2/2}$$

is an especially important Fourier transform kernel. The calculation

$$\sum_j \sum_k c_j \hat{f}(p_j - p_k) c_k^* = \int \Big| \sum_j c_j e^{ip_j^t x} \Big|^2 f(x)\, dx$$

proves positive semidefiniteness. The last example is one dimensional and involves the function $(z)_+ = \max\{z, 0\}$. If we let $z_i = p_i - c$, then positive semidefiniteness follows from the calculation

$$\sum_i \sum_j a_i (p_i - c)_+ (p_j - c)_+ a_j^* = \Big| \sum_i a_i (z_i)_+ \Big|^2 \geq 0.$$

In practice, we often start with a kernel $K(p, q)$ and construct a Hilbert space H with $K(p, q)$ as its reproducing kernel. The basic building blocks of H are linear combinations of the functions $K_p(q) = K(p, q)$. Consider two such functions $f = \sum_i a_i K_{p_i}$ and $g = \sum_i b_i K_{p_i}$. We can arrange for the two finite ranges of summation to be identical by allowing some of the coefficients a_i and b_i to be 0. The inner product of f and g is defined as

$$\langle f, g \rangle = \sum_i \sum_j a_i K(p_i, p_j) b_j^*. \tag{17.12}$$

This definition shows that the inner product is conjugate symmetric and linear in its left argument. The definition is also consistent with the norm requirement $\langle f, f \rangle \geq 0$ and the evaluation property $f(p) = \langle f, K_p \rangle$. The validity of point evaluation proves that the inner product

$$\langle f, g \rangle = \sum_j f(p_j) b_j^* = \sum_i a_i g(p_i)^*$$

depends only on the values of f and g and not on the particular linear combinations chosen to represent f and g.

Finally, we need to check that f vanishes when $\langle f, f \rangle$ vanishes. As previously mentioned, an arbitrary point p can always be included in the enumerated p_i in the representation $f = \sum_i a_i K_{p_i}$. If p is the first point in the list and M is the matrix $K(p_i, p_j)$, then $f(p)$ reduces to the quadratic form $a^t M e_1$, where $a = (a_i)$ and e_1 is the standard unit vector with first entry 1. According to Problem 15, this quadratic form satisfies the Cauchy-Schwarz inequality, which in the current circumstances translates into

$$|f(p)| \leq \sqrt{\langle f, f \rangle} \sqrt{K(p, p)}. \tag{17.13}$$

This bound simultaneously proves that f vanishes and that the evaluation functional $\mathrm{ev}_p(f)$ is bounded and therefore continuous.

The only flaw in our construction is that the inner product space may be incomplete. To make it complete, we append the limits of Cauchy sequences. Consider a Cauchy sequence f_n. The inequality

$$|f_n(p) - f_m(p)| \quad \leq \quad \|f_n - f_m\| \sqrt{K(p,p)} \qquad (17.14)$$

demonstrates that the pointwise limit $f(p)$ of the numbers $f_n(p)$ exists for every point p. The sensible strategy is to extend the inner product to such limiting functions f. Thus, suppose f_n is a Cauchy sequence converging pointwise to f and g_n is a Cauchy sequence converging pointwise to g. Let us define

$$\langle f, g \rangle \quad = \quad \lim_{n \to \infty} \langle f_n, g_n \rangle. \qquad (17.15)$$

This extended inner product is certainly conjugate symmetric, linear in its left argument, and nonnegative when the arguments agree. It is less clear (a) that the limit exists, (b) that it preserves the original inner product, (c) that it does not depend on the particular Cauchy sequences converging pointwise to f and g, and (d) that $\|f\| = 0$ implies $f(p) = 0$ for all p.

Let us prove each of these assertions in turn. (a) The inequality

$$\begin{aligned} |\langle f_m, g_m \rangle - \langle f_n, g_n \rangle| \quad &\leq \quad |\langle f_m - f_n, g_m \rangle| + |\langle f_n, g_m - g_n \rangle| \\ &\leq \quad \|f_m - f_n\| \cdot \|g_m\| + \|f_n\| \cdot \|g_m - g_n\| \end{aligned}$$

shows that the sequence $\langle f_n, g_n \rangle$ is Cauchy. Thus, the limit (17.15) exists. (b) If f and g are linear combinations of the K_p, then we can choose the constant Cauchy sequences $f_n = f$ and $g_n = g$ to represent f and g. These choices force the extended inner product to preserve the original inner product. (c) Let the Cauchy sequences $\tilde{f}_n$ and $\tilde{g}_n$ also converge pointwise to f and g. The inequality

$$|\langle f_n, g_n \rangle - \langle \tilde{f}_n, \tilde{g}_n \rangle| \quad \leq \quad \|f_n - \tilde{f}_n\| \cdot \|g_n\| + \|\tilde{f}_n\| \cdot \|g_n - \tilde{g}_n\|$$

demonstrates that in order for the two limits to agree, it suffices that the norms $\|f_n - \tilde{f}_n\|$ and $\|g_n - \tilde{g}_n\|$ converge to 0. Consider the Cauchy sequence $h_n = f_n - \tilde{f}_n$. It converges pointwise to 0, and for every $\epsilon > 0$ there is an m with $\|h_n - h_m\| < \epsilon$ for $n \geq m$. If $h_m = \sum_i a_i K_{p_i}$ and $n \geq m$, then

$$\begin{aligned} \|h_n\| \quad &= \quad \langle h_n - h_m, h_n \rangle + \langle h_m, h_n \rangle \\ &\leq \quad \epsilon \sup_k \|h_k\| + \sum_i a_i h_n(p_i)^*. \end{aligned}$$

Because ϵ is arbitrary and every $h_n(p_i)$ tends to 0 as n tends to ∞, it follows that $\|h_n\|$ tends to 0. Since the same reasoning applies to $h_n = g_n - \tilde{g}_n$, the extended inner product is well defined. (d) Finally, if f_n tends pointwise to f and $\|f\| = \lim_{n\to\infty} \|f_n\| = 0$, then inequality (17.13) applied to each f_n implies in the limit that $f(p) = 0$ for all p.

As a summary of the foregoing discussion, we have the next proposition.

Proposition 17.5.1 (Moore-Aronszajn) *Let $K(p,q)$ be a positive semidefinite function defined on a subset X of R^m. There exists one and only one Hilbert space H of functions on X with reproducing kernel K. The subspace H_0 of functions spanned by the functions $K_p(q)$ with inner product (17.12) is dense in H in the sense that every function of H is a pointwise limit of a Cauchy sequence from H_0.*

In many practical examples, the kernel $K(p,q)$ is continuous and bounded on its domain. If these conditions hold, then every linear combination $\sum_i a_i K_{p_i}$ is a continuous function, and inequality (17.14) implies that the pointwise convergence of a Cauchy sequence f_n to a function f is actually uniform. It follows that the limit f is also continuous. As well as being helpful in its own right, continuity leads to separability of the constructed Hilbert space H. This claim is verified by taking a countable dense set S in the domain of the kernel and showing that the countable set of functions $\{K_p\}_{p\in S}$ is dense in H. It suffices to prove that the only function g perpendicular to this set is the zero function. In view of the identity $g(p) = \langle g, K_p\rangle = 0$, the continuous function g vanishes on a dense set. This can only hold if g is identically 0.

Example 17.5.1 *The Sobolev Space* $\mathrm{H}^m_{\mathrm{per}}(0,1)$

Smooth periodic functions $f(x) = \sum_n a_n e^{2\pi i n x}$ have Fourier coefficients that decline rapidly. For a positive integer m, consider the subspace

$$H = \Big\{f = \sum_n a_n e^{2\pi i n x} : |a_0|^2 + \sum_{n\neq 0}(2\pi n)^{2m}|a_n|^2 < \infty\Big\}$$

of square integrable periodic functions with revised inner product

$$\langle f, g\rangle = a_0 b_0^* + \sum_{n\neq 0}(2\pi n)^{2m} a_n b_n^*$$

for $f(x) = \sum_n a_n e^{2\pi i n x}$ and $g(x) = \sum_n b_n e^{2\pi i n x}$. Point evaluation is continuous because the Cauchy-Schwarz inequality implies

$$\begin{aligned} |f(x)| &= \Big|a_0 + \sum_{n\neq 0}\frac{1}{(2\pi n)^m}(2\pi n)^m a_n e^{2\pi i n x}\Big| \\ &\leq \sqrt{1 + \sum_{n\neq 0}\frac{1}{(2\pi n)^{2m}}}\sqrt{\langle f, f\rangle} \end{aligned}$$

and boundedness is equivalent to continuity. The kernel function

$$\begin{aligned} K_x(y) &= 1 + \sum_{n\neq 0}\frac{e^{2\pi i n(y-x)}}{(2\pi n)^{2m}} \\ &= 1 + \frac{(-1)^{m-1}}{(2m)!} b_{2m}(y-x) \end{aligned}$$

involving the Bernoulli function $b_{2m}(y-x)$ clearly gives the correct evaluation

$$\langle f, K_x \rangle = \sum_n a_n e^{2\pi i n x}.$$

The new inner product and norm are designed to eliminate non-smooth functions. ■

Before introducing our second example, it is helpful to state and prove an important result that matches reproducing kernels and orthogonal decompositions of Hilbert spaces.

Proposition 17.5.2 *Let $K(p,q)$ be the reproducing kernel of a Hilbert space H_K of functions defined on a subset X of R^m. Suppose $K(p,q)$ can be decomposed as the sum $L(p,q)+M(p,q)$ of two positive semidefinite functions, whose functional pairs L_p and M_q are orthogonal vectors in H_K. Then H_K can be written as the orthogonal direct sum $H_L \oplus H_M$ of the reproducing kernel Hilbert spaces H_L and H_M corresponding to $L(p,q)$ and $M(p,q)$. Conversely, if $L(p,q)$ and $M(p,q)$ are positive semidefinite functions whose induced Hilbert spaces satisfy $H_L \cap H_M = \{\mathbf{0}\}$, then the direct sum $H_K = H_L \oplus H_M$ has reproducing kernel $K(p,q) = L(p,q) + M(p,q)$.*

Proof: Suppose $K(p,q) = L(p,q) + M(p,q)$ is the reproducing kernel of H_K with inner product $\langle \cdot, \cdot \rangle_K$. The Hilbert spaces H_L and H_M induced by $L(p,q)$ and $M(p,q)$ are subspaces of H_K with inner products $\langle \cdot, \cdot \rangle_L$ and $\langle \cdot, \cdot \rangle_M$. Because L_p and M_q are orthogonal in H_K, we have

$$\langle L_p, L_q \rangle_L = L(p,q) = \langle L_p, K_q \rangle_K = \langle L_p, L_q \rangle_K.$$

Thus, the inner products $\langle \cdot, \cdot \rangle_L$ and $\langle \cdot, \cdot \rangle_K$ agree on H_L, and H_L is a closed subspace of H_K. Similarly, the inner products $\langle \cdot, \cdot \rangle_M$ and $\langle \cdot, \cdot \rangle_K$ agree on H_M, and H_M is a closed subspace of H_K. By assumption, H_L and H_M are orthogonal. Suppose $h \in H_K$ can be written as $f_L + f_M + g$, where $f_L \in H_L$, $f_M \in H_M$, and g is perpendicular to both H_L and H_M. The calculation

$$\begin{aligned} h(p) &= \langle h, K_p \rangle_K \\ &= \langle f_L + f_M + g, L_p + M_p \rangle_K \\ &= \langle f_L, L_p \rangle_L + \langle f_M, M_p \rangle_M \\ &= f_L(p) + f_M(p) \end{aligned}$$

demonstrates that $g(p)$ is identically 0. Thus, $H_K = H_L \oplus H_M$.

For the converse, consider $f = f_L + f_M$ in $H_K = H_L \oplus H_M$. We now have

$$\begin{aligned} f(p) &= f_L(p) + f_M(p) \\ &= \langle f_L, L_p \rangle_L + \langle f_M, M_p \rangle_M \\ &= \langle f, L_p + M_p \rangle_K, \end{aligned}$$

and $K(p, q)$ is the reproducing kernel of H_K. ∎

Example 17.5.2 *A Reproducing Kernel for a Taylor Expansion*

Consider the vector space of m times differentiable functions $[0,1]$ with inner product

$$\langle f, g\rangle = \sum_{n=0}^{m-1} f^{(n)}(0)g^{(n)}(0)^* + \int_0^1 f^{(m)}(y)g^{(m)}(y)^*\, dy.$$

We do not require that $f^{(m)}(y)$ and $g^{(m)}(y)$ exist everywhere, but they should be square integrable, satisfy the fundamental theorem of calculus, and allow integration by parts. To show that the point evaluation functional $\mathrm{ev}_x(f) = f(x)$ is continuous, we use the Taylor expansion

$$f(x) = \sum_{n=0}^{m-1} f^{(n)}(0)\frac{x^n}{n!} + \int_0^1 f^{(m)}(y)\frac{(x-y)_+^{m-1}}{(m-1)!}\, dy, \quad (17.16)$$

which can be checked using repeated integration by parts. Applying the discrete and continuous versions of the Cauchy-Schwarz inequality now yields

$$\begin{aligned}|f(x)| &\le \sqrt{\sum_{n=0}^{m-1} |f^{(n)}(0)|^2 \sum_{n=0}^{m-1}\left(\frac{x^n}{n!}\right)^2} \\ &\quad + \sqrt{\int_0^1 |f^{(m)}(y)|^2 dy \int_0^1 \left[\frac{(x-y)_+^{m-1}}{(m-1)!}\right]^2 dy} \\ &\le \left\{\sqrt{\sum_{n=0}^{m-1}\left(\frac{x^n}{n!}\right)^2} + \sqrt{\int_0^1 \left[\frac{(x-y)_+^{m-1}}{(m-1)!}\right]^2 dy}\right\}\sqrt{\langle f, f\rangle}.\end{aligned}$$

Thus, the evaluation functional ev_x is bounded.

The reproducing kernel for this example can be deduced from the Taylor expansion. For instance, the sum

$$L_x(y) = \sum_{n=0}^{m-1} \frac{x^n}{n!}\frac{y^n}{n!}$$

gives the polynomial part

$$\langle f, L_x\rangle = \sum_{n=0}^{m-1} f^{(n)}(0)\frac{x^n}{n!}$$

of the expansion. The remainder term is captured by

$$M_x(y) = \int_0^1 \frac{(x-z)_+^{m-1}}{(m-1)!} \frac{(y-z)_+^{m-1}}{(m-1)!} dz.$$

According to Leibniz's rule for differentiating under an integral sign,

$$\frac{d^j}{dy^j} M_x(y) = \begin{cases} \int_0^1 \frac{(x-z)_+^{m-1}}{(m-1)!} \frac{(y-z)_+^{m-1-j}}{(m-1-j)!} dz & 0 \le j < m \\ \frac{(x-y)_+^{m-1}}{(m-1)!} & j = m \text{ and } y < x \\ 0 & j = m \text{ and } y > x. \end{cases}$$

Because $\frac{d^j}{dy^j} M_x(0) = 0$ for all $j < m$, this puts us in the position described by Proposition 17.5.2 with H_L consisting of the polynomials of degree $m-1$ or less and H_M its orthogonal complement. Thus, the reproducing kernel on the whole space boils down to the sum $K(x, y) = L(x, y) + M(x, y)$.

When $m = 2$, straightforward calculations yield

$$M(x,y) = \frac{1}{2}xy^2 - \frac{1}{6}y^3 + \frac{1}{6}(y-x)_+^3 \quad (17.17)$$

$$L(x,y) = 1 + xy. \quad (17.18)$$

A glance at Chapter 10 should convince the reader that both of these functions are cubic splines in y for x fixed. Thus, it is hardly surprising that spline regression can be rephrased in the language of reproducing kernel Hilbert spaces. ■

17.6 Application to Spline Estimation

In this section we revisit the problem of penalized estimation from the perspective of reproducing kernel Hilbert spaces. Most classical problems of estimation can be posed as minimization of a loss function subject to constraints or penalties. The most commonly employed loss function is squared error loss $[y_i - f(x_i)]^2$. Penalties are introduced in parameter estimation to steer model selection, to regularize estimation, and to enforce smoothness. As Example 17.5.2 hints, spline regression emphasizes smoothness. We therefore estimate parameters by minimizing the criterion

$$\sum_{i=1}^n [y_i - f(x_i)]^2 + \lambda \|P_M f\|^2 = \sum_{i=1}^n [y_i - f(x_i)]^2 + \lambda \int_0^1 f^{(m)}(x)^2 dx,$$

where $m = 2$, $\lambda > 0$ is a tuning constant specifying the tradeoff between fit and smoothness, and P_M is the projection of f onto the subspace H_M of functions whose derivatives of order $m-1$ or less vanish at 0.

We can represent a generic function f in H_K by

$$f(x) = \sum_{j=0}^{m-1} \alpha_j \frac{x^j}{j!} + \sum_{i=1}^{n} \beta_i M_{x_i}(x) + g(x), \tag{17.19}$$

where g is perpendicular to H_L and to all of the M_{x_i}. The combination of the reproducing property and orthogonality yields $g(x_i) = \langle g, M_{x_i} \rangle = 0$. In other words, g does not impact the loss function. The identity

$$\|P_M f\|^2 = \Big\| \sum_{i=1}^{n} \beta_i P_M M_{x_i} \Big\|^2 + \|g\|^2 \tag{17.20}$$

addressed in Problem 20 makes it clear that g should be dropped from the penalty as well. Thus, the spline estimation problem is inherently linear and finite dimensional. There is nothing about this argument that uses the specific form of the loss function, so it extends to other settings. The purely polynomial part of f plays a different role. Because the $n \times n$ matrix $Q = [M(x_i, x_j)]$ is invertible (Problem 17), there always exist coefficients $\beta_1, \ldots, \beta_n$ such that

$$f(x_j) = \sum_{i=1}^{n} \beta_i M(x_i, x_j)$$

for every j. Hence, the coefficients $\alpha_0, \ldots, \alpha_{m-1}$ also appear redundant. They are retained in the model to promote smoothness.

In practice, the representation (17.19) is not the most convenient. With cubic splines for example, it is conceptually simpler to work with linear combinations

$$f(x) = \alpha_0 + \alpha_1 x + \sum_{i=1}^{n} \beta_i (x - x_i)_+^3.$$

Natural cubic splines call for a reduced basis of lower dimension. If n is large, there is also no absolute necessity of placing a knot at every x_i [5]. In fact, the x_i and the knots can be decoupled completely. Of course, it is advisable for the distribution of the knots to mimic the distribution of the x_i. Once we make these adjustments, estimation can be reformulated as minimization of the quadratic

$$\|y - U\alpha - V\beta\|^2 + \lambda \beta^t W \beta \tag{17.21}$$

for the positive semidefinite matrix $W = [\langle P_M(x - x_i)_+^3, P_M(x - x_j)_+^3 \rangle]$. The stationary conditions

$$\begin{aligned} -2U^t(y - U\alpha - V\beta) &= \mathbf{0} \\ -2V^t(y - U\alpha - V\beta) + 2\lambda W \beta &= \mathbf{0} \end{aligned} \tag{17.22}$$

suggest block relaxation alternating the updates

$$\begin{aligned}\alpha &= (U^tU)^{-1}U^t(y - V\beta)\\ \beta &= (V^tV + \lambda W)^{-1}V^t(y - U\alpha).\end{aligned}$$

This is probably better than the complicated simultaneous solution (Problem 21), but Wahba [15] recommends a stabler numerical method.

17.7 Problems

1. Find the Fourier series of the function $|x|$ defined on $[-1/2, 1/2]$ and extended periodically to the whole real line. At what points of $[-1/2, 1/2]$ does the Fourier series converge pointwise to $|x|$?

2. Let $f(x)$ be a periodic function on the real line whose kth derivative is piecewise continuous for some positive integer k. Show that the Fourier coefficients c_n of $f(x)$ satisfy

$$|c_n| \leq \frac{\int_0^1 |f^{(k)}(x)|dx}{|2\pi n|^k}$$

for $n \neq 0$.

3. Suppose that the periodic function $f(x)$ is square integrable on $[0, 1]$. Prove the assertions: (a) $f(x)$ is an even (respectively odd) function if and only if its Fourier coefficients c_n are even (respectively odd) functions of n, (b) $f(x)$ is real and even if and only if the c_n are real and even, and (c) $f(x)$ is even (odd) if and only if it is even (odd) around 1/2. By even around 1/2 we mean $f(1/2 + x) = f(1/2 - x)$.

4. Demonstrate that

$$\frac{\pi^2}{12} = \sum_{k=1}^{\infty} \frac{(-1)^{k+1}}{k^2}, \qquad \frac{\pi^4}{90} = \sum_{k=1}^{\infty} \frac{1}{k^4}.$$

5. Show that the even Bernoulli numbers can be expressed as

$$B_{2n} = (-1)^{n+1} \frac{2(2n)!}{(2\pi)^{2n}} \left[1 + \frac{1}{2^{2n}} + \frac{1}{3^{2n}} + \frac{1}{4^{2n}} + \cdots\right].$$

Apply Stirling's formula and deduce the asymptotic relation

$$|B_{2n}| \asymp 4\sqrt{\pi n} \left(\frac{n}{\pi e}\right)^{2n}.$$

6. Show that the Bernoulli polynomials satisfy the identity

$$B_n(x) = (-1)^n B_n(1-x)$$

for all n and $x \in [0,1]$. Conclude from this identity that $B_n(1/2) = 0$ for n odd.

7. Continuing Problem 6, show inductively for $n \geq 1$ that $B_{2n}(x)$ has exactly one simple zero in $(0,1/2)$ and one in $(1/2,1)$, while $B_{2n+1}(x)$ has precisely the simple zeros 0, 1/2, and 1.

8. Demonstrate that the Bernoulli polynomials satisfy the identity

$$B_n(x+1) - B_n(x) = nx^{n-1}.$$

Use this result to verify that the sum of the nth powers of the first m integers can be expressed as

$$\sum_{k=1}^{m} k^n = \frac{1}{n+1}\Big[B_{n+1}n+1 - B_{n+1}(1)\Big].$$

(Hint: Prove the first assertion by induction or by expanding $B_n(x)$ in a Taylor series around the point 1.)

9. If μ is the probability measure associated with the binomial distribution with r trials and success probability p, then

$$\mu(\{x\}) = \binom{r}{x} p^x q^{r-x},$$

where $0 \leq x \leq r$ and $q = 1-p$. The exponential generating function

$$(1+tq)^x(1-tp)^{r-x} = \sum_{n=0}^{r} \frac{K_n^{(r,p)}(x)}{n!} t^n$$

defines the Krawtchouk polynomials. Show that

$$K_n^{(r,p)}(x) = \sum_{k=0}^{n} (-1)^k \binom{n}{k} p^k q^{n-k} x^{\underline{n-k}} (r-x)^{\underline{k}}$$

and that the normalized polynomials

$$p_n(x) = \frac{K_n^{(r,p)}(x)}{(pq)^{\frac{n}{2}} \binom{r}{n}^{\frac{1}{2}} n!}$$

constitute an orthonormal basis for the binomial distribution.

10. Let $p(x,t)$ be the exponential generating function for a sequence $p_n(x)$ of orthogonal polynomials relative to a probability measure on the real line. If A_n, B_n, and C_n are the coefficients appearing in the recurrence (17.10), then demonstrate formally that

$$\begin{aligned}
\mathrm{E}[Xp(X,s)p(X,t)] &= \sum_{m=0}^{\infty}\sum_{n=0}^{\infty} \mathrm{E}[Xp_m(X)p_n(X)]\frac{s^m t^n}{m!n!} \\
&= \sum_{n=0}^{\infty}\frac{1}{A_n}\mathrm{E}[p_{n+1}^2(X)]\frac{s^{n+1}t^n}{(n+1)!n!} \\
&\quad -\sum_{n=0}^{\infty}\frac{B_n}{A_n}\mathrm{E}[p_n^2(X)]\frac{s^n t^n}{n!n!} \\
&\quad +\sum_{n=1}^{\infty}\frac{C_n}{A_n}\mathrm{E}[p_{n-1}^2(X)]\frac{s^{n-1}t^n}{(n-1)!n!}.
\end{aligned}$$

This result can be used to calculate the coefficients A_n, B_n, and C_n. For instance, demonstrate that the exponential generating function $p(x,t)$ of the Krawtchouk polynomials of Problem 9 satisfies

$$\mathrm{E}[Xp(X,s)p(X,t)] = r(1+sq)(1+tq)p(1+stpq)^{r-1}.$$

Show that equating coefficients of $s^m t^n$ in the two formulas gives the recurrence

$$\begin{aligned}
xK_n^{(r,p)}(x) &= K_{n+1}^{(r,p)}(x) + [nq + (r-n)p]K_n^{(r,p)}(x) \\
&\quad + npq(r-n+1)K_{n-1}^{(r,p)}(x)
\end{aligned}$$

starting from $K_{-1}^{(r,p)}(x) = 0$ and $K_0^{(r,p)}(x) = 1$.

11. Let μ denote the negative binomial distribution assigning probability

$$\mu(\{x\}) = \binom{\lambda + x - 1}{x} q^\lambda p^x$$

to the nonnegative integer x, where $\lambda > 0$, $p \in (0,1)$, and $q = 1 - p$. The exponential generating function

$$\left(1 - \frac{t}{p}\right)^x (1-t)^{-x-\lambda} = \sum_{n=0}^{\infty}\frac{M_n(x;\lambda,p)}{n!}t^n$$

defines the Meixner polynomials. Show that

$$M_n(x;\lambda,p) = \sum_{k=0}^{n}(-1)^k\binom{n}{k}\frac{1}{p^k}x^{\underline{k}}(x+\lambda)^{\overline{n-k}}$$

and that the normalized polynomials

$$p_n(x) = \frac{M_n(x;\lambda,p)\sqrt{p^n}}{\sqrt{n!\lambda^{\overline{n}}}}$$

constitute an orthonormal basis for the negative binomial distribution. Derive the three-term recurrence

$$\begin{aligned} M_{n+1}(x;\lambda,p) &= \Big(1-\frac{1}{p}\Big)xM_n(x;\lambda,p) + \Big(n+\lambda+\frac{n}{p}\Big)M_n(x;\lambda,p) \\ &\quad -\frac{n}{p}(\lambda+n-1)M_{n-1}(x;\lambda,p) \end{aligned}$$

starting from $M_{-1}(x;\lambda,p) = 0$ and $M_0(x;\lambda,p) = 1$. (Hint: Apply the technique sketched in Problem 10.)

12. Show that

$$H_n(x) = (-1)^n e^{\frac{1}{2}x^2}\frac{d^n}{dx^n}e^{-\frac{1}{2}x^2}.$$

(Hint: Expand the left-hand side of the identity

$$e^{-\frac{1}{2}(x-t)^2} = \sum_{n=0}^{\infty}\frac{H_n(x)}{n!}t^n e^{-\frac{1}{2}x^2}$$

in a Taylor series and equate coefficients of t^n.)

13. Verify that

$$L_n^{(\alpha)}(x) = e^x x^{-\alpha+1}\frac{d^n}{dx^n}\Big(e^{-x}x^{n+\alpha-1}\Big).$$

14. Validate the recurrences listed in (17.11). (Hint: Note that the coefficient of x^{n-1} in $\psi_n^{(\alpha,\beta)}(x)$ is $(-1)^{n-1}n(n+\alpha-1)(2n+\alpha+\beta-3)^{\underline{n-1}}$.)

15. Let A be a positive semidefinite matrix. Prove the Cauchy-Schwarz inequality $|u^tAv|^2 \le (u^tAu)(v^tAv)$ for vectors u and v. State and prove the analogous inequality when A is Hermitian positive semidefinite. (Hints: If A is positive definite, then u^tAv defines an inner product. If A is not positive definite, then $A+\epsilon I$ is positive definite for every $\epsilon > 0$.)

16. Derive the Taylor expansion (17.16).

17. Given n distinct points $x_1 < x_2 < \cdots < x_n$ from the interval $[0,1]$, prove that the matrix $Q = [M(x_i,x_j)]$ of Example 17.5.2 is positive

definite. Also prove that the matrix $R = (x_i^j/j!)$, $0 \le j \le m-1$, has full rank whenever $n \ge m$. (Hint: Show that the quadratic form

$$v^t Q v = \Big[\frac{1}{(m-1)!}\Big]^2 \int_0^1 \Big[\sum_{i=1}^n v_i (x_i - z)_+^{m-1}\Big]^2 dz$$

is positive for every nontrivial vector v.)

18. In Example 17.5.2, verify the listed expressions for the derivatives $\frac{d^j}{dy^j} M_x(y)$ when $j \le m$.

19. Derive the expressions for $M(x,y)$ and $L(x,y)$ in equations (17.17) and (17.18) when $m = 2$.

20. Check formula (17.20) using the facts that $P_M g = g$ and $P_M x^j = \mathbf{0}$ for $j < m$.

21. Show that the system of equations (17.22) has solution

$$\begin{aligned} \hat{\alpha} &= [U^t(I - V\Sigma^{-1}V^t)U]^{-1} U^t (I - V\Sigma^{-1}V^t) y \\ \hat{\beta} &= \Sigma^{-1} V^t (y - U\hat{\alpha}), \end{aligned}$$

where $\Sigma = V^t V + \lambda W$. What is the relevance of Problem 23 of Chapter 13?

22. The Sobolev space $\mathsf{H}^1(\mathsf{R})$ has inner product

$$\langle f, g\rangle = \int_{-\infty}^{\infty} f(x) g(x)^* dx + \int_{-\infty}^{\infty} f'(x) g'(x)^* dx$$

for appropriate functions defined on the real line. Prove that

$$K(x,y) = \frac{1}{2} e^{-|y-x|}$$

is the reproducing kernel and that the functional $\mathrm{ev}_x f = f(x)$ satisfies the bound $|f(x)| \le \|f\|$. (Hints: For the first claim, integrate by parts. For the second, consider

$$|f(x)|^2 = \int_{-\infty}^{x} \frac{d}{dy} |f(y)|^2 dy,$$

and invoke the product rule of differentiation and the Cauchy-Schwarz inequality.)

17.8 References

[1] Aronszajn N (1950) Theory of reproducing kernels. *Amer Math Soc Trans* 63:337-404

[2] Berlinet A, Thomas-Agnan C (2004) *Reproducing Kernel Hilbert Spaces in Probability and Statistics.* Kluwer, Boston, MA

[3] Burges CJC (1998) A tutorial on support vector machines for pattern recognition. *Data Mining and Knowledge Discovery* 2:121–167

[4] Dym H, McKean HP (1972) *Fourier Series and Integrals.* Academic Press, New York

[5] Hastie T, Tibshirani R, Friedman J (2001) *The Elements of Statistical Learning: Data Mining, Inference, and Prediction.* Springer, New York

[6] Hewitt E, Stromberg K (1965) *Real and Abstract Analysis.* Springer, New York

[7] Hochstadt H (1986) *The Functions of Mathematical Physics.* Dover, New York

[8] Ismail MEH (2005) *Classical and Quantum Orthogonal Polynomials in One Variable.* Cambridge University Press, Cambridge

[9] Newton HJ (2002) A Conversation with Emanuel Parzen. *Stat Science* 17:357–378

[10] Parthasarathy KR (1977) *Introduction to Probability and Measure.* Springer, New York

[11] Pearce ND, Wand MP (2006) Penalised splines and reproducing kernel methods. *Amer Statistician* 60:233–240

[12] Rudin W (1973) *Functional Analysis.* McGraw-Hill, New York

[13] Schölkopf B, Smola AJ (2002) *Learning with Kernels: Support Vector Machines, Regularization, Optimization, and Beyond.* MIT Press, Cambridge, MA

[14] Vapnik V (1995) *The Nature of Statistical Learning Theory.* Springer, New York

[15] Wahba G (1990) *Spline Models for Observational Data.* SIAM, Philadelphia

18

Quadrature Methods

18.1 Introduction

The numerical calculation of one-dimensional integrals, or quadrature, is one of the oldest branches of numerical analysis. Long before calculus was invented, Archimedes found accurate approximations to π by inscribed and circumscribed polygons on a circle of unit radius. In modern applied mathematics and statistics, quadrature is so pervasive that even hand-held calculators are routinely programmed to perform it. Nonetheless, gaining a theoretical understanding of quadrature is worth the effort. In many scientific problems, large numbers of quadratures must be carried out quickly and accurately.

This chapter focuses on the two dominant methods of modern quadrature, Romberg's algorithm [6, 7] and Gaussian quadrature [6, 11, 12, 13]. To paraphrase Henrici [7], Romberg's algorithm uses an approximate knowledge of the error in integration to approximately eliminate that error. Gaussian quadrature is ideal for integration against standard probability densities such as the normal or gamma. Both methods work extremely well for good integrands such as low-degree polynomials. In fact, Gaussian quadrature is designed to give exact answers in precisely this situation. Gaussian quadrature also has the virtue of handling infinite domains of integration gracefully. In spite of these advantages, Romberg's algorithm is usually the preferred method of quadrature for a wide variety of problems. It is robust and simple to code. At its heart is the trapezoidal rule, which can be adapted to employ relatively few function evaluations for smooth regions of an integrand and many evaluations for rough regions. Whatever the method of quadrature chosen, numerical integration is an art. We briefly describe some tactics for taming bad integrands.

18.2 Euler-Maclaurin Sum Formula

As a prelude to Romberg's algorithm, we discuss the summation formula of Euler and Maclaurin [3, 5, 10, 15]. Besides providing insight into the error of the trapezoidal method of quadrature, this formula is an invaluable tool in asymptotic analysis. Our applications to harmonic series and Stirling's formula illustrate this fact.

Proposition 18.2.1 *Suppose $f(x)$ has $2m$ continuous derivatives on the*

K. Lange, *Numerical Analysis for Statisticians*, Statistics and Computing, DOI 10.1007/978-1-4419-5945-4_18, *interval* $[1,n]$ *for some positive integer* n. *Then*

$$\sum_{k=1}^{n} f(k) = \int_1^n f(x)dx + \frac{1}{2}\Big[f(n)+f(1)\Big] + \sum_{j=1}^{m} \frac{B_{2j}}{(2j)!} f^{(2j-1)}(x)\Big|_1^n - \frac{1}{(2m)!}\int_1^n b_{2m}(x) f^{(2m)}(x)dx, \quad (18.1)$$

where B_k *is a Bernoulli number and* $b_k(x)$ *is a Bernoulli function. The remainder in this expansion is bounded by*

$$\left|\frac{1}{(2m)!}\int_1^n b_{2m}(x) f^{(2m)}(x)dx\right| \leq C_{2m}\int_1^n |f^{(2m)}(x)|dx, \quad (18.2)$$

where

$$C_{2m} = \frac{2}{(2\pi)^{2m}} \sum_{k=1}^{\infty} \frac{1}{k^{2m}}.$$

Proof: Consider an arbitrary function $g(x)$ defined on $[0,1]$ with $2m$ continuous derivatives. In view of the definition of the Bernoulli polynomials in Chapter 17, repeated integration by parts gives

$$\begin{aligned}\int_0^1 g(x)dx &= \int_0^1 B_0(x)g(x)dx \\ &= B_1(x)g(x)\Big|_0^1 - \int_0^1 B_1(x)g'(x)dx \\ &= \sum_{i=1}^{2m} \frac{(-1)^{i-1}B_i(x)}{i!} g^{(i-1)}(x)\Big|_0^1 \\ &\quad + \frac{(-1)^{2m}}{(2m)!}\int_0^1 B_{2m}(x)g^{(2m)}(x)dx.\end{aligned}$$

This formula can be simplified by noting that (a) $B_{2m}(x) = b_{2m}(x)$ on $[0,1]$, (b) $B_1(x) = x - 1/2$, (c) $B_i(0) = B_i(1) = B_i$ when $i > 1$, and (d) $B_i = 0$ when $i > 1$ and i is odd. Hence,

$$\begin{aligned}\int_0^1 g(x)dx &= \frac{1}{2}\Big[g(1)+g(0)\Big] - \sum_{j=1}^{m} \frac{B_{2j}}{(2j)!} g^{(2j-1)}(x)\Big|_0^1 \\ &\quad + \frac{1}{(2m)!}\int_0^1 b_{2m}(x) g^{(2m)}(x)dx.\end{aligned}$$

If we apply this result successively to $g(x) = f(x+k)$ for $k = 1, \ldots, n-1$ and add the results, then cancellation of successive terms produces formula (18.1). The bound (18.2) follows immediately from the Fourier series representation of $b_{2m}(x)$ noted in Chapter 17. ■

Example 18.2.1 *Harmonic Series*

The harmonic series $\sum_{k=1}^{n} k^{-1}$ can be approximated by taking $f(x)$ to be x^{-1} in Proposition 18.2.1. For example with $m = 2$, we find that

$$\begin{aligned}
\sum_{k=1}^{n} \frac{1}{k} &= \int_1^n \frac{1}{x} dx + \frac{1}{2}\Big[\frac{1}{n} + 1\Big] + \frac{B_2}{2}\Big[1 - \frac{1}{n^2}\Big] \\
&\quad + \frac{B_4}{4!}\Big[3! - \frac{3!}{n^4}\Big] - \frac{1}{4!}\int_1^n b_4(x)\frac{4!}{x^5} dx \\
&= \ln n + \gamma + \frac{1}{2n} - \frac{1}{12n^2} + \frac{1}{120n^4} + \int_n^\infty b_4(x)\frac{1}{x^5} dx \\
&= \ln n + \gamma + \frac{1}{2n} - \frac{1}{12n^2} + O\Big(\frac{1}{n^4}\Big),
\end{aligned}$$

where

$$\begin{aligned}
\gamma &= \frac{1}{2} + \frac{1}{12} - \frac{1}{120} - \int_1^\infty b_4(x)\frac{1}{x^5} dx \qquad (18.3) \\
&\approx 0.5772
\end{aligned}$$

is Euler's constant. ■

Example 18.2.2 *Stirling's Formula*

If we let $f(x)$ be the function $\ln x = \frac{d}{dx}[x \ln x - x]$ and $m = 2$ in Proposition 18.2.1, then we recover Stirling's formula

$$\begin{aligned}
\ln n! &= \sum_{k=1}^{n} \ln k \\
&= \int_1^n \ln x \, dx + \frac{1}{2}\ln n + \frac{B_2}{2}\Big[\frac{1}{n} - 1\Big] \\
&\quad + \frac{B_4}{4!}\Big[\frac{2!}{n^3} - 2!\Big] + \frac{1}{4!}\int_1^n b_4(x)\frac{3!}{x^4} dx \\
&= n\ln n - n + \frac{1}{2}\ln n + s + \frac{1}{12n} - \frac{1}{360n^3} - \frac{1}{4}\int_n^\infty b_4(x)\frac{1}{x^4} dx \\
&= (n + \frac{1}{2})\ln n - n + s + \frac{1}{12n} + O\Big(\frac{1}{n^3}\Big),
\end{aligned}$$

where

$$\begin{aligned}
s &= 1 - \frac{1}{12} + \frac{1}{360} + \frac{1}{4}\int_1^\infty b_4(x)\frac{1}{x^4} dx \qquad (18.4) \\
&= \ln \sqrt{2\pi}
\end{aligned}$$

was determined in Chapter 4. ■

Example 18.2.3 *Error Bound for the Trapezoidal Rule*

The trapezoidal rule is one simple mechanism for integrating a function $f(x)$ on a finite interval $[a, b]$. If we divide the interval into n equal subintervals of length $h = (b-a)/n$, then the value of the integral of $f(x)$ between $a + kh$ and $a + (k+1)h$ is approximately $\frac{h}{2}\{f(a+kh) + f(a+[k+1]h)\}$. Summing these approximate values over all subintervals therefore gives

$$\int_a^b f(x)dx \quad \approx \quad h\Big[\frac{1}{2}g(0) + g(1) + \cdots + g(n-1) + \frac{1}{2}g(n)\Big] \qquad (18.5)$$

for $g(t) = f(a+th)$. If we abbreviate the trapezoidal approximation on the right of (18.5) by $T(h)$, then Proposition 18.2.1 implies that

$$\begin{aligned} T(h) &= h\int_0^n g(t)dt + \frac{hB_2}{2}g'(t)|_0^n + \frac{hB_4}{4!}g^{(3)}(t)|_0^n - \frac{h}{4!}\int_0^n b_4(t)g^{(4)}(t)dt \\ &= \int_a^b f(x)dx + \frac{h^2}{12}\Big[f'(b) - f'(a)\Big] - \frac{h^4}{720}\Big[f^{(3)}(b) - f^{(3)}(a)\Big] \\ &\quad - \frac{h^4}{4!}\int_a^b b_4\Big(\frac{x-a}{h}\Big)f^{(4)}(x)dx \\ &= \int_a^b f(x)dx + \frac{h^2}{12}\Big[f'(b) - f'(a)\Big] + O(h^4). \end{aligned}$$

In practice, it is inconvenient to suppose that $f'(x)$ is known, so the error committed in using the trapezoidal rule is $O(h^2)$. If $f(x)$ possesses $2k$ continuous derivatives, then a slight extension of the above argument indicates that the trapezoidal approximation satisfies

$$T(h) \;=\; \int_a^b f(x)dx + c_1h^2 + c_2h^4 + \cdots + c_{k-1}h^{2(k-1)} + O(h^{2k}) \qquad (18.6)$$

for constants $c_1, \ldots, c_{k-1}$ that depend on $f(x)$, a, and b but not on h. ■

18.3 Romberg's Algorithm

Suppose in the trapezoidal rule that we halve the integration step h. Then the error estimate (18.6) becomes

$$T\Big(\frac{1}{2}h\Big) \;=\; \int_a^b f(x)dx + \frac{c_1}{4}h^2 + \frac{c_2}{4^2}h^4 + \cdots + \frac{c_{k-1}}{4^{k-1}}h^{2(k-1)} + O(h^{2k}).$$

Romberg recognized that forming the linear combination

$$\begin{aligned} \frac{4T(\frac{1}{2}h) - T(h)}{3} &= T\Big(\frac{1}{2}h\Big) - \frac{1}{3}\Big[T(h) - T\Big(\frac{1}{2}h\Big)\Big] \qquad (18.7) \\ &= \int_a^b f(x)dx + d_2h^4 + \cdots + d_{k-1}h^{2(k-1)} + O(h^{2k}) \end{aligned}$$

eliminates the h^2 error term, where $d_2, \ldots, d_{k-1}$ are new constants that can be easily calculated. For h small, decreasing the error in estimating $\int_a^b f(x)dx$ from $O(h^2)$ to $O(h^4)$ is a striking improvement in accuracy.

Even more interesting is the fact that this tactic can be iterated. Suppose we compute the trapezoidal approximations $T_{m0} = T[2^{-m}(b-a)]$ for several consecutive integers m beginning with $m = 0$. In essence, we double the number of quadrature points and halve h at each stage. When $f(x)$ has $2k$ continuous derivatives, the natural inductive generalization of the refinement (18.7) is provided by the sequence of refinements

$$\begin{aligned} T_{mn} &= \frac{4^n T_{m,n-1} - T_{m-1,n-1}}{4^n - 1} \\ &= T_{m,n-1} - \frac{1}{4^n - 1}[T_{m-1,n-1} - T_{m,n-1}] \end{aligned} \tag{18.8}$$

for $n \leq \min\{m, k-1\}$. From this recursive definition, it follows that $\lim_{m\to\infty} T_{m0} = \int_a^b f(x)dx$ implies $\lim_{m\to\infty} T_{mn} = \int_a^b f(x)dx$ for every n. Furthermore, if

$$T_{m,n-1} = \int_a^b f(x)dx + \gamma_n 4^{-mn} + \cdots + \gamma_{k-1} 4^{-m(k-1)} + O(4^{-mk})$$

for appropriate constants $\gamma_n, \ldots, \gamma_{k-1}$, then

$$T_{mn} = \int_a^b f(x)dx + \delta_{n+1} 4^{-m(n+1)} + \cdots + \delta_{k-1} 4^{-m(k-1)} + O(4^{-mk})$$

for appropriate new constants $\delta_{n+1}, \ldots, \delta_{k-1}$. In other words, provided the condition $n + 1 \leq k$ holds, the error drops from $O(4^{-mn})$ to $O(4^{-m(n+1)})$ in going from $T_{m,n-1}$ to T_{mn}.

It is convenient to display the trapezoidal approximations to a definite integral $\int_a^b f(x)dx$ as the first column of the Romberg array

$$\begin{pmatrix} T_{00} & & & & \\ T_{10} & T_{11} & & & \\ T_{20} & T_{21} & T_{22} & & \\ T_{30} & T_{31} & T_{32} & T_{33} & \\ T_{40} & T_{41} & T_{42} & T_{43} & T_{44} \\ \vdots & \vdots & \vdots & \vdots & \vdots \end{pmatrix}.$$

Romberg's algorithm fills in an entry of this array based on the two entries immediately to the left and diagonally above and to the left of the given entry. Depending on the smoothness of the integrand, the columns to the right of the first column converge more and more rapidly to $\int_a^b f(x)dx$. In practice, convergence can occur so quickly that computing entries T_{mn} with n beyond 2 or 3 is wasted effort.

TABLE 18.1. Romberg Table for E(Y)

m	T_{m0}	T_{m1}	T_{m2}	T_{m3}
0	0.4207355			
1	0.4500805	0.4598622		
2	0.4573009	0.4597077	0.4596974	
3	0.4590990	0.4596983	0.4596977	0.4596977
∞	0.4596977	0.4596977	0.4596977	0.4596977

Example 18.3.1 *Numerical Examples of Romberg's Algorithm*

For two simple examples, let X be uniformly distributed on $[0, 1]$, and define the random variables $Y = \sin X$ and $Z = \sqrt{1 - X^2}$. The explicit values

$$
\begin{aligned}
\mathrm{E}(Y) &= \int_0^1 \sin x dx &=& \quad 1 - \cos(1) \\
\mathrm{E}(Z) &= \int_0^1 \sqrt{1 - y^2} dy &=& \quad \frac{1}{2} \arcsin(1)
\end{aligned}
$$

available for the means of Y and Z offer an opportunity to assess the performance of Romberg's algorithm. Table 18.1 clearly demonstrates the accelerated convergence possible for a smooth integrand. Because the derivative of $\sqrt{1 - y^2}$ is singular at $y = 1$, the slower convergence seen in Table 18.2 is to be expected. ■

TABLE 18.2. Romberg Table for E(Z)

m	T_{m0}	T_{m1}	T_{m2}	T_{m3}
0	0.50000			
1	0.68301	0.74402		
2	0.74893	0.77090	0.77269	
3	0.77245	0.78030	0.78092	0.78105
4	0.78081	0.78360	0.78382	0.78387
5	0.78378	0.78476	0.78484	0.78486
6	0.78482	0.78517	0.78520	0.78521
7	0.78520	0.78532	0.78533	0.78533
8	0.78533	0.78537	0.78537	0.78537
9	0.78537	0.78539	0.78539	0.78539
10	0.78539	0.78539	0.78540	0.78540
∞	0.78540	0.78540	0.78540	0.78540

18.4 Adaptive Quadrature

A crude, adaptive version of the trapezoidal rule can be easily constructed [4]. In the first stage of the adaptive algorithm, the interval of integration is split at its midpoint $c = (a+b)/2$, and the two approximations

$$\begin{aligned} S_0 &= \frac{(b-a)}{2}\Big[f(a) + f(b)\Big] \\ S_1 &= \frac{(b-a)}{4}\Big[f(a) + 2f(c) + f(b)\Big] \end{aligned}$$

are compared. If $|S_0 - S_1| < \epsilon(b-a)$ for $\epsilon > 0$ small, then $\int_a^b f(x)dx$ is set equal to S_1. If the test $|S_0 - S_1| < \epsilon(b-a)$ fails, then the integrals $\int_a^c f(x)dx$ and $\int_c^b f(x)dx$ are separately computed and the results added. This procedure is made recursive by computing an integral via the trapezoidal rule at each stage or splitting the integral for further processing. Obviously, a danger in the adaptive algorithm is that the two initial approximations S_0 and S_1 (or similar approximations at some subsequent early stage) agree by chance.

18.5 Taming Bad Integrands

We illustrate by way of example some of the usual tactics for improving integrands.

Example 18.5.1 *Subtracting off a Singularity*

Sometimes one can subtract off the singular part of an integrand and integrate that part analytically. The example

$$\begin{aligned} \int_0^1 \frac{e^x}{\sqrt{x}} dx &= \int_0^1 \frac{e^x - 1}{\sqrt{x}} dx + \int_0^1 \frac{1}{\sqrt{x}} dx \\ &= \int_0^1 \frac{e^x - 1}{\sqrt{x}} dx + 2 \end{aligned}$$

is fairly typical. The remaining integrand

$$\frac{e^x - 1}{\sqrt{x}} = \sqrt{x} \sum_{k=1}^{\infty} \frac{x^{k-1}}{k!}$$

is well behaved at $x = 0$. In the vicinity of this point, it is wise to evaluate the integrand by a few terms of its series expansion to avoid roundoff errors in the subtraction $e^x - 1$. ■

Example 18.5.2 *Splitting an Interval and Changing Variables*

Consider the problem of computing the expectation of $\ln X$ for a beta distributed random variable X. The necessary integral

$$\mathrm{E}(\ln X) = \frac{\Gamma(\alpha+\beta)}{\Gamma(\alpha)\Gamma(\beta)} \int_0^1 \ln x \, x^{\alpha-1}(1-x)^{\beta-1} dx$$

has singularities at $x = 0$ and $x = 1$. We attack this integral by splitting it into parts over $[0, 1/2]$ and $[1/2, 1]$. The integral over $[1/2, 1]$ can be tamed by making the change of variables $y^n = 1 - x$. This gives

$$\int_{\frac{1}{2}}^1 \ln x \, x^{\alpha-1}(1-x)^{\beta-1} dx = n \int_0^{\frac{1}{\sqrt[n]{2}}} \ln(1-y^n)\,(1-y^n)^{\alpha-1} y^{n\beta-1} dy.$$

Provided n is chosen so large that $n\beta - 1 \geq 0$, the transformed integral is well behaved. The integral over $[0, 1/2]$ is handled similarly by making the change of variables $y^n = x$. In this case

$$\int_0^{\frac{1}{2}} \ln x \, x^{\alpha-1}(1-x)^{\beta-1} dx = n^2 \int_0^{\frac{1}{\sqrt[n]{2}}} \ln y \, y^{n\alpha-1}(1-y^n)^{\beta-1} dy.$$

Because of the presence of the singularity in $\ln y$ at $y = 0$, it is desirable that $n\alpha - 1 > 0$. Indeed, then the factor $y^{n\alpha-1}$ yields $\lim_{y\to 0} y^{n\alpha-1} \ln y = 0$. ■

Example 18.5.3 *Infinite Integration Limit*

The integral (18.3) appearing in the definition of Euler's constant occurs over the infinite interval $[1, \infty)$. If we make the change of variable $y^{-1} = x$, then

$$\int_1^\infty b_4(x)\frac{1}{x^5} dx = \int_0^1 b_4(y^{-1}) y^3 dy.$$

The transformed integral is still challenging to evaluate because the integrand $b_4(y^{-1})y^3$ has limited smoothness and rapidly oscillates in the vicinity of $y = 0$. Fortunately, some of the sting of rapid oscillation is removed by the damping factor y^3. Problem 4 asks readers to evaluate Euler's constant by quadrature. ■

18.6 Gaussian Quadrature

Gaussian quadrature is ideal for evaluating integrals against certain probability measures μ. If $f(x)$ is a smooth function, then it is natural to consider approximations of the sort

$$\int_{-\infty}^\infty f(x) d\mu(x) \approx \sum_{i=0}^k w_i f(x_i), \tag{18.9}$$

where the finite sum ranges over fixed points x_i with attached positive weights w_i. In the trapezoidal rule, μ is a uniform measure, and the points are uniformly spaced. In Gaussian quadrature, μ is typically nonuniform, and the points cluster in regions of high probability. Since polynomials are quintessentially smooth, Gaussian quadrature requires that the approximation (18.9) be exact for all polynomials of sufficiently low degree.

If the probability measure μ possesses an orthonormal polynomial sequence $\{\psi_n(x)\}_{n=0}^{\infty}$, then the $k+1$ points $x_0, \ldots, x_k$ of formula (18.9) are taken to be the roots of the polynomial $\psi_{k+1}(x)$. Assuming for the moment that these roots are distinct and real, we have the following remarkable result.

Proposition 18.6.1 *If μ is not concentrated at a finite number of points, then there exist positive weights w_i such that the quadrature formula (18.9) is exact whenever $f(x)$ is any polynomial of degree $2k+1$ or lower.*

Proof: Let us first prove the result for a polynomial $f(x)$ of degree k or lower. If $l_i(x)$ denotes the polynomial

$$l_i(x) = \prod_{j \neq i} \frac{x - x_j}{x_i - x_j}$$

of degree k, then

$$f(x) = \sum_{i=0}^{k} l_i(x) f(x_i)$$

is the interpolating-polynomial representation of $f(x)$. The condition

$$\int f(x) d\mu(x) = \sum_{i=0}^{k} \int l_i(x) d\mu(x) f(x_i)$$

now determines the weights $w_i = \int l_i(x) d\mu(x)$, which obviously do not depend on $f(x)$.

Now let $f(x)$ be any polynomial of degree $2k+1$ or lower. The division algorithm for polynomials [2] implies that

$$f(x) = p(x)\psi_{k+1}(x) + q(x)$$

for polynomials $p(x)$ and $q(x)$ of degree k or lower. On the one hand, because $p(x)$ is orthogonal to $\psi_{k+1}(x)$,

$$\int f(x) d\mu(x) = \int q(x) d\mu(x). \tag{18.10}$$

On the other hand, because the x_i are roots of $\psi_{k+1}(x)$,

$$\sum_{i=0}^{k} w_i f(x_i) = \sum_{i=0}^{k} w_i q(x_i). \tag{18.11}$$

In view of the first part of the proof and the fact that $q(x)$ is a polynomial of degree k or lower, we also have

$$\int q(x)d\mu(x) = \sum_{i=0}^{k} w_i q(x_i). \tag{18.12}$$

Equations (18.10), (18.11), and (18.12) taken together imply that formula (18.9) is exact.

Finally, if $f(x) = l_i(x)^2$, then the calculation

$$\begin{aligned} w_i &= \sum_{j=0}^{k} w_j l_i(x_j)^2 \\ &= \int l_i(x)^2 d\mu(x) \end{aligned}$$

shows that $w_i > 0$, provided that μ is not concentrated at a finite number of points. ∎

We now make good on our implied promise concerning the roots of the polynomial $\psi_{k+1}(x)$.

Proposition 18.6.2 *Under the premises of Proposition 18.6.1, the roots of each polynomial $\psi_{k+1}(x)$ are real and distinct.*

Proof: If the contrary is true, then $\psi_{k+1}(x)$ changes sign fewer than $k+1$ times. Let the positions of the sign changes occur at the distinct roots $r_1 < \cdots < r_m$ of $\psi_{k+1}(x)$. Since the polynomial $\psi_{k+1}(x)\prod_{i=1}^{m}(x-r_i)$ is strictly negative or strictly positive except at the roots of $\psi_{k+1}(x)$, we infer that

$$\left|\int \psi_{k+1}(x)\prod_{i=1}^{m}(x-r_i)d\mu(x)\right| > 0.$$

However, $\prod_{i=1}^{m}(x-r_i)$ is a polynomial of lower degree than $\psi_{k+1}(x)$ and consequently must be orthogonal to $\psi_{k+1}(x)$. This contradiction shows that $\psi_{k+1}(x)$ must have at least $k+1$ distinct changes of sign. ∎

Good software is available for computing the roots and weights of most classical orthogonal polynomials [13]. Newton's method permits rapid computation of the roots if initial values are chosen to take advantage of the interlacing of roots from successive polynomials. The interlacing property, which we will not prove, can be stated as follows [8]: If $y_1 < \cdots < y_k$ denote the roots of the orthogonal polynomial $\psi_k(x)$, then the next orthogonal polynomial $\psi_{k+1}(x)$ has exactly one root on each of the $k+1$ intervals $(-\infty, y_1), (y_1, y_2), \ldots, (y_k, \infty)$. The weights associated with the $k+1$ roots $x_0, \ldots, x_k$ of $\psi_{k+1}(x)$ can be computed in a variety of ways.

For instance, in view of the identity $\int \psi_i(x)d\mu(x) = 0$ for $i > 0$, the set of linear equations

$$\begin{pmatrix} \psi_0(x_0) & \cdots & \psi_0(x_k) \\ \psi_1(x_0) & \cdots & \psi_1(x_k) \\ \vdots & & \vdots \\ \psi_k(x_0) & \cdots & \psi_k(x_k) \end{pmatrix} \begin{pmatrix} w_0 \\ w_1 \\ \vdots \\ w_k \end{pmatrix} = \begin{pmatrix} \int \psi_0(x)d\mu(x) \\ 0 \\ \vdots \\ 0 \end{pmatrix}$$

uniquely determines the weights.

For historical reasons, the Jacobi polynomials are employed in Gaussian quadrature rather than the beta distribution polynomials. The Jacobi polynomials are orthogonal with respect to the density $(1-x)^{\alpha-1}(1+x)^{\beta-1}$ defined on the interval $(-1,1)$. Integration against a beta distribution can be reduced to integration against a Jacobi density via the change of variables $y = (1-x)/2$. Indeed,

$$\begin{aligned} & \frac{1}{B(\alpha,\beta)} \int_0^1 f(y) y^{\alpha-1}(1-y)^{\beta-1} dy \\ = & \frac{1}{2^{\alpha+\beta-1}B(\alpha,\beta)} \int_{-1}^1 f\Big(\frac{1-x}{2}\Big)(1-x)^{\alpha-1}(1+x)^{\beta-1}dx. \end{aligned}$$

Example 18.6.1 *Variance of a Sample Median*

Consider an i.i.d. sample $X_1, \ldots, X_n$ from the standard normal distribution. Assuming n is odd and $m = \lfloor n/2 \rfloor$, the sample median $X_{(m+1)}$ has density

$$n\binom{n-1}{m}\Phi(x)^m[1-\Phi(x)]^m\phi(x),$$

where $\phi(x)$ is the standard normal density and $\Phi(x)$ is the standard normal distribution function. Since the mean of $X_{(m+1)}$ is 0, the variance of $X_{(m+1)}$ is

$$\text{Var}(X_{(m+1)}) = n\binom{n-1}{m}\int_{-\infty}^{\infty} x^2\Phi(x)^m[1-\Phi(x)]^m\phi(x)dx.$$

Table 18.3 displays the Gauss-Hermite quadrature approximations Q_{nk} to $\text{Var}(X_{(m+1)})$ for sample sizes $n = 11$ and $n = 21$ and varying numbers of quadrature points k. These results should be compared to the values 0.1428 for $n = 11$ and 0.0748 for $n = 21$ predicted by the asymptotic variance $1/[4n\phi(0)^2] = \pi/(2n)$ of the normal limit law for a sample median [14]. ■

TABLE 18.3. Quadrature Approximations to $\mathrm{Var}(X_{(m+1)})$

n	$Q_{n,2}$	$Q_{n,4}$	$Q_{n,8}$	$Q_{n,16}$	$Q_{n,32}$	$Q_{n,64}$
11	0.1175	0.2380	0.2341	0.1612	0.1379	0.1372
21	0.0070	0.0572	0.1271	0.1230	0.0840	0.0735

18.7 Problems

1. Use the Euler-Maclaurin summation formula to establish the equality

$$\frac{1}{e-1} = \sum_{n=0}^{\infty} \frac{B_n}{n!}.$$

2. Verify the asymptotic expansion

$$\sum_{k=1}^{n} k^{\alpha} = C_{\alpha} + \frac{n^{\alpha+1}}{\alpha+1} + \frac{n^{\alpha}}{2} + \sum_{j=1}^{m} \frac{B_{2j}}{2j}\binom{\alpha}{2j-1} n^{\alpha-2j+1} + O(n^{\alpha-2m-1})$$

for a real number $\alpha \neq -1$ and some constant C_{α}, which you need not determine.

3. Find asymptotic expansions for the two sums $\sum_{k=1}^{n}(n^2+k^2)^{-1}$ and $\sum_{k=1}^{n}(-1)^k/k$ valid to $O(n^{-3})$.

4. Check by quadrature that the asserted values in expressions (18.3) and (18.4) are accurate.

5. After the first application of Romberg's acceleration algorithm in equation (18.7), show that the integral $\int_a^b f(x)dx$ is approximated by

$$\frac{h}{6}\Big(f_0 + 4f_1 + 2f_2 + 4f_3 + 2f_4 + \cdots + 2f_{2n-2} + 4f_{2n-1} + f_{2n}\Big),$$

where $f_k = f[a + k(b-a)/(2n)]$. This is Simpson's rule.

6. Show that the refinement T_{mn} in Romberg's algorithm exactly equals $\int_a^b f(x)dx$ when $f(x)$ is a polynomial of degree $2n+1$ or lower.

7. For an integrand $f(x)$ with four continuous derivatives, let $Q(h)$ be the trapezoidal approximation to the integral $\int_a^b f(x)dx$ with integration step $h = (b-a)/(6n)$. Based on $Q(2h)$ and $Q(3h)$, construct a quadrature formula $R(h)$ such that

$$\int_a^b f(x)dx - R(h) = O(h^4).$$

8. Numerical differentiation can be improved by the acceleration technique employed in Romberg's algorithm. If $f(x)$ has $2k$ continuous derivatives, then the central difference formula

$$\begin{aligned} D_0(h) &= \frac{f(x+h)-f(x-h)}{2h} \\ &= \sum_{j=0}^{k-1} f^{(2j+1)}(x)\frac{h^{2j}}{(2j+1)!} + O(h^{2k-1}) \end{aligned}$$

follows from an application of Taylor's theorem. Show that the inductively defined quantities

$$\begin{aligned} D_j(h) &= \frac{4^j D_{j-1}(\frac{1}{2}h) - D_{j-1}(h)}{4^j - 1} \\ &= D_{j-1}\Big(\frac{1}{2}h\Big) - \frac{1}{4^j-1}\Big[D_{j-1}(h) - D_{j-1}\Big(\frac{1}{2}h\Big)\Big] \end{aligned}$$

satisfy

$$f'(x) = D_j(h) + O(h^{2j+2})$$

for $j = 0, \ldots, k-1$. Verify that

$$D_1(h) = \frac{1}{6}\Big[8f\Big(x+\frac{1}{2}h\Big) - 8f\Big(x-\frac{1}{2}h\Big) - f(x+h) + f(x-h)\Big].$$

Finally, try this improvement of central differencing on a few representative functions such as $\sin(x)$, e^x, and $\ln\Gamma(x)$.

9. Discuss what steps you would take to compute the following integrals accurately [1]:

$$\int_1^\infty \frac{\ln x}{x\sqrt{1+x}}dx, \quad \int_\epsilon^{\frac{\pi}{2}} \frac{\sin x}{x^2}dx, \quad \int_0^{\frac{\pi}{2}} \frac{1-\cos x}{x^2\sqrt{x}}dx.$$

In the middle integral, $\epsilon > 0$ is small.

10. For a probability measure μ concentrated on a finite interval $[a, b]$, let

$$Q_k(f) = \sum_{i=0}^{k} w_i^{(k)} f(x_i^{(k)})$$

be the sequence of Gaussian quadrature operators that exactly integrate polynomials of degree $2k+1$ or lower based on the roots of the orthonormal polynomial $\psi_{k+1}(x)$. Prove that

$$\lim_{k\to\infty} Q_k(f) = \int_a^b f(x)d\mu(x)$$

for any continuous function $f(x)$. (Hint: Apply the Weierstrass approximation theorem [9].)

11. Describe and implement Newton's method for computing the roots of the Hermite polynomials. How can you use the interlacing property of the roots and the symmetry properties of the Hermite polynomials to reduce computation time? How is Proposition 17.4.2 relevant to the computation of the derivatives required by Newton's method? To avoid overflows you should use the orthonormal version of the polynomials.

12. Continuing Problem 11, describe and implement a program for the computation of the weights in Gauss-Hermite quadrature.

13. Let $X_1, \ldots, X_n$ be an i.i.d. sample from a gamma density with scale parameter 1 and shape parameter α. Describe and implement a numerical scheme for computing the expected values of the order statistics $X_{(1)}$ and $X_{(n)}$.

14. In light of Problem 13 of Chapter 2, describe and implement a numerical scheme for computing the bivariate normal distribution function.

18.8 References

[1] Acton FS (1996) *Real Computing Made Real: Preventing Errors in Scientific and Engineering Calculations.* Princeton University Press, Princeton, NJ

[2] Birkhoff G, MacLane S (1965) *A Survey of Modern Algebra*, 3rd ed. Macmillan, New York

[3] Boas RP Jr (1977) Partial sums of infinite series, and how they grow. *Amer Math Monthly* 84:237–258

[4] Ellis TMR, Philips IR, Lahey TM (1994) *Fortran 90 Programming.* Addison-Wesley, Wokingham, England

[5] Graham RL, Knuth DE, Patashnik O (1988) *Concrete Mathematics: A Foundation for Computer Science.* Addison-Wesley, Reading, MA

[6] Hämmerlin G, Hoffmann KH (1991) *Numerical Mathematics.* Springer, New York

[7] Henrici P (1982) *Essentials of Numerical Analysis with Pocket Calculator Demonstrations.* Wiley, New York

[8] Hochstadt H (1986) *The Functions of Mathematical Physics.* Dover, New York

[9] Hoffman K (1975) *Analysis in Euclidean Space.* Prentice-Hall, Englewood Cliffs, NJ

[10] Isaacson E, Keller HB (1966) *Analysis of Numerical Methods.* Wiley, New York

[11] Körner TW (1988) *Fourier Analysis.* Cambridge University Press, Cambridge

[12] Powell MJD (1981) *Approximation Theory and Methods.* Cambridge University Press, Cambridge

[13] Press WH, Teukolsky SA, Vetterling WT, Flannery BP (1992) *Numerical Recipes in Fortran: The Art of Scientific Computing*, 2nd ed. Cambridge University Press, Cambridge

[14] Sen PK, Singer JM (1993) *Large Sample Methods in Statistics: An Introduction with Applications.* Chapman and Hall, London

[15] Wilf HS (1978) *Mathematics for the Physical Sciences.* Dover, New York

19

The Fourier Transform

19.1 Introduction

The Fourier transform is one of the most productive tools of the mathematical sciences. It crops up again and again in unexpected applications to fields as diverse as differential equations, numerical analysis, probability theory, number theory, quantum mechanics, optics, medical imaging, and signal processing [3, 5, 7, 8, 9]. One explanation for its wide utility is that it turns complex mathematical operations like differentiation and convolution into simple operations like multiplication. Readers most likely are familiar with the paradigm of transforming a mathematical equation, solving it in transform space, and then inverting the solution. Besides its operational advantages, the Fourier transform often has the illuminating physical interpretation of decomposing a temporal process into component processes with different frequencies.

In this chapter, we review the basic properties of the Fourier transform and touch lightly on its applications to Edgeworth expansions. Because of space limitations, our theoretical treatment of Fourier analysis is necessarily superficial. At this level it is difficult to be entirely rigorous without invoking some key facts from real analysis. Readers unfamiliar with the facts cited will have to take them on faith or turn to one of the many available texts on real analysis. In mitigation of these theoretical excursions, some topics from elementary probability are repeated for the sake of completeness.

19.2 Basic Properties

The Fourier transform can be defined on a variety of function spaces. For our purposes, it suffices to consider complex-valued, integrable functions whose domain is the real line. The Fourier transform of such a function $f(x)$ is defined according to the recipe

$$\hat{f}(y) = \int_{-\infty}^{\infty} e^{iyx} f(x) dx$$

for all real numbers y. Note that by the adjective "integrable" we mean $\int_{-\infty}^{\infty} |f(x)| dx < \infty$. In the sequel we usually omit the limits of integration. If $f(x)$ is a probability density, then the Fourier transform $\hat{f}(y)$ coincides with the characteristic function of $f(x)$.

K. Lange, *Numerical Analysis for Statisticians*, Statistics and Computing,
DOI 10.1007/978-1-4419-5945-4_19, © Springer Science+Business Media, LLC 2010

TABLE 19.1. Fourier Transform Pairs

Function	Transform	Function	Transform
(a) $af(x) + bg(x)$	$a\hat{f}(y) + b\hat{g}(y)$	(e) $f(x)^*$	$\hat{f}(-y)^*$
(b) $f(x - x_0)$	$e^{iyx_0}\hat{f}(y)$	(f) $ixf(x)$	$\frac{d}{dy}\hat{f}(y)$
(c) $e^{iy_0x}f(x)$	$\hat{f}(y + y_0)$	(g) $\frac{d}{dx}f(x)$	$-iy\hat{f}(y)$
(d) $f(\frac{x}{a})$	$\lvert a\rvert\hat{f}(ay)$	(h) $f * g(x)$	$\hat{f}(y)\hat{g}(y)$

Proposition 19.2.1 *Table 19.1 summarizes the operational properties of the Fourier transform. In the table, a, b, x_0, and y_0 are constants, and the functions $f(x)$, $xf(x)$, $\frac{d}{dx}f(x)$, and $g(x)$ are assumed integrable as needed. In entry (g), $f(x)$ is taken to be absolutely continuous.*

Proof: All entries in the table except (f) through (h) are straightforward to verify. To prove (f), note that $\frac{d}{dy}\hat{f}(y)$ is the limit of the difference quotients

$$\frac{\hat{f}(y+u) - \hat{f}(y)}{u} = \int \frac{e^{iux} - 1}{u} e^{iyx} f(x)dx.$$

The integrand on the right is bounded above in absolute value by

$$\begin{aligned} \left|\frac{e^{iux} - 1}{u} e^{iyx} f(x)\right| &= \left|\int_0^x e^{iuz}dz\right| |f(x)| \\ &\leq |xf(x)|. \end{aligned}$$

Hence, the dominated convergence theorem permits one to interchange limit and integral signs as u tends to 0.

To verify property (g), we first observe that the absolute continuity of $f(x)$ is a technical condition permitting integration by parts and application of the fundamental theorem of calculus. Now the integration-by-parts formula

$$\int_c^d e^{iyx} \frac{d}{dx} f(x)dx = e^{iyx} f(x)|_c^d - iy \int_c^d e^{iyx} f(x)dx.$$

proves property (g) provided we can demonstrate that

$$\lim_{c \to -\infty} f(c) = \lim_{d \to \infty} f(d) = 0.$$

Since $\frac{d}{dx}f(x)$ is assumed integrable, the reconstruction

$$f(d) - f(0) = \int_0^d \frac{d}{dx} f(x)dx$$

implies that $\lim_{d\to\infty} f(d)$ exists. This right limit is necessarily 0 because $f(x)$ is integrable. The left limit $\lim_{c\to-\infty} f(c) = 0$ follows by the same argument. We defer the proof of property (h) until we define convolution. ■

19.3 Examples

Before developing the theory of the Fourier transform further, it is useful to pause and calculate some specific transforms.

Example 19.3.1 *Uniform Distribution*

If $f(x)$ is the uniform density on the interval $[a,b]$, then

$$\begin{aligned}\hat{f}(y) &= \frac{1}{b-a}\int_a^b e^{iyx}\,dx \\ &= \frac{e^{iyx}}{(b-a)iy}\bigg|_a^b \\ &= \frac{e^{i\frac{1}{2}(a+b)y}\left[e^{\frac{1}{2}i(b-a)y} - e^{-\frac{1}{2}i(b-a)y}\right]}{\frac{1}{2}(b-a)y2i} \\ &= e^{i\frac{1}{2}(a+b)y}\frac{\sin[\frac{1}{2}(b-a)y]}{\frac{1}{2}(b-a)y}.\end{aligned}$$

When $a=-b$, this reduces to $\hat{f}(y) = \sin(by)/(by)$. ■

Example 19.3.2 *Gaussian Distribution*

To find the Fourier transform of the Gaussian density $f(x)$ with mean 0 and variance σ^2, we derive and solve a differential equation. Indeed, integration by parts and property (f) of Table 19.1 imply that

$$\begin{aligned}\frac{d}{dy}\hat{f}(y) &= \frac{1}{\sqrt{2\pi\sigma^2}}\int e^{iyx}ixe^{-\frac{x^2}{2\sigma^2}}\,dx \\ &= \frac{1}{\sqrt{2\pi\sigma^2}}\int e^{iyx}(-i\sigma^2)\frac{d}{dx}e^{-\frac{x^2}{2\sigma^2}}\,dx \\ &= \frac{-i\sigma^2}{\sqrt{2\pi\sigma^2}}e^{iyx}e^{-\frac{x^2}{2\sigma^2}}\bigg|_{-\infty}^{\infty} - \frac{\sigma^2 y}{\sqrt{2\pi\sigma^2}}\int e^{iyx}e^{-\frac{x^2}{2\sigma^2}}\,dx \\ &= -\sigma^2 y\hat{f}(y).\end{aligned}$$

The unique solution to this differential equation with initial value $\hat{f}(0) = 1$ is $\hat{f}(y) = e^{-\sigma^2y^2/2}$. ■

Example 19.3.3 *Gamma Distribution*

We can also derive the Fourier transform of the gamma density $f(x)$ with shape parameter α and scale parameter β by solving a differential equation. Now integration by parts and property (f) yield

$$\begin{aligned}
& \frac{d}{dy}\hat{f}(y) \\
&= \frac{\beta^\alpha}{\Gamma(\alpha)}\int_0^\infty e^{(iy-\beta)x} ix x^{\alpha-1}\,dx \\
&= \frac{i\beta^\alpha}{(iy-\beta)\Gamma(\alpha)}\int_0^\infty x^\alpha \frac{d}{dx}e^{(iy-\beta)x}\,dx \\
&= \frac{i\beta^\alpha}{(iy-\beta)\Gamma(\alpha)} x^\alpha e^{(iy-\beta)x}\Big|_0^\infty - \frac{i\alpha\beta^\alpha}{(iy-\beta)\Gamma(\alpha)}\int_0^\infty e^{(iy-\beta)x}x^{\alpha-1}\,dx \\
&= -\frac{i\alpha}{(iy-\beta)}\hat{f}(y).
\end{aligned}$$

The solution to this differential equation with initial condition $\hat{f}(0) = 1$ is clearly $\hat{f}(y) = [\beta/(\beta - iy)]^\alpha$. ■

Example 19.3.4 *Bilateral Exponential*

The exponential density $f(x) = e^{-x}1_{[0,\infty)}(x)$ reduces to the special case $\alpha = \beta = 1$ of the last example. Since the bilateral exponential density $e^{-|x|}/2$ can be expressed as $[f(x) + f(-x)]/2$, property (d) of Table 19.1 shows that it has Fourier transform

$$\begin{aligned}
\frac{1}{2}\left[\hat{f}(y) + \hat{f}(-y)\right] &= \frac{1}{2(1-iy)} + \frac{1}{2(1+iy)} \\
&= \frac{1}{1+y^2}.
\end{aligned}$$

Up to a normalizing constant, this is the standard Cauchy density. ■

Example 19.3.5 *Hermite Polynomials*

The Hermite polynomial $H_n(x)$ can be expressed as

$$H_n(x) = (-1)^n e^{\frac{1}{2}x^2}\frac{d^n}{dx^n}e^{-\frac{1}{2}x^2}. \tag{19.1}$$

Indeed, if we expand the left-hand side of the identity

$$e^{-\frac{1}{2}(x-t)^2} = e^{-\frac{1}{2}x^2}e^{xt-\frac{1}{2}t^2} = \sum_{n=0}^\infty \frac{H_n(x)e^{-\frac{1}{2}x^2}}{n!}t^n$$

in a Taylor series about $t = 0$, then the coefficient of $t^n/n!$ is

$$\begin{aligned} H_n(x)e^{-\frac{1}{2}x^2} &= \frac{d^n}{dt^n}e^{-\frac{1}{2}(x-t)^2}\Big|_{t=0} \\ &= (-1)^n\frac{d^n}{dx^n}e^{-\frac{1}{2}(x-t)^2}\Big|_{t=0} \\ &= (-1)^n\frac{d^n}{dx^n}e^{-\frac{1}{2}x^2}. \end{aligned}$$

Example 19.3.2 and repeated application of property (g) of Table 19.1 therefore yield

$$\begin{aligned} \frac{1}{\sqrt{2\pi}}\int e^{iyx}H_n(x)e^{-\frac{1}{2}x^2}dx &= \frac{1}{\sqrt{2\pi}}\int e^{iyx}(-1)^n\frac{d^n}{dx^n}e^{-\frac{1}{2}x^2}dx \quad (19.2) \\ &= (iy)^n e^{-\frac{1}{2}y^2}. \end{aligned}$$

This Fourier transform will appear in our subsequent discussion of Edgeworth expansions. ■

19.4 Further Theory

We now delve more deeply into the theory of the Fourier transform.

Proposition 19.4.1 (Riemann-Lebesgue) *If the function $f(x)$ is integrable, then its Fourier transform $\hat{f}(y)$ is bounded, continuous, and tends to 0 as $|y|$ tends to ∞.*

Proof: The transform $\hat{f}(y)$ is bounded because

$$\begin{aligned} |\hat{f}(y)| &= \Big|\int e^{iyx}f(x)dx\Big| \\ &\le \int |e^{iyx}||f(x)|dx \quad (19.3) \\ &= \int |f(x)|dx. \end{aligned}$$

To prove continuity, let $\lim_{n\to\infty} y_n = y$. Then the sequence of functions $g_n(x) = e^{iy_nx}f(x)$ is bounded in absolute value by $|f(x)|$ and satisfies

$$\lim_{n\to\infty} g_n(x) = e^{iyx}f(x).$$

Hence the dominated convergence theorem implies that

$$\lim_{n\to\infty}\int g_n(x)dx = \int e^{iyx}f(x)dx.$$

To prove the last assertion, we use the fact that the space of step functions with bounded support is dense in the space of integrable functions. Thus, given any $\epsilon > 0$, there exists a step function

$$g(x) = \sum_{j=1}^{m} c_j 1_{[x_{j-1},x_j)}(x)$$

vanishing off some finite interval and satisfying $\int |f(x) - g(x)|dx < \epsilon$. The Fourier transform $\hat{g}(y)$ has the requisite behavior at ∞ because Example 19.3.1 allows us to calculate

$$\hat{g}(y) = \sum_{j=1}^{m} c_j e^{i\frac{1}{2}(x_{j-1}+x_j)y} \frac{\sin[\frac{1}{2}(x_j - x_{j-1})y]}{\frac{1}{2}y},$$

and this finite sum clearly tends to 0 as $|y|$ tends to ∞. The original transform $\hat{f}(y)$ exhibits the same behavior because the bound (19.3) entails the inequality

$$\begin{aligned} |\hat{f}(y)| &\leq |\hat{f}(y) - \hat{g}(y)| + |\hat{g}(y)| \\ &\leq \epsilon + |\hat{g}(y)|. \end{aligned}$$

This completes the proof. ■

The Fourier transform can be inverted. If $g(x)$ is integrable, then

$$\check{g}(y) = \frac{1}{2\pi} \int e^{-iyx} g(x)dx$$

supplies the inverse Fourier transform of $g(x)$. This terminology is justified by the next proposition.

Proposition 19.4.2 *Let $f(x)$ be a bounded, continuous function. If $f(x)$ and $\hat{f}(y)$ are both integrable, then*

$$f(x) = \frac{1}{2\pi} \int e^{-iyx} \hat{f}(y)dy. \tag{19.4}$$

Proof: Consider the identities

$$\begin{aligned} \frac{1}{2\pi} \int e^{-iyx} e^{-\frac{y^2}{2\sigma^2}} \hat{f}(y)dy &= \frac{1}{2\pi} \int e^{-iyx} e^{-\frac{y^2}{2\sigma^2}} \int e^{iyu} f(u)du\, dy \\ &= \int f(u) \frac{1}{2\pi} \int e^{iy(u-x)} e^{-\frac{y^2}{2\sigma^2}} dy\, du \\ &= \int f(u) \frac{\sigma}{\sqrt{2\pi}} e^{-\frac{\sigma^2(u-x)^2}{2}} du \\ &= \frac{1}{\sqrt{2\pi}} \int f(x + \frac{v}{\sigma}) e^{-\frac{v^2}{2}} dv, \end{aligned}$$

which involve Example 19.3.2 and the change of variables $u = x + v/\sigma$. As σ tends to ∞, the last integral tends to

$$\frac{1}{\sqrt{2\pi}} \int f(x) e^{-\frac{v^2}{2}} dv \quad = \quad f(x),$$

while the original integral tends to

$$\frac{1}{2\pi} \int e^{-iyx} \lim_{\sigma \to \infty} e^{-\frac{y^2}{2\sigma^2}} \hat{f}(y) dy \quad = \quad \frac{1}{2\pi} \int e^{-iyx} \hat{f}(y) dy.$$

Equating these two limits yields the inversion formula (19.4). ■

Example 19.4.1 *Cauchy Distribution*

If $f(x)$ is the bilateral exponential density, then Proposition 19.4.2 and Example 19.3.4 show that the Cauchy density $1/[\pi(1 + x^2)]$ has Fourier transform $e^{-|y|}$. ■

Proposition 19.4.3 (Parseval-Plancherel) *If either of the integrable functions $f(x)$ or $g(x)$ obeys the further assumptions of Proposition 19.4.2, then*

$$\int f(x) g(x)^* dx \quad = \quad \frac{1}{2\pi} \int \hat{f}(y) \hat{g}(y)^* dy. \tag{19.5}$$

In particular, when $f(x) = g(x)$,

$$\int |f(x)|^2 dx \quad = \quad \frac{1}{2\pi} \int |\hat{f}(y)|^2 dy. \tag{19.6}$$

Proof: If $\hat{f}(y)$ satisfies the assumptions of Proposition 19.4.2, then

$$\begin{aligned}
\int f(x) g(x)^* dx &= \int \frac{1}{2\pi} \int e^{-iyx} \hat{f}(y) dy \, g(x)^* dx \\
&= \int \hat{f}(y) \frac{1}{2\pi} \int e^{-iyx} g(x)^* dx \, dy \\
&= \int \hat{f}(y) \frac{1}{2\pi} \Big[\int e^{iyx} g(x) dx \Big]^* dy \\
&= \frac{1}{2\pi} \int \hat{f}(y) \hat{g}(y)^* dy.
\end{aligned}$$

With obvious modifications the same proof works if $\hat{g}(y)$ satisfies the assumptions of Proposition 19.4.2. ■

Example 19.4.2 *Computation of an Integral*

Let $f(x) = 1_{[-b,b]}(x)/(2b)$ be the uniform density. Then the Parseval-Plancherel relation (19.6) implies

$$\frac{1}{2b} = \frac{1}{(2b)^2}\int_{-b}^{b} 1\,dx = \frac{1}{2\pi}\int \left[\frac{\sin(by)}{by}\right]^2 dy.$$

Evaluation of the second integral is not so obvious. ■

There are several definitions of the Fourier transform that differ only in how the factor of 2π and the sign of the argument of the complex exponential are assigned. We have chosen the definition that coincides with the characteristic function of a probability density. For some purposes the alternative definitions

$$\begin{aligned}\hat{f}(y) &= \frac{1}{\sqrt{2\pi}}\int e^{iyx} f(x)dx \\ \breve{g}(x) &= \frac{1}{\sqrt{2\pi}}\int e^{-iyx} g(y)dy\end{aligned}$$

of the Fourier transform and its inverse are better. Obviously, the transform and its inverse are now more symmetrical. Also, the Parseval-Plancherel relation (19.5) simplifies to

$$\int f(x)g(x)^* dx = \int \hat{f}(y)\hat{g}(y)^* dy.$$

In other words, the Fourier transform now preserves inner products and norms on a subspace of the Hilbert space $\mathrm{L}^2(-\infty,\infty)$ of square-integrable functions. Such a transformation is said to be unitary. One can show that this subspace is dense in $\mathrm{L}^2(-\infty,\infty)$ and therefore that the Fourier transform extends uniquely to a unitary transformation from $\mathrm{L}^2(-\infty,\infty)$ onto itself [3, 4, 9]. Proof of these theoretical niceties would take us too far afield. Let us just add that norm preservation forces the Fourier transform of a function to be unique.

Our final theoretical topic is convolution. If $f(x)$ and $g(x)$ are integrable functions, then their convolution $f * g(x)$ is defined by

$$f * g(x) = \int f(x-u)g(u)du.$$

Doubtless readers will recall that if $f(x)$ and $g(x)$ are the densities of independent random variables U and V, then $f * g(x)$ is the density of the sum $U+V$. The fundamental properties of convolution valid in the context of density functions carry over to the more general setting of integrable functions.

Proposition 19.4.4 *The convolution of two integrable functions $f(x)$ and $g(x)$ is integrable with Fourier transform $\hat{f}(y)\hat{g}(y)$. Furthermore, convolution is a commutative, associative, and linear operation.*

Proof: Integrability of $f * g(x)$ is a consequence of the calculation

$$\begin{aligned}
\int |f * g(x)| dx &= \int \left| \int f(x-u)g(u)du \right| dx \\
&\leq \int \int |f(x-u)||g(u)|du\, dx \\
&= \int \int |f(x-u)|dx|g(u)|du \\
&= \int \int |f(x)|dx|g(u)|du \\
&= \int |f(x)|dx \int |g(u)|du.
\end{aligned}$$

The product form of the Fourier transform follows from

$$\begin{aligned}
\int e^{iyx} f * g(x)dx &= \int e^{iyx} \int f(x-u)g(u)du\, dx \\
&= \int \int e^{iy(x-u)} f(x-u)dx e^{iyu} g(u)du \\
&= \int \int e^{iyx} f(x)dx e^{iyu} g(u)du \\
&= \hat{f}(y)\hat{g}(y).
\end{aligned}$$

The remaining assertions are easy consequences of the result just established and the uniqueness of the Fourier transform. ∎

Example 19.4.3 *Convolution of Cauchy Densities*

Let $c_1, \ldots, c_n$ be positive constants and $X_1, \ldots, X_n$ an i.i.d. sequence of random variables with standard Cauchy density $1/[\pi(1+x^2)]$ and characteristic function $e^{-|y|}$. Then the sum $c_1 X_1 + \cdots + c_n X_n$ has characteristic function $e^{-c|y|}$ and Cauchy density $c/[\pi(c^2+x^2)]$, where $c = c_1 + \cdots + c_n$. When $c = 1$ we retrieve the standard Cauchy density. ∎

19.5 Edgeworth Expansions

An Edgeworth expansion is an asymptotic approximation to a density or distribution function [1, 4, 6]. The main ideas can be best illustrated by considering the proof of the central limit theorem for i.i.d. random variables

$X_1, X_2, \ldots$ with common mean μ, variance σ^2, and density $f(x)$. Assuming that $f(x)$ possesses a moment generating function, we can write its characteristic function as

$$\begin{aligned}\hat{f}(y) &= \sum_{j=0}^{\infty} \frac{\mu_j}{j!}(iy)^j \\ &= \exp\left[\sum_{j=1}^{\infty} \frac{\kappa_j}{j!}(iy)^j\right],\end{aligned}$$

where μ_j and κ_j are the jth moment and jth cumulant of $f(x)$, respectively. The moment series for $\hat{f}(y)$ follows from repeated application of property (f) of Table 19.1 with y set equal to 0. Cumulants are particularly handy in this context because Proposition 19.4.4 implies that the jth cumulant of the sum $S_n = \sum_{i=1}^n X_i$ is just $n\kappa_j$. The identities $\kappa_1 = \mu_1$ and $\kappa_2 = \sigma^2$ hold in general. For notational convenience, we let $\rho_j = \kappa_j/\sigma^j$.

Owing to properties (b), (d), and (h) of Table 19.1, the characteristic function of the standardized sum $T_n = (S_n - n\mu)/(\sigma\sqrt{n})$ reduces to

$$\begin{aligned}& e^{-\frac{i\sqrt{n}\mu y}{\sigma}} \hat{f}\left(\frac{y}{\sigma\sqrt{n}}\right)^n \\ &= \exp\left[-\frac{y^2}{2} + n\sum_{j=3}^{\infty} \frac{\rho_j}{n^{\frac{j}{2}} j!}(iy)^j\right] \\ &= e^{-\frac{y^2}{2}} e^{\frac{\rho_3 (iy)^3}{6\sqrt{n}} + \frac{\rho_4 (iy)^4}{24n} + O(n^{-3/2})} \\ &= e^{-\frac{y^2}{2}}\left[1 + \frac{\rho_3 (iy)^3}{6\sqrt{n}} + \frac{\rho_4 (iy)^4}{24n} + \frac{\rho_3^2 (iy)^6}{72n} + O(n^{-\frac{3}{2}})\right].\end{aligned}$$

Formal inversion of this Fourier transform taking into account equation (19.2) yields the asymptotic expansion

$$\phi(x)\left[1 + \frac{\rho_3 H_3(x)}{6\sqrt{n}} + \frac{\rho_4 H_4(x)}{24n} + \frac{\rho_3^2 H_6(x)}{72n} + O(n^{-\frac{3}{2}})\right] \tag{19.7}$$

for the density of T_n. Here $\phi(x)$ denotes the standard normal density. To approximate the distribution function of T_n in terms of the standard normal distribution function $\Phi(x)$, note that integration of the Hermite polynomial identity (19.1) gives

$$\int_{-\infty}^{x} \phi(u) H_n(u)\,du = -\phi(x) H_{n-1}(x).$$

Applying this fact to the integration of expression (19.7) yields the asymptotic expansion

$$\Phi(x) - \phi(x)\left[\frac{\rho_3 H_2(x)}{6\sqrt{n}} + \frac{\rho_4 H_3(x)}{24n} + \frac{\rho_3^2 H_5(x)}{72n} + O(n^{-\frac{3}{2}})\right] \tag{19.8}$$

for the distribution function of T_n. It is worth stressing that the formal manipulations leading to the expansions (19.7) and (19.8) can be made rigorous [4, 6].

Both of the expansions (19.7) and (19.8) suggest that the rate of convergence of T_n to normality is governed by the $O(n^{-1/2})$ correction term. However, this pessimistic impression is misleading at $x = 0$ because $H_3(0) = 0$. (A quick glance at the recurrence relation (17.11) of Chapter 17 confirms that $H_n(x)$ is even for n even and odd for n odd.) The device known as exponential tilting exploits this peculiarity [1].

If we let $K(t) = \sum_{j=1}^{\infty} \kappa_j t^j/j!$ be the cumulant generating function of $f(x)$ and $g(x)$ be the density of S_n, then we tilt $g(x)$ to the density $e^{tx-nK(t)}g(x)$. Because $e^{nK(t)}$ is the moment generating function of S_n, we find that

$$\int e^{sx}e^{tx-nK(t)}g(x)dx \quad = \quad e^{nK(s+t)-nK(t)}.$$

This calculation confirms that $e^{tx-nK(t)}g(x)$ is a probability density with moment generating function $e^{nK(s+t)-nK(t)}$. The tilted density has mean $nK'(t)$ and variance $nK''(t)$. We can achieve an arbitrary mean x_0 by choosing t_0 to be the solution of the equation $nK'(t_0) = x_0$. In general, this equation must be solved numerically. Once t_0 is chosen, we can approximate the standardized tilted density

$$\sqrt{nK''(t_0)}e^{t_0[\sqrt{nK''(t_0)}x+x_0]-nK(t_0)}g\Big(\sqrt{nK''(t_0)}x + x_0\Big)$$

at $x = 0$ by the asymptotic expansion (19.7). To order $O(n^{-1})$ this gives

$$\sqrt{nK''(t_0)}e^{t_0x_0-nK(t_0)}g(x_0) \quad = \quad \phi(0)\Big[1 + O(n^{-1})\Big]$$

or

$$g(x_0) \quad = \quad \frac{e^{-t_0x_0+nK(t_0)}}{\sqrt{2\pi nK''(t_0)}}\Big[1 + O(n^{-1})\Big]. \tag{19.9}$$

This result is also called a saddle point approximation.

Further terms can be included in the saddle point approximation if we substitute the appropriate normalized cumulants

$$\rho_j(t_0) \quad = \quad \frac{nK^{(j)}(t_0)}{[nK''(t_0)]^{\frac{j}{2}}}$$

of the tilted density in the Edgeworth expansion (19.7). Once we determine the required coefficients $H_3(0) = 0$, $H_4(0) = 3$, and $H_6(0) = -15$ from recurrence relation (17.11) of Chapter 17, it is obvious that

$$g(x_0) \quad = \quad \frac{e^{-t_0x_0+nK(t_0)}}{\sqrt{2\pi nK''(t_0)}}\Big[1 + \frac{3\rho_4(t_0) - 5\rho_3^2(t_0)}{24n} + O(n^{-\frac{3}{2}})\Big]. \tag{19.10}$$

Example 19.5.1 *Spread of a Random Sample from the Exponential*

If $X_1, \ldots, X_{n+1}$ are independent, exponentially distributed random variables with common mean 1, then the spread $X_{(n+1)} - X_{(1)}$ has density

$$n \sum_{k=1}^{n} (-1)^{k-1} \binom{n-1}{k-1} e^{-kx}. \qquad (19.11)$$

This is also the density of the sum $Y_1 + \cdots + Y_n$ of independent, exponentially distributed random variables Y_j with respective means $1, 1/2, \ldots, 1/n$. (A proof of these obscure facts is sketched in Problem I.13.13 of [4]). For the sake of comparison, we compute the Edgeworth approximation (19.7) and the two saddle point approximations (19.9) and (19.10) to the exact density (19.11). Because the Y_j have widely different variances, a naive normal approximation based on the central limit theorem is apt to be poor.

TABLE 19.2. Saddle Point Approximations to the Spread Density

x_0	Exact $g(x_0)$	Error (19.7)	Error (19.9)	Error (19.10)
.50000	.00137	-.04295	-.00001	-.00001
1.00000	.05928	-.04010	-.00027	-.00024
1.50000	.22998	.04979	.00008	.00004
2.00000	.36563	.09938	.00244	.00211
2.50000	.37974	.05913	.00496	.00439
3.00000	.31442	-.00245	.00550	.00499
3.50000	.22915	-.03429	.00423	.00394
4.00000	.15508	-.03588	.00242	.00235
4.50000	.10046	-.02356	.00098	.00103
5.00000	.06340	-.00992	.00012	.00022
5.50000	.03939	-.00136	-.00026	-.00017
6.00000	.02424	.00167	-.00037	-.00029
6.50000	.01483	.00209	-.00035	-.00029
7.00000	.00904	.00213	-.00028	-.00024
7.50000	.00550	.00220	-.00021	-.00018
8.00000	.00334	.00203	-.00014	-.00013
8.50000	.00203	.00160	-.00010	-.00009
9.00000	.00123	.00112	-.00006	-.00006

A brief calculation shows that the cumulant generating function of the sum $Y_1 + \cdots + Y_n$ is $nK(t) = -\sum_{j=1}^{n} \ln(1 - t/j)$. The equation $nK'(t_0) = x_0$ becomes $\sum_{j=1}^{n} 1/(j - t_0) = x_0$, which obviously must be solved numerically. The kth cumulant of the tilted density is

$$nK^{(k)}(t_0) = (k-1)! \sum_{j=1}^{n} \frac{1}{(j - t_0)^k}.$$

Table 19.2 displays the exact density (19.11) and the errors (exact values minus approximate values) committed in using the Edgeworth expansion (19.7) and the two saddle point expansions (19.9) and (19.10) when $n = 10$. Note that we apply equation (19.7) at the standardized point

$$x = [x_0 - nK'(0)]/\sqrt{nK''(0)}$$

and divide the result of the approximation to the standardized density by $\sqrt{nK''(0)}$.

Both saddle point expansions clearly outperform the Edgeworth expansion except very close to the mean $\sum_{j=1}^{10} 1/j \approx 2.93$. Indeed, it is remarkable how well the saddle point expansions do considering how far this example is from the ideal of a sum of i.i.d. random variables. The refined saddle point expansion (19.10) is an improvement over the ordinary saddle point expansion (19.9) in the tails of the density but not necessarily in the center. Daniels [2] considers variations on this problem involving pure birth processes. ■

19.6 Problems

1. Verify the first five entries in Table 19.1.

2. For an even integrable function $f(x)$, show that

$$\hat{f}(y) = 2\int_0^\infty \cos(yx)f(x)dx,$$

and for an odd integrable function $g(x)$, show that

$$\hat{g}(y) = 2i\int_0^\infty \sin(yx)g(x)dx.$$

Conclude that (a) $\hat{f}(y)$ is even, (b) $\hat{f}(y)$ is real if $f(x)$ is real, and (c) $\hat{g}(y)$ is odd.

3. If $f(x)$ is integrable, then define

$$\mathrm{S}f(x) = f(\ln x)$$

for $x > 0$. Show that S is a linear mapping satisfying

   (a) $\int_0^\infty |\mathrm{S}f(x)|x^{-1}dx < \infty$,

   (b) $\mathrm{S}(f * g)(x) = \int_0^\infty \mathrm{S}f(xz^{-1})\mathrm{S}g(z)z^{-1}dz$,

   (c) $\hat{f}(y) = \int_0^\infty \mathrm{S}f(x)x^{iy}x^{-1}dx$.

The function $\int_0^\infty h(x)x^{iy}x^{-1}dx$ defines the Mellin transform of $h(x)$.

4. Suppose $f(x)$ is integrable and $\hat{f}(y) = c\sqrt{2\pi}f(y)$ for some constant c. Prove that either $f(x) = 0$ for all x or c is drawn from the set $\{1, i, -1, -i\}$ of fourth roots of unity. (Hint: Take the Fourier transform of $f(x)$.)

5. Compute

$$\lim_{\alpha\to 0+} \int e^{-\alpha x^2} \frac{\sin(\lambda x)}{x} dx$$

for λ real. (Hint: Use the Parseval-Plancherel relation.)

6. Find a random variable X symmetrically distributed around 0 such that X cannot be represented as $X = Y - Z$ for i.i.d. random variables Y and Z. (Hint: Assuming Y and Z possess a density function, demonstrate that the Fourier transform of the density of X must be nonnegative.)

7. Let $X_1, X_2, \ldots$ be a sequence of i.i.d. random variables that has common density $f(x)$ and is independent of the integer-valued random variable $N \geq 0$. If N has generating function

$$G(s) = \sum_{n=0}^{\infty} \Pr(N = n)s^n,$$

then show that the density of the random sum $\sum_{i=1}^{N} X_i$ has Fourier transform $G[\hat{f}(y)]$.

8. Let $X_1, \ldots, X_n$ be a random sample from a normal distribution with mean μ and variance σ^2. Show that the saddle point approximation (19.9) to the density of $S_n = \sum_{j=1}^{n} X_j$ is exact.

9. Let $X_1, \ldots, X_n$ be a random sample from an exponential distribution with mean 1. Show that the saddle point approximation (19.9) to the density of $S_n = \sum_{j=1}^{n} X_j$ is exact up to Stirling's approximation.

10. Let $X_1, \ldots, X_n$ be a random sample from a member of an exponential family of densities $f(x|\theta) = h(x)e^{\theta u(x) - \gamma(\theta)}$. Show that the saddle point approximation (19.9) to the density of $S_n = \sum_{j=1}^{n} u(X_j)$ at x_0 reduces to

$$\frac{e^{n[\gamma(\theta_0+\theta)-\gamma(\theta)]-\theta_0 x_0}}{\sqrt{2\pi n\gamma''(\theta_0+\theta)}}\Big[1 + O(n^{-1})\Big],$$

where θ_0 satisfies the equation $n\gamma'(\theta_0 + \theta) = x_0$.

11. Compute the Edgeworth approximation (19.8) to the distribution function of a sum of i.i.d. Poisson random variables with unit means. Compare your results for sample sizes $n = 4$ and $n = 8$ to the exact distribution function and, if available, to the values in Table 4.2 of [1]. Note that in computing $\Pr(S_n \leq z)$ in this discrete case it is wise to incorporate a continuity correction by applying the Edgeworth approximation at the point $x = (z - n\mu + 1/2)/(\sigma\sqrt{n})$.

19.7 References

[1] Barndorff-Nielsen OE, Cox DR (1989) *Asymptotic Techniques for Use in Statistics.* Chapman and Hall, London

[2] Daniels HE (1982) The saddlepoint approximation for a general birth process. *J Appl Prob* 19:20–28

[3] Dym H, McKean HP (1972) *Fourier Series and Integrals.* Academic Press, New York

[4] Feller W (1971) *An Introduction to Probability Theory and Its Applications, Volume 2*, 2nd ed. Wiley, New York

[5] Folland GB (1992) *Fourier Analysis and its Applications.* Wadsworth and Brooks/Cole, Pacific Grove, CA

[6] Hall P (1992) *The Bootstrap and Edgeworth Expansion.* Springer, New York

[7] Körner TW (1988) *Fourier Analysis.* Cambridge University Press, Cambridge

[8] Lighthill MJ (1958) *An Introduction to Fourier Analysis and Generalized Functions.* Cambridge University Press, Cambridge

[9] Rudin W (1973) *Functional Analysis.* McGraw-Hill, New York

20

The Finite Fourier Transform

20.1 Introduction

In previous chapters we have met Fourier series and the Fourier transform. These are both incarnations of Fourier analysis on a commutative group, namely the unit circle and the real line under addition. In this chapter we study Fourier analysis in the even simpler setting of the additive group of integers modulo a fixed positive integer n [6, 12]. Here, for obvious reasons, the Fourier transform is called the finite Fourier transform. Although the finite Fourier transform has many interesting applications in abstract algebra, combinatorics, number theory, and complex variables [8], we view it mainly as a tool for approximating Fourier series. Computation of finite Fourier transforms is done efficiently by an algorithm known as the fast Fourier transform [1, 3, 5, 9, 13, 15]. Although it was discovered by Gauss, the fast Fourier transform has come into prominence only with the advent of modern computing. As an indication of its critical role in many scientific and engineering applications, it is often implemented in hardware rather than software.

In this chapter we first study the operational properties of the finite Fourier transform. With minor differences these parallel the properties of Fourier series and the ordinary Fourier transform. We then derive the fast Fourier transform for any highly composite number n. In many applications n is a power of 2, but this choice is hardly necessary. Once we have developed the fast Fourier transform, we discuss applications to time series [1, 2, 4, 9, 11] and other areas of statistics.

20.2 Basic Properties

Periodic sequences $\{c_j\}_{j=-\infty}^{\infty}$ of period n constitute the natural domain of the finite Fourier transform. The transform of such a sequence is defined by

$$\hat{c}_k = \frac{1}{n}\sum_{j=0}^{n-1} c_j e^{-2\pi i \frac{jk}{n}}. \tag{20.1}$$

From this definition it follows immediately that the finite Fourier transform is linear and maps periodic sequences into periodic sequences with the same

K. Lange, *Numerical Analysis for Statisticians*, Statistics and Computing,
DOI 10.1007/978-1-4419-5945-4_20, © Springer Science+Business Media, LLC 2010

period. The inverse transform turns out to be

$$\check{d}_j = \sum_{k=0}^{n-1} d_k e^{2\pi i \frac{jk}{n}}. \tag{20.2}$$

It is fruitful to view each of these operations as a matrix times vector multiplication. Thus, if we let $u_n = e^{2\pi i/n}$ denote the principal nth root of unity, then the finite Fourier transform represents multiplication by the matrix (u_n^{-kj}/n) and the inverse transform multiplication by the matrix (u_n^{jk}). To warrant the name "inverse transform", the second matrix should be the inverse of the first. Indeed, we have

$$\begin{aligned}
\sum_{l=0}^{n-1} u_n^{jl} \frac{1}{n} u_n^{-kl} &= \frac{1}{n} \sum_{l=0}^{n-1} u_n^{(j-k)l} \\
&= \begin{cases} \frac{1}{n} \frac{1-u_n^{(j-k)n}}{1-u_n^{j-k}} & j \neq k \bmod n \\ \frac{1}{n} n & j = k \bmod n \end{cases} \\
&= \begin{cases} 0 & j \neq k \bmod n \\ 1 & j = k \bmod n. \end{cases}
\end{aligned}$$

More symmetry in the finite Fourier transform (20.1) and its inverse (20.2) can be achieved by replacing the factor $1/n$ in the transform by the factor $1/\sqrt{n}$. The inverse transform then includes the $1/\sqrt{n}$ factor as well, and the matrix $(u_n^{-kj}/\sqrt{n})$ is unitary.

We modify periodic sequences of period n by convolution, translation, reversion, and stretching. The convolution of two periodic sequences c_j and d_j is the sequence

$$c * d_k = \sum_{j=0}^{n-1} c_{k-j} d_j = \sum_{j=0}^{n-1} c_j d_{k-j}$$

with the same period. The translate of the periodic sequence c_j by index r is the periodic sequence $T_r c_j$ defined by $T_r c_j = c_{j-r}$. Thus, the operator T_r translates a sequence r places to the right. The reversion operator R takes a sequence c_j into $Rc_j = c_{-j}$. Finally, the stretch operator S_r interpolates $r-1$ zeros between every pair of adjacent entries of a sequence c_j. In symbols,

$$S_r c_j = \begin{cases} c_{\frac{j}{r}} & r \mid j \\ 0 & r \not\mid j, \end{cases}$$

where $r \mid j$ indicates r divides j without remainder. The sequence $S_r c_j$ has period rn, not n. For instance, if $n = 2$ and $r = 2$, the periodic sequence $\ldots, 1, 2, 1, 2 \ldots$ becomes $\ldots, 1, 0, 2, 0, 1, 0, 2, 0, \ldots$.

Proposition 20.2.1 *The finite Fourier transform satisfies the rules:*

1. *(a)* $\widehat{c*d}_k = n\hat{c}_k\hat{d}_k$
2. *(b)* $\widehat{T_r c}_k = u_n^{-rk}\hat{c}_k$
3. *(c)* $\widehat{Rc}_k = R\hat{c}_k = \widehat{c^*}^*_k$
4. *(d)* $\widehat{S_r c}_k = \frac{\hat{c}_k}{r}$.

In (d) the transform on the left has period rn.

Proof: To prove rule (d), note that

$$\begin{aligned}\widehat{S_r c}_k &= \frac{1}{rn}\sum_{j=0}^{rn-1} S_r c_j u_{rn}^{-jk} \\ &= \frac{1}{rn}\sum_{l=0}^{n-1} c_{\frac{lr}{r}} u_{rn}^{-lrk} \\ &= \frac{1}{rn}\sum_{l=0}^{n-1} c_l u_n^{-lk} \\ &= \frac{\hat{c}_k}{r}.\end{aligned}$$

Verification of rules (a) through (c) is left to the reader. ■

20.3 Derivation of the Fast Fourier Transform

The naive approach to computing the finite Fourier transform (20.1) takes $3n^2$ arithmetic operations (additions, multiplications, and complex exponentiations). The fast Fourier transform accomplishes the same task in $O(n \log n)$ operations when n is a power of 2. Proposition 20.2.1 lays the foundation for deriving this useful and clever result.

Consider a sequence c_j of period n, and suppose n factors as $n = rs$. For $k = 0, 1, \ldots, r-1$, define related sequences $c_j^{(k)}$ according to the recipe $c_j^{(k)} = c_{jr+k}$. Each of these secondary sequences has period s. We now argue that we can recover the primary sequence through

$$c_j = \sum_{k=0}^{r-1} T_k S_r c_j^{(k)}. \tag{20.3}$$

In fact, $T_k S_r c_j^{(k)} = 0$ unless $r \mid j-k$. The condition $r \mid j-k$ occurs for exactly one value of k between 0 and $r-1$. For the chosen k,

$$T_k S_r c_j^{(k)} = c_{\frac{j-k}{r}}^{(k)}$$

$$\begin{aligned} &= c_{\frac{j-k}{r}r+k} \\ &= c_j. \end{aligned}$$

In view of properties (b) and (d) of Proposition 20.2.1, taking the finite Fourier transform of equation (20.3) gives

$$\begin{aligned} \hat{c}_j &= \sum_{k=0}^{r-1} u_n^{-kj} \widehat{S_r c^{(k)}}_j \\ &= \sum_{k=0}^{r-1} u_n^{-kj} \frac{1}{r} \widehat{c^{(k)}}_j. \end{aligned} \tag{20.4}$$

Now let $Op(n)$ denote the number of operations necessary to compute a finite Fourier transform of period n. From equation (20.4) it evidently takes $3r$ operations to compute each $\hat{c}_j$ once the $\widehat{c^{(k)}}_j$ are computed. Since there are n numbers $\hat{c}_j$ to compute and r sequences $c_j^{(k)}$, it follows that

$$Op(n) = 3nr + rOp(s). \tag{20.5}$$

If r is prime but s is not, then the same procedure can be repeated on each $c_j^{(k)}$ to further reduce the amount of arithmetic. A simple inductive argument based on (20.5) that splits off one prime factor at a time yields

$$Op(n) = 3n(p_1 + \cdots + p_d),$$

where $n = p_1 \cdots p_d$ is the prime factorization of n. In particular, if $n = 2^d$, then $Op(n) = 6n \log_2 n$. In this case, it is noteworthy that all computations can be done in place without requiring computer storage beyond that allotted to the original vector $(c_0, \ldots, c_{n-1})^t$ [3, 9, 13, 15].

20.4 Approximation of Fourier Series Coefficients

The finite Fourier transform can furnish approximations to the Fourier coefficients of a periodic function $f(x)$. If $f(x)$ has period 1, then its kth Fourier coefficient c_k can be approximated by

$$\begin{aligned} c_k &= \int_0^1 f(x) e^{-2\pi i k x} dx \\ &\approx \frac{1}{n} \sum_{j=0}^{n-1} f\Big(\frac{j}{n}\Big) e^{-2\pi i \frac{jk}{n}} \\ &= \hat{b}_k, \end{aligned}$$

where $b_j = f(j/n)$ and n is some large, positive integer. Because the transformed values $\hat{b}_k$ are periodic, only n of them are distinct, say $\hat{b}_{-n/2}$ through $\hat{b}_{n/2-1}$ for n even.

An important question is how well $\hat{b}_k$ approximates c_k. To assess the error, suppose that $\sum_k |c_k| < \infty$ and that the Fourier series of $f(x)$ converges to $f(x)$ at the points j/n for $j = 0, \ldots, n-1$. The calculation

$$\begin{aligned}
\hat{b}_k &= \frac{1}{n}\sum_{j=0}^{n-1} u_n^{-jk} f\Big(\frac{j}{n}\Big) \\
&= \frac{1}{n}\sum_{j=0}^{n-1} u_n^{-jk} \sum_m c_m u_n^{jm} \\
&= \sum_m c_m \frac{1}{n}\sum_{j=0}^{n-1} u_n^{j(m-k)} \\
&= \sum_m c_m \begin{cases} 1 & m = k \bmod n \\ 0 & m \neq k \bmod n \end{cases}
\end{aligned}$$

implies that

$$\begin{aligned}
\hat{b}_k - c_k &= \sum_{l \neq 0} c_{ln+k} \qquad (20.6) \\
&= \cdots + c_{-2n+k} + c_{-n+k} + c_{n+k} + c_{2n+k} + \cdots.
\end{aligned}$$

If the Fourier coefficients c_j decline sufficiently rapidly to 0 as $|j|$ tends to ∞, then the error $\hat{b}_k - c_k$ will be small for $-n/2 \leq k \leq n/2 - 1$. Problems (6), (7), and (8) explore this question in more depth.

Example 20.4.1 *Number of Particles in a Branching Process*

In a branching process the probability that there are k particles at generation j is given by the coefficient p_{jk} of s^k in the probability generating function $P_j(s) = \sum_{k=0}^{\infty} p_{jk} s^k$ [10]. The generating function $P_j(s)$ is calculated from an initial progeny generating function $P_1(s) = P(s) = \sum_{k=0}^{\infty} p_k s^k$ by taking its j-fold functional composition

$$P_j(s) = \overbrace{P(P(\cdots(P(}^{j \text{ times}}s))\cdots)). \qquad (20.7)$$

The progeny generating function $P(s)$ summarizes the distribution of the number of progeny left at generation 1 by the single ancestral particle at generation 0. In general, it is impossible to give explicit expressions for the p_{jk}. However, these can be easily computed numerically by the finite Fourier transform. If we extend $P_j(s)$ to the unit circle by the formula

$$P_j(e^{2\pi i t}) = \sum_{k=0}^{\infty} p_{jk} e^{2\pi i k t},$$

then we can view the p_{jk} as Fourier series coefficients and recover them as discussed above. Fortunately, evaluation of $P_j(e^{2\pi it})$ at the points $t = m/n$, $m = 0, \ldots, n-1$, is straightforward under the functional composition rule (20.7). As a special case, consider $P(s) = \frac{1}{2} + \frac{s^2}{2}$. Then the algebraically formidable

$$\begin{aligned} P_4(s) &= \frac{24,305}{32,768} + \frac{445}{4,096}s^2 + \frac{723}{8,192}s^4 + \frac{159}{4,096}s^6 + \frac{267}{16,384}s^8 \\ &\quad + \frac{19}{4,096}s^{10} + \frac{11}{8,192}s^{12} + \frac{1}{4,096}s^{14} + \frac{1}{32,768}s^{16} \end{aligned}$$

can be derived by symbolic algebra programs such as Maple. Alternatively, the finite Fourier transform approximation

$$\begin{aligned} P_4(s) &= 0.74172974 + 0.10864258s^2 + 0.08825684s^4 + 0.03881836s^6 \\ &\quad + 0.01629639s^8 + 0.00463867s^{10} + 0.00134277s^{12} \\ &\quad + 0.00024414s^{14} + 0.00003052s^{16} \end{aligned}$$

is trivial to compute and is exact up to machine precision if we take the period $n > 16$. ■

Example 20.4.2 *Differentiation of an Analytic Function*

If the function $f(x)$ has a power series expansion $\sum_{j=0}^{\infty} a_j x^j$ converging in a disc $\{x : |x| < r\}$ centered at 0 in the complex plane, then we can approximate the derivatives $f^{(j)}(0) = j!a_j$ by evaluating $f(x)$ on the boundary of a small circle of radius $h < r$. This is accomplished by noting that the periodic function $t \to f(he^{2\pi it})$ has Fourier series expansion

$$f(he^{2\pi it}) = \sum_{j=0}^{\infty} a_j h^j e^{2\pi ijt}.$$

Thus, if we take the finite Fourier transform $\hat{b}_k$ of the sequence $b_j = f(hu_n^j)$, equation (20.6) mutates into

$$\hat{b}_k - a_k h^k = \sum_{l=1}^{\infty} a_{ln+k} h^{ln+k} = O\Big(h^{n+k}\Big)$$

for $0 \le k \le n-1$ under fairly mild conditions on the coefficients a_j. Rearranging this equation gives the derivative approximation

$$f^{(k)}(0) = \frac{k!\hat{b}_k}{h^k} + O\Big(h^n\Big) \qquad (20.8)$$

highlighted in [8].

The two special cases

$$
\begin{aligned}
f'(0) &= \frac{1}{2h}\Big[f(h) - f(-h)\Big] + O\Big(h^2\Big) \\
f''(0) &= \frac{1}{2h^2}\Big[f(h) - f(ih) + f(-h) - f(-ih)\Big] + O\Big(h^4\Big)
\end{aligned}
$$

of equation (20.8) when $n = 2$ and $n = 4$, respectively, deserve special mention. If h is too small, subtractive cancellation causes roundoff error in both of these equations. For a real analytic function $f(x)$, there is an elegant variation of the central difference approximation $f'(0) \approx \frac{1}{2h}\Big[f(h) - f(-h)\Big]$ that eliminates roundoff error. To derive this improved approximation, we define $g(x) = f(ix)$ and exploit the fact that $g'(0) = if'(0)$. Because the coefficients a_j are real, the identity $f(-ih) = f(ih)^*$ holds and allows us to deduce that

$$
\begin{aligned}
f'(0) &= \frac{1}{i}g'(0) \\
&= \frac{1}{2ih}\Big[g(h) - g(-h)\Big] + O\Big(h^2\Big) \\
&= \frac{1}{2ih}\Big[f(ih) - f(-ih)\Big] + O\Big(h^2\Big) \\
&= \frac{1}{2ih}\Big[f(ih) - f(ih)^*\Big] + O\Big(h^2\Big) \\
&= \frac{1}{h}\mathrm{Im} f(ih) + O\Big(h^2\Big).
\end{aligned}
$$

The approximation $f'(0) \approx \frac{1}{h}\mathrm{Im} f(ih)$ not only eliminates the roundoff error jeopardizing the central difference approximation, but it also requires one less function evaluation. Of course, the latter advantage is partially offset by the necessity of using complex arithmetic.

TABLE 20.1. Numerical Derivatives of $f(x) = e^x/(\sin^3 x + \cos^3 x)$

h	$\frac{1}{2h}[f(x+h) - f(x-h)]$	$\frac{1}{h}\mathrm{Im} f(x+ih)$
10^{-2}	3.62298	3.62109
10^{-3}	3.62229	3.62202
10^{-4}	3.62158	3.62203
10^{-5}	3.60012	3.62203
10^{-6}	3.57628	3.62203
10^{-7}	4.76837	3.62203
10^{-8}	0.00000	3.62203
10^{-9}	0.00000	3.62203

As an example, consider the problem of differentiating the analytic function $f(x) = e^x/(\sin^3 x + \cos^3 x)$ at $x = 1.5$. Table 20.1 reproduces a single precision numerical experiment from reference [14] and shows the lethal effects of roundoff in the central difference formula. The formula $f'(x) \approx \frac{1}{h}\mathrm{Im} f(x+ih)$ approximates the true value $f'(1.5) = 3.62203$ extremely well and is stable even for small values of h. This stability makes it possible to circumvent the delicate question of finding the right h to balance truncation and roundoff errors. ■

20.5 Convolution

Proposition 20.2.1 suggests a fast method of computing the convolution of two sequences c_j and d_j of period n; namely, compute the transforms $\hat{c}_k$ and $\hat{d}_k$ via the fast Fourier transform, multiply pointwise to form the product transform $n\hat{c}_k\hat{d}_k$, and then invert the product transform via the fast inverse Fourier transform. This procedure requires on the order of $O(n \ln n)$ operations, whereas the naive evaluation of a convolution requires on the order of n^2 operations unless one of the sequences consists mostly of zeros. Here are some examples where fast convolution is useful.

Example 20.5.1 *Repeated Differencing*

The classical finite difference $\Delta c_j = c_{j+1} - c_j$ corresponds to convolution against the sequence

$$d_j = \begin{cases} 1 & j = -1 \bmod n \\ -1 & j = 0 \bmod n \\ 0 & \text{otherwise.} \end{cases}$$

Hence, the sequence $\Delta^r c_j$ is sent into the sequence $(u_n^k - 1)^r \hat{c}_k$ under the finite Fourier transform. ■

Example 20.5.2 *Data Smoothing*

In many statistical applications, observations $x_0, \ldots, x_{n-1}$ are smoothed by a linear filter w_j. Smoothing creates a new sequence y_j according to the recipe

$$y_j = w_r x_{j-r} + w_{r-1} x_{j-r+1} + \cdots + w_{-r+1} x_{j+r-1} + w_{-r} x_{j+r}.$$

For instance, $y_j = \frac{1}{3}(x_{j-1} + x_j + x_{j+1})$ replaces x_j by a moving average of x_j and its two nearest neighbors. For the convolution paradigm to make sense, we must extend x_j and w_j to be periodic sequences of period n and pad w_j with zeros so that $w_{r+1} = \cdots = w_{n-r-1} = 0$. In many situations it is natural to require the weights to satisfy $w_j \geq 0$ and $\sum_{j=-r}^{r} w_j = 1$. Problem 3 provides the finite Fourier transforms of two popular smoothing sequences. ■

Example 20.5.3 *Multiplication of Generating Functions*

One can write the generating function $R(s)$ of the sum $X+Y$ of two independent, nonnegative, integer-valued random variables X and Y as the product $R(s) = P(s)Q(s)$ of the generating function $P(s) = \sum_{j=0}^{\infty} p_j s^j$ of X and the generating function $Q(s) = \sum_{j=0}^{\infty} q_j s^j$ of Y. The coefficients of $R(s)$ are given by the convolution formula

$$r_k = \sum_{j=0}^{k} p_j q_{k-j}.$$

Assuming that the p_j and q_j are 0 or negligible for $j \geq m$, we can view the two sequences as having period $n = 2m$ provided we set $p_j = q_j = 0$ for $j = m, \ldots, n-1$. Introducing these extra zeros makes it possible to write

$$r_k = \sum_{j=0}^{n-1} p_j q_{k-j} \tag{20.9}$$

without embarrassment. The r_j returned by the suggested procedure are correct in the range $0 \leq j \leq m-1$. Clearly, the same process works if $P(s)$ and $Q(s)$ are arbitrary polynomials of degree $m-1$ or less. ■

Example 20.5.4 *Multiplication of Large Integers*

If p and q are large integers, then we can express them in base b as

$$\begin{aligned} p &= p_0 + p_1 b + \cdots + p_{m-1} b^{m-1} \\ q &= q_0 + q_1 b + \cdots + q_{m-1} b^{m-1}, \end{aligned}$$

where each $0 \leq p_j \leq b-1$ and $0 \leq q_j \leq b-1$. We can represent the product $r = pq$ as $r = \sum_{k=0}^{n-1} r_k b^k$ with the r_k given by equation (20.9) and $n = 2m$. Although a given r_k may not satisfy the constraint $r_k \leq b-1$, once we replace it by its representation in base b and add and carry appropriately, we quickly recover the base b representation of r. For very large integers, computing r via the fast Fourier transform represents a large savings. ■

Example 20.5.5 *Fast Solution of a Renewal Equation*

The discrete renewal equation

$$u_n = a_n + \sum_{m=0}^{n} f_m u_{n-m} \tag{20.10}$$

arises in many applications of probability theory [7]. Here f_n is a known discrete probability density with $f_0 = 0$, and a_n is a known sequence with partial sums converging absolutely to $\sum_{n=0}^{\infty} a_n = a$. Beginning with the initial value $u_0 = a_0$, it takes on the order of n^2 operations to compute $u_0, \ldots, u_n$ recursively via the convolution equation (20.10).

If we multiply both sides of (20.10) by s^n and sum on n, then we get the equation

$$U(s) = A(s) + F(s)U(s) \tag{20.11}$$

involving the generating functions

$$U(s) = \sum_{n=0}^{\infty} u_n s^n, \quad A(s) = \sum_{n=0}^{\infty} a_n s^n, \quad F(s) = \sum_{n=0}^{\infty} f_n s^n.$$

The solution

$$U(s) = \frac{A(s)}{1 - F(s)}$$

of equation (20.11) has a singularity at $s = 1$. This phenomenon is merely a reflection of the fact that the u_n do not tend to 0 as n tends to ∞. Indeed, under a mild hypothesis on the coefficients f_n, one can show that $\lim_{n\to\infty} u_n = a/\mu$, where $\mu = \sum_{n=0}^{\infty} nf_n$ [7]. The required hypothesis on the f_n says that the set $\{n\colon f_n > 0\}$ has greatest common divisor 1. Equivalently, the only complex number s satisfying both $F(s) = 1$ and $|s| = 1$ is $s = 1$. (See Problem 12.)

These observations suggest that it would be better to estimate the coefficients $v_n = u_n - a/\mu$ of the generating function

$$\begin{aligned} V(s) &= U(s) - \frac{a}{\mu(1-s)} \\ &= \frac{A(s)\mu(1-s) - a[1-F(s)]}{[1-F(s)]\mu(1-s)}. \end{aligned}$$

A double application of l'Hôpital's rule implies that

$$\lim_{s\to 1} V(s) = \frac{aF''(1)}{2\mu^2} - \frac{A'(1)}{\mu}.$$

In other words, we have removed the singularity of $U(s)$ in forming $V(s)$. Provided $F(s)$ satisfies the greatest common divisor hypothesis, we can now recover the coefficients v_n by the approximate Fourier series method of Section 20.4. The advantage of this oblique attack on the problem is that it takes on the order of only $n \ln n$ operations to compute $u_0, \ldots, u_{n-1}$.

As a concrete illustration of the proposed method, consider the classical problem of computing the probability u_n of a new run of r heads ending at trial n in a sequence of coin-tossing trials. If p and $q = 1-p$ are the head and tail probabilities per trial, respectively, then in this case the appropriate renewal equation has $A(s) = 1$ and

$$F(s) = \frac{p^r s^r (1 - ps)}{1 - s + qp^r s^{r+1}}. \tag{20.12}$$

TABLE 20.2. Renewal Probabilities in a Coin Tossing Example

n	u_n	n	u_n	n	u_n
0	1.0000	5	0.1563	10	0.1670
1	0.0000	6	0.1719	11	0.1665
2	0.2500	7	0.1641	12	0.1667
3	0.1250	8	0.1680	13	0.1666
4	0.1875	9	0.1660	∞	0.1667

(See reference [7] or Problem 13.) A brief but tedious calculation shows that $F(s)$ has mean and variance

$$\mu = \frac{1-p^r}{qp^r}, \quad \sigma^2 = \frac{1}{(qp^r)^2} - \frac{2r+1}{qp^r} - \frac{p}{q^2},$$

which may be combined to give $F''(1) = \sigma^2 + \mu^2 - \mu$. Fourier transforming $n = 32$ values of $V(s)$ on the boundary of the unit circle when $r = 2$ and $p = 1/2$ yields the renewal probabilities displayed in Table 20.2. In this example, convergence to the limiting value occurs so rapidly that the value of introducing the finite Fourier transform is debatable. Other renewal equations exhibit less rapid convergence. ■

20.6 Time Series

The canonical example of a time series is a stationary sequence $Z_0, Z_1, \ldots$ of real, square-integrable random variables. The sample average $\frac{1}{n}\sum_{j=0}^{n-1} Z_j$ over the n data points collected is the natural estimator of the common theoretical mean μ of the Z_j. Of considerably more interest is the autocovariance sequence

$$c_k = \text{Cov}(Z_j, Z_{j+k}) = c_{-k}.$$

Since we can subtract from each Z_j the sample mean, let us assume that each Z_j has mean 0. Given this simplification, the natural estimator of c_k is

$$d_k = \frac{1}{n}\sum_{j=0}^{n-k-1} Z_j Z_{j+k}.$$

If the finite sequence $Z_0, \ldots, Z_{n-1}$ is padded with n extra zeros and extended to a periodic sequence e_j of period $2n$, then

$$d_k = \frac{2}{2n}\sum_{j=0}^{2n-1} e_j e_{j+k}$$

$$= \frac{2}{2n} \sum_{j=0}^{2n-1} e_{j-k} e_j.$$

According to properties (a) and (c) of Proposition 20.2.1, $d_0, \ldots, d_{n-1}$ can be quickly computed by inverting the finite Fourier transform $2|\hat{e}_k|^2$.

If the terms c_k of the autocovariance sequence decline sufficiently rapidly, then $\sum_k |c_k| < \infty$, and the Fourier series $\sum_k c_k e^{2\pi i k x}$ converges absolutely to a continuous function $f(x)$ called the spectral density of the time series. One of the goals of time series analysis is to estimate the periodic function $f(x)$. The periodogram

$$I_n(x) = \frac{1}{n}\left|\sum_{j=0}^{n-1} Z_j e^{-2\pi i j x}\right|^2$$

provides an asymptotically unbiased estimator of $f(x)$. Indeed, the dominated convergence theorem and the premise $\sum_k |c_k| < \infty$ together imply

$$\begin{aligned}
\lim_{n\to\infty} \mathrm{E}[I_n(x)] &= \lim_{n\to\infty} \frac{1}{n} \sum_{j=0}^{n-1}\sum_{k=0}^{n-1} \mathrm{E}(Z_j Z_k) e^{2\pi i (k-j) x} \\
&= \lim_{n\to\infty} \sum_{m=-n+1}^{n-1} \left(1 - \frac{|m|}{n}\right) c_m e^{2\pi i m x} \\
&= \sum_m c_m e^{2\pi i m x} \\
&= f(x).
\end{aligned}$$

As a by-product of this convergence proof, we see that $f(x) \geq 0$. In view of the fact that the c_k are even, Problem 6 of Chapter 17 indicates that $f(x)$ is also even around both 0 and 1/2.

Unfortunately, the sequence of periodogram estimators $I_n(x)$ is not consistent. Suppose we take two sequences l_n and m_n with $\lim_{n\to\infty} l_n/n = x$ and $\lim_{n\to\infty} m_n/n = y$. Then one can show that $\lim_{n\to\infty} \mathrm{Var}[I_n(l_n/n)]$ is proportional to $f(x)^2$ and that $\lim_{n\to\infty} \mathrm{Cov}[I_n(l_n/n), I_n(m_n/n)] = 0$ for $x \pm y \neq 0 \bmod 1$ [11]. The inconsistency of the periodogram has prompted statisticians to replace $I_n(k/n)$ by the smoothed estimator

$$\sum_{j=-r}^{r} w_j I_n\left(\frac{j+k}{n}\right)$$

with positive weights w_j satisfying $w_{-j} = w_j$ and $\sum_{j=-r}^{r} w_j = 1$. The smoothed periodogram decreases mean square error at the expense of increasing bias slightly. This kind of compromise occurs throughout statistics. Of course, the value of the fast Fourier transform in computing the finite Fourier transforms $\frac{1}{n}\sum_{j=0}^{n-1} Z_j e^{-2\pi i j k/n}$ and smoothing the periodogram should be obvious. Here, as elsewhere, speed of computation dictates much of statistical practice.

20.7 Problems

1. Explicitly calculate the finite Fourier transforms of the four sequences $c_j = 1$, $c_j = 1_{\{0\}}$, $c_j = (-1)^j$, and $c_j = 1_{\{0,1,\ldots,n/2-1\}}$ defined on $\{0, 1, \ldots, n-1\}$. For the last two sequences assume that n is even.

2. Show that the sequence $c_j = j$ on $\{0, 1, \ldots, n-1\}$ has finite Fourier transform

$$\hat{c}_k = \begin{cases} \frac{n-1}{2} & k = 0 \\ -\frac{1}{2} + \frac{i}{2}\cot\frac{k\pi}{n} & k \neq 0. \end{cases}$$

3. For $0 \leq r < n/2$, define the rectangular and triangular smoothing sequences

$$\begin{aligned} c_j &= \frac{1}{2r+1} 1_{\{-r \leq j \leq r\}} \\ d_j &= \frac{1}{r} 1_{\{-r \leq j \leq r\}} \left(1 - \frac{|j|}{r}\right) \end{aligned}$$

and extend them to have period n. Show that

$$\begin{aligned} \hat{c}_k &= \frac{1}{n(2r+1)} \frac{\sin\frac{(2r+1)k\pi}{n}}{\sin\frac{k\pi}{n}} \\ \hat{d}_k &= \frac{1}{nr^2}\left(\frac{\sin\frac{rk\pi}{n}}{\sin\frac{k\pi}{n}}\right)^2. \end{aligned}$$

4. Prove parts (a) through (c) of Proposition 20.2.1.

5. From a periodic sequence c_k with period n, form the circulant matrix

$$C = \begin{pmatrix} c_0 & c_{n-1} & c_{n-2} & \cdots & c_1 \\ c_1 & c_0 & c_{n-1} & \cdots & c_2 \\ \vdots & \vdots & \vdots & & \vdots \\ c_{n-1} & c_{n-2} & c_{n-3} & \cdots & c_0 \end{pmatrix}.$$

For $u_n = e^{2\pi i/n}$ and m satisfying $0 \leq m \leq n-1$, show that the vector $(u_n^{0m}, u_n^{1m}, \ldots, u_n^{(n-1)m})^t$ is an eigenvector of C with eigenvalue $n\hat{c}_m$. From this fact deduce that the circulant matrix C can be written in the diagonal form $C = UDU^*$, where D is the diagonal matrix with kth diagonal entry $n\hat{c}_{k-1}$, U is the unitary matrix with entry $u_n^{(j-1)(k-1)}/\sqrt{n}$ in row j and column k, and U^* is the conjugate transpose of U.

6. For $0 \leq m \leq n-1$ and a periodic function $f(x)$ on [0,1], define the sequence $b_m = f(m/n)$. If $\hat{b}_k$ is the finite Fourier transform of the

sequence b_m, then we can approximate $f(x)$ by $\sum_{k=-\lfloor n/2\rfloor}^{\lfloor n/2\rfloor} \hat{b}_k e^{2\pi ikx}$. Show that this approximation is exact when $f(x)$ is equal to $e^{2\pi ijx}$, $\cos(2\pi jx)$, or $\sin(2\pi jx)$ for j satisfying $0 \le |j| < \lfloor n/2\rfloor$.

7. Continuing Problem 6, let c_k be the kth Fourier series coefficient of a general periodic function $f(x)$. If $|c_k| \le ar^{|k|}$ for constants $a \ge 0$ and $0 \le r < 1$, then verify using equation (20.6) that

$$|\hat{b}_k - c_k| \quad \le \quad ar^n \frac{r^k + r^{-k}}{1 - r^n}$$

for $|k| < n$. Functions analytic around 0 automatically possess Fourier coefficients satisfying the bound $|c_k| \le ar^{|k|}$.

8. Continuing Problems 6 and 7, suppose a constant $a \ge 0$ and positive integer p exist such that

$$|c_k| \quad \le \quad \frac{a}{|k|^{p+1}}$$

for all $k \neq 0$. (As Problem 2 of Chapter 17 shows, this criterion holds if $f^{(p+1)}(x)$ is piecewise continuous.) Verify the inequality

$$|\hat{b}_k - c_k| \quad \le \quad \frac{a}{n^{p+1}} \sum_{j=1}^{\infty} \left[\frac{1}{\left(j + \frac{k}{n}\right)^{p+1}} + \frac{1}{\left(j - \frac{k}{n}\right)^{p+1}} \right]$$

when $|k| < n/2$. To simplify this inequality, demonstrate that

$$\begin{aligned} \sum_{j=1}^{\infty} \frac{1}{(j+\alpha)^{p+1}} \quad &< \quad \int_{\frac{1}{2}}^{\infty} (x+\alpha)^{-p-1} dx \\ &= \quad \frac{1}{p\left(\frac{1}{2} + \alpha\right)^p} \end{aligned}$$

for $\alpha > -1/2$. Finally, conclude that

$$|\hat{b}_k - c_k| \quad < \quad \frac{a}{pn^{p+1}} \left[\frac{1}{\left(\frac{1}{2} + \frac{k}{n}\right)^p} + \frac{1}{\left(\frac{1}{2} - \frac{k}{n}\right)^p} \right].$$

9. For a complex number c with $|c| > 1$, show that the periodic function $f(x) = (c - e^{2\pi ix})^{-1}$ has the simple Fourier series coefficients $c_k = c^{-k-1} 1_{\{k \ge 0\}}$. Argue from equation (20.6) that the finite Fourier transform approximation $\hat{b}_k$ to c_k is

$$\hat{b}_k \quad = \quad \begin{cases} c^{-k-1} \frac{1}{1-c^{-n}} & 0 \le k \le \frac{n}{2} - 1 \\ c^{-n-k-1} \frac{1}{1-c^{-n}} & -\frac{n}{2} \le k \le 0. \end{cases}$$

10. For some purposes it is preferable to have a purely real transform. If $c_1, \ldots, c_{n-1}$ is a finite sequence of real numbers, then we define its finite sine transform by

$$\hat{c}_k = \frac{2}{n}\sum_{j=1}^{n-1} c_j \sin\left(\frac{\pi k j}{n}\right).$$

Show that this transform has inverse

$$\check{d}_j = \sum_{k=1}^{n-1} d_k \sin\left(\frac{\pi k j}{n}\right).$$

(Hint: It is helpful to consider $c_1, \ldots, c_{n-1}$ as part of a sequence of period $2n$ that is odd about n.)

11. From a real sequence c_k of period $2n$ we can concoct a complex sequence of period n according to the recipe $d_k = c_{2k} + ic_{2k+1}$. Because it is quicker to take the finite Fourier transform of the sequence d_k than that of the sequence c_k, it is desirable to have a simple method of constructing $\hat{c}_k$ from $\hat{d}_k$. Show that

$$\hat{c}_k = \frac{1}{4}\left(\hat{d}_k + \hat{d}^*_{n-k}\right) - \frac{i}{4}\left(\hat{d}_k - \hat{d}^*_{n-k}\right)e^{-\frac{\pi i k}{n}}.$$

12. Let $F(s) = \sum_{n=1}^{\infty} f_n s^n$ be a probability generating function. Show that the equation $F(s) = 1$ has only the solution $s = 1$ on $|s| = 1$ if and only if the set $\{n: f_n > 0\}$ has greatest common divisor 1.

13. Let W be the waiting time until the first run of r heads in a coin-tossing experiment. If heads occur with probability p, and tails occur with probability $q = 1 - p$ per trial, then show that W has the generating function displayed in equation (20.12). (Hint: Argue that either $W = r$ or $W = k + 1 + W_k$, where $0 \leq k \leq r - 1$ is the initial number of heads and W_k is a probabilistic replica of W.)

14. Consider a power series $f(x) = \sum_{m=0}^{\infty} c_m x^m$ with radius of convergence $r > 0$. Prove that

$$\sum_{m=k \bmod n}^{\infty} c_m x^m = \frac{1}{n}\sum_{j=0}^{n-1} u_n^{-jk} f(u_n^j x)$$

for any x with $|x| < r$. As a special case, verify the identity

$$\sum_{m=k \bmod n}^{\infty} \binom{p}{m} = \frac{2^p}{n}\sum_{j=0}^{n-1} \cos\left[\frac{(p-2k)j\pi}{n}\right] \cos^p\left[\frac{j\pi}{n}\right]$$

for any positive integer p.

15. For a fixed positive integer n, we define the segmental functions ${}_n\alpha_j(x)$ of x as the finite Fourier transform coefficients

$$ {}_n\alpha_j(x) = \frac{1}{n}\sum_{k=0}^{n-1} e^{xu_n^k} u_n^{-jk}. $$

These functions generalize the hyperbolic trig functions $\cosh(x)$ and $\sinh(x)$. Prove the following assertions:

(a) ${}_n\alpha_j(x) = {}_n\alpha_{j+n}(x)$.

(b) ${}_n\alpha_j(x+y) = \sum_{k=0}^{n-1} {}_n\alpha_k(x)\,{}_n\alpha_{j-k}(y)$.

(c) ${}_n\alpha_j(x) = \sum_{k=0}^{\infty} x^{j+kn}/(j+kn)!$ for $0 \le j \le n-1$.

(d) $\frac{d}{dx}\left[{}_n\alpha_j(x)\right] = {}_n\alpha_{j-1}(x)$.

(e) Consider the differential equation $\frac{d^n}{dx^n}f(x) = kf(x)$ with initial conditions $\frac{d^j}{dx^j}f(0) = c_j$ for $0 \le j \le n-1$, where k and the c_j are constants. Show that

$$ f(x) = \sum_{j=0}^{n-1} c_j k^{-\frac{j}{n}}\,{}_n\alpha_j(k^{\frac{1}{n}}x). $$

(f) The differential equation $\frac{d^n}{dx^n}f(x) = kf(x) + g(x)$ with initial conditions $\frac{d^j}{dx^j}f(0) = c_j$ for $0 \le j \le n-1$ has solution

$$ \begin{aligned} f(x) &= \int_0^x k^{-\frac{n-1}{n}}\,{}_n\alpha_{n-1}[k^{\frac{1}{n}}(x-y)]g(y)\,dy \\ &\quad + \sum_{j=0}^{n-1} c_j k^{-\frac{j}{n}}\,{}_n\alpha_j(k^{\frac{1}{n}}x). \end{aligned} $$

(g) $\lim_{x\to\infty} e^{-x}\,{}_n\alpha_j(x) = 1/n$.

(h) In a Poisson process of intensity 1, $e^{-x}\,{}_n\alpha_j(x)$ is the probability that the number of random points on $[0,x]$ equals j modulo n.

(i) Relative to this Poisson process, let N_x count every nth random point on $[0,x]$. Then N_x has probability generating function

$$ P(s) = e^{-x}\sum_{j=0}^{n-1} s^{-\frac{j}{n}}\,{}_n\alpha_j(s^{\frac{1}{n}}x). $$

(j) Furthermore, N_x has mean

$$ \mathrm{E}(N_x) = \frac{x}{n} - \frac{e^{-x}}{n}\sum_{j=0}^{n-1} j\,{}_n\alpha_j(x). $$

(k) $\lim_{x\to\infty}\left[\mathrm{E}(N_x) - x/n\right] = -(n-1)/(2n)$.

20.8 References

[1] Blackman RB, Tukey JW (1959) *The Measurement of Power Spectra.* Dover, New York

[2] Bloomfield P (1976) *Fourier Analysis of Time Series: An Introduction.* Wiley, New York

[3] Brigham EO (1974) *The Fast Fourier Transform.* Prentice-Hall, Englewood Cliffs, NJ

[4] Brillinger D (1975) *Time Series: Data Analysis and Theory.* Holt, Rinehart, and Winston, New York

[5] Cooley JW, Lewis PAW, Welch PD (1969) The finite Fourier transform. *IEEE Trans Audio Electroacoustics* AU-17:77–85

[6] Dym H, McKean HP (1972) *Fourier Series and Integrals.* Academic Press, New York

[7] Feller W (1968) *An Introduction to Probability Theory and Its Applications, Volume 1*, 3rd ed. Wiley, New York

[8] Henrici P (1979) Fast Fourier transform methods in computational complex analysis. *SIAM Review* 21:481–527

[9] Henrici P (1982) *Essentials of Numerical Analysis with Pocket Calculator Demonstrations.* Wiley, New York

[10] Karlin S, Taylor HM (1975) *A First Course in Stochastic Processes*, 2nd ed. Academic Press, New York

[11] Koopmans LH (1974) *The Spectral Analysis of Time Series.* Academic Press, New York

[12] Körner TW (1988) *Fourier Analysis.* Cambridge University Press, Cambridge

[13] Press WH, Teukolsky SA, Vetterling WT, Flannery BP (1992) *Numerical Recipes in Fortran: The Art of Scientific Computing*, 2nd ed. Cambridge University Press, Cambridge

[14] Squire W, Trapp G (1998) Using complex variables to estimate derivatives of real functions. *SIAM Review* 40:110–112

[15] Wilf HS (1986) *Algorithms and Complexity.* Prentice-Hall, Englewood Cliffs, NJ

21

Wavelets

21.1 Introduction

Wavelets are just beginning to enter statistical theory and practice [2, 5, 7, 10, 12]. The pace of discovery is still swift, and except for orthogonal wavelets, the theory has yet to mature. However, the advantages of wavelets are already obvious in application areas such as image compression. Wavelets are more localized in space than the competing sines and cosines of Fourier series. They also use fewer coefficients in representing images, and they pick up edge effects better. The secret behind these successes is the capacity of wavelets to account for image variation on many different scales.

In this chapter we develop a small fraction of the relevant theory and briefly describe applications of wavelets to density estimation and image compression. For motivational purposes, we begin with the discontinuous wavelets of Haar. These wavelets are easy to understand but have limited utility. The recent continuous wavelets of Daubechies are both more subtle and more practical. Daubechies' wavelets fortunately lend themselves to fast computation. By analogy to the fast Fourier transform, there is even a fast wavelet transform [9]. The challenge to applied mathematicians, computer scientists, engineers, and statisticians is to find new applications that exploit wavelets. The edited volume [1] and the articles [6, 3] describe some opening moves by statisticians.

21.2 Haar's Wavelets

Orthonormal bases are not unique. For example, ordinary Fourier series and the beta distribution polynomials $\phi_n^{(1,1)}$ studied in Chapter 11 both provide bases for the space $\mathrm{L}^2[0,1]$ of square-integrable functions relative to the uniform distribution on $[0,1]$. Shortly after the turn of the twentieth century, Haar introduced yet another orthonormal basis for $\mathrm{L}^2[0,1]$. His construction anticipated much of the modern development of wavelets.

We commence our discussion of Haar's contribution with the indicator function $h_0(x) = 1_{[0,1)}(x)$ of the unit interval. This function satisfies the identities $\int h_0(x)dx = \int h_0(x)^2 dx = 1$. It can also be rescaled and translated to give the indicator function $h_0(2^j x - k)$ of the interval $[k/2^j, (k+1)/2^j)$. If we want to stay within the unit interval $[0,1]$, then

K. Lange, *Numerical Analysis for Statisticians*, Statistics and Computing,
DOI 10.1007/978-1-4419-5945-4_21, © Springer Science+Business Media, LLC 2010

we restrict j to be a nonnegative integer and k to be an integer between 0 and $2^j - 1$. If we prefer to range over the whole real line, then k can be any integer. For the sake of simplicity, let us focus on $[0, 1]$. Since step functions are dense in $L^2[0, 1]$, we can approximate any square-integrable function by a linear combination of the $h_0(2^j x - k)$. Within a fixed level j, two different translates $h_0(2^j x - k)$ and $h_0(2^j x - l)$ are orthogonal, but across levels orthogonality fails. Thus, the normalized functions $2^{j/2} h_0(2^j x - k)$ do not provide an orthonormal basis.

Haar turned the scaling identity

$$h_0(x) = h_0(2x) + h_0(2x - 1) \tag{21.1}$$

around to construct a second function

$$w(x) = h_0(2x) - h_0(2x - 1), \tag{21.2}$$

which is 1 on $[0, 1/2)$ and -1 on $[1/2, 1)$. In modern terminology, $h_0(x)$ is called the scaling function and $w(x)$ the mother wavelet. We subject $w(x)$ to dilation and translation and construct a sequence of functions

$$h_n(x) = 2^{\frac{j}{2}} w(2^j x - k)$$

to supplement $h_0(x)$. Here $n > 0$ and j and k are uniquely determined by writing $n = 2^j + k$ subject to the constraint $0 \leq k < 2^j$. As with the corresponding dilated and translated version of $h_0(x)$, the function $h_n(x)$ has support on the interval $[k/2^j, (k+1)/2^j) \subset [0, 1)$. We claim that the sequence $\{h_n(x)\}_{n=0}^{\infty}$ constitutes an orthonormal basis of $L^2[0, 1]$.

To prove the claim, first note that

$$\begin{aligned}
\int_0^1 h_n^2(x)dx &= \int_0^1 \left[2^{\frac{j}{2}} w(2^j x - k)\right]^2 dx \\
&= \int_0^1 w(y)^2 dy \\
&= 1.
\end{aligned}$$

Second, observe that

$$\begin{aligned}
\int_0^1 h_0(x) h_n(x)dx &= \int_0^1 h_n(x)dx \\
&= 0
\end{aligned}$$

for any $n \geq 1$ because of the balancing positive and negative parts of $h_n(x)$. If $0 < m = 2^r + s < n$ for $0 \leq s < 2^r$, then

$$\begin{aligned}
\int_0^1 h_m(x) h_n(x)dx &= 2^{\frac{r}{2}} 2^{\frac{j}{2}} \int_0^1 w(2^r x - s) w(2^j x - k)dx \\
&= 2^{\frac{j-r}{2}} \int w(y - s) w(2^{j-r} y - k)dy. \qquad (21.3)
\end{aligned}$$

If $r = j$ in the integral (21.3), then the support $[k, k+1)$ of the right integrand is disjoint from the support $[s, s+1)$ of the left integrand, and the integral is trivially 0. If $r < j$, then the support $[k/2^{j-r}, (k+1)/2^{j-r})$ of the right integrand is disjoint from the interval $[s, s+1)$ or wholly contained within either $[s, s+1/2)$ or $[s+1/2, s+1)$. If the two supports are disjoint, then again the integral is trivially 0. If they intersect, then the positive and negative contributions of the integral exactly cancel. This proves that the Haar functions $\{h_n(x)\}_{n=0}^{\infty}$ form an orthonormal sequence.

To verify completeness, it suffices to show that the indicator function $h_0(2^j x - k)$ of an arbitrary dyadic interval $[k/2^j, (k+1)/2^j) \subset [0,1)$ can be written as a finite linear combination $\sum_n c_n h_n(x)$. For example,

$$\begin{aligned} 1_{[0,\frac{1}{2})}(x) &= h_0(2x) = \frac{1}{2}[h_0(x) + w(x)] \\ 1_{[\frac{1}{2},1)}(x) &= h_0(2x-1) = \frac{1}{2}[h_0(x) - w(x)] \end{aligned}$$

are immediate consequences of equations (21.1) and (21.2). The general case follows by induction from the analogous identities

$$\begin{aligned} h_0(2^j x - 2k) &= \frac{1}{2}[h_0(2^{j-1}x - k) + w(2^{j-1}x - k)] \\ h_0(2^j x - 2k - 1) &= \frac{1}{2}[h_0(2^{j-1}x - k) - w(2^{j-1}x - k)]. \end{aligned}$$

Obvious extensions of the above arguments show that we can construct an orthonormal basis for $L^2(-\infty, \infty)$ from the functions $h_0(x-k)$ and $2^{j/2}w(2^j x - k)$, where j ranges over the nonnegative integers and k over all integers. In this basis, it is always possible to express the indicator function $h_0(2^r x - s)$ of an interval $[s/2^r, (s+1)/2^r)$ as a finite linear combination of the $h_0(x-k)$ and the $2^{j/2}w(2^j x - k)$ for $0 \le j < r$.

21.3 Histogram Estimators

One application of the Haar functions is in estimating the common density function $f(x)$ of an i.i.d. sequence $X_1, \ldots, X_n$ of random variables. For j large and fixed, we can approximate $f(x)$ accurately in $L^2(-\infty, \infty)$ by a linear combination of the orthonormal functions $g_k(x) = 2^{j/2}h_0(2^j x - k)$. The best choice of the coefficient c_k in the approximate expansion

$$f(x) \approx \sum_k c_k g_k(x)$$

is $c_k = \int g_k(z) f(z) dz = \mathrm{E}[g_k(X_1)]$. This suggests that we replace the expectation c_k by the sample average

$$\bar{c}_k = \frac{1}{n} \sum_{i=1}^{n} g_k(X_i).$$

The resulting estimator $\sum_k \bar{c}_k g_k(x)$ of $f(x)$ is called a histogram estimator.

If we let

$$a_{jk} = 2^j \int_{\frac{k}{2^j}}^{\frac{k+1}{2^j}} f(z)dz,$$

then we can express the expectation of the histogram estimator as

$$\begin{aligned} \mathrm{E}\Big[\sum_k \bar{c}_k g_k(x)\Big] &= \sum_k c_k g_k(x) \\ &= \sum_k a_{jk} 1_{[\frac{k}{2^j}, \frac{k+1}{2^j})}(x). \end{aligned}$$

If $f(x)$ is continuous, this sum tends to $f(x)$ as j tends to ∞. Since $g_k(x)g_l(x) = 0$ for $k \neq l$, the variance of the histogram estimator amounts to

$$\begin{aligned} \frac{1}{n}\mathrm{Var}\Big[\sum_k g_k(X_1)g_k(x)\Big] &= \frac{1}{n}\sum_k \mathrm{Var}\left[g_k(X_1)\right] g_k(x)^2 \\ &= \frac{2^j}{n}\sum_k a_{jk}\Big(1 - \frac{1}{2^j}a_{jk}\Big)1_{[\frac{k}{2^j}, \frac{k+1}{2^j})}(x), \end{aligned}$$

which is small when the ratio $2^j/n$ is small. We can minimize the mean square error of the histogram estimator by taking some intermediate value for j and balancing bias against variance. Clearly, this general procedure extends to density estimators based on other orthonormal sequences [12].

21.4 Daubechies' Wavelets

Our point of departure in developing Daubechies' lovely generalization of the Haar functions is the scaling equation [2, 8, 11]

$$\psi(x) = \sum_{k=0}^{n-1} c_k \psi(2x - k). \tag{21.4}$$

When $n = 2$ and $c_0 = c_1 = 1$, the indicator function of the unit interval solves equation (21.4); this solution generates the Haar functions. Now we look for a continuous solution of (21.4) that leads to an orthogonal wavelet sequence on the real line instead of the unit interval. For the sake of simplicity, we limit our search to the special value $n = 4$. In fact, there exists a solution to (21.4) for every even $n > 0$. These higher-order Daubechies' scaling functions generate progressively smoother and less localized wavelet sequences.

In addition to continuity, we require that the scaling function $\psi(x)$ have bounded support. If the support of $\psi(x)$ is the interval $[a, b]$, then the support of $\psi(2x-k)$ is $[(a+k)/2, (b+k)/2]$. Thus, the right-hand side of equation (21.4) implies that $\psi(x)$ has support between $a/2$ and $(b+n-1)/2$. Equating these to a and b yields $a = 0$ and $b = n-1$. Because $\psi(x)$ vanishes outside $[0, n-1]$, continuity dictates that $\psi(0) = 0$ and $\psi(n-1) = 0$. Therefore, when $n = 4$, the only integers k permitting $\psi(k) \neq 0$ are $k = 1$ and 2. The scaling equation (21.4) determines the ratio $\psi(1)/\psi(2)$ through the eigenvector equation

$$\begin{pmatrix} \psi(1) \\ \psi(2) \end{pmatrix} = \begin{pmatrix} c_1 & c_0 \\ c_3 & c_2 \end{pmatrix} \begin{pmatrix} \psi(1) \\ \psi(2) \end{pmatrix}. \tag{21.5}$$

If we take $\psi(1) > 0$, then $\psi(1)$ and $\psi(2)$ are uniquely determined either by the convention $\int \psi(x)dx = 1$ or by the convention $\int |\psi(x)|^2 dx = 1$. As we will see later, these constraints can be simultaneously met. Once we have determined $\psi(x)$ for all integer values k, then these values determine $\psi(x)$ for all half-integer values $k/2$ through the scaling equation (21.4). The half-integer values $k/2$ determine the quarter-integer values $k/4$ and so forth. Since any real number is a limit of dyadic rationals $k/2^j$, the postulated continuity of $\psi(x)$ completely determines all values of $\psi(x)$. The scaling equation truly is a potent device.

Only certain values of the coefficients c_k are compatible with our objective of constructing an orthonormal basis for $\mathrm{L}^2(-\infty, \infty)$. To determine these values, we first note that the scaling equation (21.4) implies

$$\begin{aligned} 2\int \psi(x)dx &= 2\sum_k c_k \int \psi(2x-k)dx \\ &= \sum_k c_k \int \psi(z)dz. \end{aligned}$$

(Here and in the following, we omit limits of summation by defining $c_k = 0$ for k outside $0, \ldots, n-1$.) Assuming that $\int \psi(z)dz \neq 0$, we find

$$2 = \sum_k c_k. \tag{21.6}$$

If we impose the orthogonality constraints

$$1_{\{m=0\}} = \int \psi(x)\psi(x-m)^* dx$$

on the integer translates of the current unknown scaling function $\psi(x)$, then the scaling equation (21.4) implies

$$1_{\{m=0\}} = \int \psi(x)\psi(x-m)^* dx$$

$$\begin{aligned}
&= \sum_k \sum_l c_k c_l^* \int \psi(2x-k)\psi(2x-2m-l)^*\,dx \\
&= \frac{1}{2}\sum_k \sum_l c_k c_l^* \int \psi(z)\psi(z+k-2m-l)^*\,dz \qquad (21.7)\\
&= \frac{1}{2}\sum_k c_k c_{k-2m}^*.
\end{aligned}$$

For reasons that will soon be apparent, we now define the mother wavelet

$$w(x) = \sum_k (-1)^k c_{1-k}\psi(2x-k). \qquad (21.8)$$

In the case of the Haar functions, $w(x)$ satisfies $\int w(x)dx = 0$. In view of definition (21.8), imposing this constraint yields

$$\begin{aligned}
0 &= \sum_k (-1)^k c_{1-k} \int \psi(2x-k)dx \\
&= \frac{1}{2}\sum_k (-1)^k c_{1-k}. \qquad (21.9)
\end{aligned}$$

We can restate this result by taking the Fourier transform of equation (21.8). This gives

$$\hat{w}(y) = Q\Big(\frac{y}{2}\Big)\hat{\psi}\Big(\frac{y}{2}\Big),$$

where

$$Q(y) = \frac{1}{2}\sum_k (-1)^k c_{1-k} e^{iky}.$$

From the identity $\int w(x)dx = \hat{w}(0) = 0$, we deduce that $Q(0) = 0$. Finally, the constraint $\int xw(x)dx = \frac{d}{idy}\hat{w}(0) = 0$, which is false for the Haar mother wavelet, ensures a limited amount of symmetry in the current $w(x)$. This final constraint on the coefficients c_k amounts to

$$\begin{aligned}
0 &= \frac{d}{dy}\hat{w}(0) \\
&= \frac{1}{2}\Big[\frac{d}{dy}Q(0)\Big]\hat{\psi}(0) + \frac{1}{2}Q(0)\frac{d}{dy}\hat{\psi}(0) \qquad (21.10)\\
&= \frac{1}{2}\Big[\frac{d}{dy}Q(0)\Big]\int \psi(x)dx \\
&= \frac{1}{4}\sum_k (-1)^k ikc_{1-k}.
\end{aligned}$$

Our findings (21.6), (21.7), (21.9), and (21.10) can be summarized for $n = 4$ by the system of equations

$$\begin{aligned} c_0 + c_1 + c_2 + c_3 &= 2 \\ |c_0|^2 + |c_1|^2 + |c_2|^2 + |c_3|^2 &= 2 \\ c_0 c_2^* + c_1 c_3^* &= 0 \\ -c_0 + c_1 - c_2 + c_3 &= 0 \\ -c_0 + c_2 - 2c_3 &= 0. \end{aligned} \tag{21.11}$$

The first four of these equations are redundant and have general solution

$$\begin{pmatrix} c_0 \\ c_1 \\ c_2 \\ c_3 \end{pmatrix} = \frac{1}{c^2+1} \begin{pmatrix} c(c-1) \\ 1-c \\ c+1 \\ c(c+1) \end{pmatrix}$$

for some real constant c. The last equation determines $c = \pm 1/\sqrt{3}$. With the coefficients c_k in hand, we return to equation (21.5) and identify the eigenvector $\binom{c-1}{c+1}$ determining the ratio of $\psi(1)$ to $\psi(2)$. By virtue of the fact that the coefficients c_k are real, choosing either $\psi(1)$ or $\psi(2)$ to be real forces $\psi(x)$ to be real for all x. This in turn compels $w(x)$ to be real. It follows that we can safely omit complex conjugate signs in calculating inner products.

Figures 21.1 and 21.2 plot Daubechies' $\psi(x)$ and $w(x)$ when $n = 4$ for the choices $c = -1/\sqrt{3}$, $\psi(1) = (1+\sqrt{3})/2$, and $\psi(2) = (1-\sqrt{3})/2$, which incidentally give the correct ratio $\psi(1)/\psi(2) = (c-1)/(c+1)$. The functions $\psi(x)$ and $w(x)$ are like no other special functions of applied mathematics. Despite our inability to express $\psi(x)$ explicitly, the scaling equation offers an effective means of computing its values on the dyadic rationals. Continuity fills in the holes.

The choices $c = -1/\sqrt{3}$, $\psi(1) = (1+\sqrt{3})/2$, and $\psi(2) = (1-\sqrt{3})/2$ also yield the partition of unity property

$$\sum_l \psi(x-l) = 1 \tag{21.12}$$

at any integer x. To prove that this property extends to all real numbers, let $\mathrm{ev}(x) = \sum_m \psi(x-2m)$, and consider a half-integer x. The scaling relation (21.4), induction, and the first and fourth equations in (21.11) imply

$$\begin{aligned} \sum_l \psi(x-l) &= \sum_l \sum_k c_k \psi(2x - 2l - k) \\ &= c_0 \mathrm{ev}(2x) + c_1[1 - \mathrm{ev}(2x)] + c_2 \mathrm{ev}(2x) + c_3[1 - \mathrm{ev}(2x)] \\ &= c_1 + c_3 \\ &= 1. \end{aligned}$$

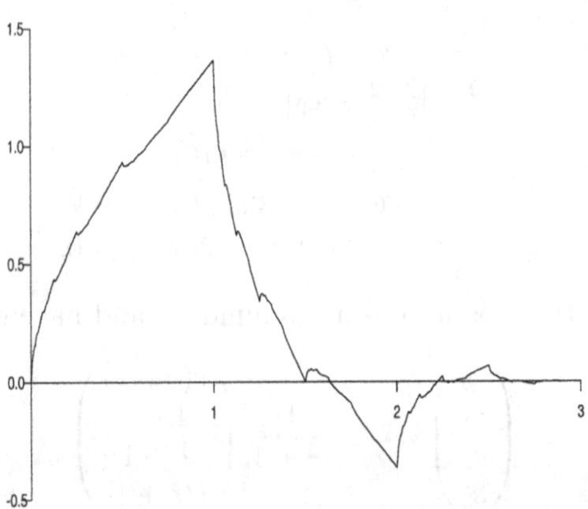

FIGURE 21.1. Plot of Daubechies' $\psi(x)$

Induction extends the partition-of-unity property (21.12) beyond half integers to all dyadic rationals, and continuity extends it from there to all real numbers.

At first glance it is not obvious that the choices $\psi(1) = (1 + \sqrt{3})/2$ and $\psi(2) = (1 - \sqrt{3})/2$ are compatible with the conventions $\int \psi(x)dx = 1$ and $\int \psi(x)^2 dx = 1$. Since $\psi(x)$ has bounded support and satisfies the partition-of-unity property, the first convention follows from

$$\begin{aligned} 1 &= \lim_{n\to\infty} \frac{1}{2n+1} \int \sum_{k=-n}^{n} \psi(x-k)dx \\ &= \int \psi(x)dx. \end{aligned}$$

Here we use the fact that $\sum_{k=-n}^{n} \psi(x-k) = 1_{[-n,n]}(x)$ except for small intervals around $-n$ and n. The orthogonality of the different $\psi(x-k)$ and the partition-of-unity property now justify the calculation

$$\begin{aligned} 1 &= \int \psi(x)dx \\ &= \int \psi(x) \sum_k \psi(x-k)dx \\ &= \int \psi(x)^2 dx. \end{aligned}$$

The scaling function $\psi(x)$ and the mother wavelet $w(x)$ together generate

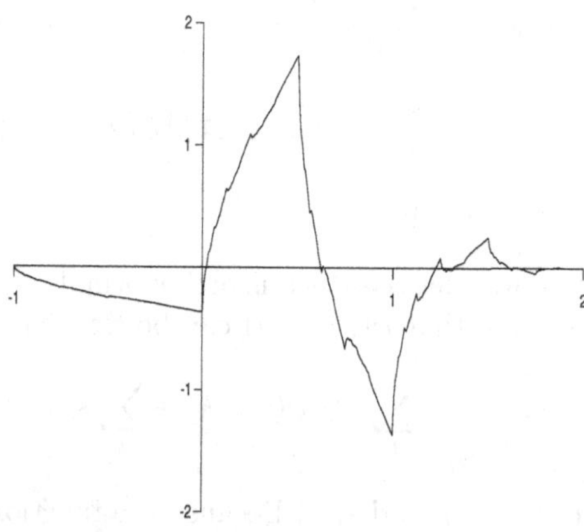

FIGURE 21.2. Plot of Daubechies' $w(x)$

a wavelet basis for $L^2(-\infty, \infty)$ consisting of all translates $\psi(x-m)$ of $\psi(x)$ plus all translated dilates $w_{jk}(x) = 2^{j/2}w(2^j x - k)$ of $w(x)$. Here m and k are arbitrary integers while j is an arbitrary nonnegative integer. The orthonormality of the translates $\psi(x-m)$ is built into the definition of $\psi(x)$. By the same reasoning that led to equation (21.7), we deduce that

$$\int w(x)\psi(x-m)dx = \frac{1}{2}\sum_k (-1)^k c_{1-k}c_{k-2m}.$$

This sum vanishes because the term $(-1)^k c_{1-k}c_{k-2m}$ exactly cancels the term $(-1)^{1-k+2m}c_{1-(1-k+2m)}c_{1-k+2m-2m}$ in which $1-k+2m$ replaces k. Now we see the purpose of the strange definition of $w(x)$. Orthonormality of the translates of $w(x)$ comes down to the constraints

$$\begin{aligned}
1_{\{m=0\}} &= \int w(x)w(x-m)dx \\
&= \frac{1}{2}\sum_k (-1)^k c_{1-k}(-1)^{k-2m}c_{1-k+2m}\int \psi(z)^2 dz \\
&= \frac{1}{2}\sum_l c_l c_{l+2m}
\end{aligned}$$

already imposed on the c_k in equation (21.7).

The coefficient $2^{j/2}$ is chosen to make $\int w_{jk}(x)^2 dx = 1$. The orthogonality conditions $\int \psi(x)w_{jk}(x)dx = 0$ and $\int w_{jk}(x)w_{lm}(x)dx = 0$ for pairs

$(j,k) \neq (l,m)$ follow by induction. For instance, induction implies

$$\begin{aligned}\int \psi(x)w_{jk}(x)dx &= \sum_l c_l \int \psi(2x-l)2^{\frac{j}{2}}w(2^j x-k)dx \\ &= \sum_l c_l \int \psi(z)2^{\frac{j}{2}-1}w(2^{j-1}z+2^{j-1}l-k)dz \\ &= 0\end{aligned}$$

for $j>0$. Thus, the wavelet sequence is orthonormal.

We next demonstrate that $\psi(2x-m)$ can be written as a finite sum

$$\psi(2x-m) = \sum_k r_{mk}\psi(x-k) + \sum_k s_{mk}w(x-k) \qquad (21.13)$$

for certain coefficients r_{mk} and s_{mk}. Because the functions on the right of the representation (21.13) are orthonormal, we calculate r_{mk} as

$$\begin{aligned}\int \psi(2x-m)\psi(x-k)dx &= \sum_j c_j \int \psi(2x-m)\psi(2x-2k-j)dx \\ &= \frac{1}{2}c_{m-2k}\end{aligned}$$

and s_{mk} as

$$\begin{aligned}&\int \psi(2x-m)w(x-k)dx \\ &= \sum_j (-1)^j c_{1-j}\int \psi(2x-m)\psi(2x-2k-j)dx \\ &= \frac{(-1)^{m-2k}}{2}c_{1-m+2k}.\end{aligned}$$

In light of the second identity in (21.11), we conclude that

$$\begin{aligned}\int \psi(2x-m)^2 dx &= \frac{1}{2} \\ &= \frac{1}{4}\sum_k c_{m-2k}^2 + \frac{1}{4}\sum_k c_{1-m+2k}^2 \\ &= \sum_k r_{mk}^2 + \sum_k s_{mk}^2.\end{aligned}$$

Bessel's equality (17.3) of Chapter 17 now yields equation (21.13). Induction and substitution of $2^{l-1}x$ for $2x$ in equation (21.13) demonstrate that every function $\psi(2^l x-m)$ can be written as a finite linear combination of the functions $\psi(x-k)$ and the $w_{jk}(x)$ with $j<l$.

Finally, we address the completeness of the orthonormal wavelet sequence. Observe first that the sequence of functions $\psi_{jk}(x) = 2^{j/2}\psi(2^j x - k)$ is orthonormal for j fixed. To prove completeness, consider the projection

$$P_j f(x) = \sum_k \psi_{jk}(x) \int f(y)\psi_{jk}(y)dy$$

of a square-integrable function $f(x)$ onto the subspace spanned by the sequence $\{\psi_{jk}(x)\}_k$. It suffices to show that these projections converge to $f(x)$ in $\mathrm{L}^2(-\infty,\infty)$. Due to Bessel's inequality (17.2) in Chapter 17, convergence in $\mathrm{L}^2(-\infty,\infty)$ is equivalent to $\lim_{j\to\infty}\|P_j f\|^2 = \|f\|^2$. Since step functions are dense in $\mathrm{L}^2(-\infty,\infty)$, we can further reduce the problem to the case where $f(x)$ is the indicator function $1_{[a,b]}(x)$ of an interval. Making the change of variables $y = 2^j x$, we calculate

$$\begin{aligned}
\|P_j 1_{[a,b]}\|^2 &= \sum_k \Big[\int 1_{[a,b]}(x) 2^{\frac{j}{2}}\psi(2^j x - k)dx\Big]^2 \\
&= \sum_k \Big[\int 1_{[2^j a, 2^j b]}(y)\psi(y-k)dy\Big]^2 2^{-j}.
\end{aligned}$$

For the vast majority of indices k when j is large, the support of $\psi(y-k)$ is wholly contained within or wholly disjoint from $[2^j a, 2^j b]$. Hence, the condition $\int \psi(y)dy = 1$ implies that

$$\begin{aligned}
\lim_{j\to\infty}\|P_j 1_{[a,b]}\|^2 &= \lim_{j\to\infty} 2^{-j}\#\{k : k \in [2^j a, 2^j b]\} \\
&= b - a \\
&= \int 1_{[a,b]}(y)^2 dy.
\end{aligned}$$

This proves completeness.

Doubtless the reader has noticed that we have never proved that $\psi(x)$ exists and is continuous for all real numbers. For the sake of brevity, we refer interested readers to Problems 10 and 11 for a sketch of one attack on these thorny questions [8]. Although continuity of $\psi(x)$ is assured, differentiability is not. It turns out that $\psi(x)$ is left differentiable, but not right differentiable, at each dyadic rational of $[0,3]$. Problem 9 makes a start on the issue of differentiability.

21.5 Multiresolution Analysis

It is now time to step back and look at the larger landscape. At the coarsest level of detail, we have a (closed) subspace V_0 of $\mathrm{L}^2(-\infty,\infty)$ spanned by the translates $\psi(x-k)$ of the scaling function. This is the first of a hierarchy of closed subspaces V_j constructed from the translates $\psi_{jk}(x)$ of the dilated functions $2^{j/2}\psi(2^j x)$. The scaling equation (21.4) tells us that $V_j \subset V_{j+1}$ for every j, and completeness tells us that $\mathrm{L}^2(-\infty,\infty)$ is the closed span of the union $\cup_{j=0}^{\infty} V_j$. This favorable state of affairs is marred by the fact that the functions $\psi_{jk}(x)$ are only orthogonal within a level j and not across levels. The remedy is to introduce the wavelets $w_{jk}(x)$. These are designed so that V_j is spanned by a basis of V_{j-1} plus the translates $w_{j-1,k}(x)$ of $w_{j-1,0}(x)$. This fact follows from the obvious generalization

$$\begin{aligned} & \psi_{jk}(x) \\ = {} & \frac{1}{\sqrt{2}}\sum_l c_{k-2l}\psi_{j-1,l}(x) + \frac{1}{\sqrt{2}}\sum_l (-1)^{k-2l} c_{1-k+2l} w_{j-1,l}(x) \end{aligned} \tag{21.14}$$

of equation (21.13). Representation (21.14) permits us to express the fine distinctions of V_j partially in terms of the coarser distinctions of V_{j-1}. The restatements

$$\begin{aligned} \psi_{jk}(x) &= \frac{1}{\sqrt{2}}\sum_l c_l \psi_{j+1,2k+l}(x) \\ w_{jk}(x) &= \frac{1}{\sqrt{2}}\sum_l (-1)^l c_{1-l}\psi_{j+1,2k+l}(x) \end{aligned} \tag{21.15}$$

of the scaling equation and the definition of $w(x)$ allow us to move in the opposite direction from coarse to fine.

21.6 Image Compression and the Fast Wavelet Transform

One of the major successes of wavelets is image compression. For the sake of simplicity, we will discuss the compression of one-dimensional images. Our remarks are immediately pertinent to acoustic recordings [13] and, with minor changes, to visual images. The fact that most images are finite in extent suggests that we should be using periodic wavelets rather than ordinary wavelets. It is possible to periodize wavelets by defining

$$\overline{w}_{jk}(x) = \sum_l w_{jk}(x-l) = \sum_l 2^{\frac{j}{2}} w(2^j x - k - l2^j) \tag{21.16}$$

for $j \geq 0$ and $0 \leq k < 2^j$. The reader is asked in Problem 12 to show that these functions of period 1 together with the constant 1 form an orthonormal basis for the space $\mathrm{L}^2[0,1]$. The constant 1 enters because $1 = \sum_l \psi(x-l)$.

To the subspace V_j in $\mathrm{L}^2(-\infty, \infty)$ corresponds the subspace $\overline{V}_j$ in $\mathrm{L}^2[0,1]$ spanned by the 2^j periodic functions

$$\overline{\psi}_{jk}(x) = \sum_l \psi_{jk}(x-l) = \sum_l 2^{\frac{j}{2}}\psi(2^j x - k - l2^j).$$

It is possible to pass between the $\overline{\psi}_{jk}(x)$ at level j and the basis functions $\overline{w}_{lm}(x)$ for $0 \le l < j$ via the analogs

$$\begin{aligned}
\overline{\psi}_{jk}(x) &= \frac{1}{\sqrt{2}}\sum_l c_{k-2l}\overline{\psi}_{j-1,l}(x) + \frac{1}{\sqrt{2}}\sum_l (-1)^{k-2l} c_{1-k+2l}\overline{w}_{j-1,l}(x) \\
\overline{\psi}_{jk}(x) &= \frac{1}{\sqrt{2}}\sum_l c_l \overline{\psi}_{j+1,2k+l}(x) \qquad (21.17) \\
\overline{w}_{jk}(x) &= \frac{1}{\sqrt{2}}\sum_l (-1)^l c_{1-l}\overline{\psi}_{j+1,2k+l}(x)
\end{aligned}$$

of equations (21.14) and (21.15).

If we start with a linear approximation

$$f(x) \approx \sum_{k=0}^{2^j-1} r_k \overline{\psi}_{jk}(x) \qquad (21.18)$$

to a given function $f(x)$ by the basis functions of $\overline{V}_j$, it is clear that the first of the recurrences in (21.17) permits us to replace this approximation by an equivalent linear approximation

$$f(x) \approx \sum_{k=0}^{2^{j-1}-1} s_k \overline{\psi}_{j-1,k}(x) + \sum_{k=0}^{2^{j-1}-1} t_k \overline{w}_{j-1,k}(x) \qquad (21.19)$$

involving the basis functions of $\overline{V}_{j-1}$. We can then substitute

$$\sum_{k=0}^{2^{j-1}-1} s_k \overline{\psi}_{j-1,k}(x) = \sum_{k=0}^{2^{j-2}-1} u_k \overline{\psi}_{j-2,k}(x) + \sum_{k=0}^{2^{j-2}-1} v_k \overline{w}_{j-2,k}(x)$$

in equation (21.19) and so forth. This recursive procedure constitutes the fast wavelet transform. It is efficient because in the case of Daubechies' wavelets, only four coefficients c_k are involved, and because only half of the basis functions must be replaced at each level.

In image compression a function $f(x)$ is observed on an interval $[a,b]$. Extending the function slightly, we can easily make it periodic. We can also arrange that $[a,b] = [0,1]$. If we choose 2^j sufficiently large, then the approximation (21.18) will be good provided each coefficient r_k satisfies $r_k = \int_0^1 f(x)\overline{\psi}_{jk}(x)dx$. We omit the practical details of how these integrations are done. Once the linear approximation (21.18) is computed, we can apply the fast wavelet transform to reduce the approximation to one involving the $\overline{w}_{l,k}(x)$ of order $l \leq j-1$ and the constant 1. Image compression is achieved by throwing away any terms in the final expansion having coefficients smaller in absolute value than some threshold $\epsilon > 0$. If we store the image as the list of remaining coefficients, then we can readily reconstruct the image by forming the appropriate linear combination of basis functions. If we want to return to the basis of $\overline{V}_j$ furnished by the $\overline{\psi}_{jk}(x)$, then the second and third recurrences of (21.17) make this possible.

As mentioned in the introduction to this chapter, the value of wavelet expansions derives from their ability to capture data at many different scales. A periodic wavelet $\overline{w}_{jk}(x)$ is quite localized if j is even moderately large. In regions of an image where there is little variation, the coefficients of the pertinent higher-order wavelets $\overline{w}_{jk}(x)$ are practically zero because $\int w(x)dx = 0$. Where edges or rapid oscillations occur, the higher-order wavelets $\overline{w}_{jk}(x)$ are retained.

21.7 Problems

1. Let $X_1, \ldots, X_n$ be a random sample from a well-behaved density $f(x)$. If $\{g_k(x)\}_{k=1}^{\infty}$ is a real, orthonormal basis for $\mathrm{L}^2(-\infty, \infty)$, then a natural estimator of $f(x)$ is furnished by

$$\begin{aligned} \bar{f}(x) &= \sum_{k=1}^{\infty} \bar{c}_k g_k(x) \\ \bar{c}_k &= \frac{1}{n} \sum_{i=1}^{n} g_k(X_i). \end{aligned}$$

Show formally that

$$\begin{aligned} \mathrm{E}[\bar{f}(x)] &= f(x) \\ \mathrm{Var}[\bar{f}(x)] &= \frac{1}{n} \sum_{k=1}^{\infty} \sum_{l=1}^{\infty} \int g_k(z) g_l(z) f(z) dz g_k(x) g_l(x) - \frac{1}{n} f(x)^2, \end{aligned}$$

provided the orthogonal expansion of $f(x)$ converges pointwise to $f(x)$.

2. Let $C_1(x)$ be the uniform density on $[0, 1)$. The cardinal B-spline $C_m(x)$ of order m is the m-fold convolution of $C_1(x)$ with itself. Prove that this function satisfies the scaling equation

$$C_m(x) = \frac{1}{2^{m-1}} \sum_{k=0}^{m} \binom{m}{k} C_m(2x - k).$$

(Hint: Show that both sides have the same Fourier transform.)

3. For the choice $c = -1/\sqrt{3}$, show that $\begin{pmatrix} c-1 \\ c+1 \end{pmatrix}$ is the eigenvector sought in equation (21.5).

4. Write software to evaluate and graph Daubechies' scaling function and mother wavelet for $n = 4$ and $c = -1/\sqrt{3}$.

5. Suppose $\psi(x)$ is a continuous function with bounded support that satisfies the scaling equation (21.4) and the condition $\int \psi(x)dx = 1$. If the coefficients satisfy $\sum_k c_k = 2$, then show that

$$2\int x\psi(x)dx = \sum_k kc_k.$$

6. Show that the Fourier transform of Daubechies' scaling function satisfies $\hat{\psi}(y) = P(y/2)\hat{\psi}(y/2)$, where $P(y) = (\sum_k c_k e^{iky})/2$. Conclude that $\hat{\psi}(y) = \prod_{k=1}^{\infty} P(y/2^k)$ holds.

7. Verify the identity

$$1 = \sum_k |\hat{\psi}(2\pi y + 2\pi k)|^2$$

satisfied by Daubechies' scaling function. (Hint: Apply the Parseval-Plancherel identity to the first line of (21.7), interpret the result as a Fourier series, and use the fact that the Fourier series of an integrable function determines the function almost everywhere.)

8. Demonstrate the identities

$$\begin{aligned} 1 &= |P(\pi y)|^2 + |P(\pi y + \pi)|^2 \\ 1 &= |Q(\pi y)|^2 + |Q(\pi y + \pi)|^2 \\ 0 &= P(\pi y)Q(\pi y)^* + P(\pi y + \pi)Q(\pi y + \pi)^* \end{aligned}$$

involving Daubechies' functions

$$\begin{aligned} P(y) &= \frac{1}{2}\sum_k c_k e^{iky} \\ Q(y) &= \frac{1}{2}\sum_k (-1)^k c_{1-k} e^{iky}. \end{aligned}$$

(Hint: For the first identity, see Problems 6 and 7.)

9. Show that Daubechies' scaling function $\psi(x)$ is left differentiable at $x = 3$ but not right differentiable at $x = 0$ when $n = 4$ and $c = -1/\sqrt{3}$. (Hint: Take difference quotients, and invoke the scaling equation (21.4).)

10. This problem and the next deal with Pollen's [8] proof of the existence of a unique continuous solution to the scaling equation (21.4). Readers are assumed to be familiar with some results from functional analysis [4]. Let $a = (1+\sqrt{3})/4$ and $\overline{a} = (1-\sqrt{3})/4$. If $f(x)$ is a function defined on $[0,3]$, then we map it to a new function $M(f)(x)$ defined on $[0,3]$ according to the piecewise formulas

$$\begin{aligned}
M(f)\Big(\frac{0+x}{2}\Big) &= af(x) \\
M(f)\Big(\frac{1+x}{2}\Big) &= \overline{a}f(x) + ax + \frac{2+\sqrt{3}}{4} \\
M(f)\Big(\frac{2+x}{2}\Big) &= af(1+x) + \overline{a}x + \frac{\sqrt{3}}{4} \\
M(f)\Big(\frac{3+x}{2}\Big) &= \overline{a}f(1+x) - ax + \frac{1}{4} \\
M(f)\Big(\frac{4+x}{2}\Big) &= af(2+x) - \overline{a}x + \frac{3-2\sqrt{3}}{4} \\
M(f)\Big(\frac{5+x}{2}\Big) &= \overline{a}f(2+x)
\end{aligned}$$

for $x \in [0,1]$. To ensure that the transformation $M(f)(x)$ is well defined at the half-integers, we postulate that $f(x)$ takes the values $f(0) = f(3) = 0$, $f(1) = 2a$, and $f(2) = 2\overline{a}$. Show first that $M(f)(x) = f(x)$ at these particular points. Now consider the functional identities

$$\begin{aligned}
2f(x) + f(1+x) &= x + \frac{1+\sqrt{3}}{2} \\
2f(2+x) + f(1+x) &= -x + \frac{3-\sqrt{3}}{2}
\end{aligned}$$

for $x \in [0,1]$. If $f(x)$ satisfies these two identities, then show that $M(f)(x)$ does as well. The set S of continuous functions $f(x)$ that have the values 0, $2a$, $2\overline{a}$, and 0 at 0, 1, 2, and 3 and that satisfy the two functional identities is nonempty. Indeed, prove that S contains the function that takes the required values and is linear between successive integers on $[0,3]$. Also show that S is a closed, convex subset of the Banach space (complete normed linear space) of continuous functions on $[0,3]$ under the norm $\|f\| = \sup_{x\in[0,3]} |f(x)|$. Given this fact, prove that $M(f)$ is a contraction mapping on S and therefore has a unique fixed point $\psi(x)$ [4]. Here it is helpful to note that $|\overline{a}| \leq |a| < 1$.

11. Continuing Problem 10, suppose we extend the continuous, fixed-point function $\psi(x)$ of the contraction map M to the entire real line by setting $\psi(x) = 0$ for $x \notin [0,3]$. Show that $\psi(x)$ satisfies the scaling

equation (21.4). (Hint: You will have to use the two functional identities imposed on S as well as the functional identities implied by the fixed-point property of M.)

12. Demonstrate that the constant 1 plus the periodic wavelets defined by equation (21.16) constitute an orthonormal basis for $L^2[0,1]$.

21.8 References

[1] Antoniadis A, Oppenheim G (1995) *Wavelets and Statistics.* Springer, New York

[2] Daubechies I (1992) *Ten Lectures on Wavelets.* SIAM, Philadelphia

[3] Donoho DL, Johnstone IM (1995) Adapting to unknown smoothness via wavelet shrinkage. *J Amer Stat Assoc* 90:1200–1224

[4] Hoffman K (1975) *Analysis in Euclidean Space.* Prentice-Hall, Englewood Cliffs, NJ

[5] Jawerth B, Sweldens W (1994) An overview of wavelet based multiresolution analysis. *SIAM Review* 36:377–412

[6] Kolaczyk ED (1996) A wavelet shrinkage approach to tomographic image reconstruction. *J Amer Stat Assoc* 91:1079–1090

[7] Meyer Y (1993) *Wavelets: Algorithms and Applications.* Ryan RD, translator, SIAM, Philadelphia

[8] Pollen D (1992) Daubechies's scaling function on [0,3]. In *Wavelets: A Tutorial in Theory and Applications,* Chui CK, editor, Academic Press, New York, pp 3–13

[9] Press WH, Teukolsky SA, Vetterling WT, Flannery BP (1992) *Numerical Recipes in Fortran: The Art of Scientific Computing,* 2nd ed. Cambridge University Press, Cambridge

[10] Strang G (1989) Wavelets and dilation equations: a brief introduction. *SIAM Review* 31:614–627

[11] Strichartz RS (1993) How to make wavelets. *Amer Math Monthly* 100:539–556

[12] Walter GG (1994) *Wavelets and Other Orthogonal Systems with Applications.* CRC Press, Boca Raton, FL

[13] Wickerhauser MV (1992) Acoustic signal compression with wavelet packets. In *Wavelets: A Tutorial in Theory and Applications,* Chui CK, editor, Academic Press, New York, pp 679–700

22

Generating Random Deviates

22.1 Introduction

Statisticians rely on a combination of mathematical theory and statistical simulation to develop new methods. Because simulations are often conducted on a massive scale, it is crucial that they be efficiently executed. In the current chapter, we investigate techniques for producing random samples from univariate and multivariate distributions. These techniques stand behind every successful simulation and play a critical role in Monte Carlo integration. Exceptionally fast code for simulations almost always depends on using a lower-level computer language such as C or Fortran. This limitation forces the statistician to write custom software. Mastering techniques for generating random variables (or deviates in this context) is accordingly a useful survival skill.

Almost all lower-level computer languages fortunately have facilities for computing a random sample from the uniform distribution on $[0, 1]$. Although there are important philosophical, mathematical, and statistical issues involved in whether and to what extent a deterministic computer can deliver independent uniform deviates [18, 25, 11], we take the relaxed attitude that this problem has been solved for all practical purposes. For the sake of completeness, the next section discusses how to implement a simple portable random number generator. Our focus in later sections is on fast methods for turning uniform deviates into more complicated random samples [4, 5, 11, 15, 16, 18, 23, 25, 26, 27]. Because statisticians must constantly strike a balance between programming costs and machine efficiency, we stress methods that are straightforward to implement.

22.2 Portable Random Number Generators

A carefully tuned multiplicative random number generator

$$I_{n+1} = mI_n \bmod p$$

can perform exceptionally well [22, 23]. Here the modulus p is a large prime number, the multiplier m is an integer between 2 and $p-1$, and I_n and I_{n+1} are two successive random numbers output by the generator. The Mersenne prime $p = 2^{31} - 1 = 214783647$ is the most widely used modulus. Note that the function $I \mapsto mI \bmod p$ maps the set $\{1, \ldots, p-1\}$ onto

K. Lange, *Numerical Analysis for Statisticians*, Statistics and Computing,
DOI 10.1007/978-1-4419-5945-4_22,

itself. Indeed, if $mI = mJ \bmod p$, then $m(I - J)$ must be a multiple of p. This is impossible unless $I - J$ is a multiple of p.

The period n of the generator is the minimal positive number such that $m^n = 1 \bmod p$. In this circumstance,

$$I_n = m^n \bmod p = I_0 \bmod p,$$

and the generator repeats itself every n steps under any seed I_0 between 1 and $p - 1$. Fermat's little theorem says that $m^{p-1} = 1 \bmod p$, so the period satisfies $n \leq p - 1$. The multipliers meeting the test $n = p - 1$ are called primitive. Although these are relatively rare, the surprising choice $m = 7^5 = 16807$ is primitive. The generator $I_{n+1} = 7^5 I_n \bmod (2^{31} - 1)$ can be easily implemented in any modern computer language with 64-bit integers. The output is reduced to a real number on (0,1) by dividing by $2^{31} - 1$. In addition to the Mersenne prime and its multiplier, the less impressive primes 30269 and 30323 and corresponding primitive multipliers 171 and 170 enjoy wide usage [30].

Some of the more sophisticated generators are constructed by adding the output of several independent generators [30]. This procedure is justified by the simple observation that the sum $U_1 + U_2 \bmod 1$ of two independent uniform deviates U_1 and U_2 on $[0, 1]$ is also uniform on $[0, 1]$. If this tactic is adopted, then the prime moduli used to generate U_1 and U_2 should be different. With a common prime modulus p and a common period $p - 1$, the sum $U_1 + U_2 \bmod 1$ also has period $p - 1$. Problem 1 asks the reader to write code implementing these ideas.

22.3 The Inverse Method

The inverse method embodied in the next proposition is one of the simplest and most natural methods of generating non-uniform random deviates [1].

Proposition 22.3.1 *Let X be a random variable with distribution function $F(x)$.*

(a) If $F(x)$ is continuous, then $U = F(X)$ is uniformly distributed on $[0, 1]$.

(b) Even if $F(x)$ is not continuous, the inequality $\Pr[F(X) \leq t] \leq t$ is still true for all $t \in [0, 1]$.

(c) If $F^{[-1]}(y) = \inf\{x : F(x) \geq y\}$ for any $0 < y < 1$, and if U is uniform on $[0, 1]$, then $F^{[-1]}(U)$ has distribution function $F(x)$.

Proof: Let us first demonstrate that

$$\Pr[F(X) \leq F(t)] = F(t). \tag{22.1}$$

To prove this assertion, note that $\{X > t\} \cap \{F(X) < F(t)\} = \emptyset$ and $\{X \leq t\} \cap \{F(X) > F(t)\} = \emptyset$ together entail

$$\{F(X) \leq F(t)\} = \{X \leq t\} \cup \{F(X) = F(t),\ X > t\}.$$

However, the event $\{F(X) = F(t),\ X > t\}$ maps under X to an interval of constancy of $F(x)$ and therefore has probability 0. Equation (22.1) follows immediately.

For part (a) let $u \in (0,1)$. Because $F(x)$ is continuous, there exists t with $F(t) = u$. In view of equation (22.1),

$$\Pr[F(X) \leq u] = \Pr[F(X) \leq F(t)] = u.$$

Part (c) follows if we can show that the events $u \leq F(t)$ and $F^{[-1]}(u) \leq t$ are identical for both u and $F(t)$ in $(0,1)$. Assume that $F^{[-1]}(u) \leq t$. Because $F(x)$ is increasing and right continuous, the set $\{x : u \leq F(x)\}$ is an interval containing its left endpoint. Hence, $u \leq F(t)$. Conversely, if $u \leq F(t)$, then $F^{[-1]}(u) \leq t$ by definition. Finally for part (b), apply part (c) and write $X = F^{[-1]}(U)$ for U uniform on $[0,1]$. Then the inequality $U \leq F(X)$ implies

$$\Pr[F(X) \leq t] \leq \Pr(U \leq t) = t.$$

This completes the proof. ■

Example 22.3.1 *Exponential Distribution*

If X is exponentially distributed with mean 1, then $F(x) = 1 - e^{-x}$, and $F^{[-1]}(u) = -\ln(1-u)$. Because $1 - U$ is uniform on $[0,1]$ when U is uniform on $[0,1]$, the random variable $-\ln U$ is distributed as X. The positive multiple $Y = -\mu \ln U$ is exponentially distributed with mean μ. ■

Example 22.3.2 *Cauchy Distribution*

The distribution function $F(x) = 1/2 + \arctan(x)/\pi$ of a standard Cauchy random variable X has inverse $F^{[-1]}(u) = \tan[\pi(u - 1/2)]$. To generate a Cauchy random variable $Y = \sigma X + \mu$ with location and scale parameters μ and σ, simply take $Y = \sigma \tan[\pi(U - 1/2)] + \mu$ for U uniform on $[0,1]$. ■

Example 22.3.3 *Probability Plots*

If $X_{(1)} < X_{(2)} < \cdots < X_{(n)}$ are the order statistics of a random sample from a continuous distribution function $F(x)$, then taking $U_i = F(X_i)$ generates the order statistics $U_{(1)} < U_{(2)} < \cdots < U_{(n)}$ of a random sample from the uniform distribution. The fact that $\mathrm{E}[U_{(i)}] = i/(n+1)$ suggests that a plot of the points $(i/[n+1], F[X_{(i)}])$ should fall approximately on a straight line. This is the motivation for the diagnostic tool of probability plotting [24]. ■

Example 22.3.4 *Discrete Uniform*

One can sample a number uniformly from the set $\{1, 2, \ldots, n\}$ by taking $\lfloor nU \rfloor + 1$, where U is uniform on $[0, 1]$ and $\lfloor r \rfloor$ denotes the greatest integer less than or equal to the real number r. ■

Example 22.3.5 *Geometric*

In a Bernoulli sampling scheme with success probability p, the number of trials N until the first success follows a geometric distribution. If we choose λ so that $q = 1 - p = e^{-\lambda}$, then N can be represented as $N = \lfloor X \rfloor + 1$, where X is exponentially distributed with intensity λ. Indeed,

$$\begin{aligned} \Pr(N = k+1) &= \Pr(k \le X < k+1) \\ &= e^{-\lambda k} - e^{-\lambda(k+1)} \\ &= q^k - q^{k+1}. \end{aligned}$$

In light of Example 22.3.1, $N = \lfloor -\ln(U)/\lambda \rfloor + 1$, where U is uniform on $[0, 1]$. For the geometric that counts total failures until success rather than total trials, we replace N by $N - 1$. ■

22.4 Normal Random Deviates

Although in principle normal random deviates can be generated by the inverse method, the two preferred methods involve substantially less computation. Both the Box and Muller [2] and the Marsaglia methods generate two independent, standard normal deviates X and Y at a time starting from two independent, uniform deviates U and V. The Box and Muller method transforms the random Cartesian coordinates (X, Y) in the plane to random polar coordinates (Θ, R). It is clear from their joint density $e^{-\frac{r^2}{2}} r/(2\pi)$ that Θ and R are independent, with Θ uniformly distributed on $[0, 2\pi]$ and R^2 exponentially distributed with mean 2. Example 22.3.1 says we can generate Θ and R^2 by taking $\Theta = 2\pi U$ and $R^2 = -2\ln V$. Transforming from polar coordinates back to Cartesian coordinates, we define $X = R\cos\Theta$ and $Y = R\sin\Theta$.

In Marsaglia's polar method, a random point (U, V) in the unit square is transformed into a random point (S, T) in the square $[-1, 1] \times [-1, 1]$ by taking $S = 2U - 1$ and $T = 2V - 1$. If $W^2 = S^2 + T^2 > 1$, then the random point (S, T) falls outside the unit circle. When this occurs, the current U and V are discarded and resampled. If $W^2 = S^2 + T^2 \le 1$, then the point (S, T) generates a uniformly distributed angle Θ with $\cos\Theta = S/W$ and $\sin\Theta = T/W$. Furthermore, the distribution of the random variable $Z = -2\ln W^2$ is

$$\Pr(Z \le z) = \Pr(W \ge e^{-\frac{z}{4}})$$

$$\begin{aligned} &= 1 - \frac{\pi(e^{-\frac{z}{4}})^2}{\pi} \\ &= 1 - e^{-\frac{z}{2}}, \end{aligned}$$

which implies that Z is distributed as R^2 in the Box and Muller method. Thus, we need only set

$$\begin{aligned} X &= \sqrt{-2\ln W^2}\frac{S}{W} \\ Y &= \sqrt{-2\ln W^2}\frac{T}{W} \end{aligned}$$

to recover the normally distributed pair (X, Y).

The polar method avoids the trigonometric function evaluations of the Box and Muller method but uses $4/\pi$ as many random pairs (U, V) on average. Both methods generate normal deviates with mean μ and variance σ^2 by replacing X and Y by $\sigma X + \mu$ and $\sigma Y + \mu$.

22.5 Acceptance-Rejection Method

The acceptance-rejection method is predicated on a notion of majorization [10] more primitive than that used in the MM algorithm. Suppose we want to sample from a complicated probability density $f(x)$ that is majorized by a simple probability density $g(x)$ in the sense that $f(x) \leq cg(x) = h(x)$ for all x and some constant $c > 1$. If we sample a deviate X distributed according to $g(x)$, then we can accept or reject X as a representative of $f(x)$. John von Neumann [29] suggested making this decision based on sampling a uniform deviate U and accepting X if and only if $U \leq f(X)/h(X)$. This procedure gives the probability of generating an accepted value in the interval $(x, x + dx)$ as proportional to

$$g(x)dx\frac{f(x)}{h(x)} = \frac{1}{c}f(x)dx.$$

In other words, the density function of the accepted deviates is precisely $f(x)$. The fraction of sampled deviates accepted is $1/c$. The density $g(x)$ is called the instrumental density. The first stage of the polar method provides a simple geometric example of the acceptance-rejection method.

As we have seen in Example 22.3.1, generating exponential deviates is computationally quick. This fact suggests exploiting exponential curves as majorizing functions in the acceptance-rejection method [5]. On a log scale an exponential curve is a straight line. If a density $f(x)$ is log-concave, then any line tangent to $\ln f(x)$ will lie above $\ln f(x)$. Thus, log-concave densities are ideally suited to acceptance-rejection sampling with piecewise exponential envelopes. Commonly encountered log-concave densities

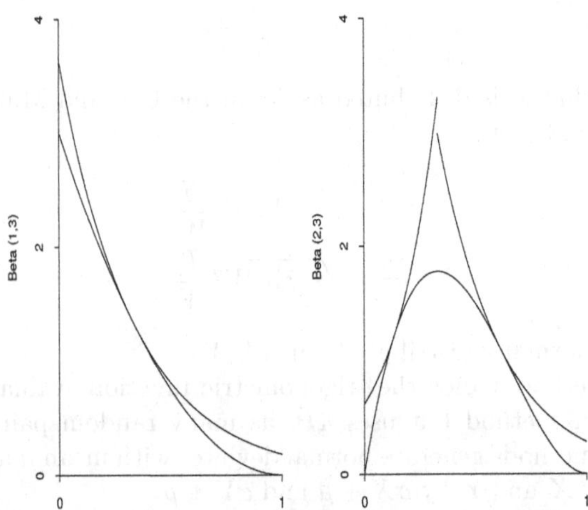

FIGURE 22.1. Exponential Envelopes for Two Beta Densities

include the normal, the gamma with shape parameter $\alpha \geq 1$, the beta with parameters α and $\beta \geq 1$, the exponential power density, and Fisher's z density. The reader can easily check log-concavity in each of these examples and in the three additional examples mentioned in Problem 9 by showing that $\frac{d^2}{dx^2} \ln f(x) \leq 0$ on the support of $f(x)$.

A strictly log-concave density $f(x)$ defined on an interval is unimodal. The mode m of $f(x)$ may occur at either endpoint or on the interior of the interval. In the former case, we suggest using a single exponential envelope; in the latter case, we recommend two exponential envelopes oriented in opposite directions from the mode m. Figure 22.1 depicts the two situations. With different left and right envelopes, the appropriate majorizing function is

$$h(x) = \begin{cases} c_l \lambda_l e^{-\lambda_l(m-x)} & x < m \\ c_r \lambda_r e^{-\lambda_r(x-m)} & x \geq m. \end{cases}$$

Note that $h(x)$ has total mass $c = c_l + c_r$. The proposal density is the admixture density

$$g(x) = \begin{cases} \frac{c_l}{c_l+c_r} \lambda_l e^{-\lambda_l(m-x)} & x < m \\ \frac{c_r}{c_l+c_r} \lambda_r e^{-\lambda_r(x-m)} & x \geq m. \end{cases}$$

To maximize the acceptance rate $1/(c_l + c_r)$ and the efficiency of sampling, we minimize the mass constants c_l and c_r. Geometrically this is accomplished by choosing optimal tangent points x_l and x_r. The tangency condition for the right envelope amounts to

$$\begin{aligned} f(x_r) &= c_r\lambda_r e^{-\lambda_r(x_r-m)} \\ f'(x_r) &= -c_r\lambda_r^2 e^{-\lambda_r(x_r-m)}. \end{aligned} \tag{22.2}$$

These equations allow us to solve for λ_r as $-f'(x_r)/f(x_r)$ and then for c_r as

$$c_r(x_r) = -\frac{f(x_r)^2}{f'(x_r)} e^{-\frac{f'(x_r)}{f(x_r)}(x_r-m)}.$$

Finding x_r to minimize c_r is now a matter of calculus. A similar calculation for the left envelope shows that $c_l(x_l) = -c_r(x_l)$.

Example 22.5.1 *Exponential Power Density*

The exponential power density

$$f(x) = \frac{e^{-|x|^\alpha}}{2\Gamma(1+\frac{1}{\alpha})}, \qquad \alpha \geq 1,$$

has mode $m = 0$. For $x_r \geq 0$ we have

$$\begin{aligned} \lambda_r &= \alpha x_r^{\alpha-1} \\ c_r(x_r) &= \frac{e^{(\alpha-1)x_r^\alpha}}{2\Gamma(1+\frac{1}{\alpha})\alpha x_r^{\alpha-1}}. \end{aligned}$$

The equation $\frac{d}{dx}c_r(x) = 0$ has solution $-x_l = x_r = \alpha^{-1/\alpha}$. This allows us to calculate the acceptance probability

$$\frac{1}{2c_r(x_r)} = \Gamma\Big(1+\frac{1}{\alpha}\Big)\alpha^{\frac{1}{\alpha}} e^{\frac{1}{\alpha}-1},$$

which ranges from 1 at $\alpha = 1$ (the double or bilateral exponential distribution) to $e^{-1} = .368$ as α tends to ∞. For a normal density ($\alpha = 2$), the acceptance probability reduces to $\sqrt{\pi/2e} \approx .76$. In practical implementations, the acceptance-rejection method for normal deviates is slightly less efficient than the polar method. ■

A completely analogous development holds for a discrete density $f(x)$ defined and positive on an interval of integers. Now, however, we substitute the easily generated geometric distribution for the exponential distribution [6, 14]. In extending the notion of log-concavity to a discrete density $f(x)$, we linearly interpolate $\ln f(x)$ between supporting integers as shown in Figure 22.2. If the linearly interpolated function is concave, then $f(x)$ is said to be log-concave. Analytically, log-concavity of $f(x)$ is equivalent to the inequality

$$\ln f(x) \geq \frac{1}{2}[\ln f(x-1) + \ln f(x+1)]$$

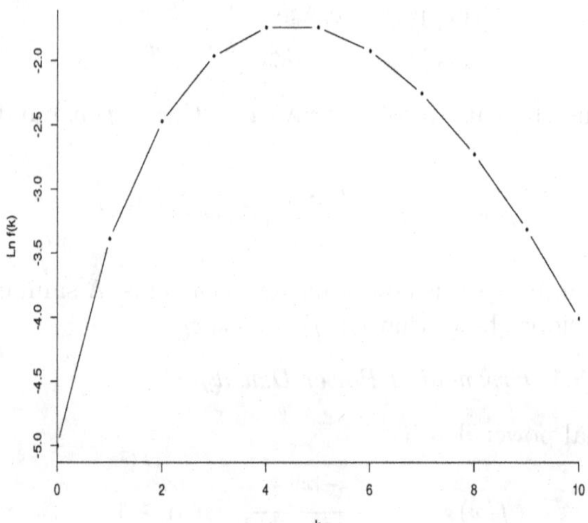

FIGURE 22.2. Linearly Interpolated Log Poisson Density

for all supporting integers x. This inequality is in turn equivalent to the inequality $f(x)^2 \geq f(x-1)f(x+1)$ for all integers x.

For a discrete density with an interior mode m, the majorizing function

$$h(x) = \begin{cases} c_l(1-q_l)q_l^{m-1-x} & x < m \\ c_r(1-q_r)q_r^{x-m} & x \geq m \end{cases}$$

consists of two geometric envelopes oriented in opposite directions from the mode m. The analog of the tangency condition (22.2) is

$$\begin{aligned} f(x_r) &= c_r(1-q_r)q_r^{x_r-m} \\ f(x_r+1) &= c_r(1-q_r)q_r^{x_r+1-m}. \end{aligned}$$

Solving these two equations gives $q_r = f(x_r+1)/f(x_r)$ and

$$c_r(x_r) = \frac{f(x_r)}{1-\frac{f(x_r+1)}{f(x_r)}}\left[\frac{f(x_r+1)}{f(x_r)}\right]^{m-x_r}.$$

We now minimize the mass constant $c_r(x_r)$ by adjusting x_r. To the left of the mode, a similar calculation yields

$$\begin{aligned} c_l(x_l) &= \frac{f(x_l)}{1-\frac{f(x_l)}{f(x_l+1)}}\left[\frac{f(x_l)}{f(x_l+1)}\right]^{x_l+1-m} \\ &= -c_r(x_l). \end{aligned}$$

Example 22.5.2 *Poisson Distribution*

For the Poisson density $f(x) = \lambda^x e^{-\lambda}/x!$, the mode $m = \lfloor \lambda \rfloor$ because $f(x+1)/f(x) = \lambda/(x+1)$. It follows that the mass constant

$$c_r(x_r) = \frac{\frac{\lambda^{x_r} e^{-\lambda}}{x_r!}\left(\frac{\lambda}{x_r+1}\right)^{m-x_r}}{1-\frac{\lambda}{x_r+1}}.$$

To minimize $c_r(x_r)$, we treat x_r as a continuous variable and invoke Stirling's asymptotic approximation in the form

$$\begin{aligned}\ln x! &= \ln(x+1)! - \ln(x+1) \\ &= \left(x+\frac{3}{2}\right)\ln(x+1) - (x+1) + \ln\sqrt{2\pi} + - \ln(x+1) \\ &= \left(x+\frac{1}{2}\right)\ln(x+1) - (x+1) + \ln\sqrt{2\pi}.\end{aligned}$$

Substitution of this expression in the expansion of $\ln c_r(x_r)$ produces

$$\begin{aligned}\ln c_r(x_r) &= x_r \ln\lambda - \lambda - \left(x+\frac{1}{2}\right)\ln(x_r+1) + (x_r+1) - \ln\sqrt{2\pi} \\ &\quad + (m-x_r)\ln\lambda - (m-x_r)\ln(x_r+1) - \ln\left(1-\frac{\lambda}{x_r+1}\right).\end{aligned}$$

Hence,

$$\begin{aligned}\frac{d}{dx_r}\ln c_r(x_r) &= \ln\lambda - \ln(x_r+1) - \frac{x_r+\frac{1}{2}}{x_r+1} + 1 - \ln\lambda \\ &\quad + \ln(x_r+1) - \frac{m-x_r}{x_r+1} - \frac{\frac{\lambda}{(x_r+1)^2}}{1-\frac{\lambda}{x_r+1}} \\ &= \frac{x_r^2 + \left(\frac{3}{2}-m-\lambda\right)x_r + \frac{1}{2} - m - \frac{3}{2}\lambda + m\lambda}{(x_r+1)(x_r+1-\lambda)}.\end{aligned}$$

Setting this derivative equal to 0 identifies x_r and x_l as the two roots of the quadratic equation $x^2 + (3/2 - m - \lambda)x + 1/2 - m - 3\lambda/2 + m\lambda = 0$. In practice, this acceptance-rejection method is faster than the alternative acceptance-rejection method featured in [23]. ■

The efficiency of the acceptance-rejection method depends heavily on programming details. For instance, the initial choice of the left or right envelope in a two-envelope problem involves comparing a uniform random variable U to the ratio $r = c_l/(c_l + c_r)$. Once this choice is made, U can be reused to generate the appropriate exponential deviate. Indeed, given $U < r$, the random variable $V = U/r$ is uniformly distributed on $[0, 1]$ and independent of U. Similarly, given the complementary event $U \geq r$, the random variable $W = (U - r)/(1 - r)$ is also uniformly distributed on $[0, 1]$ and independent of U. In the continuous case, the acceptance step is based on the ratio $f(x)/h(x)$. Usually it is more efficient to base acceptance on the log ratio

$$\ln \frac{f(x)}{h(x)} = \begin{cases} \ln f(x) - \ln f(x_l) + \lambda_l(x_l - x) & x < m \\ \ln f(x) - \ln f(x_r) + \lambda_r(x - x_r) & x \geq m, \end{cases}$$

from which one cancels as many common terms as possible. Going over to log ratios would appear to require computation of $\ln U$ for the uniform deviate U used in deciding acceptance. In view of the inequality

$$\ln u \leq \frac{1}{2}(u - 1)(3 - u)$$

for $u \in (0, 1]$, whenever $\frac{1}{2}(U - 1)(3 - U) \leq \ln[f(x)/h(x)]$, we can accept the proposed deviate without actually computing $\ln U$. This is an example of Marsaglia's squeeze principle [21].

The next section takes the idea of piecewise exponential envelopes to its logical extreme. Of course, such envelopes are not the only majorizing functions possible. Problems 17, 18, and 21 provide some alternative examples.

22.6 Adaptive Acceptance-Rejection Sampling

The acceptance-rejection method just sketched has the disadvantage of requiring a detailed mathematical analysis for each family of distributions. In Gibbs sampling exotic distributions arise naturally that defy easy analysis. This dilemma prompted Gilks and Wild [12, 13] to devise an adaptive method of acceptance-rejection sampling that handles any log-concave density. The method comes in tangent and secant versions. We restrict our attention to the secant version because it avoids the cumbersome calculation of derivatives.

The method begins with p points $x_1 < x_2 \cdots < x_p$ scattered over the domain of the target density $f(x)$ and p corresponding values $y_i = \ln f(x_i)$. The vast majority of the mass of $f(x)$ should occur between x_1 and x_p. If the domain is bounded, then x_1 and x_p should fall near the boundaries. If the domain is unbounded, then x_1 and x_p should fall far out in the left and right tails of $f(x)$. The exact positions of the interior points $x_2, \ldots, x_{p-1}$ are less relevant because we add interior points as we go to improve sampling efficiency. A good choice is $p = 5$ with y_3 the largest value.

When $f(x)$ is log-concave, the chords between the successive points of $\ln f(x)$ lie below $\ln f(x)$ as depicted in Figure 22.3 for the case $p = 5$. These chords are crucial in implementing the squeeze principle. Extending the chords beyond their endpoints yields line segments majorizing $\ln f(x)$.

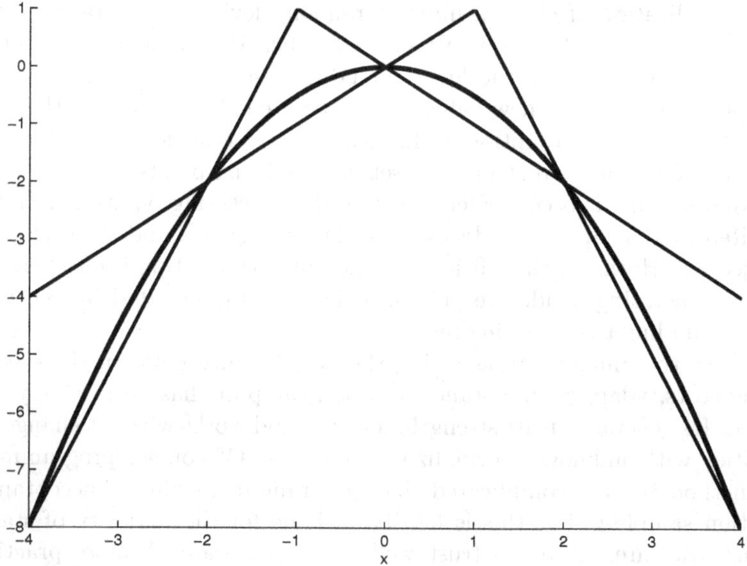

FIGURE 22.3. Lower and Upper Envelopes for the Log Density $-\frac{1}{2}x^2$

In Figure 22.3 a single extended chord majorizes $\ln f(x)$ on the leftmost interval $[x_1, x_2]$; likewise, a single extended chord majorizes $\ln f(x)$ on the rightmost interval $[x_{p-1}, x_p]$. On an interior interval, the majorizing function is a tent formed by taking the minimum of the chords coming in from the left and right.

Let $u(x)$ and $l(x)$ denote the upper and lower envelopes majorizing and minorizing $\ln f(x)$. Because $u(x)$ is piecewise linear, the total area lying below $e^{u(x)}$ is readily calculated interval by interval via the formula

$$\int_a^b e^{c+dx} dx = \frac{1}{d}(e^{c+db} - e^{c+da}).$$

To sample from the normalized version of $e^{u(x)}$, one first chooses an interval (a, b) with the appropriate probability. The possible choices for (a, b) are (x_1, x_2), (x_{p-1}, x_p), (x_i, c_i), and (c_i, x_{i+1}) for $1 < i < p-1$, where c_i is the abscissa under the peak of the tent on (x_i, x_{i+1}). Once the interval (a, b) is chosen, it is possible to apply inverse sampling. In this regard, the distribution function and its inverse

$$z = \frac{e^{c+dx} - e^{c+da}}{e^{c+db} - e^{c+da}}, \quad x = \frac{1}{d}\ln\left[ze^{db} + (1-z)e^{da}\right]$$

come into play.

Implementation of adaptive acceptance-rejection sampling follows the usual well-trod path. In addition to the random deviate X drawn from the normalization of $e^{u(x)}$, a uniform random deviate U is drawn. If the squeeze test $\ln U \leq l(X) - u(X)$ succeeds, then the point X is accepted. If the squeeze test fails, the less demanding test $\ln U \leq \ln f(X) - u(X)$ is applied. If it fails, a new value X is generated. If it succeeds, then the point X is accepted. Adaption to the shape of $f(x)$ is achieved by adding an accepted X to the current ordered set of bracketing points $x_1 < \cdots < x_p$. To avoid sorting, it is convenient to store the bracketing points as a linked list. Repeated sampling produces larger bracketing sets and lower chances of rejection. However, the efficiency gained by low rejection is offset by the cost of generating candidate points. This cost is dominated by the time taken in finding the sampling interval (a, b).

Despite the difficulty in assessing the best balance between these competing costs, adaptive acceptance-rejection sampling has been widely embraced. Its generality is its strength. The method works without change for densities with unknown normalizing constants. Of course, programming the method is more complicated than programming ordinary acceptance-rejection sampling, but this is hardly an issue for the majority of statisticians, who are willing to trust well-crafted software. A more practical matter is mathematical verification of log-concavity of typical densities $f(x)$. The knee-jerk reaction is to apply the second derivative test. Often the closure properties of log-concave functions lead to the conclusion of log-concavity with less pain; see Problems 7 through 10.

22.7 Ratio Method

The ratio method is a kind of generalization of the polar method. Suppose that $f(x)$ is a probability density and $h(x) = cf(x)$ for $c > 0$. Consider the set $S_h = \{(u, v) : 0 < u \leq \sqrt{h(v/u)}\}$ in the plane. If this set is bounded, then we can enclose it in a well-behaved set such as a rectangle and sample uniformly from the enclosing set. The next proposition shows how this leads to a method for sampling from $f(x)$.

Proposition 22.7.1 *Suppose $k_u = \sup_x \sqrt{h(x)}$ and $k_v = \sup_x |x|\sqrt{h(x)}$ are finite. Then the rectangle $[0, k_u] \times [-k_v, k_v]$ encloses S_h. If $h(x) = 0$ for $x < 0$, then the rectangle $[0, k_u] \times [0, k_v]$ encloses S_h. Finally, if the point (U, V) sampled uniformly from the enclosing set falls within S_h, then the ratio $X = V/U$ is distributed according to $f(x)$.*

Proof: From the definition of S_h it is clear that the permitted u lie in $[0, k_u]$. Multiplying the inequality $|v|u/|v| \leq \sqrt{h(v/u)}$ by $|v|/u$ implies that $|v| \leq k_v$. If $h(x) = 0$ for $x < 0$, then no $v < 0$ yields a pair (u, v) in S_h. Finally, note that the transformation $(u, v) \to (u, v/u)$ has Jacobian u^{-1}.

Hence,

$$\begin{aligned}\int\int 1_{\{\frac{v}{u}\le x_0\}}1_{\{0<u\le\sqrt{h(\frac{v}{u})}\}}du\,dv &= \int\int 1_{\{x\le x_0\}}1_{\{0<u\le\sqrt{h(x)}\}}u du\,dx \\ &= \int_{-\infty}^{x_0}\frac{1}{2}h(x)dx\end{aligned}$$

is the distribution function of the accepted random variable X up to a normalizing constant. ■

Example 22.7.1 *Gamma with Shape Parameter $\alpha > 1$*

Here we take $h(x) = x^{\alpha-1}e^{-x}1_{(0,\infty)}(x)$. The maximum of $\sqrt{h(x)}$ occurs at $x = \alpha - 1$ and equals $k_u = [(\alpha-1)/e]^{(\alpha-1)/2}$. Likewise, the maximum of $x\sqrt{h(x)}$ occurs at $x = \alpha + 1$ and equals $k_v = [(\alpha+1)/e]^{(\alpha+1)/2}$. To carry out the ratio method, we sample uniformly from the rectangular region $[0, k_u] \times [0, k_v]$ by multiplying two independent, uniform deviates U and V by k_u and k_v, respectively. The ratio $X = k_vV/(k_uU)$ is accepted as a random deviate from $f(x)$ if and only if

$$k_uU \le X^{\frac{\alpha-1}{2}}e^{-\frac{X}{2}},$$

which simplifies to

$$\frac{2}{\alpha-1}\ln U - 1 - \ln W + W \le 0$$

for $W = X/(\alpha - 1)$. Problem 22 describes an alternative algorithm that is faster and just as easy to program. ■

22.8 Deviates by Definition

In many cases we can generate a random variable by exploiting its definition in terms of simpler random variables or familiar stochastic processes. Here are some examples.

Example 22.8.1 *Admixture Distributions*

Consider the admixture distribution

$$F(x) = \sum_{j=1}^{k} p_jF_j(x)$$

with admixture proportions p_j. If it is easy to sample from each of the distributions $F_j(x)$, then the obvious strategy is to randomly choose an index j with probability p_j. Once this choice is made, generate the required random deviate X from $F_j(x)$. For example, to draw X from the admixture

$$F(x) = p_1(1 - e^{-\alpha_1 x}) + p_2(1 - e^{-\alpha_2 x})$$

of two exponentials, generate a random deviate U and decide to sample from the first exponential when $U \le p_1$ and from the second exponential when $U > p_1$. Given the choice of the index j, generate an exponential deviate according to the inverse method by sampling a second uniform deviate V and setting $X = -\alpha_j \ln V$. ■

Example 22.8.2 *Binomial*

For a small number of trials n, a binomial deviate S_n can be quickly generated by taking n independent, uniform deviates $U_1, \ldots, U_n$ and setting $S_n = \sum_{i=1}^n 1_{\{U_i \le p\}}$, where p is the success probability per trial. When n is large, the acceptance-rejection method featured in Problem 13 is more efficient. ■

Example 22.8.3 *Negative Binomial*

Consider a Bernoulli sampling process with success probability p. The number of failures S_n until the nth success follows a negative binomial distribution. When $n = 1$, we recover the geometric distribution. Adding n independent geometric deviates gives the negative binomial. In view of the ease with which we can generate geometric deviates (Example 22.3.5), we can sample S_n quickly for n small. When n is large or fails to be an integer, one can apply the acceptance-rejection method or exploit the fact that the negative binomial distribution can be represented as a gamma mixture of Poisson distributions. See Problems 13 and 14. ■

Example 22.8.4 *Poisson*

To generate a Poisson deviate X with mean λ, consider a Poisson process with unit intensity. The number of random points falling on the interval $[0, \lambda]$ follows a Poisson distribution with mean λ. Furthermore, the waiting times between successive random points are independent, exponentially distributed random variables with common mean 1. If we generate a sequence $Z_1, Z_2, \ldots$ of independent, exponential deviates and stop with Z_j satisfying $\sum_{i=1}^{j-1} Z_i \le \lambda < \sum_{i=1}^{j} Z_i$, then $X = j - 1$. Rephrasing the stopping condition as $\prod_{i=1}^{j-1} e^{-Z_i} \ge e^{-\lambda} > \prod_{i=1}^{j} e^{-Z_i}$ allows us to use uniform deviates $U_1, \ldots, U_j$ since Example 22.3.1 implies that U_i and e^{-Z_i} are identically distributed. This procedure is more efficient for small λ than the acceptance-rejection method discussed in Example 22.5.2. ■

Example 22.8.5 *Lognormal*

If a random variable X can be represented as a function $f(Y)$ of another random variable that is easy to generate, then obviously we should sample Y and compute $f(Y)$ to generate X. For example, if X is a standard normal deviate, then $e^{\sigma X+\mu}$ is a lognormal deviate for all choices of the mean μ and standard deviation σ of the normal deviate $\sigma X + \mu$. ■

Example 22.8.6 *Chi-square*

A chi-square distribution with n degrees of freedom is a gamma distribution with shape parameter $\alpha = n/2$ and scale parameter $\beta = 2$. The acceptance-rejection method sketched in Problem 21 or the ratio method discussed in Example 22.7.1 delivers a gamma deviate with shape parameter α and scale parameter 1. Doubling their output when $\alpha = n/2$ gives a χ_n^2 deviate. Alternatively for small n, we can exploit the definition of χ_n^2 as a sum of squares of n independent, standard normal deviates. Once we have generated a χ_n^2 deviate, we can compute derived deviates such as the inverse chi-square, the inverse chi, and the log chi-square by forming $1/\chi_n^2$, $1/\chi_n$, and $\ln \chi_n^2$, respectively.

Generating deviates from the noncentral chi-square distribution is another matter. If X has a noncentral chi-square distribution with n degrees of freedom and noncentrality parameter μ^2, then by definition we can write $X = (Y+\mu)^2+\chi_{n-1}^2$, where the standard normal deviate Y is independent of the chi-square deviate χ_{n-1}^2. In view of our previous constructions, this representation yields a convenient method of simulating X. ■

Example 22.8.7 *F Distribution*

If χ_m^2 and χ_n^2 are independent chi-square random variables with m and n degrees of freedom, respectively, then the ratio

$$F_{mn} = \frac{\frac{1}{m}\chi_m^2}{\frac{1}{n}\chi_n^2}$$

follows an F distribution. Since we can readily generate chi-square deviates, this definition provides a convenient method for generating F deviates. ■

Example 22.8.8 *Student's t Distribution*

If X is a standard normal deviate and χ_n^2 is an independent chi-square deviate, then the definition $T_n = X/\sqrt{\frac{1}{n}\chi_n^2}$ gives a convenient method of generating t deviates. ■

Example 22.8.9 *Beta*

If X_α and X_β are independent gamma deviates with shape parameters α and β and scale parameter 1, then the ratio $X_\alpha/(X_\alpha+X_\beta)$ is by definition a beta deviate with parameter pair (α, β). ■

22.9 Multivariate Deviates

We make no attempt to be systematic in presenting the following examples.

Example 22.9.1 *Multinomial*

The multinomial distribution, like the binomial distribution, involves random assignment to a finite number of categories. Instead of two categories with probabilities p and $1 - p$, we have k categories with probabilities $p_1, \ldots, p_k$. For a small number of trials n, the simplest approach to simulation is to conduct the independent trials and add the number of successes in each category. For large n this procedure is cumbersome. Brown and Bromberg [3] suggest an alternative that exploits the intimate relationship between multinomial sampling and independent Poisson sampling. Suppose that we choose a positive constant λ and sample k independent Poisson deviates $X_1, \ldots, X_k$ with means $\lambda p_1, \ldots, \lambda p_k$. If the total number of successes $S = \sum_{i=1}^k X_i$ equals n, then the random vector $X = (X_1, \ldots, X_k)^t$ follows the multinomial distribution just described [19]. Of course, we must be fortunate for the Poisson deviate S to exactly equal n.

Given $S = m < n$, we have a multinomial sample with m trials. If $n - m$ is small, then we can easily work our way up to n trials by conducting an additional $n - m$ trials in the naive manner. When $m > n$ we are out of luck and must redo the entire Poisson sample. These considerations suggest that we take $\lambda = \sum_{i=1}^k \lambda p_i$ slightly less than n. Brown and Bromberg recommend the choices

$$\lambda = \begin{cases} n - k & \text{for } k \le \sqrt{n} \\ n - \sqrt{n} - (k - \sqrt{n})^{1/2} & \text{for } k > \sqrt{n}\,. \end{cases}$$

Because a Poisson random variable with mean λ has standard deviation $\sqrt{\lambda}$, most of the time a single round of Poisson sampling suffices. ■

Example 22.9.2 *Multivariate Normal*

In R^n the simplest multivariate normal random vector X has n independent, standard normal components. To generate a multivariate normal deviate Y with mean vector μ and covariance matrix Ω, we first generate X and then form $Y = \Omega^{1/2}X + \mu$. Any square root $\Omega^{1/2}$ will do; for instance, we can use the Cholesky decomposition of Ω. ■

Example 22.9.3 *Multivariate t*

Let X be a multivariate normal deviate with mean vector $\mathbf{0}$ and covariance matrix Ω, and let χ^2_ν be an independent chi-square deviate with possibly noninteger degrees of freedom ν. The translated ratio

$$T_\nu = \frac{X}{\sqrt{\chi^2_\nu/\nu}} + \mu \tag{22.3}$$

follows a multivariate t distribution with location vector μ, scale matrix Ω, and degree of freedom ν [8]. For ν small, the t distribution has fatter tails than the normal distribution and offers the opportunity of estimating location and scale parameters robustly [20]. As ν tends to ∞, the t distribution tends to the normal with mean vector μ and covariance matrix Ω. Again, the most natural way of generating T_ν is via its definition (22.3). ■

Example 22.9.4 *Wishart*

The Wishart distribution $\mathcal{W}_m(n, \Sigma)$ resides on the space of $m \times m$ positive semidefinite matrices. If the vectors $X_1, \ldots, X_n$ constitute a random sample from the multivariate normal with m components, mean vector $\mathbf{0}$, and variance matrix Σ, then the sum of outer products $\sum_{i=1}^n X_i X_i^t$ follows a $\mathcal{W}_m(n, \Sigma)$ distribution. This is yet another case where the definition offers a viable method of simulation. ■

Example 22.9.5 *Multivariate Uniform*

In R^n there are many sets of finite Lebesgue measure from which we might want to sample uniformly. The rectangle $[a, b] = \prod_{i=1}^n [a_i, b_i]$ is the simplest. In this case we take n independent uniform deviates $U_1, \ldots, U_n$ and construct the vector V with ith component $V_i = (b_i - a_i)U_i + a_i$. To sample uniformly from the unit sphere $S_n = \{x : \|x\|_2 \leq 1\}$, we sample a standard, multivariate normal random vector X and note that $V = X/\|X\|_2$ is uniformly distributed on the surface of S_n. We then choose an independent radius $R \leq 1$ and form the contracted point RV within S_n. Since the volume of a sphere depends on the nth power of its radius, we construct R by the inverse method employing the distribution function $F(r) = r^n$ on $[0, 1]$. More complicated sets can be accommodated by enclosing them within a rectangle or sphere and using a rejection procedure. ■

Example 22.9.6 *Dirichlet*

The Dirichlet density is the natural generalization of the beta density [17]. To generate a Dirichlet deviate $Y = (Y_1, \ldots, Y_n)^t$, we take n independent gamma deviates $X_1, \ldots, X_n$ with shape parameters α_i and scale parameters 1 and form the ratios

$$Y_i = \frac{X_i}{\sum_{i=1}^n X_i}.$$

The random vector Y lives on the simplex

$$\Delta_n = \Big\{(y_1, \ldots, y_n)^t : y_1 > 0\,, \ldots, y_n > 0, \sum_{i=1}^n y_i = 1\Big\}$$

and has density

$$f(y) = \frac{\Gamma(\sum_{i=1}^n \alpha_i)}{\prod_{i=1}^n \Gamma(\alpha_i)} \prod_{i=1}^n y^{\alpha_i - 1}$$

there relative to the uniform measure. When each $\alpha_i = 1$, the deviate Y is uniformly distributed on Δ_n, and the reduced deviate $(Y_1, \ldots, Y_{n-1})^t$ derived by deleting Y_n is uniformly distributed over the reduced simplex

$$\Big\{(y_1, \ldots, y_{n-1})^t : y_1 > 0\,, \ldots, y_{n-1} > 0,\ \sum_{i=1}^{n-1} y_i \leq 1\Big\}.$$

When $n = 3$, the reduced simplex is just the triangle in the plane with vertices $(0,0)$, $(1,0)$, and $(0,1)$. Given the ability to randomly sample this special triangle, one can randomly sample from any triangle. ■

Example 22.9.7 *Order Statistics*

At first glance, generating order statistics $Y_{(1)}, \ldots, Y_{(n)}$ from a distribution $F(y)$ appears easy. However, if n is large and we are interested only in a few order statistics at the beginning or end of the sequence, we can do better than generate all n deviates Y_i and order them in $O(n \ln n)$ steps. Consider the special case of exponential deviates X_i with mean 1. From the calculation

$$\Pr(X_{(1)} \geq x) \quad = \quad \prod_{i=1}^{n} \Pr(X_i \geq x) \quad = \quad e^{-nx},$$

we find that $X_{(1)}$ is exponentially distributed with intensity n. Because of the lack of memory property of the exponential, the $n-1$ random points to the right of $X_{(1)}$ provide an exponentially distributed sample of size $n-1$ starting at $X_{(1)}$. Duplicating our argument for $X_{(1)}$, we find that the difference $X_{(2)} - X_{(1)}$ is independent of $X_{(1)}$ and exponentially distributed with intensity $n-1$. Arguing inductively, we now see that $Z_1 = X_{(1)}$, that the differences $Z_{i+1} = X_{(i+1)} - X_{(i)}$ are independent, and that Z_i is exponentially distributed with intensity $n - i + 1$. This result, which is proved more rigorously in [9], suggests that we can sample the Z_i and add them to get the $X_{(i)}$. If we are interested only in the $X_{(i)}$ for $i \leq j$, then we omit generating $Z_{j+1}, \ldots, Z_n$.

To capitalize on this special case, note that Example 22.3.1 permits us to generate the n order statistics from the uniform distribution by defining $U_{(i)} = e^{-X_{(n-i+1)}}$. The order statistics are reversed here because e^{-x} is strictly decreasing. Given the fact that $1 - U$ is uniform when U is uniform, we can equally well define $U_{(i)} = 1 - e^{-X_{(i)}}$. If we desire the order statistics $Y_{(i)}$ from a general, continuous distribution function $F(y)$ with inverse $F^{[-1]}(u)$, then we apply the inverse method and set

$$Y_{(i)} \quad = \quad F^{[-1]}(e^{-X_{(n-i+1)}})$$

or

$$Y_{(i)} \quad = \quad F^{[-1]}(1 - e^{-X_{(i)}}).$$

The first construction is more convenient if we want only the last j order statistics, and the second construction is more convenient if we want only the first j order statistics. In both cases we generate only $X_{(1)}, \ldots, X_{(j)}$. ■

Example 22.9.8 *Random Orthogonal Matrix*

The easiest way to generate an $n \times n$ random orthogonal matrix is to subject n vectors pointing in random directions to the Gram-Schmidt orthogonalization process. The n vectors can be generated as multivariate normal deviates with mean vector $\mathbf{0}$ and covariance matrix I. In Example 25.2.4 we consider the related problem of randomly sampling rotation matrices. The articles [7, 28] discuss even more efficient schemes. ■

22.10 Sequential Sampling

In many cases we can sample a random vector $X = (X_1, \ldots, X_k)$ by sampling each component in turn. For the sake of notational convenience, suppose that X is discretely distributed. Then the decomposition

$$\begin{aligned} & \Pr(X_1 = x_1, \ldots, X_k = x_k) \\ = \; & \Pr(X_1 = x_1) \prod_{j=2}^{k} \Pr(X_j = x_j \mid X_1 = x_1, \ldots, X_{j-1} = x_{j-1}) \end{aligned}$$

provides the rationale for sequential sampling.

Example 22.10.1 *Multinomial*

Let us revisit the problem of multinomial sampling. Once again suppose that X_j represents the number of outcomes of type j over n trials. It is obvious that X_1 follows a binomial distribution with success probability p_1 over n trials. Given we sample $X_1 = x_1$, the next component X_2 is binomially distributed with success probability $p_2/(1 - p_1)$ over $n - x_1$ trials. In general, given $X_1 = x_1, \ldots, X_{j-1} = x_{j-1}$, the next component X_j is binomially distributed with success probability $p_j/(1-p_1-\cdots-p_{j-1})$ over $n - x_1 - \cdots - x_{j-1}$ trials. ■

Example 22.10.2 *Conditional Poisson-Binomial*

Let $X_1, \ldots, X_n$ be independent Bernoulli random variables with a possibly different success probability p_k for each X_k. In Section 1.7 we showed how to compute the distribution $q_n(i) = \Pr(S_n = i)$ of the Poisson-binomial sum $S_n = \sum_{k=1}^{n} X_k$. These probabilities come in handy when we want to sample the vector $(X_1, \ldots, X_n)$ conditional on $S_n = i$. Indeed, X_n follows a Bernoulli distribution with success probability $p_n q_{n-1}(i-1)/q_n(i)$. Once we sample $X_n = x_n$, the value of the Poisson-binomial sum S_{n-1} is determined as $i - x_n$. Thus, we can sample X_{n-1} and so forth recursively. Fortunately, the algorithm for computing $q_n(i)$ puts all of the required probabilities $q_m(j)$ with $m \leq n$ and $j \leq i$ at our fingertips. ■

Example 22.10.3 *Multivariate Normal*

Let X be a normally distributed random vector in R^p with mean vector μ and covariance matrix Ω. The first component X_1 is univariate normal with mean μ_1 and variance Ω_{11}. Once we sample X_1, we must sample $(X_2, \ldots, X_p)$ conditional on X_1. As suggested in Section 7.3, the necessary ingredients for computing the conditional mean vector and covariance matrix emerge after sweeping on Ω_{11}. This allows us to sample X_2, and successive sweeps down the diagonal of Ω allow us to sample X_3 through X_p similarly. This process is more intuitive but less efficient than sampling via the Cholesky decomposition. ■

22.11 Problems

1. Write and test a uniform random number generator that adds the output modulo 1 of three multiplicative generators with different prime moduli and primitive multipliers.

2. Discuss how you would use the inverse method to generate a random variable with (a) the continuous logistic density

$$f(x|\mu,\sigma) = \frac{e^{-\frac{x-\mu}{\sigma}}}{\sigma[1+e^{-\frac{x-\mu}{\sigma}}]^2},$$

(b) the Pareto density

$$f(x|\alpha,\beta) = \frac{\beta\alpha^\beta}{x^{\beta+1}}1_{(\alpha,\infty)}(x),$$

and (c) the Weibull density

$$f(x|\delta,\gamma) = \frac{\gamma}{\delta}x^{\gamma-1}e^{-\frac{x^\gamma}{\delta}}1_{(0,\infty)}(x),$$

where α, β, γ, δ, and σ are taken positive.

3. Continuing Problem 2, discuss how the inverse method applies to (d) the Gumbel density

$$f(x) = e^{-x}e^{-e^{-x}},$$

(e) the arcsine density

$$f(x) = \frac{1}{\pi\sqrt{x(1-x)}}1_{(0,1)}(x),$$

and (f) the slash density

$$f(x) = \alpha x^{\alpha-1} 1_{(0,1)}(x),$$

where $\alpha > 0$.

4. Demonstrate how Examples 22.3.4 and 22.3.5 follow from Proposition 22.3.1.

5. Suppose the random variable X has distribution function $F(x)$. If U is uniformly distributed on $[0, 1]$, then show that

$$Y = F^{[-1]}[UF(t)]$$

is distributed as X conditional on $X \leq t$ and that

$$Z = F^{[-1]}\{F(t) + U[1 - F(t]\}$$

is distributed as X conditional on $X > t$.

6. One can implement the inverse method even when the inverse distribution function $F^{[-1]}(y)$ is not explicitly available. Demonstrate that Newton's method for solving $F(x) = y$ has the update

$$x_{n+1} = x_n - \frac{F(x_n) - y}{f(x_n)}$$

when $F(x)$ has density $f(x) = F'(x)$. Observe that if $F(x_n) > y$, then $x_{n+1} < x_n$, and if $F(x_n) < y$, then $x_{n+1} > x_n$. Prove that x_n approaches the solution from above if $F(x)$ is convex and from below if $F(x)$ is concave. Implement Newton's method for the standard normal distribution, and describe its behavior.

7. A positive function $f(x)$ is said to be log-convex if and only if $\ln f(x)$ is convex. Demonstrate the following properties:

   (a) (a) If $f(x)$ is log-convex, then $f(x)$ is convex.

   (b) (b) If $f(x)$ is convex and $g(x)$ is log-convex and increasing, then the functional composition $g \circ f(x)$ is log-convex.

   (c) (c) If $f(x)$ is log-convex, then the functional composition $f(Ax + b)$ of $f(x)$ with an affine function $Ax + b$ is log-convex.

   (d) (d) If $f(x)$ is log-convex, then $f(x)^\alpha$ and $\alpha f(x)$ are log-convex for any $\alpha > 0$.

   (e) (e) If $f(x)$ and $g(x)$ are log-convex, then $f(x) + g(x)$, $f(x)g(x)$, and $\max\{f(x), g(x)\}$ are log-convex.

   (f) (f) If $f_n(x)$ is a sequence of log-convex functions, then $\lim_{n\to\infty} f_n(x)$ is log-convex whenever it exists and is positive.

(Hint: To prove that the sum of log-convex functions is log-convex, apply Hölder's inequality.)

8. Show that the normal distribution, the gamma distribution with shape parameter $\alpha \geq 1$, the beta distribution with parameters α and $\beta \geq 1$, the exponential power distribution with parameter $\alpha \geq 1$, and Fisher's z distribution of Problem 12 all have log-concave densities.

9. Verify that the logistic and Weibull ($\gamma \geq 1$) densities of Problem 2 and the Gumbel density of Problem 3 are log-concave. Prove that the Cauchy density is not log-concave.

10. Check that the Poisson, binomial, negative binomial, and hypergeometric distributions have discrete log-concave densities.

11. Verify that the gamma distribution with shape parameter $\alpha > 1$ and scale parameter $\beta = 1$ has mode $m = \alpha - 1$ and that the beta distribution with parameters $\alpha > 1$ and $\beta > 1$ has mode

$$m = \frac{\alpha - 1}{\alpha + \beta - 2}.$$

Demonstrate that the corresponding optimal tangency points of the acceptance-rejection method of Section 22.5 are the roots of the respective quadratics

$$\begin{aligned} & m(m-x)^2 - x \\ = \;& (\alpha - 1)(\alpha - 1 - x)^2 - x \\ & (\alpha + \beta - 1)x^2 + (2m - \alpha - \alpha m - \beta m)x + \alpha m - m \\ = \;& (\alpha + \beta - 1)x^2 + (1 - 2\alpha)x + \frac{(1-\alpha)^2}{\alpha + \beta - 2}. \end{aligned}$$

(Hints: You may want to use a computer algebra program such as Maple. The beta distribution involves a quartic polynomial, one of whose quadratic factors has imaginary roots.)

12. If X has an F distribution with m and n degrees of freedom, then $\ln(X)/2$ has Fisher's Z distribution. Show that $\ln X$ has density

$$f(x) = \frac{m^{\frac{m}{2}} n^{\frac{n}{2}} \Gamma(\frac{m}{2} + \frac{n}{2}) e^{\frac{mx}{2}}}{\Gamma(\frac{m}{2})\Gamma(\frac{n}{2})(n + me^x)^{\frac{m}{2} + \frac{n}{2}}}.$$

Prove that $f(x)$ has mode $m = 0$ and that the corresponding optimal tangency points x_l and x_r of the acceptance-rejection method of Section 22.5 are the roots of the transcendental equation

$$m(nx - 2)e^x = 2n + mnx.$$

(Hint: You may want to use a computer algebra program such as Maple.)

13. Demonstrate that the binomial density $f(x) = \binom{n}{x}p^x q^{n-x}$ has mode $\lfloor (n+1)p \rfloor$ and that the negative binomial density $f(x) = \binom{x+n-1}{n-1}p^n q^x$ has mode $\lfloor (n-1)q/p \rfloor$. Verify that the corresponding optimal tangency points x_l and x_r of the acceptance-rejection method of Section 22.5 are the roots of the respective quadratic equations

$$\begin{aligned} 2x^2 + (3 - 2m - 2np)x + 1 - 2m - p + 2mp - 3np + 2mnp &= 0, \\ 2px^2 + (3p - 2mp - 2nq)x + 1 - 2m - 3nq + 2mnq &= 0. \end{aligned}$$

(Hints: Use Stirling's formula and a computer algebra program such as Maple.)

14. Verify the identity

$$\frac{\Gamma(r+k)}{k!\,\Gamma(r)}p^r(1-p)^k = \int_0^\infty \frac{\lambda^k}{k!}e^{-\lambda}\frac{\lambda^{r-1}e^{-\lambda p/(1-p)}}{\Gamma(r)[(1-p)/p]^r}d\lambda$$

demonstrating that the negative binomial distribution is a mixture of Poisson distributions whose intensities are gamma distributed.

15. Check that the hypergeometric density

$$f(x) = \frac{\binom{R}{x}\binom{N-R}{n-x}}{\binom{N}{n}}$$

has mode $m = \lfloor (R+1)(n+1)/(2+N) \rfloor$. Prove that the corresponding optimal tangency points x_l and x_r of the acceptance-rejection method of Section 22.5 are the roots of the quadratic

$$\begin{aligned} 0 &= (2N+2)x^2 + (3N - 2mN - 4m + 4 - 2nR)x \\ &\quad + (2m-1)(R - N + n - 1 + nR) - 2nR. \end{aligned}$$

(Hints: Use Stirling's formula and a computer algebra program such as Maple. This quadratic is one of two possible quadratics. Why can you discard the other one?)

16. Suppose the random variable X follows a standard normal distribution. Conditional on the event $X \geq c > 0$, what is the density of X? Construct the optimal exponential envelope

$$h(x) = 1_{\{x \geq c\}} b\lambda e^{-\lambda(x-c)}$$

for acceptance-rejection sampling of the conditional density. In particular, demonstrate that $\lambda = (c + \sqrt{c^2+4})/2$.

17. The von Mises distribution for a random angle Θ has density

$$f(\theta) = \frac{e^{\kappa \cos\theta}}{I_0(\kappa)},$$

where $\theta \in [0, 2\pi]$, $\kappa > 0$, and $I_0(\kappa)$ is a Bessel function. Devise an acceptance-rejection method for generating random deviates from $f(\theta)$.

18. Devise an acceptance-rejection method for generating beta deviates based on the inequality $x^{\alpha-1}(1-x)^{\beta-1} \leq x^{\alpha-1} + (1-x)^{\beta-1}$.

19. For $\alpha \in (0,1)$ let Y be beta distributed with parameters α and $1-\alpha$. If Z is independent of Y and exponentially distributed with mean 1, then prove that $X = YZ$ is gamma distributed with parameters α and 1. In conjunction with the previous problem, this gives a method of generating gamma deviates.

20. Suppose Y is gamma distributed with parameters $\alpha+1$ and 1 and Z is uniformly distributed on $(0,1)$. If Y and Z are independent, then demonstrate that $X = YZ^{1/\alpha}$ is gamma distributed with parameters α and 1.

21. When $\alpha < 1$, show that the gamma density

$$f(x) = \frac{x^{\alpha-1}}{\Gamma(\alpha)} e^{-x} 1_{(0,\infty)}(x)$$

with scale 1 is majorized by the mixture density

$$g(x) = \frac{e}{e+\alpha} \alpha x^{\alpha-1} 1_{(0,1)}(x) + \frac{\alpha}{e+\alpha} e^{1-x} 1_{[1,\infty)}(x)$$

with mass constant $c = (e+\alpha)/[e\Gamma(\alpha)\alpha]$. Give a detailed algorithm for implementing the acceptance-rejection method employing the majorizing function $h(x) = cg(x)$.

22. The gamma deviate method of Marsaglia and Tsang [21] combines speed and simplicity. This acceptance-rejection algorithm can be derived by the following steps.

    (a) Let X be a random variable with density

    $$\frac{1}{\Gamma(\alpha)} f(x)^{\alpha-1} e^{-f(x)} f'(x)$$

    for some nonnegative function $f(x)$ and $\alpha \geq 1$. Show that the random variable $Y = f(X)$ has gamma density with shape parameter α and scale parameter 1.

    (b) Let c and d be positive constants to be specified. For the choice

    $$f(x) = \begin{cases} d(1+cx)^3 & x \geq -c^{-1} \\ 0 & x < -c^{-1}, \end{cases}$$

define a function $g(x)$ and constant $a > 0$ by the conditions

$$ae^{g(x)} = \frac{1}{\Gamma(\alpha)} f(x)^{\alpha-1} e^{-f(x)} f'(x)$$

and $g(0) = 0$. If $d = \alpha - \frac{1}{3}$ and $c^2 d = \frac{1}{9}$, then prove that

$$g(x) = d\ln(1+cx)^3 - d(1+cx)^3 + d.$$

(c) To generate a deviate X with the density $ae^{g(x)}$, one can use acceptance-rejection with a truncated standard normal deviate Z, conditioning on the event $Z \geq -c^{-1}$. Explain how the inequality

$$e^{g(x)} \leq e^{-\frac{1}{2}x^2} \tag{22.4}$$

can be used to sample X. What fraction of standard normal deviates Z satisfy $Z \geq -c^{-1}$? Of these, what proportion are accepted as valid X's?

(d) Demonstrate inequality (22.4) by showing that the derivatives of $w(x) = -\frac{1}{2}x^2 - g(x)$ satisfy

$$\begin{aligned}(1+cx)w'(x) &= \frac{c^2}{3}x^3 \\ (1+cx)^2 w''(x) &= c^2x^2\Big(1+\frac{2}{3}cx\Big).\end{aligned}$$

23. Program and test the gamma deviate algorithm described in Problem 22. Check that the sample mean and variance match the theoretical mean and variance of the gamma distribution.

24. Let S and T be independent deviates sampled from the uniform distribution on $[-1,1]$. Show that conditional on the event $S^2 + T^2 \leq 1$ the ratio S/T is Cauchy.

25. Specify the ratio method for generating normal deviates starting from the multiple $h(x) = e^{-x^2/2}$ of the standard normal density. Show that the smallest enclosing rectangle is defined by the inequalities $0 \leq u \leq 1$ and $v^2 \leq 2/e$.

26. Describe and implement in computer code an algorithm for sampling from the hypergeometric distribution. Use the "deviate by definition" method of Section 22.8.

27. Describe how to generate deviates from the noncentral F and noncentral t distributions. Implement one of these algorithms in computer code.

28. Three vertices v_1, v_2, and v_3 in the plane define a triangle. To generate a random point T within the triangle, sample three independent exponential deviates X, Y, and Z with mean 1 and put

$$T = \frac{X}{X+Y+Z} v_1 + \frac{Y}{X+Y+Z} v_2 + \frac{Z}{X+Y+Z} v_3.$$

Implement this sampling method in code. How can it be justified in theory? How does the method generalize to "triangles" in higher dimensions?

29. Suppose the n-dimensional random deviate X is uniformly distributed within the unit sphere $S_n = \{x : \|x\| \le 1\}$. If Ω is a covariance matrix with square root $\Omega^{1/2}$, then show that $Y = \Omega^{1/2}X$ is uniformly distributed within the ellipsoid $\{y : y^t\Omega^{-1}y \le 1\}$.

22.12 References

[1] Angus J (1994) The probability integral transform and related results. *SIAM Review* 36:652–654

[2] Box GEP, Muller ME (1958) A note on the generation of random normal deviates. *Ann Math Stat* 29:610–611

[3] Brown MB, Bromberg J (1984) An efficient two-stage procedure for generating variates from the multinomial distribution. *Amer Statistician* 38:216–219

[4] Dagpunar J (1988) *Principles of Random Variate Generation.* Oxford University Press, Oxford

[5] Devroye L (1986) *Non-Uniform Random Variate Generation.* Springer, New York

[6] Devroye L (1987) A simple generator for discrete log-concave distributions. *Computing* 39:87–91

[7] Diaconis P, Mehrdad S (1987) The subgroup algorithm for generating uniform random variables. *Prob in Engineering and Info Sci* 1:15-32

[8] Fang KT, Kotz S, Ng KW (1990) *Symmetric Multivariate and Related Distributions.* Chapman & Hall, London

[9] Feller W (1971) *An Introduction to Probability Theory and Its Applications, Volume 2,* 2nd ed. Wiley, New York

[10] Flury BD (1990) Acceptance-rejection sampling made easy. *SIAM Review* 32:474–476

[11] Gentle JE (2003) *Random Number Generation and Monte Carlo Methods,* 2nd ed. Springer, New York

[12] Gilks WR (1992) Derivative-free adaptive rejection sampling for Gibbs sampling. In *Bayesian Statistics 4*, Bernardo JM, Berger JO, Dawid AP, Smith AFM, editors, Oxford University Press, Oxford, pp 641–649

[13] Gilks WR, Wild P (1992) Adaptive rejection sampling for Gibbs sampling. *Appl Stat* 41:337–348

[14] Hörmann W (1994) A universal generator for discrete log-concave distributions. *Computing* 52:89–96

[15] Kalos MH, Whitlock PA (1986) *Monte Carlo Methods, Volume 1: Basics.* Wiley, New York

[16] Kennedy WJ Jr, Gentle JE (1980) *Statistical Computing.* Marcel Dekker, New York

[17] Kingman JFC (1993) *Poisson Processes.* Oxford University Press, Oxford

[18] Knuth D (1981) *The Art of Computer Programming, 2: Seminumerical Algorithms*, 2nd ed. Addison-Wesley, Reading MA

[19] Lange K (2003) *Applied Probability.* Springer, New York

[20] Lange K, Little RJA, Taylor JMG (1989) Robust modeling using the t distribution. *J Amer Stat Assoc* 84:881–896

[21] Marsaglia G, Tsang WW (2000) A simple method for generating gamma deviates. *ACM Trans Math Software* 26:363–372

[22] Park SK, Miller KW (1988) Random number generators: good ones are hard to find. *Communications ACM* 31:1192–1201

[23] Press WH, Teukolsky SA, Vetterling WT, Flannery BP (1992) *Numerical Recipes in Fortran: The Art of Scientific Computing*, 2nd ed. Cambridge University Press, Cambridge

[24] Rice JA (1995) *Mathematical Statistics and Data Analysis*, 2nd ed. Duxbury Press, Belmont, CA

[25] Ripley BD (1983) Computer Generation of Random Variables. *International Stat Review* 51:301–319

[26] Robert CP, Casella G (1999) *Monte Carlo Statistical Methods.* Springer, New York

[27] Rubinstein RY (1981) *Simulation and the Monte Carlo Method.* Wiley, New York

[28] Stewart GW (1980) The efficient generation of random orthogonal matrices with an application to condition estimation. *SIAM J Numer Anal* 17:403-409

[29] von Neumann J (1951) Various techniques used in connection with random digits. Monte Carlo methods. *Nat Bureau Standards* 12:36-38

[30] Wichman BA, Hill ID (1982) Algorithm AS 183: an efficient and portable pseudo-random number generator. *Appl Stat* 31:188–190

23

Independent Monte Carlo

23.1 Introduction

Monte Carlo integration is a rough and ready technique for calculating high-dimensional integrals and dealing with nonsmooth integrands [4, 5, 6, 8, 9, 10, 11, 12, 13]. Although quadrature methods can be extended to multiple dimensions, these deterministic techniques are almost invariably defeated by the curse of dimensionality. For example, if a quadrature method relies on n quadrature points in one dimension, then its product extension to d dimensions relies on n^d quadrature points. Even in one dimension, quadrature methods perform best for smooth functions. Both Romberg acceleration and Gaussian quadrature certainly exploit smoothness.

Monte Carlo techniques ignore smoothness and substitute random points for fixed quadrature points. If we wish to approximate the integral

$$\mathrm{E}[f(X)] = \int f(x)d\mu(x)$$

of an arbitrary integrand $f(x)$ against a probability measure μ, then we can take an i.i.d. sample $X_1, \ldots, X_n$ from μ and estimate $\int f(x)d\mu(x)$ by the sample average $\frac{1}{n}\sum_{i=1}^n f(X_i)$. The law of large numbers implies that these Monte Carlo estimates converge to $\mathrm{E}[f(X)]$ as n tends to ∞. If $f(x)$ is square integrable, then the central limit theorem allows us to refine this conclusion by asserting that the estimator $\frac{1}{n}\sum_{i=1}^n f(X_i)$ is approximately normally distributed around $\mathrm{E}[f(X)]$ with standard deviation $\sqrt{\mathrm{Var}[f(X)]/n}$. In practice, we estimate the order of the Monte Carlo error as $\sqrt{v/n}$, where

$$v = \frac{1}{n-1}\sum_{i=1}^n \Big[f(X_i) - \frac{1}{n}\sum_{j=1}^n f(X_j)\Big]^2$$

is the usual unbiased estimator of $\mathrm{Var}[f(X)]$.

The central limit theorem perspective also forces on us two conclusions. First, the error estimate does not depend directly on the dimensionality of the underlying space. This happy conclusion is balanced by the disappointing realization that the error in estimating $\mathrm{E}[f(X)]$ declines at the slow rate $n^{-1/2}$. In contrast, the errors encountered in quadrature formulas with n quadrature points typically vary as $O(n^{-k})$ for k at least 2. Rather than bemoan the $n^{-1/2}$ rate of convergence in Monte Carlo integration, practitioners now attempt to reduce the $\mathrm{Var}[f(X)]$ part of the standard error

K. Lange, *Numerical Analysis for Statisticians*, Statistics and Computing,
DOI 10.1007/978-1-4419-5945-4_23,

formula $\sqrt{\mathrm{Var}[f(X)]/n}$. Our limited overview accordingly stresses variance reduction methods. Several of these are predicated on the common sense principle of substituting an exact partial result for an approximate partial result. We begin our survey by discussing importance sampling, a supple technique that is often combined with acceptance-rejection and sequential sampling.

23.2 Importance Sampling

Importance sampling is one technique for variance reduction. Suppose that the probability measure μ is determined by a density $g(x)$ relative to a measure ν such as Lebesgue measure or counting measure. If $h(x)$ is another density relative to ν with $h(x) > 0$ when $f(x)g(x) \neq 0$, then we can write

$$\int f(x)g(x)d\nu(x) = \int \frac{f(x)g(x)}{h(x)}h(x)d\nu(x).$$

Thus, if $Y_1, \ldots, Y_n$ is an i.i.d. sample from $h(x)$, then the sample average

$$\frac{1}{n}\sum_{i=1}^{n} f(Y_i)\frac{g(Y_i)}{h(Y_i)} = \frac{1}{n}\sum_{i=1}^{n} f(Y_i)w(Y_i) \tag{23.1}$$

offers an alternative unbiased estimator of $\int f(x)g(x)d\nu(x)$. The ratios $w(Y_i) = g(Y_i)/h(Y_i)$ are called importance weights.

The weighted estimator (23.1) has smaller variance than the naive estimator $\frac{1}{n}\sum_{i=1}^{n} f(X_i)$ if and only if the second moments of the two sampling distributions satisfy

$$\int \Big[\frac{f(x)g(x)}{h(x)}\Big]^2 h(x)d\nu(x) \le \int f(x)^2 g(x)d\nu(x).$$

If we choose $h(x) = |f(x)|g(x)/\int |f(z)|g(z)d\nu(z)$, then the Cauchy-Schwarz inequality implies

$$\begin{aligned}\int \Big[\frac{f(x)g(x)}{h(x)}\Big]^2 h(x)d\nu(x) &= \Big[\int |f(x)|g(x)d\nu(x)\Big]^2 \\ &\le \int f(x)^2 g(x)d\nu(x)\int g(x)d\nu(x) \\ &= \int f(x)^2 g(x)d\nu(x).\end{aligned}$$

Equality occurs if and only if $|f(x)|$ is constant with probability 1 relative to $g(x)d\nu(x)$. If $f(x)$ is nonnegative and $h(x)$ is chosen according to the above recipe, then the variance of the estimator $\frac{1}{n}\sum_{i=1}^{n} f(Y_i)g(Y_i)/h(Y_i)$ reduces to 0.

This elegant result is slightly irrelevant since $\int |f(x)|g(x)d\nu(x)$ is unknown. However, it suggests that the variance of the sample average estimator will be reduced if $h(x)$ resembles $|f(x)|g(x)$. When this is the case, random points tend to be sampled where they are most needed to achieve accuracy.

In practice, one or both of the densities $g(x)$ and $h(x)$ may only be known up to an unspecified constant. This means the true importance weight is $cw(Y_i)$ for some constant $c > 0$ rather than $w(Y_i)$. Although formula (23.1) is no longer available, there is a simple fix. We now approximate $\mathrm{E}[f(X)]$ by

$$\frac{\sum_{i=1}^n f(Y_i)w(Y_i)}{\sum_{i=1}^n w(Y_i)} = \frac{\frac{1}{n}\sum_{i=1}^n f(X_i)cw(Y_i)}{\frac{1}{n}\sum_{i=1}^n cw(Y_i)}. \tag{23.2}$$

A double application of the law of large numbers to the right-hand side of equation (23.2) dictates that the numerator tends almost surely to $\mathrm{E}[f(X)]$ and that the denominator tends almost surely to 1.

Example 23.2.1 *Binomial Tail Probabilities*

Suppose the random variable X is binomially distributed with m trials and success probability p. For x much larger than mp, the right-tail probability $\Pr(X \geq x)$ is very small and estimating its value by taking independent replicates of X is ill advised. If we use importance sampling with random binomial deviates with the same number of trials m but a higher success probability q, then we are apt to be more successful. In the earlier notation of this section, importance sampling now depends on the following functions

$$\begin{aligned}
f(y) &= 1_{\{y \geq x\}} \\
g(y) &= \binom{m}{y} p^y (1-p)^{m-y} \\
h(y) &= \binom{m}{y} q^y (1-q)^{m-y} \\
w(y) &= \left(\frac{p}{q}\right)^y \left(\frac{1-p}{1-q}\right)^{m-y}.
\end{aligned}$$

As a guess, we might choose q to satisfy $mq = x$. This puts most of the mass of the sampling distribution near x. To rationalize this guess, we need to minimize the variance of the corresponding importance estimator. As noted earlier, this amounts to minimizing the second moment

$$\sum_{y \geq x} \binom{m}{y} \frac{[p^y(1-p)^{m-y}]^2}{q^y(1-q)^{m-y}}.$$

If we assume that the terms in this sum are rapidly decreasing and replace the sum by its first term, then it suffices to minimize $q^{-x}(1-q)^{x-m}$. A quick calculation shows that $mq = x$ is optimal. ∎

Example 23.2.2 *Tilted Importance Sampling*

The notion of tilting introduced in Section 19.5 is also relevant to approximating a tail probability $\Pr(X \geq x)$. If the random variable X with density $g(x)$ has cumulant generating function $K(t)$, then tilting by t gives a new density $h_t(x) = e^{xt-K(t)}g(x)$. The importance weight associated with an observation x drawn from $h_t(x)$ is $w(t) = e^{-xt+K(t)}$. If E_t denotes expectation with respect to $h_t(x)$, then the optimal tilt minimizes the second moment

$$\begin{aligned} \mathrm{E}_t(1_{\{X\geq x\}}e^{-2Xt+2K(t)}) &= \mathrm{E}_0(1_{\{X\geq x\}}e^{-Xt+K(t)}) \\ &\leq e^{-xt+K(t)}. \end{aligned}$$

It is far simpler to minimize the displayed bound than the second moment. Differentiation with respect to t shows that the minimum is attained when $K'(t) = x$. For example, when X is normally distributed with mean μ and variance σ^2, the cumulant generating function is $K(t) = \mu t + \frac{1}{2}\sigma^2 t^2$. For a given x, a good tilt is therefore $t = (x-\mu)/\sigma^2$. ■

Example 23.2.3 *Expected Returns to the Origin*

In a three-dimensional, symmetric random walk on the integer lattice [3], the expected number of returns to the origin equals

$$\frac{1}{2^3}\int_{-1}^{1}\int_{-1}^{1}\int_{-1}^{1}\frac{3}{3-\cos(\pi x_1)-\cos(\pi x_2)-\cos(\pi x_3)}dx_1dx_2dx_3.$$

Detailed analytic calculations too lengthy to present here show that this integral approximately equals 1.516. A crude Monte Carlo estimate based on 10,000 uniform deviates from the cube $[-1,1]^3$ is 1.478 ± 0.036. The singularity of the integrand at the origin $\mathbf{0}$ explains the inaccuracy and implies that the estimator has infinite variance. Thus, the standard error 0.036 attached to the estimate 1.478 is bogus.

We can improve the estimate by importance sampling. Let

$$S_3 = \{(x_1,x_2,x_3) : r \leq 1\}$$

be the unit sphere in R^3, where $r = \sqrt{x_1^2+x_2^2+x_3^2}$. Since the singularity near the origin behaves like a multiple of r^{-2}, we decompose the integral as

$$\begin{aligned} &\frac{1}{2^3}\int_{-1}^{1}\int_{-1}^{1}\int_{-1}^{1}\frac{3}{3-\cos(\pi x_1)-\cos(\pi x_2)-\cos(\pi x_3)}dx_1dx_2dx_3 \\ &= \frac{1}{(2^3-\frac{4\pi}{3})}\int_{[-1,1]^3\setminus S_3}\frac{3(2^3-\frac{4\pi}{3})/2^3}{3-\cos(\pi x_1)-\cos(\pi x_2)-\cos(\pi x_3)}dx_1dx_2dx_3 \\ &\quad + \int_{S_3}\frac{3(4\pi r^2)/2^3}{3-\cos(\pi x_1)-\cos(\pi x_2)-\cos(\pi x_3)}\frac{1}{4\pi r^2}dx_1dx_2dx_3. \end{aligned} \tag{23.3}$$

Sampling from the density $1/(4\pi r^2)$ on S_3 concentrates random points near the origin. As the brief calculation

$$\Pr(R^* \leq r) = \frac{1}{4\pi}\int_0^r \int_0^{2\pi} \int_0^{\pi} \frac{1}{s^2} s^2 \sin\phi \, d\phi \, d\theta \, ds = r$$

with spherical coordinates shows, the radius R^* under such sampling is uniformly distributed on $[0, 1]$. This contrasts with the nonuniform distribution

$$\Pr(R \leq r) = \frac{3}{4\pi}\int_0^r \int_0^{2\pi} \int_0^{\pi} s^2 \sin\phi \, d\phi \, d\theta \, ds = r^3$$

of the radius R under uniform sampling on S_3. These formulas demonstrate that we can generate R^* by taking a uniform sample from S_3 and setting $R^* = R^3$.

A strategy for computing the expected number of returns is now clear. We sample a point (x_1, x_2, x_3) uniformly from the cube $[-1, 1]^3$. If $r > 1$, then the point is uniform in $[-1, 1]^3 \setminus S_3$, and we use it to compute a Monte Carlo estimate of the first integral on the right of (23.3). If $r \leq 1$, then we have a uniform point in S_3. If we replace r by r^3 and adjust (x_1, x_2, x_3) accordingly, then we use the revised point in S_3 to compute a Monte Carlo estimate of the second integral on the right of (23.3) based on the density $1/(4\pi r^2)$. Carrying out this procedure with 10,000 random points from $[-1, 1]^3$ and adding the two Monte Carlo estimates produces the improved estimate 1.513 ± 0.030 of the expected number of returns. ■

23.3 Stratified Sampling

In stratified sampling, we partition the domain of integration S of an expectation $\mathrm{E}[f(X)] = \int_S f(x) d\mu(x)$ into $m > 1$ disjoint subsets S_i and sample a fixed number of points $X_{i1}, \ldots, X_{in_i}$ from each S_i according to the conditional probability measure $\mu(A \mid S_i)$. If we estimate the conditional expectation $\mathrm{E}[f(X) \mid X \in S_i]$ by $\frac{1}{n_i}\sum_{j=1}^{n_i} f(X_{ij})$, then the weighted estimator

$$\sum_{i=1}^{m} \mu(S_i) \frac{1}{n_i} \sum_{j=1}^{n_i} f(X_{ij}) \qquad (23.4)$$

is unbiased for $\mathrm{E}[f(X)]$. If the n_i are chosen carefully, then the variance $\sum_{i=1}^m \mu(S_i)^2 \operatorname{Var}[f(X) \mid X \in S_i]/n_i$ of this estimator will be smaller than the variance $\operatorname{Var}[f(X)]/n$ of the sample average Monte Carlo estimator with the same number of points $n = \sum_{i=1}^m n_i$ drawn randomly from S.

For instance, if we take $n_i = n\mu(S_i)$, then the variance of the stratified estimator (23.4) reduces to

$$\frac{1}{n}\sum_{i=1}^{m}\mu(S_i)\operatorname{Var}[f(X) \mid X \in S_i] \quad = \quad \frac{1}{n}\operatorname{E}\{\operatorname{Var}[f(X) \mid Z]\},$$

where Z is a random variable satisfying $Z = i$ when the random point X drawn from μ falls in S_i. Since we can write

$$\operatorname{Var}[f(X)] \quad = \quad \operatorname{E}\{\operatorname{Var}[f(X) \mid Z]\} + \operatorname{Var}\{\operatorname{E}[f(X) \mid Z]\},$$

it is clear that the stratified estimator has smaller variance than the sample average estimator.

In principle, one can improve on proportional sampling. To minimize the variance of the stratified estimator, we treat the n_i as continuous variables, introduce a Lagrange multiplier λ, and look for a stationary point of the Lagrangian

$$\sum_{i=1}^{m}\mu(S_i)^2\frac{1}{n_i}\operatorname{Var}[f(X) \mid X \in S_i] + \lambda\Big(n - \sum_{i=1}^{m} n_i\Big).$$

Equating its partial derivative with respect to n_i to zero and taking into account the constraint $\sum_{i=1}^{m} n_i = n$ yields

$$n_i \quad = \quad n\frac{\mu(S_i)\sqrt{\operatorname{Var}[f(X) \mid X \in S_i]}}{\sum_{k=1}^{m}\mu(S_k)\sqrt{\operatorname{Var}[f(X) \mid X \in S_k]}}.$$

Although the values of the conditional variances $\operatorname{Var}[f(X) \mid X \in S_i]$ are inaccessible in practice, we can estimate them using a small pilot sample of points from each S_i. Once this is done, we can collect a more intelligent, final stratified sample that puts more points where $f(x)$ shows more variation. Obviously, it is harder to give general advice about how to choose the strata S_i and compute their probabilities $\mu(S_i)$ in the first place.

23.4 Antithetic Variates

In the method of antithetic variates, we look for unbiased estimators V and W of an integral that are negatively correlated rather than independent. The average $(V + W)/2$ is also unbiased, and its variance

$$\operatorname{Var}\Big(\frac{V+W}{2}\Big) \quad = \quad \frac{1}{4}\operatorname{Var}(V) + \frac{1}{4}\operatorname{Var}(W) + \frac{1}{2}\operatorname{Cov}(V, W)$$

is reduced compared to what it would be if V and W were independent. The next proposition provides a sufficient condition for achieving negative correlation. Its proof exploits coupled random variables; by definition these reside on the same probability space [7].

Proposition 23.4.1 *Suppose X is a random variable and the functions $f(x)$ and $g(x)$ are both increasing or both decreasing. If the random variables $f(X)$ and $g(X)$ have finite second moments, then*

$$\text{Cov}[f(X), g(X)] \ \geq \ 0.$$

If $f(x)$ is increasing and $g(x)$ is decreasing, or vice versa, then the reverse inequality holds.

Proof: Consider a second random variable Y independent of X but having the same distribution. If $f(x)$ and $g(x)$ are both increasing or both decreasing, then the product $[f(X) - f(Y)][g(X) - g(Y)] \geq 0$. Hence,

$$\begin{aligned} 0 &\leq \text{E}\{[f(X) - f(Y)][g(X) - g(Y)]\} \\ &= \text{E}[f(X)g(X)] + \text{E}[f(Y)g(Y)] - \text{E}[f(X)]\,\text{E}[g(Y)] - \text{E}[f(Y)]\,\text{E}[g(X)] \\ &= 2\,\text{Cov}[f(X), g(X)]. \end{aligned}$$

The same proof with obvious modifications holds when one of the two functions is increasing and the other is decreasing. ■

Example 23.4.1 *Antithetic Uniform Estimators*

Consider the integral $\int f(x)g(x)\,dx$, where $f(x)$ is increasing and the density $g(x)$ has distribution function $G(x)$ with inverse $G^{[-1]}(u)$. If U is uniformly distributed on $[0, 1]$, then in view of Proposition 22.3.1, the random variables $f[G^{[-1]}(U)]$ and $f[G^{[-1]}(1 - U)]$ are both unbiased estimators of $\int f(x)g(x)\,dx$. According to Proposition 23.4.1, they are also negatively correlated. It follows that for a random uniform sample $U_1, \ldots, U_{2n}$, the first of the two estimators

$$\begin{aligned} A_1 &= \frac{1}{2n}\sum_{i=1}^{n}\{f[G^{[-1]}(U_i)] + f[G^{[-1]}(1 - U_i)]\} \\ A_2 &= \frac{1}{2n}\sum_{i=1}^{2n} f[G^{[-1]}(U_i)] \end{aligned}$$

has smaller variance.

The inverse distribution function is not always readily available. As a substitute, one can exploit the symmetry of $g(x)$. If X is distributed as $g(x)$, and $g(x)$ is symmetric around the point μ, then $X - \mu$ has the same distribution as $\mu - X$. In other words, $2\mu - X$ is distributed as X. Therefore, one can substitute $f(X_i)$ for $f[G^{[-1]}(U_i)]$ and $f(2\mu - X_i)$ for $f[G^{[-1]}(1 - U_i)]$ in the above sample averages and reach the same conclusion. ■

23.5 Control Variates

In computing $\text{E}[f(X)]$, suppose that we can calculate exactly the expectation $\text{E}[g(X)]$ for a function $g(x)$ close to $f(x)$. Then it makes sense to write

$$\mathrm{E}[f(X)] = \mathrm{E}[f(X) - g(X)] + \mathrm{E}[g(X)]$$

and approximate $\mathrm{E}[f(X) - g(X)]$ by a Monte Carlo estimate rather than $\mathrm{E}[f(X)]$. Example 23.2.3 provides a test case of this tactic. Near the origin the integrand

$$\begin{aligned} f(x) &= \frac{3}{3 - \cos(\pi x_1) - \cos(\pi x_2) - \cos(\pi x_3)} \\ &\approx \frac{6}{\pi^2 r^2} \\ &= g(x). \end{aligned}$$

By transforming to spherical coordinates it is straightforward to calculate

$$\frac{1}{\frac{4\pi}{3}} \int_{S_3} g(x) dx_1 dx_2 dx_3 = \frac{18}{\pi^2}.$$

In Example 23.2.3 we can avoid importance sampling by forming a Monte Carlo estimate of the conditional expectation $\mathrm{E}[f(X) - g(X) \mid X \in S_3]$ and adding it to the exact conditional expectation

$$\mathrm{E}[g(X) \mid X \in S_3] = \frac{18}{\pi^2}.$$

Making this change but proceeding otherwise precisely as sketched in Example 23.2.3 yields the impressive approximation 1.517 ± 0.015 for the expected number of returns to the origin. In this case, the method of control variates operates by subtracting off the singularity of the integrand and performs better than importance sampling.

23.6 Rao-Blackwellization

The Rao-Blackwell theorem in statistics takes an unbiased estimator and replaces it by its conditional expectation given a sufficient statistic. This process reduces the variance of the statistic while retaining its unbiasedness. The same principle applies in Monte Carlo integration. The next example illustrates the idea.

Example 23.6.1 *Covariance of the Bivariate Exponential*

Let U, V, and W be independent exponentially distributed random variables with intensities λ, μ, and ν, respectively. The two random variables $X = \min\{U, W\}$ and $Y = \min\{V, W\}$ are exponentially distributed with intensities $\lambda + \nu$ and $\mu + \nu$. The random vector (X, Y) follows the bivariate exponential distribution. Calculation of the covariance $\text{Cov}(X, Y)$ is an interesting theoretical exercise. Problem 17 asks the reader to demonstrate that

$$\text{Cov}(X, Y) = \frac{\nu}{(\lambda + \nu)(\mu + \nu)(\lambda + \mu + \nu)}. \tag{23.5}$$

If one ignores this exact result and wants to approximate $\text{Cov}(X, Y)$ numerically, the natural approach is to generate a trivariate sample (U_i, V_i, W_i) for $1 \le i \le n$, form the bivariate sample (X_i, Y_i) with $X_i = \min\{U_i, W_i\}$ and $Y_i = \min\{V_i, W_i\}$, and compute the latter's sample covariance. Let us call this method 1.

Method 2 is less direct but more accurate. It exploits the formula

$$\text{Cov}(X, Y) = \text{Cov}[\text{E}(X \mid W), \text{E}(Y \mid W)] + \text{E}[\text{Cov}(X, Y \mid W)].$$

The second term on the right vanishes because X and Y are independent given W. The first term can be simplified by noting that

$$\begin{aligned} \text{E}(X \mid W) &= \int_0^\infty \Pr(\min\{U, W\} \ge x \mid W)\, dx \\ &= \int_0^W e^{-\lambda x} dx \\ &= \frac{1 - e^{-\lambda W}}{\lambda}, \end{aligned}$$

and similarly for $\text{E}(Y \mid W)$. (See Problem 18.) Thus, method 2 takes a random sample $W_1, \ldots, W_n$ and computes the sample covariance of the conditional expectations $\text{E}(X_i \mid W_i)$ and $\text{E}(Y_i \mid W_i)$. As a test of the two methods, we set $\lambda = 1$, $\mu = 2$, and $\nu = 3$. The exact covariance is then 0.02500. With 10,000 trials, method 1 yields the approximation 0.02466 and method 2 the approximation 0.02490. ■

Rao-Blackwellization also illuminates integration via acceptance-rejection sampling [1]. Recall that in the acceptance-rejection method we sample from an envelope density $h(x)$ that satisfies an inequality $g(x) \le ch(x)$ relative to a target density $g(x)$. We accept a sampled point X drawn from the density $h(x)$ based on an independent uniform deviate U. If $U \le V = g(X)/[ch(X)]$, then we accept X; otherwise, we reject X. The accepted points conform to the target density $g(x)$.

In computing an expectation $\text{E}[f(X)] = \int f(x)g(x)dx$ by the Monte Carlo method, suppose we generate n points $X_1, \ldots, X_n$ according to $h(x)$ and accept precisely m of them using the uniform deviates $U_1, \ldots, U_n$. The Monte Carlo estimate of $\text{E}[f(X)]$ is

$$\text{E}[f(X)] \approx \frac{1}{m} \sum_{i=1}^n f(X_i) 1_{\{U_i \le V_i\}}. \tag{23.6}$$

The conditional expectation

$$\begin{aligned}
&\frac{1}{m}\mathrm{E}\Big[\sum_{i=1}^n f(X_i)1_{\{U_i\le V_i\}} \mid X_1,\ldots,X_n,\sum_{j=1}^n 1_{\{U_j\le V_j\}}=m\Big] \\
&= \frac{1}{m}\sum_{i=1}^n f(X_i)\,\mathrm{E}\Big[1_{\{U_i\le V_i\}} \mid X_1,\ldots,X_n,\sum_{j=1}^n 1_{\{U_j\le V_j\}}=m\Big]
\end{aligned}$$

retains the unbiased character of the Monte Carlo estimate while reducing its variance. The revised estimate achieves this trick by using both the rejected and the accepted points with appropriate weights for each.

To make this scheme viable, we must compute the conditional probability R_i that the ith deviate X_i is accepted, given its success probability

$$V_i = \frac{g(X_i)}{ch(X_i)}$$

and the fact that there are m successes in n trials. Fortunately, Example 22.10.2 of Chapter 22 deals with exactly this problem. Once the R_i are in hand, the improved approximation

$$\mathrm{E}[f(X)] \approx \frac{1}{m}\sum_{i=1}^n f(X_i)R_i \tag{23.7}$$

is available.

When n is large, R_i is apt to be close to V_i, and m is apt to be close to $\sum_{i=1}^n V_i$. If we drop the factor c^{-1} from V_i and define the revised weight $w(X_i) = g(X_i)/h(X_i)$, then formula (23.7) reduces under these approximations to the alternative importance sampling formula (23.2). Chen [2] argues that formula (23.2) is generally preferable to the acceptance-rejection formula (23.6). This conclusion is hardly surprising given the Rao-Blackwell connection. In addition to leading to more precise estimates, the importance sampling perspective avoids explicit mention of the constant c and the necessity of generating the uniform deviates $U_1,\ldots,U_n$.

23.7 Sequential Importance Sampling

As its name implies, sequential importance sampling combines sequential sampling with importance sampling. In practice this involves sequential sampling of the importance density $h(x)$. Suppose that x has d components and that we factor $h(x)$ as

$$h(x) = h_1(x_1)\prod_{j=2}^d h_j(x_j \mid x_1,\ldots,x_{j-1}).$$

To randomly draw from $h(x)$, we draw the first component according to $h_1(x_1)$. Conditional on the first component, we then draw the second component according to $h_2(x_2 \mid x_1)$ and so forth until the entire random vector X is filled out. In this process, there is nothing to prevent x_j from being a subvector instead of a single component. Often the importance weight $w(X)$ is computed recursively as the components of X are added. Two examples treated more fully in [8] illustrate the general approach.

Example 23.7.1 *Self-Avoiding Symmetric Random Walks*

This famous problem of polymer growth attracted the interest of applied mathematicians and physicists in the mid 20*th* century. For the sake of concreteness, suppose the walker starts at the origin of the plane and currently occupies the integer lattice point (j, k). At the next step he moves randomly to one of the four adjacent lattice points $(j-1, k)$, $(j+1, k)$, $(j, k-1)$, or $(j, k+1)$ with the uniform probability of $\frac{1}{4}$. When the walker visits a point already visited, the walk fails to model polymer growth well. Let z_d be the number of paths of d steps that do not cross themselves. The most obvious question is how to compute z_d. For d small, direct enumeration is possible, but as d grows it becomes more and more difficult to check each of the 4^{d-1} possible paths for self-avoidance. Among the z_d permitted paths, one would like to know the average distance from the origin and other statistics.

The primary obstacle to computing these quantities by Monte Carlo summation is that the fraction of self-avoiding paths declines quickly as d grows. Naive sampling is therefore terribly inefficient. A better alternative is to generate paths that tend to avoid themselves. One can then correct for biased sampling by introducing importance weights. A reasonable possibility is to modify the move of the walker from the lattice point (j, k) defining position $m-1$. If we scan the points already visited, then some of the four nearest neighbors of (j, k) will be among those points. Let c_{m-1} count the number of neighboring points not visited in the first $m-1$ steps of the walk. In the redesigned walk, the walker chooses the next destination among these points with the uniform probability $1/c_{m-1}$. Thus, an entire path P of length d without self-intersection is chosen with probability $\prod_{j=1}^{d-1} c_j^{-1}$. A partial path that stops prematurely at step e is assigned probability $\prod_{j=1}^{e-1} c_j^{-1}$. The sample space for the revised sampling scheme contains both partial and full paths. Figure 23.1 depicts a sample random walk of $d = 30$ steps constructed under the revised scheme.

The importance weight

$$w(P) = \begin{cases} z_d^{-1} \prod_{j=1}^{d-1} c_j(P), & P \text{ a successful path} \\ 0, & P \text{ an unsuccessful path} \end{cases}$$

can clearly be computed recursively up to the unknown constant z_d. Because $\mathrm{E}[w(P)] = 1$ under the revised scheme, this gives an opening to

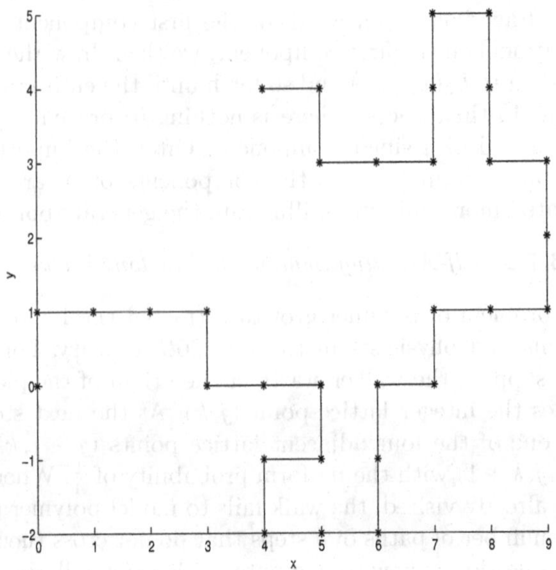

FIGURE 23.1. A Random Walk Biased toward Self-Avoidance

compute z_d. If we generate independent paths $P_1, \ldots, P_n$ under the revised scheme, then

$$1 \approx \frac{1}{n}\sum_{j=1}^{n} w(P_j),$$

which we rearrange to yield the estimate

$$\hat{z}_d = \frac{1}{n}\sum_{i=1}^{n} 1_{\{e(P_i)=d\}} \prod_{j=1}^{e(P_i)-1} c_j(P_i).$$

If R_d is the distance from the origin conditional on self-avoidance, then we estimate $\mathrm{E}(R_d)$ by

$$\hat{\mathrm{E}}(R_d) = \frac{\sum_{j=1}^{n} R_d(P_j)w(P_j)}{\sum_{j=1}^{n} w(P_j)}.$$

In this case, the unknown partition constant z_d cancels, and we can employ revised weights omitting it.

To get a feel for the size of z_d and R_d, we conducted a Monte Carlo experiment with $d = 100$ steps and n =100,000 trials. The importance sampling formulas just given yield $\hat{z}_d = 4.3 \times 10^{42}$, $\hat{\mathrm{E}}(R_d) = 26.0$, and $\hat{\mathrm{Var}}(R_d) = 85.5$. The first of these numbers makes it clear that only the miniscule fraction 1.1×10^{-17} of all realizations under ordinary sampling avoid self-intersection. ■

Example 23.7.2 *Permanents of 0-1 Matrices*

The permanent of a $k \times k$ matrix $M = (m_{ij})$ is defined by the formula

$$\operatorname{perm}(M) = \sum_{\sigma \in S_k} \prod_{i=1}^{k} m_{i,\sigma(i)} = \sum_{\sigma \in S_k} \prod_{i=1}^{k} m_{\sigma(i),i},$$

where S_k is the set of all permutations of $\{1, \ldots, k\}$. This is precisely the definition of the determinant of M with the alternating signs removed. Permanents find applications in combinatorics and permutation tests. Here we will be concerned with computing the permanent of a sparse matrix with all entries equal to 0 or 1. In this setting $\operatorname{perm}(M)$ counts the number of permutations with restricted positions. For example, the permanent of the matrix

$$M = \begin{pmatrix} 1 & 0 & 0 & 1 & 0 \\ 0 & 0 & 1 & 0 & 1 \\ 0 & 1 & 0 & 0 & 0 \\ 1 & 1 & 1 & 1 & 0 \\ 0 & 1 & 1 & 0 & 0 \end{pmatrix} \tag{23.8}$$

equals 2. This can be rephrased by considering random permutations σ and defining the random variable $X(\sigma) = \prod_{i=1}^{k} m_{i,\sigma(i)}$. It is then clear that $\mathrm{E}(X) = \frac{2}{5!} = 0.0167$.

In larger matrices, exhaustive enumeration of all permutations is out of the question. If we randomly sample permutations and compute X for a sparse matrix, then most of the time $X = 0$. To overcome this problem, we need to sample from a distribution that is enriched for permutations σ satisfying $X(\sigma) = 1$. Consider sampling one column at a time. In column 1 there must be a non-zero entry m_{i1}; otherwise, $\operatorname{perm}(M) = 0$. If T_1 is the set $\{i : m_{i1} = 1\}$, then one possibility is to sample $\sigma(1)$ uniformly from T_1. Unfortunately, this tactic is not ideal. To understand why, let r_i count the number of 1's in row i. If we have $r_i = 1$ for some $i \in T_1$, then we are forced to choose $\sigma(1) = i$. Failure to do so compels us to pick a 0 entry in row i at a later stage. In fact, we should concentrate on using the sparse rows early in the hope that the dense rows will still have enough non-zero entries when they are encountered to create a full permutation σ with $X(\sigma) = 1$. When all $r_i > 1$ for $i \in T_1$, Liu and Chen [8] suggest choosing i with probability $(r_i - 1)^{-1} d_1^{-1}$, where

$$d_1 = \sum_{j \in T_1} \frac{1}{r_j - 1}.$$

This favors the sparse rows.

Once $\sigma(1) = i$ is selected, we reset the entries in column 1 and row i of M to 0. We then move on to column 2 and choose $\sigma(2)$ by the same process. This sequential construction either produces a full permutation σ with $X(\sigma) = 1$ or terminates prematurely with a partial permutation σ with $X(\sigma) = 0$. Either kind of outcome has a well-defined probability equaling the product of the choice probabilities along the way. We start the importance weight $w(\sigma)$ attached to a permutation σ at 1 and replace it at stage i by $w(\sigma)(r_{\sigma(i)} - 1)d_i$. Because we have omitted the normalizing constant $\frac{1}{k!}$ from the weight, a random sample $\sigma_1, \ldots, \sigma_n$ of permutations from the importance sampling distribution gives the estimate

$$\widehat{\text{perm}}(M) = \frac{1}{n}\sum_{j=1}^{n} X(\sigma_j)w(\sigma_j).$$

We estimate the permanent of the toy matrix (23.8) as 1.95 using 1000 trials of importance sampling and as 2.64 using 1000 trials of ordinary uniform sampling. ■

23.8 Problems

1. Consider the integral $\int_0^1 \cos(\pi x/2)\,dx = 2/\pi$ [6]. Interpreting this as an expectation relative to the uniform distribution on $[0, 1]$, show that

$$\text{Var}\left[\cos\left(\frac{\pi X}{2}\right)\right] = \frac{1}{2} - \left(\frac{2}{\pi}\right)^2 \approx 0.095.$$

   The importance density $h(x) = 3(1 - x^2)/2$ roughly resembles the integrand $\cos(\pi x/2)$. Demonstrate numerically that

$$\text{Var}\left[\frac{2\cos\left(\frac{\pi Y}{2}\right)}{3(1-Y^2)}\right] \approx 0.00099$$

   when Y is sampled from $h(y)$. Thus, importance sampling reduces the variance of the corresponding Monte Carlo estimator by almost a factor of 100.

2. Continuing Problem 1, devise a ratio method for sampling from the density $h(x) = 3(1 - x^2)/2$ on $[0, 1]$. Show that $[0, \sqrt{3/2}] \times [0, \sqrt{3/8}]$ is an enclosing rectangle.

3. Suppose X is binomially distributed with 100 trials and success probability 0.25. Compute the right-tail probability $\Pr(X \geq 75)$ by importance sampling as suggested in Example 23.2.1.

4. Let Y be a random variable with density $g(y)$ relative to a measure ν. Suppose that the random variable $Z = f(Y)$ satisfies $\text{E}_g(Z^2) < \infty$. It

is helpful for the sample average (23.1) used in estimating $\mathrm{E}_g(Z)$ to have finite variance. Show that this condition is fulfilled whenever the importance weight $g(y)/h(y)$ is bounded on the set $\{y : f(y) \neq 0\}$. If $f(y)$ never vanishes, then this is just the boundedness requirement imposed in the acceptance-rejection algorithm for sampling from $g(y)$ via $h(y)$.

5. In conjunction with Example 23.2.2, calculate the cumulant generating functions $K(t)$ of the exponential and Poisson distributions. Solve the equation $K'(t) = x$ for t.

6. Suppose X follows a standard normal distribution. Write and test a program to approximate the right-tail probability $\Pr(X \geq x)$ by tilted importance sampling. Assume that x is large and positive.

7. Monte Carlo simulation also provides a device for sampling from a marginal distribution based on a joint distribution. Suppose the random vector $Y = (U, V)^t$ has joint density $f_{UV}(u, v)$ relative to the product measure $\mu \times \nu$. To approximate the marginal density $f_U(u) = \int f_{UV}(u, v)\, d\nu(v)$, we take a random sample $y_1, \ldots, y_n$ from Y and form the average

$$A_n = \frac{1}{n} \sum_{i=1}^{n} \frac{f_{UV}(u, v_i) g(u_i)}{f_{UV}(u_i, v_i)}.$$

Here $g(w)$ is any probability density relative to μ. Demonstrate that A_n is an unbiased estimator of $f_U(u)$ and that the variance of A_n is minimized by choosing

$$g(w) = \frac{f_{UV}(w, v)}{\int f_{UV}(u, v)\, d\mu(u)},$$

the conditional density of U given $V = v$. (Hint: Apply the Cauchy-Schwarz inequality.)

8. Continuing Problem 7, suppose U is Poisson distributed with mean V and V is gamma distributed as described in Problem 14 of Chapter 22. The marginal distribution U is negative binomial. Write a program to compute the marginal distribution of U by sampling from the joint distribution of U and V. Test several densities $g(w)$ as described in Problem 7 and see which one works best in practice.

9. Write a program to carry out the naive Monte Carlo method, the importance sampling method, and the control variate method for estimating the random walk integral discussed in Example 23.2.3 and Section 23.5. Show analytically that about 52.4 percent of the points generated in the cube $[-1, 1]^3$ fall within S_3.

10. Suppose you compute by stratified sampling from the two subintervals $[0, \sqrt{3}/2)$ and $[\sqrt{3}/2, 1]$ a Monte Carlo estimate of the integral $\int_0^1 \sqrt{1-x^2}dx$. If you employ n points overall, how many points should you allot to each subinterval to achieve the greatest variance reduction [9]?

11. In the stratified sampling method, suppose the domain of integration splits into two subsets S_1 and S_2 such that

$$\begin{aligned}\Pr(X \in S_1) &= \Pr(X \in S_2)\\ \mathrm{E}[f(X) \mid X \in S_1] &= \mathrm{E}[f(X) \mid X \in S_2]\\ \mathrm{Var}[f(X) \mid X \in S_1] &= \mathrm{Var}[f(X) \mid X \in S_2].\end{aligned}$$

Show that the stratified estimate of $\mathrm{E}[f(X)]$ with $3n/4$ points drawn randomly from S_1 and $n/4$ points drawn randomly from S_2 is actually worse than the sample mean estimate with all n points drawn randomly from $S_1 \cup S_2$.

12. Consider Monte Carlo estimation of the integral $\int_0^1 e^x dx$. Calculate the reduction in variance achieved by applying antithetic variables.

13. Proposition 23.4.1 demonstrates what a powerful tool coupling of random variables is in proving inequalities. Here is another example involving the monotonicity of power functions in hypothesis testing. Let X and Y follow binomial distributions with n trials and success probabilities $p < q$, respectively. We can realize X and Y on the same probability space by scattering n points randomly on the unit interval. Interpret X as the number of points less than or equal to the cutoff p, and interpret Y as the number of points less than or equal to the cutoff q. Show that this interpretation leads to an immediate proof of the inequality $\Pr(X \geq k) \leq \Pr(Y \geq k)$ for all k.

14. Continuing Problem 13, suppose that X follows the hypergeometric distribution

$$\Pr(X = i) = \frac{\binom{r}{i}\binom{n-r}{m-i}}{\binom{n}{m}}.$$

Let Y follow the same hypergeometric distribution except that $r+1$ replaces r. Prove that $\Pr(X \geq k) \leq \Pr(Y \geq k)$ for all k. (Hint: Consider an urn with r red balls, 1 white ball, and $n-r-1$ black balls. If we draw m balls from the urn without replacement, then X is the number of red balls drawn, and Y is the number of red or white balls drawn.)

15. The method of control variates can be used to estimate the moments of the sample median $X_{(n)}$ from a random sample of size $2n-1$ from

a symmetric distribution. Because we expect the difference $X_{(n)} - \bar{X}$ between the sample median and the sample mean to be small, the moments of $\bar{X}$ serve as a first approximation to the moments of $X_{(n)}$. Put this insight into practice by writing a Monte Carlo program to compute $\text{Var}(X_{(n)})$ for a sample from the standard normal distribution.

16. Suppose that X and Y are independent random variables with distribution functions $F(x)$ and $G(y)$. Sketch a naive simulation tactic for estimating $\Pr(X + Y \le z)$. Describe how you can improve on the naive tactic by conditioning on either X or Y.

17. Consider the bivariate exponential distribution defined in Example 23.6.1. Demonstrate that:

    (a) X and Y are exponentially distributed with the stated means.

    (b) (X, Y) has the tail probability

    $$\Pr(X \ge x, Y \ge y) = e^{-\lambda x - \mu y - \nu \max\{x,y\}}.$$

    (c) The probabilities

    $$\begin{aligned}\Pr(X < Y) &= \frac{\lambda}{\lambda + \mu + \nu} \\ \Pr(Y < X) &= \frac{\mu}{\lambda + \mu + \nu} \\ \Pr(X = Y) &= \frac{\nu}{\lambda + \mu + \nu}\end{aligned}$$

    are valid.

    (d) (X, Y) possesses the density

    $$f(x, y) = \begin{cases} \lambda(\mu + \nu)e^{-\lambda x - (\mu+\nu)y} & x < y \\ \mu(\lambda + \nu)e^{-(\lambda+\nu)x - \mu y} & x > y \end{cases}$$

    off the line $y = x$.

    (e) The covariance expression (23.5) holds.

18. Let Z be a nonnegative random variable with distribution function $G(z)$. Demonstrate that

    $$\text{E}(Z) = \int_0^\infty [1 - G(z)]dz$$

    by interchanging the order of integration in the implied double integral.

19. Program the naive method and the sequential importance sampling method for either of the two examples 23.7.1 and 23.7.2. Compare your findings with the results in the text.

23.9 References

[1] Casella G, Robert CP (1996) Rao-Blackwellisation of sampling schemes. *Biometrika* 83:81–94

[2] Chen Y, (2005) Another look at rejection sampling through importance sampling. *Prob Stat Letters* 72:277–283

[3] Doyle PG, Snell JL (1984) *Random Walks and Electrical Networks.* The Mathematical Association of America, Washington, DC

[4] Gentle JE (2003) *Random Number Generation and Monte Carlo Methods*, 2nd ed. Springer, New York

[5] Hammersley JM, Handscomb DC (1964) *Monte Carlo Methods.* Methuen, London

[6] Kalos MH, Whitlock PA (1986) *Monte Carlo Methods, Volume 1: Basics.* Wiley, New York

[7] Lindvall T (1992) *Lectures on the Coupling Method.* Wiley, New York

[8] Liu JS (2001) *Monte Carlo Strategies in Scientific Computing.* Springer, New York

[9] Morgan BJT (1984) *Elements of Simulation.* Chapman & Hall, London

[10] Press WH, Teukolsky SA, Vetterling WT, Flannery BP (1992) *Numerical Recipes in Fortran: The Art of Scientific Computing*, 2nd ed. Cambridge University Press, Cambridge

[11] Robert CP, Casella G (1999) *Monte Carlo Statistical Methods.* Springer, New York

[12] Ross SM (2002) *Simulation,* 3rd ed. Academic Press, San Diego

[13] Rubinstein RY (1981) *Simulation and the Monte Carlo Method.* Wiley, New York

24

Permutation Tests and the Bootstrap

24.1 Introduction

In this chapter we discuss two techniques, permutation testing and the bootstrap, of immense practical value. Both techniques involve random resampling of observed data and liberate statisticians from dubious model assumptions and large sample requirements. Both techniques initially met with considerable intellectual resistance. The notion that one can conduct hypothesis tests or learn something useful about the properties of estimators and confidence intervals by resampling data was alien to most statisticians of the past. The computational demands of permutation testing and bootstrapping alone made them unthinkable. These philosophical and practical objections began to crumble with the advent of modern computing.

Permutation testing is one of the many seminal ideas in statistics introduced by Fisher [11]. Despite the fact that permutation tests have flourished in biomedical research, they receive little mention in most statistics courses. This educational oversight stems from three sources. First, permutation tests involve little elegant mathematics and rely on brute force computation. Second, the exchangeability hypothesis on which they depend is often false. Third, as pure hypothesis tests, they forge only weak connections to parameter estimation and confidence intervals. Despite these drawbacks, permutation methods belong in the statistics curriculum. They are simple to explain, require almost no distributional assumptions, and can be nearly as powerful as parametric analogs in ideal settings.

Part of the mystery of permutation tests revolves around the distinction between population sampling and randomization [10, 19]. Statistics courses emphasize population sampling. This is attractive because conclusions drawn from the sample can be generalized to the entire population. In many biomedical applications, data sets are small, and a nonrandom sample is purposely randomized to protect against hidden biases.

The bootstrap is an even more versatile tool than permutation testing. Because it avoids normality assumptions and large sample theory, it perfectly embodies the healthy skepticism of many statisticians. The introduction of the jackknife by Quenouille [25] and Tukey [29] demonstrated some of the virtues of data resampling. In the last quarter century, Efron's bootstrap has largely supplanted the jackknife [7].

In essence, the bootstrap sets up a miniature world parallel to the real

K. Lange, *Numerical Analysis for Statisticians*, Statistics and Computing,
DOI 10.1007/978-1-4419-5945-4_24, © Springer Science+Business Media, LLC 2010

world where the data originate. Inferences in the bootstrap world are made by resampling the data with replacement. This enables one to explore the properties of an estimator such as bias and variance. The bootstrap correspondence principle allows one to transfer these findings from the bootstrap world back to the real world. In the early days of the bootstrap when computing was still relatively expensive, Efron and others set guidelines for the minimum number of bootstrap samples for safe inference. In view of the speed of current computers, it seems better to take a more experimental attitude and continue simulating until inferences stabilize. For this reason, we refrain from recommending bootstrap sample sizes.

24.2 Permutation Tests

In designing permutation tests, statisticians such as Fisher and Pitman [22, 23, 24] were guided by their experience with parametric models. There are many permutation analogs of likelihood ratio tests. Permutation tests differ from parametric tests in how the distribution of the test statistic is evaluated. Asymptotic distributions are dropped in favor of exact distributions conditional on the sample. The idea is best introduced in the concrete setting of the two-sample t-test popular in classical statistics. The t-distribution arises when the measured variable is normally distributed in each sampled population. If the variable is far from normally distributed, a nonparametric test is a good alternative. Tests based on ranks are often recommended, but the permutation t-test can be more powerful.

Suppose the observations from population 1 are labeled $x_1, \ldots, x_m$, and the observations from population 2 are labeled $y_1, \ldots, y_n$. The sample means $\bar{x}$ and $\bar{y}$ should not differ much if the variable has the same distribution in the two populations. Accordingly, the difference $T = \bar{x} - \bar{y}$ is a reasonable test statistic. In permutation testing, we concatenate the observations into one long sequence $z_1, \ldots, z_{m+n}$ and assume under the null hypothesis that all permutations of the data vector z are equally likely. These permutations generate $\binom{m+n}{m}$ equally likely versions of the test statistic T. In evaluating T, $z_1, \ldots, z_m$ replace $x_1, \ldots, x_m$, and $z_{m+1}, \ldots, z_{m+n}$ replace $y_1, \ldots, y_n$. The p-value attached to the observed value T_{obs} is just the fraction of permuted vectors z with $|T| \geq |T_{\text{obs}}|$. If the number of combinations $\binom{m+n}{m}$ is not too large, one can enumerate all of them and compute the exact p-value; otherwise, one can randomly sample permutations and approximate the exact p-value by the sample proportion.

When the simulation route is taken, one should indicate the precision of the approximate p-value. If k random permutations are sampled and p is the true p-value, then the sample proportion $\hat{p}$ has mean p and variance $\sigma_k^2 = k^{-1}p(1-p)$. For k large, the coverage probability of the interval $(\hat{p} - 2\hat{\sigma}_k, \hat{p} + 2\hat{\sigma}_k)$ is about 95%. In interpreting p-values, two cautions should be borne in mind. First, the set of possible p-values is discrete with jumps of size $1/\binom{m+n}{m}$. Obviously, this limitation is more of an issue for small samples. Second, the width of the 95% confidence interval is inversely proportional to $\sqrt{k}$. This fact makes it difficult to approximate very small p-values accurately. Many practitioners of permutation testing include the observed statistic T_{obs} as one of the sample statistics in computing $\hat{p}$. If the data conform to the null hypothesis, this action is certainly legitimate. However, if the data do not conform, then the p-value is slightly inflated, and the test is conservative. Inclusion or exclusion is largely a matter of taste. Our examples exclude the observed statistic.

Because permutation tests rely so heavily on random permutations, it is worth mentioning how to generate a random permutation π on $\{1,\ldots,n\}$. The standard algorithm starts with an arbitrary permutation π. It then redefines π_1 by choosing a random integer j between 1 and n and swapping π_1 and π_j. Once $\pi_1,\ldots,\pi_i$ have been redefined, it chooses a random integer j between $i+1$ and n and swaps π_{i+1} and π_j. A total of $n-1$ such random swaps completely redefine π.

TABLE 24.1. Midge Morphological Data

Amerohelea fasciata		*Amerohelea pseudofasciata*	
Wing Length	**Antenna Length**	**Wing Length**	**Antenna Length**
1.72	1.24	1.78	1.14
1.64	1.38	1.86	1.20
1.74	1.36	1.96	1.30
1.70	1.40	2.00	1.26
1.82	1.38	2.00	1.28
1.82	1.48	1.96	1.18
1.90	1.38		
1.82	1.54		
2.08	1.56		

Example 24.2.1 *Distinguishing Two Midge Species*

Our first example considers the midge data of Grogan and Wirth [14]. In an attempt to differentiate the two related species *Amerohelea fasciata* and *Amerohelea pseudofasciata*, Grogan and Wirth measured several morphological features, including the wing lengths and antenna lengths featured in Table 24.1. The permutation t-tests for wing length and antenna length have approximate p-values of 0.0697 and 0.0023, respectively, based on 100,000 random permutations. The corresponding 95% confidence intervals are $(0.0681, 0.0713)$ and $(0.0020, 0.0026)$. In this case, there are only $\binom{15}{9} = 5005$ possible combinations, and complete enumeration of all relevant combinations is clearly possible. Chapter 3 of the reference [20] discusses efficient algorithms for this task.

Antenna length alone discriminates between the two species, but the borderline significance of wing length suggests that a bivariate test might be even more significant. Because the two measurements are correlated, the appropriate classical test employs Hotelling's two-sample T^2 statistic. Assuming normally distributed observations, the test statistic follows an F distribution under the null hypothesis. Passing to the corresponding permutation test obviates the normality assumption. Up to an irrelevant constant, the test statistic is now defined by $T^2 = (\bar{x} - \bar{y})^t W^{-1}(\bar{x} - \bar{y})$, where

$$W = \sum_{i=1}^{m}(x_i - \bar{x})(x_i - \bar{x})^t + \sum_{j=1}^{n}(y_j - \bar{y})(y_j - \bar{y})^t$$

and each observation x_i or y_i has two components. The approximate p-value falls to 0.00015 for the permutation version of Hotelling's test with 100,000 random permutations. The corresponding 95% confidence interval $(0.00007, 0.00023)$ covers the number $1/5005 = 0.00020$, and we can safely conclude that the natural order of the data gives the most extreme test statistic. ■

Example 24.2.2 *The One-Way Layout and Reading Speeds*

In the one-way layout, there are k samples of sizes $n_1, \ldots, n_k$ with unknown means $\mu_1, \ldots, \mu_k$ and a common unknown variance σ^2. The null hypothesis postulates that the means coincide as well. Let y_{ij} denote the jth observation from population i and $n = \sum_{i=1}^{n}$ the total sample size. Assuming the y_{ij} are normally distributed, the likelihood ratio test of the null hypothesis takes the difference of the two statistics

$$\min_{\mu} \sum_{i=1}^{k}\sum_{j=1}^{n_i}(y_{ij} - \mu)^2 = \sum_{i=1}^{k}\sum_{j=1}^{n_i} y_{ij}^2 - n\bar{y}^2$$

and

$$\sum_{i=1}^{k}\min_{\mu_i}\sum_{j=1}^{n_i}(y_{ij} - \mu_i)^2 = \sum_{i=1}^{k}\sum_{j=1}^{n_i} y_{ij}^2 - \sum_{i=1}^{k} n_i\bar{y}_i^2,$$

where $\bar{y}_i = n_i^{-1}\sum_{j=1}^{n_i} y_{ij}$. In the permutation setting, we omit irrelevant constants and define the test statistic

$$S = \sum_{i=1}^{k} n_i\bar{y}_i^2.$$

The sample space consists of all permutations of the vector z obtained by concatenating the outcome vectors for the k different populations. In recomputing S, the first n_1 entries of z replace $y_{11}, \ldots, y_{1n_1}$, the next n_2 entries replace $y_{21}, \ldots, y_{2n_2}$, and so forth. Large values of S reject the null hypothesis.

TABLE 24.2. Reading Speeds and Typeface Styles

Typeface Style		
1	**2**	**3**
135	175	105
91	130	147
111	514	159
87	283	107
122		194

Table 24.2 presents data of Bradley [3] on the relationship between reading speed and typeface style. The measured variables are reading speeds for 14 subjects randomly assigned to three typeface styles. With 100,000 random permutations, we estimate the p-value of no effect of typeface as 0.0111 with a 95% confidence interval of $(0.0104, 0.0117)$. ■

Example 24.2.3 *Exact Tests of Independence in Contingency Tables*

To illustrate permutation testing in a discrete setting, we now turn to the problem of testing independence in large sparse contingency tables [18]. Consider a table with n multivariate observations scored on m factors. We will denote a typical cell of the table by a multi-index $\mathbf{i} = (i_1, \ldots, i_m)$, where i_j represents the level of factor j in the cell. If the probability associated with level k of factor j is p_{jk}, then under the assumption of independence of the various factors, a multivariate observation falls in cell $\mathbf{i} = (i_1, \ldots, i_m)$ with probability

$$p_{\mathbf{i}} = \prod_{j=1}^{m} p_{ji_j}.$$

Furthermore, the cell counts $\{n_{\mathbf{i}}\}$ from the sample follow a multinomial distribution with parameters $(n, \{p_{\mathbf{i}}\})$.

For the purposes of testing independence, the probabilities p_{jk} are nuisance parameters. In exact inference, one conditions on the marginal counts $\{n_{jk}\}_k$ of each factor j. For j fixed these follow a multinomial distribution with parameters $(n, \{p_{jk}\}_k)$. Because marginal counts are independent from factor to factor under the null hypothesis, the conditional distribution of the cell counts is

$$\Pr(\{n_{\mathbf{i}}\} \mid \{n_{jk}\}) \quad = \quad \frac{\binom{n}{\{n_{\mathbf{i}}\}} \prod_{\mathbf{i}} p_{\mathbf{i}}^{n_{\mathbf{i}}}}{\prod_{j=1}^{m} \binom{n}{\{n_{jk}\}_k} \prod_k (p_{jk})^{n_{jk}}}$$

$$= \frac{\binom{n}{\{n_i\}}}{\prod_{j=1}^{m}\binom{n}{\{n_{jk}\}_k}}. \tag{24.1}$$

One of the pleasant facts of exact inference is that the multivariate Fisher-Yates distribution (24.1) does not depend on the marginal probabilities p_{jk}. Problem 4 indicates how to compute the moments of the distribution (24.1).

We can also derive the Fisher-Yates distribution by a counting argument involving a related but different sample space. Consider an $m \times n$ matrix whose rows correspond to factors and whose columns correspond to the multivariate observations attributed to the different cells. For instance with three factors, the column vector $(a_1, b_3, c_2)^t$ represents an observation with level a_1 at factor 1, level b_3 at factor 2, and level c_2 at factor 3. By assumption at factor j there are n_{jk} observations representing level k. If we uniquely label each of these $n = \sum_k n_{jk}$ observations, then there are $n!$ distinguishable permutations of the level labels in row j. The uniform sample space consists of the $(n!)^m$ matrices derived from the $n!$ permutations of each of the m rows. Each such matrix is assigned probability $1/(n!)^m$. For instance, if we distinguish duplicate labels by a superscript $*$, then the 3×4 matrix

$$\begin{pmatrix} a_1 & a_2 & a_1^* & a_2^* \\ b_3 & b_1 & b_1^* & b_2 \\ c_2 & c_1 & c_3 & c_2^* \end{pmatrix} \tag{24.2}$$

for $m = 3$ factors and $n = 4$ multivariate observations represents one out of $(4!)^3$ equally likely matrices and yields the nonzero cell counts

$$\begin{aligned} n_{a_1b_3c_2} &= 1 \\ n_{a_2b_1c_1} &= 1 \\ n_{a_1b_1c_3} &= 1 \\ n_{a_2b_2c_2} &= 1. \end{aligned}$$

To count the number of matrices consistent with a cell count vector $\{n_i\}$, note that the n multivariate observations can be assigned to the columns of a typical matrix from the uniform space in $\binom{n}{\{n_i\}}$ ways. Within each such assignment, there are $\prod_k n_{jk}!$ consistent permutations of the labels at level j; over all levels, there are $\prod_{j=1}^m \prod_k n_{jk}!$ consistent permutations. It follows that the cell count vector $\{n_i\}$ has probability

$$\begin{aligned} \Pr(\{n_i\}) &= \frac{\binom{n}{\{n_i\}} \prod_{j=1}^m \prod_k n_{jk}!}{(n!)^m} \\ &= \frac{\binom{n}{\{n_i\}}}{\prod_{j=1}^{m}\binom{n}{\{n_{jk}\}_k}}. \end{aligned}$$

In other words, we recover the Fisher-Yates distribution.

The uniform sample space also suggests a device for random sampling from the Fisher-Yates distribution [18]. If we arrange our n multivariate observations in an $m \times n$ matrix as just described and randomly permute the entries within each row, then we get a new matrix whose cell counts are drawn from the Fisher-Yates distribution. For example, appropriate permutations within each row of the matrix (24.2) produce the matrix

$$\begin{pmatrix} a_1 & a_1^* & a_2 & a_2^* \\ b_1 & b_1^* & b_2 & b_3 \\ c_2 & c_2^* & c_1 & c_3 \end{pmatrix}$$

with nonzero cell counts

$$\begin{aligned} n_{a_1 b_1 c_2} &= 2 \\ n_{a_2 b_2 c_1} &= 1 \\ n_{a_2 b_3 c_3} &= 1. \end{aligned}$$

Iterating this permutation procedure r times generates an independent, random sample $Z_1, \ldots, Z_r$ of tables from the Fisher-Yates distribution. In practice, it suffices to permute all rows except the bottom row m because cell counts do not depend on the order of the columns in a matrix such as (24.2). Given the observed value T_{obs} of a test statistic T for independence, we estimate the corresponding p-value by the sample average $\frac{1}{r}\sum_{l=1}^{r} 1_{\{T(Z_l) \geq T_{\text{obs}}\}}$.

In Fisher's exact test, the statistic T is the negative of the Fisher-Yates probability (24.1). Thus, the null hypothesis of independence is rejected if the observed Fisher-Yates probability is too low. The chi-square statistic $\sum_{\mathbf{i}}[n_{\mathbf{i}} - \mathrm{E}(n_{\mathbf{i}})]^2/\mathrm{E}(n_{\mathbf{i}})$ is also reasonable for testing independence, provided we estimate its p-value by random sampling and do not foolishly rely on the standard chi-square approximation. As noted in Problem 4, the expectation $\mathrm{E}(n_{\mathbf{i}}) = n\prod_{j=1}^{m}(n_{ji_j}/n)$.

Lazzeroni and Lange [18] apply this Monte Carlo method to test for linkage equilibrium (independence) at six linked genetic markers on chromosome 11. Here each marker locus is considered a factor, and each allele of a marker is considered a level. An observation of a chromosome with a particular sequence of alleles at the marker loci is called a haplotype. With only 180 random haplotypes and 2, 2, 10, 5, 3, and 2 alleles at the six loci, the resulting contingency table qualifies as sparse. Large sample chi-square tests strongly suggest linkage disequilibrium (dependence). Monte Carlo exact tests do not reach significance at the 0.1 level and correct this misimpression. ■

24.3 The Bootstrap

Like most good ideas, the principle behind the bootstrap is simple. Suppose for the sake of argument that we draw an i.i.d. sample $\mathbf{x} = (x_1, \ldots, x_n)$ from some unknown probability distribution $F(x)$. If we want to understand the sampling properties of a complicated estimator $T(\mathbf{x})$ of a parameter (or functional) $t(F)$ of $F(x)$, then we study the properties of the corresponding estimator $T(\mathbf{x}^*)$ on the space of i.i.d. samples $\mathbf{x}^* = (x_1^*, \ldots, x_n^*)$ drawn from a data-based approximation $F_n^*(x)$ to $F(x)$. In the case of the nonparametric bootstrap, $F_n^*(x)$ is the empirical distribution function, putting weight $1/n$ on each of the n observed points $x_1, \ldots, x_n$. In the case of the parametric bootstrap, we assume a parametric form for $F_\alpha(x)$, estimate the parameter α from the data by $\hat{\alpha}$, and then sample from $F_n^*(x) = F_{\hat{\alpha}}(x)$. The bootstrap correspondence principle suggests that not only do $T(\mathbf{x})$ and $T(\mathbf{x}^*)$ have similar distributions, but equally important in practice, that $T(\mathbf{x}) - t(F)$ and $T(\mathbf{x}^*) - t(F_n^*)$ have similar distributions. In many examples, the identity $t(F_n^*) = T(\mathbf{x})$ holds. In ordinary English, the functional t applied to the approximate distribution F_n^* equals the estimator T applied to the actual sample.

These insights are helpful, but finding the theoretical sampling distribution of $T(\mathbf{x}^*)$ is usually impossible. Undeterred by this fact, Efron [7] suggested that we approximate the distribution and moments of $T(\mathbf{x}^*)$ by independent Monte Carlo sampling, in effect substituting computing brawn for mathematical weakness. As we have seen in Chapter 22, there are inefficient and efficient ways of carrying out Monte Carlo estimation. One of the themes of this chapter is efficient Monte Carlo estimation for the nonparametric bootstrap, the more interesting and widely applied version of the bootstrap. Limitations of space prevent us from delving more deeply into the theoretical justifications of the bootstrap. The underlying theory involves contiguity of probability measures and Edgeworth expansions. The papers [1, 2] and books [16, 27] are good places to start for mathematically sophisticated readers. Efron and Tibshirani [9] and Davison and Hinkley [5] provide practical guides to the bootstrap that go well beyond the abbreviated account presented here.

24.3.1 Range of Applications

In the bootstrap literature, the scope of the word "parameter" is broadened to include any functional $t(F)$ of a probability distribution $F(x)$. The moments and central moments

$$\begin{aligned} \mu_k(F) &= \int x^k dF(x) \\ \omega_k(F) &= \int [x - \mu_1(F)]^k dF(x) \end{aligned}$$

are typical parameters whenever they exist. The pth quantile

$$\xi_p(F) = \inf\{x: F(x) \geq p\}$$

is another commonly encountered parameter. If we eschew explicit parametric models, then the natural estimators of these parameters are the corresponding sample statistics

$$\begin{aligned} \hat{\mu}_k(\mathbf{x}) &= \mu_k(F_n^*) = \frac{1}{n}\sum_{i=1}^n x_i^k \\ \hat{\omega}_k(\mathbf{x}) &= \omega_k(F_n^*) = \frac{1}{n}\sum_{i=1}^n (x_k - \bar{x})^k \\ \hat{\xi}_p(\mathbf{x}) &= \xi_p(F_n^*) = \inf\{x: F_n^*(x) \geq p\}. \end{aligned}$$

By construction, these estimators obey the rule $T(\mathbf{x}) = t(F_n^*)$ and consequently are called "plug-in" estimators by Efron and Tibshirani [9]. The unbiased version of the sample variance,

$$s_n^2 = \frac{1}{n-1}\sum_{i=1}^n (x_i - \bar{x})^2 = \frac{n}{n-1}\hat{\omega}_2(x),$$

fails to qualify as a plug-in estimator. In this chapter we consider only plug-in estimators.

24.3.2 Estimation of Standard Errors

As a motivating example for the bootstrap, Efron [7] discusses calculation of the variance of the sample median $\hat{\xi}_{1/2}(\mathbf{x}) = \xi_{1/2}(F_n^*)$. Classical large sample theory implies that

$$\mathrm{Var}[\xi_{\frac{1}{2}}(F_n^*)] \asymp \frac{1}{4nf(\xi_{\frac{1}{2}})^2},$$

where $f(x) = F'(x)$ is the density of $F(x)$ [26]. However, $f(x)$ is hard to estimate accurately even in large samples. The bootstrap provides a way out of this dilemma that avoids direct estimation of $f(\xi_{1/2})$.

In general, the bootstrap correspondence principle suggests that we estimate the variance of an estimator $T(\mathbf{x})$ by the variance of the corresponding estimator $T(\mathbf{x}^*)$ on the bootstrap sample space. Because the variance $\mathrm{Var}(T)^*$ of $T(\mathbf{x}^*)$ is usually difficult to calculate exactly, we take independent bootstrap samples $\mathbf{x}_b^* = (x_{b1}^*, \ldots, x_{bn}^*)$ for $b = 1, \ldots, B$ and approximate $\mathrm{Var}(T)^*$ by

$$\widehat{\mathrm{Var}}(T)^* = \frac{1}{B-1}\sum_{b=1}^B [T(\mathbf{x}_b^*) - \hat{\mathrm{E}}(T)^*]^2,$$

TABLE 24.3. Law School Admission Data

LSAT	GPA	LSAT	GPA
576	3.39	651	3.36
635	3.30	605	3.13
558	2.81	653	3.12
578	3.03	575	2.74
666	3.44	545	2.76
580	3.07	572	2.88
555	3.00	594	2.96
661	3.43		

where

$$\hat{\mathrm{E}}(T)^* = \frac{1}{B}\sum_{b=1}^{B} T(\mathbf{x}_b^*).$$

If we can calculate $\mathrm{E}(T)^*$ exactly, then in the approximation $\widehat{\mathrm{Var}}(T)^*$ we substitute $\mathrm{E}(T)^*$ for the sample mean $\hat{\mathrm{E}}(T)^*$ and replace the divisor $B-1$ by B. Of course, to derive the estimated standard deviation, one simply takes the square root of the estimated variance.

For example, Table 24.3 lists average LSAT and GPA admission scores for 15 representative American law schools [9]. The median LSAT score is 580.0. Straightforward random sampling with replacement generates the median histogram of Figure 24.1. The 1000 bootstrap values depicted there have mean 590.6 and standard deviation 20.5. Hence, under the correspondence principle, the approximate standard deviation of the sample median is also 20.5.

24.3.3 Bias Reduction

Provided users bear in mind that it can increase the variability of an estimator, bias reduction is another valuable application of the bootstrap. We approximate the bootstrap bias

$$\mathrm{bias}^* = \mathrm{E}[T(\mathbf{x}^*)] - t(F_n^*)$$

by the Monte Carlo average

$$\widehat{\mathrm{bias}}_B^* = \frac{1}{B}\sum_{b=1}^{B} T(\mathbf{x}_b^*) - t(F_n^*). \tag{24.3}$$

In accord with the bootstrap correspondence principle, the revised estimator

$$T(\mathbf{x}) - \widehat{\mathrm{bias}}_B^* = 2T(\mathbf{x}) - \frac{1}{B}\sum_{b=1}^{B} T(\mathbf{x}_b^*) \tag{24.4}$$

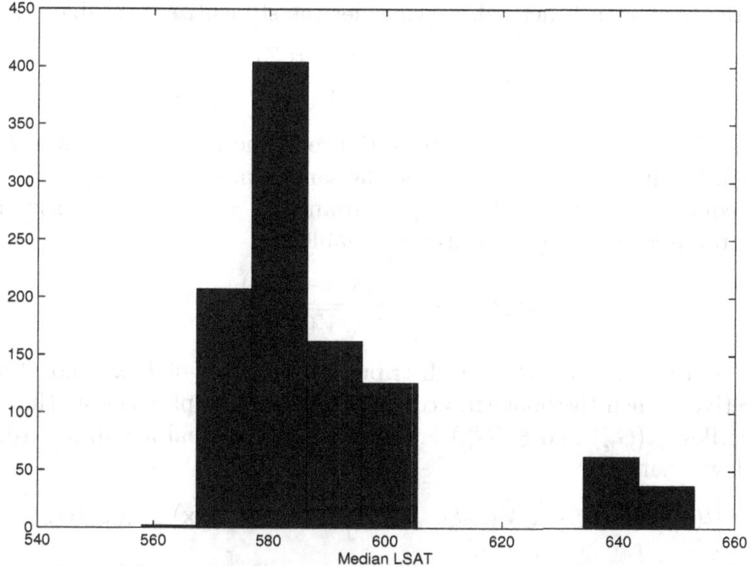

FIGURE 24.1. Histogram of 1000 Bootstrap Medians of 15 LSAT Scores

usually has much less bias than $T(\mathbf{x})$. For the sample median in the law school data, the bootstrap bias (24.3) is 10.1, and the bias-adjusted estimate (24.4) is 569.9. As a second example, the sample correlation between LSAT and GPA is $\hat{\rho} = 0.776$. The 1000 bootstrap samples suggest that $\hat{\rho}$ has the small bias -0.004 and the standard deviation 0.140.

24.3.4 Confidence Intervals

Our third application of the bootstrap involves confidence intervals. Recall that a set $C(\mathbf{x}) = C(x_1, \ldots, x_n)$ is a $1 - \alpha$ level confidence set for a parameter $t(F)$ if $\Pr[t(F) \in C(\mathbf{x})] \geq 1 - \alpha$. A good share of classical theoretical statistics is devoted to the construction and interpretation of confidence intervals. In his initial development of the bootstrap, Efron [7] introduced the bootstrap percentile confidence interval. Let G_n and G_n^* denote the distribution functions of $T(\mathbf{x})$ and $T(\mathbf{x}^*)$, respectively. The bootstrap correspondence principle suggests that the percentiles $\xi_\alpha(G_n)$ and $\xi_\alpha(G_n^*)$ are approximately equal for all $\alpha \in (0, 1)$. Thus, the interval

$$[\xi_{\frac{\alpha}{2}}(G_n^*), \xi_{1-\frac{\alpha}{2}}(G_n^*)]$$

should serve as a crude α-level confidence interval for $t(F)$. In practice, the probability that the bootstrap percentile interval actually covers $t(F)$ can deviate markedly from $1 - \alpha$.

In the bootstrap-t method, we consider the studentized variable

$$R(\mathbf{x}) = \frac{T(\mathbf{x}) - t(F)}{\sqrt{V(\mathbf{x})}},$$

where $V(\mathbf{x})$ is an estimator of the variance of the plug-in estimator $T(\mathbf{x})$ of $t(F)$. For instance, $T(\mathbf{x})$ could be the sample mean $\bar{\mathbf{x}} = \frac{1}{n}\sum_{i=1}^n x_i$, and $V(\mathbf{x})$ could be n^{-1} times the sample variance $S_n^2 = \frac{1}{n}\sum_{i=1}^n (x_i - \bar{\mathbf{x}})^2$. The bootstrap analog of the studentized variable is

$$R(\mathbf{x}^*) = \frac{T(\mathbf{x}^*) - T(\mathbf{x})}{\sqrt{V(\mathbf{x}^*)}}.$$

If G_n and G_n^* now denote the distribution functions of $R(\mathbf{x})$ and $R(\mathbf{x}^*)$, respectively, then the bootstrap correspondence principle suggests that the percentiles $\xi_\alpha(G_n)$ and $\xi_\alpha(G_n^*)$ are approximately equal for all $\alpha \in (0,1)$. It follows that

$$\begin{aligned}
\Pr\left[t(F) \geq T(\mathbf{x}) - \sqrt{V(\mathbf{x})}\xi_{1-\alpha}(G_n^*)\right] &= \Pr\left[R(\mathbf{x}) \leq \xi_{1-\alpha}(G_n^*)\right] \\
&\approx \Pr\left[R(\mathbf{x}^*) \leq \xi_{1-\alpha}(G_n^*)\right] \\
&= 1 - \alpha.
\end{aligned}$$

In other words, $T(\mathbf{x}) - \sqrt{V(\mathbf{x})}\xi_{1-\alpha}(G_n^*)$ is an approximate level $1-\alpha$ lower confidence bound for $t(F)$. Unfortunately, the bootstrap-t is also not a panacea. To use it, we need a good estimator $V(\mathbf{x})$ of the variance of $T(\mathbf{x})$. Such estimators are not always available. Best results obtain when $R(\mathbf{x}^*) = [T(\mathbf{x}^*) - T(\mathbf{x})]/\sqrt{V(\mathbf{x}^*)}$ is nearly pivotal, that is, depends only weakly on $F(x)$.

In response to the deficiencies of the percentile and t confidence intervals, Efron [8] crafted an improved version of the percentile interval called the BC$_a$ (bias-corrected and accelerated) interval. We now briefly discuss this interval, giving only enough details for implementation. Mathematical proofs and elaborations of the method can be found in the references [8, 9, 16, 27]. To motivate the method, let $\theta = t(F)$ and $\hat{\theta} = T(\mathbf{x})$ and assume that there is an increasing differentiable transformation $\phi = m(\theta)$ such that the random variable

$$\frac{\hat{\phi} - \phi}{\tau[1 + a(\phi - \phi_0)]} + z_0$$

is approximately standard normal. Here z_0 is the bias correction, a is the acceleration constant, τ is a scaling constant, and ϕ_0 is a convenient reference point. We will discuss how to estimate z_0 and a in a moment. Fortunately, $m(\theta)$, τ, and ϕ_0 fade into the background, and we can formulate the confidence interval solely in terms of the function

$$p(c) = \Phi\Big\{z_o + \frac{z_0 + \Phi^{[-1]}(c)}{1 - a[z_0 + \Phi^{[-1]}(c)]}\Big\},$$

where $\Phi(x)$ and $\Phi^{[-1]}(x)$ are the standard normal distribution function and its inverse. In this notation the BC_a confidence interval for $\theta = t(F)$ can be expressed as

$$[\xi_{p(\frac{\alpha}{2})}(G_n^*), \xi_{p(1-\frac{\alpha}{2})}(G_n^*)],$$

where G_n^* is the distribution function of the bootstrap statistic $\hat{\theta}^* = T(\mathbf{x}^*)$. One typically sets z_0 equal to the normal percentile point corresponding to the fraction of bootstrap statistics falling below the original statistic; in symbols,

$$z_0 = \Phi^{[-1]}\Big[\#(\hat{\theta}^* < \hat{\theta})/B\Big].$$

The acceleration is related to the skewness of the jackknife estimates of θ. Thus, we take

$$a = -\frac{\sum_{i=1}^n (\theta_{(i)} - \bar{\theta})^3}{6[\sum_{i=1}^n (\theta_{(i)} - \bar{\theta})^2]^{3/2}},$$

where $\theta_{(i)}$ is the jackknife estimate of θ leaving out observation i and $\bar{\theta}$ is the average of these n estimates.

Efron and Tibshirani [9] analyze the spatial ability scores A_i displayed in Table 24.4 for 26 neurologically impaired children. To illustrate the various bootstrap confidence intervals, they consider the statistic $\theta = \text{Var}(A)$. The plug-in estimator

$$\hat{\theta} = \frac{1}{26}\sum_{i=1}^{26}(A_i - \bar{A})^2, \quad \bar{A} = \frac{1}{26}\sum_{i=1}^{26} A_i$$

has value 171.5, which is biased downward because we divide by 26 rather than 25. A bootstrap sample of 10,000 suggests a bias of -7.1. The bias-adjusted estimate of 178.6 is in remarkable agreement with the usual unbiased estimate of 178.4 that divides by 25. Efron and Tibshirani compute 90% confidence intervals for ϕ based on 2000 bootstraps. Their approximate percentile, bootstrap-t, and BC_a intervals are (100.8,233.6), (112.3,314.8), and (115.8,259.6), respectively. They state a clear preference for the BC_a interval for reasons that we will not pursue here.

TABLE 24.4. Spatial Ability Scores of 26 Neurologically Impaired Children

48	36	20	29	42	42	20	42
22	41	45	14	6	0	33	28
34	4	32	24	47	41	24	26
30	41						

24.3.5 Applications in Regression

In the linear regression model $y = X\beta + u$, bootstrapping is conducted by resampling either cases (y_i, x_i) or residuals $r_i = y_i - \hat{y}_i$. In bootstrapping residuals, we sample $\mathbf{r}^* = (r_1^*, \ldots, r_n^*)$ from $\mathbf{r} = (r_1, \ldots, r_n)$ with replacement and construct the new vector of dependent (or response) variables $y^* = X\hat{\beta} + \mathbf{r}^*$. Recall that the least squares estimator $\hat{\beta} = (X^t X)^{-1} X^t y$ is unbiased for β provided the errors u_i satisfy $\mathrm{E}(u_i) = 0$. If, in addition, the errors are uncorrelated with common variance $\mathrm{Var}(u_i) = \sigma^2$, then $\mathrm{Var}(\hat{\beta}) = \sigma^2 (X^t X)^{-1}$. The predicted value of y is obviously defined as $\hat{y} = X\hat{\beta}$.

For the least squares estimator to qualify as a plug-in estimator in bootstrapping residuals, the condition $\mathrm{E}(\hat{\beta}^*) = \hat{\beta}$ must hold. Problem 11 asks the reader to verify this condition for the intercept model $X = (\mathbf{1}, Z)$ under the above hypotheses. Bootstrap-t confidence intervals for the components of β are based on the studentized variables $(\hat{\beta}_i - \beta_i)/(\hat{\sigma}\sqrt{w_{ii}})$ and their bootstrap analogs, where $\hat{\sigma}^2 = \frac{1}{n}\sum_{i=1}^n (y_i - \hat{y}_i)^2$ and w_{ii} is the ith diagonal entry of $(X^t X)^{-1}$. Bootstrapping cases is more robust to violations of model assumptions than bootstrapping residuals. Resampling of residuals can be misleading if there are gross differences in the error variance across the spectrum of predictor vectors.

In generalized linear models, the parametric bootstrap is the natural recourse for computing standard errors and confidence intervals. The parametric bootstrap entails estimating regression coefficients by maximum likelihood and resampling responses. For instance, in logistic regression the estimated regression coefficients determine for each observation the probability that it is a success or failure. For a bootstrap replicate, one simply resamples a success or failure with the estimated probability. In Poisson regression, the estimated regression coefficients supply an estimated mean for each observation, and for a bootstrap replicate one samples a Poisson deviate with the given mean.

As a concrete example, consider the AIDS model of Section 14.7. Recall that the intercept and slope of the model had estimates of 0.3396 and 0.2565, respectively. The large sample standard errors of these estimates based on inverting the expected information matrix are 0.2512 and 0.0220. The classical 95% confidence intervals for the two parameters are therefore (-0.1528,0.8320) and (0.2134,0.2996), respectively. By contrast, 1000 bootstrap replicates suggest a standard error of 0.2676 and a 95% BC$_a$ confidence interval of (-0.2264,0.7774) for the intercept. For the slope, the bootstrap standard error is 0.0234, and the BC$_a$ confidence interval is (0.2169,0.3044). The classical and bootstrap confidence intervals are nearly the same length, but the bootstrap intercept interval is shifted left, and the bootstrap slope interval is shifted right.

24.4 Efficient Bootstrap Simulations

24.4.1 The Balanced Bootstrap

We now turn to methods of reducing Monte Carlo sampling error. The first of these methods, balanced bootstrap resampling [6], is best illustrated by an example where simulation is unnecessary. According to the bootstrap correspondence principle, we estimate the bias of the sample mean $\bar{x}$ of n i.i.d. observations $\mathbf{x} = (x_1, \ldots, x_n)$ by the Monte Carlo difference

$$\frac{1}{B}\sum_{b=1}^{B} \overline{\mathbf{x}_b^*} - \bar{x}.$$

By chance alone, this difference often is nonzero. We can eliminate this artificial bias by adopting nonindependent Monte Carlo sampling. Balanced bootstrap resampling retains as much randomness in the bootstrap resamples $\mathbf{x}_b$ as possible while forcing each original observation x_i to appear exactly B times in the bootstrap resamples. The naive implementation of the balanced bootstrap involves concatenating B copies of the data $(x_1, \ldots, x_n)$, randomly permuting the resulting data vector v of length nB, and then taking successive blocks of size n from v for the B bootstrap resamples.

A drawback of this permutation method of producing balanced bootstrap resamples is that it requires storage of v. Gleason [13] proposed an acceptance-rejection algorithm that sequentially creates only one block of v at a time and therefore minimizes computer storage. We can visualize Gleason's algorithm by imagining n urns, with urn i initially filled with B replicates of observation x_i. In the first step of the algorithm, we simply choose an urn at random and extract one of its replicate observations. This forms the first observation of the first bootstrap resample. In filling out the first resample and constructing subsequent ones, we must adjust our sampling procedure to reflect the fact that the urns may contain different numbers of observations. Let c_i be the number of observations currently left in urn i, and set $c = \max_i c_i$. If there are k nonempty urns, we choose one of these k urns at random, say the ith, and propose extracting a replicate x_i. Our decision to accept or reject the replicate is determined by selecting a random uniform deviate U. If $U \leq c_i/c$, then we accept the replicate; otherwise, we reject the replicate. When a replicate x_i is accepted, c_i is reduced by 1, and the maximum c is recomputed. In practice, recomputation of c is the most time-consuming step of the algorithm. Gleason [13] suggests some remedies that require less frequent updating of c.

The balanced bootstrap is best suited to bias estimation. One can impose additional constraints on balanced resampling such as the requirement that every pair of observations occur the same number of times in the B bootstrap resamples. These higher-order balanced bootstraps are harder to generate, and limited evidence suggests that they are less effective than other methods of reducing Monte Carlo error [9].

24.4.2 The Antithetic Bootstrap

The method of antithetic resampling discussed in Chapter 23 is also pertinent to the bootstrap. Hall [15, 16] suggests implementing antithetic resampling by replacing scalar i.i.d. observations $(x_1, \ldots, x_n)$ by their order statistics $x_{(1)} \leq \cdots \leq x_{(n)}$. He then defines the permutation $\pi(i) = n-i+1$ that reverses the order statistics; in other words, $x_{(\pi[1])} \geq \cdots \geq x_{(\pi[n])}$. For any bootstrap resample $\mathbf{x}^* = (x_1^*, \ldots, x_n^*)$, we can construct a corresponding antithetic bootstrap resample $\mathbf{x}^{**}$ by substituting $x_{(\pi[i])}$ for every appearance of $x_{(i)}$ in $\mathbf{x}^*$. Often it is intuitively clear that the identically distributed statistics $T^* = T(\mathbf{x}^*)$ and $T^{**} = T(\mathbf{x}^{**})$ are negatively correlated. Negative correlation makes it advantageous to approximate the mean $\mathrm{E}(T^*)$ by taking sample averages of the statistic $(T^* + T^{**})/2$.

If $T(\mathbf{x})$ is a monotonic function of the sample mean $\bar{x}$, then negative correlation is likely to occur. For example, when $T^* = T(\bar{x^*})$ is symmetric about its mean, the residuals $T^* - \mathrm{E}(T^*)$ and $T^{**} - \mathrm{E}(T^*)$ are always of opposite sign, and a large positive value of one residual is matched by a large negative value of the other residual. This favorable state of affairs for the antithetic bootstrap persists for statistics $T^* = T(\bar{x^*})$ having low skewness.

For a vector sample $(x_1, \ldots, x_n)$, the order statistics $x_{(1)} \leq \cdots \leq x_{(n)}$ no longer exist. However, for a statistic $T(\bar{x})$ depending only on the sample mean, we can order the observations by their deflection of $T(\bar{x^*})$ from $T(\bar{x})$. To a first approximation, this deflection is measured by

$$T(x_i) - T(\bar{x}) \quad \approx \quad dT(\bar{x})(x_i - \bar{x}),$$

where $dT(x)$ is the differential of $T(x)$. Hall [15, 16] recommends implementing antithetic resampling after ordering the x_i by their approximate deflections $dT(\bar{x})(x_i - \bar{x})$.

24.4.3 Importance Resampling

In standard bootstrap resampling, each observation x_i is resampled uniformly with probability $1/n$. In some applications it is helpful to implement importance sampling by assigning different resampling probabilities p_i to the different observations x_i [4, 17]. For instance, with univariate observations $(x_1, \ldots, x_n)$, we may want to emphasize one of the tails of the empirical distribution. If we elect to resample nonuniformly according to the multinomial distribution with proportions $p = (p_1, \ldots, p_n)^t$, then the equality

$$\begin{aligned}
\mathrm{E}[T(\mathbf{x}^*)] &= \mathrm{E}_p\left[T(\mathbf{x}^*)\frac{\binom{n}{m_1^* \cdots m_n^*}\left(\frac{1}{n}\right)^n}{\binom{n}{m_1^* \cdots m_n^*}\prod_{i=1}^n p_i^{m_i^*}}\right] \\
&= \mathrm{E}_p\left[T(\mathbf{x}^*)\prod_{i=1}^n (np_i)^{-m_i^*}\right]
\end{aligned}$$

connects the uniform expectation and the importance expectation on the bootstrap resampling space. Here m_i^* represents the number of times sample point x_i appears in $\mathbf{x}^*$. Thus, we can approximate the mean $\mathrm{E}[T(\mathbf{x}^*)]$ by taking a bootstrap average

$$\frac{1}{B}\sum_{b=1}^B T(\mathbf{x}_b^*)\prod_{i=1}^n (np_i)^{-m_{bi}^*} \tag{24.5}$$

with multinomial sampling relative to p. This Monte Carlo approximation has variance

$$\frac{1}{B}\left\{\mathrm{E}_p\left[T(\mathbf{x}^*)^2\prod_{i=1}^n (np_i)^{-2m_{bi}^*}\right] - \mathrm{E}\left[T(\mathbf{x}^*)\right]^2\right\},$$

which we can minimize with respect to p by minimizing the second moment $\mathrm{E}_p\left[T(\mathbf{x}^*)^2\prod_{i=1}^n (np_i)^{-2m_{bi}^*}\right]$.

Hall [16] suggests approximately minimizing the second moment by taking a preliminary uniform bootstrap sample of size B_1. Based on the preliminary resample, we approximate $\mathrm{E}_p\left[T(\mathbf{x}^*)^2\prod_{i=1}^n (np_i)^{-2m_{bi}^*}\right]$ by the Monte Carlo average

$$\begin{aligned}
s(p) &= \frac{1}{B_1}\sum_{b=1}^{B_1} T(\mathbf{x}_b^*)^2\prod_{i=1}^n (np_i)^{-2m_{bi}^*}\prod_{i=1}^n (np_i)^{m_{bi}^*} \\
&= \frac{1}{B_1}\sum_{b=1}^{B_1} T(\mathbf{x}_b^*)^2\prod_{i=1}^n (np_i)^{-m_{bi}^*}.
\end{aligned} \tag{24.6}$$

The function $s(p)$ serves as a surrogate for $\mathrm{E}_p\left[T(\mathbf{x}^*)^2\prod_{i=1}^n (np_i)^{-2m_{bi}^*}\right]$. It is possible to minimize $s(p)$ on the open unit simplex

$$U = \left\{p : \sum_{i=1}^n p_i = 1,\ \ p_i > 0,\ \ 1 \le i \le n\right\}$$

by standard methods.

For instance, we can apply the adaptive barrier method sketched in Chapter 11 for moderately sized n. The method of geometric programming offers another approach [21]. For large n the combination of recursive quadratic programming and majorization sketched in Problem 16 works well. All methods are facilitated by the convexity of $s(p)$. Convexity is evident from the form

$$\begin{aligned} d^2 s(p) &= \frac{1}{B_1}\sum_{b=1}^{B_1} T(\mathbf{x}_b^*)^2 \prod_{i=1}^{n}(np_i)^{-m_{bi}^*} \qquad (24.7) \\ &\times \left\{ \begin{pmatrix} \frac{m_{b1}^*}{p_1} \\ \vdots \\ \frac{m_{bn}^*}{p_n} \end{pmatrix} \begin{pmatrix} \frac{m_{b1}^*}{p_1} & \cdots & \frac{m_{bn}^*}{p_n} \end{pmatrix} + \begin{pmatrix} \frac{m_{b1}^*}{p_1^2} & \cdots & 0 \\ \vdots & \ddots & \vdots \\ 0 & \cdots & \frac{m_{bn}^*}{p_n^2} \end{pmatrix} \right\} \end{aligned}$$

of the second differential. Problems 14 and 15 ask the reader to verify this formula and check that $s(p)$ is strictly convex and attains its minimum on U provided there exists for each index i an index b such that $T(\mathbf{x}_b^*) \neq 0$ and $m_{bi}^* \neq 0$. When this condition fails, one or more optimal p_i are estimated as 0. This makes it impossible to resample some observations and suggests that the constraint $p_i \geq \epsilon > 0$ be imposed for all i and some small ϵ.

Once the optimal vector p_{opt} is calculated, we take a second bootstrap resample of size B_2 with multinomial probabilities p_{opt} and compute the sample average (24.5) with B_2 replacing B. Call the result A_2. Given the outcomes of the first bootstrap, we can also easily compute a sample average A_1 approximating $\mathrm{E}[T(\mathbf{x}^*)]$ under uniform resampling. Each of these sample averages has an attached sample variance V_i. The convex combination

$$\frac{V_2}{V_1+V_2}A_1 + \frac{V_1}{V_1+V_2}A_2$$

is unbiased for $\mathrm{E}[T(\mathbf{x}^*)]$ and should have nearly minimal variance. (See Problem 17.)

The obvious strengths of Hall's strategy of importance sampling are its generality, its adaptive nature, and its use of stage-one resampling for approximating both the importance density and the sought-after expectation. In practice, the strategy has at least two drawbacks. First, it entails solving a nontrivial optimization problem as an intermediate step. Second, Hall [16] argues theoretically that the method offers little advantage in approximating certain expectations such as central moments. However, Hall's method appears to yield substantial dividends in approximating distribution functions and tail probabilities.

To illustrate importance sampling in action, we now consider Monte Carlo approximation of the left-tail probability $\Pr(\rho^* \leq 0.4)$ for the correlation coefficient in the law school data. In $B = 500$ ordinary bootstrap replicates, the event $\rho^* \leq 0.4$ occurs only 11 times. Thus, we are fairly far out in the left tail. Minimizing the importance criterion (24.6) for these 500 replicates yields an optimal importance probability vector with minimum

0.0134 (school 5) and maximum 0.2085 (school 1). Figure 24.2 plots running Monte Carlo averages approximating the left-tail probability $\Pr(\rho^* \leq 0.4)$ based on the ordinary bootstrap and the importance bootstrap. Each curve in the figure extends from $b = 10^3$ to $b = 10^6$ bootstrap resamples. Although importance sampling (dark line) appears to converge faster, it is unclear whether the sample averages of either method have fully stabilized. Other examples offer a more dramatic contrast.

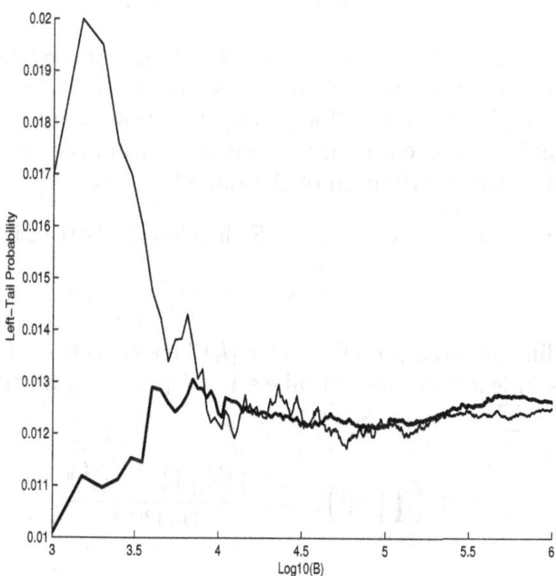

FIGURE 24.2. Running Monte Carlo Averages for $\Pr(\rho^* \leq 0.4)$

24.5 Problems

1. Consider a random sample $(x_1, y_1), \ldots, (x_n, y_n)$ from a bivariate distribution $F(x, y)$. Devise a permutation test of the null hypothesis that the two components are independent. Program the test and apply it to real or simulated data. (Hint: Use the sample covariance function.)

2. The permutation t-test emphasizes sample means and ignores sample variances. Devise a permutation test to check for equality of variances. Program the test and apply it to real or simulated data.

3. The two-way layout without repetition deals with an $m \times n$ matrix of observations (y_{ij}) from independent random variables Y_{ij} having

common variance σ^2 and means $\mathrm{E}(Y_{ij}) = \mu + \alpha_i + \beta_j$ subject to the constraints $\sum_i \alpha_i = \sum_j \beta_j = 0$. The obvious estimate of μ is the sample mean $\bar{y}$, so without loss of generality drop μ from the model and assume $\bar{y} = 0$. Using known facts from the analysis of variance, argue that

$$T = \sum_{i=1}^{m} y_{i.}^2 = \sum_{i=1}^{m}\Big[\sum_{j=1}^{n} y_{ij}\Big]^2$$

is a reasonable statistic for testing the null hypothesis that all $\alpha_i = 0$. Program a permutation test based on this statistic, and apply it to real or simulated data. (Hint: If the deviations from the means are identically distributed, then permutation within each column should not change the distribution of the table.)

4. To compute moments under the Fisher-Yates distribution (24.1), let

$$u^{\underline{r}} = u(u-1)\cdots(u-r+1)$$

be a falling factorial power, and let $\{l_{\mathbf{i}}\}$ be a collection of nonnegative integers indexed by the cell labels $\mathbf{i} = (i_1 \ldots, i_m)$. Setting $l = \sum_{\mathbf{i}} l_{\mathbf{i}}$ and $l_{jk} = \sum_{\mathbf{i}} 1_{\{i_j = k\}} l_{\mathbf{i}}$, show that

$$\mathrm{E}\Big(\prod_{\mathbf{i}} n_{\mathbf{i}}^{\underline{l_{\mathbf{i}}}}\Big) = \frac{\prod_{j=1}^{m}\prod_k (n_{jk})^{\underline{l_{jk}}}}{(n^{\underline{l}})^{m-1}}.$$

In particular, verify that $\mathrm{E}(n_{\mathbf{i}}) = n\prod_{j=1}^{m}(n_{ji_j}/n)$.

5. Table 24.5 gives tallies of eye color and hair color on 592 people [28]. Test for independence of the two attributes in this contingency table by implementing a permutation version of either Fisher's exact test or a chi-square test.

TABLE 24.5. Eye Color versus Hair Color for 592 People

Eye Color	Black	Brunette	Red	Blond
Brown	68	119	26	7
Blue	20	84	17	94
Hazel	15	54	14	10
Green	5	29	14	16

6. If the n observations $\{x_1, \ldots, x_n\}$ are distinct, then we can represent a bootstrap resample by an n-tuple $(m_1, \ldots, m_n)$ of nonnegative integers, where m_i indicates the number of repetitions of x_i in the

resample. Show that there are $\binom{2n-1}{n}$ such n-tuples. If the n-tuple $(m_1, \ldots, m_n)$ is assigned multinomial probability $\binom{n}{m_1 \cdots m_n} n^{-n}$, then further demonstrate that the most probable n-tuple is $(1, \ldots, 1)$. Apply Stirling's formula and show that

$$\begin{aligned} \binom{2n-1}{n} &\asymp \frac{2^{2n-1}}{\sqrt{n\pi}} \\ \binom{n}{1 \cdots 1} \left(\frac{1}{n}\right)^n &\asymp \sqrt{2n\pi} e^{-n}. \end{aligned}$$

Finally, prove that a given x_i appears in a bootstrap resample with approximate probability $1 - e^{-1} \approx 0.632$.

7. Problem 6 indicates that the most likely bootstrap resample from n distinct observations $\{x_1, \ldots, x_n\}$ has probability $p_n \asymp \sqrt{2n\pi} e^{-n}$. Show that b bootstrap samples are all distinct with probability

$$\begin{aligned} q_{nb} &\geq (1 - p_n)(1 - 2p_n) \cdots (1 - [b-1]p_n) \\ &\geq 1 - \frac{1}{2} b(b-1) p_n. \end{aligned}$$

For $n = 20$ and $b = 2000$, compute the bound $q_{nb} \geq 0.954$ [16].

8. Replicate any of the bootstrapping numerical examples discussed in this chapter. Your answers will differ because of random sampling. Either Matlab or R is a good environment for bootstrapping.

9. Suppose $X = (Y, Z)$ is a bivariate random vector with independent components Y and Z. If you are given n independent realizations $x_1, \ldots, x_n$ of X, what alternative distribution would be a reasonable substitute for the empirical distribution in bootstrap resampling?

10. In a certain medical experiment summarized in Table 24.6, mice were assigned to either a treatment or control group, and their survival times recorded in days [9]. Compute a 90% bootstrap-t confidence interval for the mean of each group.

11. In the linear regression model of Section 24.3.5, prove that the condition $\mathrm{E}(\hat{\beta}^*) = \hat{\beta}$ holds for the intercept model $X = (\mathbf{1}, Z)$ under bootstrapping residuals. (Hint: Show that $\mathbf{1}^t \hat{\mathbf{r}} = 0$, where $\hat{\mathbf{r}}$ is the residual vector.)

12. If $n = 2m - 1$ is odd, and the observations $\{x_1, \ldots, x_n\}$ are distinct, then the sample median $x_{(m)}$ is uniquely defined. Prove that a bootstrap resample $\{x_1^*, \ldots, x_n^*\}$ has median $x_{(m)}^*$ with distribution function

$$G_n^*(x_{(k)}) = \Pr(x_{(m)}^* \leq x_{(k)}) = \sum_{j=m}^{n} \binom{n}{j} \frac{k^j (n-k)^{n-j}}{n^n}.$$

TABLE 24.6. Mouse Survival Data

Group	Survival Time	Group	Survival Time
Treatment	94	Control	10
Treatment	38	Control	40
Treatment	23	Control	104
Treatment	197	Control	51
Treatment	99	Control	27
Treatment	16	Control	146
Treatment	141	Control	30
Control	52	Control	46

By definition, the quantile $\xi_\alpha(G_n^*)$ of $x_{(m)}^*$ is the smallest order statistic $x_{(k)}$ satisfying $G_n^*(x_{(k)}) \geq \alpha$. The bootstrap percentile method gives $\xi_\alpha(G_n^*)$ as an approximate $1-\alpha$ lower confidence bound for the true median.

13. The bootstrap may yield nonsense results in some circumstances. Generate a random sample from the standard Cauchy distribution. Estimate the standard error and bias of the sample mean. Alternatively, generate a random sample from the uniform distribution on $(0,\theta)$. Estimate the standard error and bias of the sample maximum [12].

14. Calculate the second differential (24.7) of the function $s(p)$ defined in equation (24.6).

15. Show that the function $s(p)$ in equation (24.6) is convex on the open simplex $U = \{p : \sum_{i=1}^n p_i = 1,\ p_i > 0,\ i = 1,\ldots,n\}$. Prove in addition that $s(p)$ is strictly convex and attains its minimum on U provided there exists for each index i an index b such that $T(\mathbf{x}_b^*) \neq 0$ and $m_{bi}^* \neq 0$.

16. When the number of observations n is large, a combination of methods works best for minimizing the importance function $s(p)$ defined by equation (24.6). Let q be the current point in an iterative search for the minimum, and define

$$
\begin{aligned}
c_b &= T(x_b^*)^2 \prod_{j=1}^n (nq_j)^{-m_{b_j}^*} \\
v_b &= \Big(\frac{m_{b_1}^*}{q_1}, \ldots, \frac{m_{b_n}^*}{q_n}\Big)^t \\
D_b &= \operatorname{diag}\Big(\frac{m_{b_1}^*}{q_1^2}, \ldots, \frac{m_{b_n}^*}{q_n^2}\Big).
\end{aligned}
$$

In this notation, $s(p)$ has the second-order Taylor expansion

$$s(p) \approx s(q) + ds(q)(p-q) + \frac{1}{2}(p-q)^t d^2 s(q)(p-q) \quad (24.8)$$

around q. Apply the Cauchy-Schwarz inequality and show that the quadratic (24.8) is majorized by the simpler quadratic

$$\frac{1}{2}p^t Dp - (u + Dq)^t p + e,$$

where $u = \sum_b c_b v_b$, $D = \sum_b c_b(D_b + \|v_b\|_2^2 I_n)$, and e is an irrelevant constant. The strategy is now to minimize this surrogate function subject to the additional safeguards $p_i \geq \epsilon$ for all i. Demonstrate that the surrogate function can be recast as

$$\frac{1}{2}\|r - (D^{-1/2}u + D^{1/2}q)\|_2^2$$

by making the change of variable $r = D^{1/2}p$. In the new coordinates, the constraints are $\mathbf{1}^t D^{-1/2} r = 1$ and $r_i \geq D_{ii}^{1/2}\epsilon$ for all i. To find the next iterate p, one should minimize the surrogate as a function of r and transform the result back to p. Explain how Michelot's algorithm described in Problem 13 of Chapter 16 applies.

17. Suppose two independent random variables Y_1 and Y_2 have the same mean but different variances v_1 and v_2. Demonstrate that the convex combination $\beta Y_1 + (1-\beta)Y_2$ with minimal variance is achieved by taking $\beta = v_2/(v_1 + v_2)$.

24.6 References

[1] Babu GJ, Singh K (1984) On a one term Edgeworth correction for Efron's bootstrap. *Sankhyā A* 46:219–232

[2] Bickel PJ, Freedman DA (1981) Some asymptotics for the bootstrap. *Ann Stat* 9:1196–1217

[3] Bradley JV (1968). *Distribution-Free Statistical Tests.* Prentice-Hall, Englewood Cliffs, NJ

[4] Davison AC (1988) Discussion of papers by DV Hinkley and TJ DiCiccio and JP Romano. *J Roy Stat Soc B* 50:356–357

[5] Davison AC, Hinkley DV (1997) *Bootstrap Methods and Their Applications.* Cambridge University Press, Cambridge

[6] Davison AC, Hinkley DV, Schechtman E (1986) Efficient bootstrap simulation. *Biometrika* 73:555–566

[7] Efron B (1979) Bootstrap methods: Another look at the jackknife. *Ann Stat* 7:1–26

[8] Efron B (1987) Better bootstrap confidence intervals (with discussion). *J Amer Stat Assoc* 82:171–200

[9] Efron B, Tibshirani RJ (1993) *An Introduction to the Bootstrap.* Chapman & Hall, New York

[10] Ernst MD (2004) Permutation methods. *Stat Sci* 19:676–685

[11] Fisher RA (1935) *The Design of Experiments.* Hafner, New York

[12] Givens GH, Hoeting JA (2005) *Computational Statistics.* Wiley, Hoboken, NJ

[13] Gleason JR (1988) Algorithms for balanced bootstrap simulations. *Amer Statistician* 42:263–266

[14] Grogan WL, Wirth WW (1981) A new American genus of predaceous midges related to *Palpomyia* and *Bezzia* (Diptera: Ceratopogonidae). *Proc Biol Soc Wash* 94:1279–1305

[15] Hall P (1989) Antithetic resampling for the bootstrap. *Biometrika* 76:713–724

[16] Hall P (1992) *The Bootstrap and Edgeworth Expansion.* Springer, New York

[17] Johns MV Jr (1988) Importance sampling for bootstrap confidence intervals. *J Amer Stat Assoc* 83:709–714

[18] Lazzeroni LC, Lange K (1997) Markov chains for Monte Carlo tests of genetic equilibrium in multidimensional contingency tables. *Ann Stat* 25:138–168

[19] Ludbrook J, Dudley H (1998) Why permutation tests are superior to t and F tests in biomedical research. Amer Statistician 52:127–132

[20] Nijenhuis A, Wilf HS (1978) *Combinatorial Algorithms for Computers and Calculators*, 2nd ed. Academic Press, New York

[21] Peressini AL, Sullivan FE, Uhl JJ Jr (1988) *The Mathematics of Nonlinear Programming.* Springer, New York

[22] Pitman EJG (1937) Significance tests which may be applied to samples from any population. *J Roy Stat Soc Suppl* 4:119–130

[23] Pitman EJG (1937) Significance tests which may be applied to samples from any population. II. The correlation coefficient test. *J Roy Stat Soc Suppl* 4:225–232

[24] Pitman EJG (1938) Significance tests which may be applied to samples from any population. III. The analysis of variance test. *Biometrika* 29:322–335

[25] Quenouille M (1949) Approximate tests of correlation in time series. *J Roy Stat Soc Ser B* 11:18–44

[26] Serfling RJ (1980) *Approximation Theorems in Mathematical Statistics.* Wiley, New York

[27] Shao J, Tu D (1995) *The Jackknife and Bootstrap.* Springer, New York

[28] Snee RD (1974) Graphical display of two-way contingency tables. *Amer Statistician* 38:9–12

[29] Tukey JW (1958) Bias and confidence in not quite large samples. (Abstract) *Ann Math Stat* 29:614

25

Finite-State Markov Chains

25.1 Introduction

Applied probability and statistics thrive on models. Markov chains are one of the richest sources of good models for capturing dynamical behavior with a large stochastic component [2, 3, 7, 9, 13, 18, 19, 21]. Certainly, every research statistician should be comfortable formulating and manipulating Markov chains. In this chapter we give a quick overview of some of the relevant theory of Markov chains in the simple context of finite-state chains. We cover both discrete-time and continuous-time chains in what we hope is a lively blend of applied probability, graph theory, linear algebra, and differential equations. Since this may be a first account for many readers, we stress intuitive explanations and computational techniques rather than mathematical rigor.

To convince readers of the statistical utility of Markov chains, we introduce the topic of hidden Markov chains [1, 5, 27]. This brings in Baum's forward and backward algorithms and inhomogeneous chains. Limitations of space prevent us from considering specific applications. Interested readers can consult our listed references on speech recognition [27], physiological models of single-ion channels [10], gene mapping by radiation hybrids [23], and alignment of multiple DNA sequences [31].

25.2 Discrete-Time Markov Chains

For the sake of simplicity, we will only consider chains with a finite state space [2, 9, 13, 18, 19]. The movement of such a chain from epoch to epoch (equivalently, generation to generation) is governed by its transition probability matrix $P = (p_{ij})$. If Z_n denotes the state of the chain at epoch n, then $p_{ij} = \Pr(Z_n = j \mid Z_{n-1} = i)$. As a consequence, every entry of P satisfies $p_{ij} \geq 0$, and every row of P satisfies $\sum_j p_{ij} = 1$. Implicit in the definition of p_{ij} is the fact that the future of the chain is determined by its present regardless of its past. This Markovian property is expressed formally by the equation

$$\Pr(Z_n = i_n \mid Z_{n-1} = i_{n-1}, \ldots, Z_0 = i_0) \quad = \quad \Pr(Z_n = i_n \mid Z_{n-1} = i_{n-1}).$$

The n-step transition probability $p_{ij}^{(n)} = \Pr(Z_n = j \mid Z_0 = i)$ is given by the entry in row i and column j of the matrix power P^n. This follows

K. Lange, *Numerical Analysis for Statisticians*, Statistics and Computing,
DOI 10.1007/978-1-4419-5945-4_25, © Springer Science+Business Media, LLC 2010

because the decomposition

$$p_{ij}^{(n)} = \sum_{i_1} \cdots \sum_{i_{n-1}} p_{ii_1} \cdots p_{i_{n-1}j}$$

over all paths $i \to i_1 \to \cdots \to i_{n-1} \to j$ corresponds to matrix multiplication. A question of fundamental theoretical importance is whether the matrix powers P^n converge. If the chain eventually forgets its starting state, then the limit should have identical rows. Denoting the common limiting row by π, we deduce that $\pi = \pi P$ from the calculation

$$\begin{aligned} \begin{pmatrix} \pi \\ \vdots \\ \pi \end{pmatrix} &= \lim_{n\to\infty} P^{n+1} \\ &= \Big(\lim_{n\to\infty} P^n\Big) P \\ &= \begin{pmatrix} \pi \\ \vdots \\ \pi \end{pmatrix} P. \end{aligned}$$

Any probability distribution π on the states of the chain satisfying the condition $\pi = \pi P$ is termed an equilibrium (or stationary) distribution of the chain. For finite-state chains, equilibrium distributions always exist [9, 13]. The real issue is uniqueness.

Mathematicians have attacked the uniqueness problem by defining appropriate ergodic conditions. For finite-state Markov chains, two ergodic assumptions are invoked. The first is aperiodicity; this means that the greatest common divisor of the set $\{n \geq 1 : p_{ii}^{(n)} > 0\}$ is 1 for every state i. Aperiodicity trivially holds when $p_{ii} > 0$ for all i. The second ergodic assumption is irreducibility; this means that for every pair of states (i, j), there exists a positive integer n_{ij} such that $p_{ij}^{(n_{ij})} > 0$. In other words, every state is reachable from every other state. Said yet another way, all states communicate. For an irreducible chain, Problem 1 states that the integer n_{ij} can be chosen independently of the particular pair (i, j) if and only if the chain is also aperiodic. Thus, we can merge the two ergodic assumptions into the single assumption that some power P^n has all entries positive. Under this single ergodic condition, we showed in Chapter 6 that a unique equilibrium distribution π exists and that $\lim_{n\to\infty} p_{ij}^{(n)} = \pi_j$. Because all states communicate, the entries of π are necessarily positive.

Equally important is the ergodic theorem [9, 13]. This theorem permits one to run a chain and approximate theoretical means by sample means. More precisely, let $f(z)$ be some function defined on the states of an ergodic chain. Then $\lim_{n\to\infty} \frac{1}{n} \sum_{i=0}^{n-1} f(Z_i)$ exists and equals the theoretical mean $\mathrm{E}_\pi[f(Z)] = \sum_z \pi_z f(z)$ of $f(Z)$ under the equilibrium distribution π. This result generalizes the law of large numbers for independent sampling.

The equilibrium condition $\pi = \pi P$ can be restated as the system of equations

$$\pi_j = \sum_i \pi_i p_{ij} \tag{25.1}$$

for all j. In many Markov chain models, the stronger condition

$$\pi_j p_{ji} = \pi_i p_{ij} \tag{25.2}$$

holds for all pairs (i, j). If this is the case, then the probability distribution π is said to satisfy detailed balance. Summing equation (25.2) over i yields the equilibrium condition (25.1). An irreducible Markov chain with equilibrium distribution π satisfying detailed balance is said to be reversible. Irreducibility is imposed to guarantee that π is unique and has positive entries.

If $i_1, \ldots, i_m$ is any sequence of states in a reversible chain, then detailed balance implies

$$\begin{aligned}
\pi_{i_1} p_{i_1 i_2} &= \pi_{i_2} p_{i_2 i_1} \\
\pi_{i_2} p_{i_2 i_3} &= \pi_{i_3} p_{i_3 i_2} \\
&\vdots \\
\pi_{i_{m-1}} p_{i_{m-1} i_m} &= \pi_{i_m} p_{i_m i_{m-1}} \\
\pi_{i_m} p_{i_m i_1} &= \pi_{i_1} p_{i_1 i_m}.
\end{aligned}$$

Multiplying these equations together and cancelling the common positive factor $\pi_{i_1} \cdots \pi_{i_m}$ from both sides of the resulting equality give Kolmogorov's circulation criterion [20]

$$p_{i_1 i_2} p_{i_2 i_3} \cdots p_{i_{m-1} i_m} p_{i_m i_1} = p_{i_1 i_m} p_{i_m i_{m-1}} \cdots p_{i_3 i_2} p_{i_2 i_1}. \tag{25.3}$$

Conversely, if an irreducible Markov chain satisfies Kolmogorov's criterion, then the chain is reversible. This fact can be demonstrated by explicitly constructing the equilibrium distribution and showing that it satisfies detailed balance. The idea behind the construction is to choose some arbitrary reference state i and to pretend that π_i is given. If j is another state, let $i \to i_1 \to \cdots \to i_m \to j$ be any path leading from i to j. Then the formula

$$\pi_j = \pi_i \frac{p_{ii_1} p_{i_1 i_2} \cdots p_{i_m j}}{p_{j i_m} p_{i_m i_{m-1}} \cdots p_{i_1 i}} \tag{25.4}$$

defines π_j. A straightforward application of Kolmogorov's criterion (25.3) shows that the definition (25.4) does not depend on the particular path chosen from i to j. To validate detailed balance, suppose that k is adjacent to j. Then $i \to i_1 \to \cdots \to i_m \to j \to k$ furnishes a path from i to k

through j. It follows from (25.4) that $\pi_k = \pi_j p_{jk}/p_{kj}$, which is obviously equivalent to detailed balance. In general, the value of π_i is not known beforehand. Setting $\pi_i = 1$ produces the equilibrium distribution up to a normalizing constant.

Example 25.2.1 *Random Walk on a Graph*

Consider a connected graph with vertex set V and edge set E. The number of edges $d(v)$ incident on a given vertex v is called the degree of v. Due to the connectedness assumption, $d(v) > 0$ for all $v \in V$. Now define the transition probability matrix $P = (p_{uv})$ by

$$p_{uv} = \begin{cases} \frac{1}{d(u)} & \text{for } \{u, v\} \in E \\ 0 & \text{for } \{u, v\} \notin E . \end{cases}$$

This Markov chain is irreducible because of the connectedness assumption; it is also aperiodic unless the graph is bipartite. (A graph is said to be bipartite if we can partition its vertex set into two disjoint subsets F and M, say females and males, such that each edge has one vertex in F and the other vertex in M.) If V has m edges, then the equilibrium distribution π of the chain has components $\pi_v = d(v)/(2m)$. It is trivial to show that this choice of π satisfies detailed balance.

For instance, consider the $n!$ different permutations

$$\sigma = (\sigma_1, \ldots, \sigma_n)$$

of the set $\{1, \ldots, n\}$ equipped with the uniform distribution $\pi_\sigma = 1/n!$ [6]. Declare a permutation ω to be a neighbor of σ if there exist two indices $i \neq j$ such that $\omega_i = \sigma_j$, $\omega_j = \sigma_i$, and $\omega_k = \sigma_k$ for $k \notin \{i, j\}$. Evidently, each permutation has $\binom{n}{2}$ neighbors. If we put $p_{\sigma\omega} = 1/\binom{n}{2}$ for each neighbor ω of σ, then $p_{\sigma\omega}$ satisfies detailed balance. Thus, randomly choosing a pair of indices $i \neq j$ and switching σ_i with σ_j produces a Markov chain on the set of permutations. This chain has period 2. It can be made aperiodic by allowing the randomly chosen indices i and j to coincide. ■

Example 25.2.2 *Wright's Model of Genetic Drift*

Consider a population of m organisms from some animal or plant species. Each member of this population carries two genes at some genetic locus, and these genes assume two forms (or alleles) labeled a_1 and a_2. At each generation, the population reproduces itself by sampling $2m$ genes with replacement from the current pool of $2m$ genes. If Z_n denotes the number of a_1 alleles at generation n, then it is clear that the Z_n constitute a Markov chain with transition probability matrix

$$p_{jk} = \binom{2m}{k}\left(\frac{j}{2m}\right)^k\left(1 - \frac{j}{2m}\right)^{2m-k}.$$

This chain is reducible because once one of the states 0 or $2m$ is reached, then the corresponding allele is fixed in the population, and no further variation is possible. An infinite number of equilibrium distributions exist determined by $\pi_0 = \alpha$ and $\pi_{2m} = 1 - \alpha$ for $\alpha \in [0, 1]$. ■

Example 25.2.3 *Ehrenfest's Model of Diffusion*

Consider a box with m gas molecules. Suppose the box is divided in half by a rigid partition with a very small hole. Molecules drift aimlessly around each half until one molecule encounters the hole and passes through. Let Z_n be the number of molecules in the left half of the box at epoch n. If epochs are timed to coincide with molecular passages, then the transition matrix of the chain is

$$p_{jk} = \begin{cases} 1 - \frac{j}{m} & \text{for } k = j + 1 \\ \frac{j}{m} & \text{for } k = j - 1 \\ 0 & \text{otherwise.} \end{cases}$$

This chain is periodic with period 2, irreducible, and reversible with equilibrium distribution $\pi_j = \binom{m}{j} 2^{-m}$. ■

Example 25.2.4 *A Chain with a Continuous-State Space*

Although this chapter is devoted to discrete-state Markov chains, many chains have continuous state spaces. Hastings [14] suggests an interesting chain on the space of $n \times n$ orthogonal matrices R with $\det(R) = 1$. These multidimensional rotations form a compact subgroup of the set of all $n \times n$ matrices and consequently possess an invariant Haar probability measure [26]. The proposal stage of Hastings's algorithm consists of choosing at random two indices $i \neq j$ and an angle $\theta \in [-\pi, \pi]$. The proposed replacement for the current rotation matrix R is then $S = E_{ij}(\theta)R$, where the matrix $E_{ij}(\theta)$ is a rotation in the (i, j) plane through angle θ. Note that $E_{ij}(\theta) = (e_{kl})$ coincides with the identity matrix except for entries $e_{ii} = e_{jj} = \cos\theta$, $e_{ij} = \sin\theta$, and $e_{ji} = -\sin\theta$. Since $E_{ij}(\theta)^{-1} = E_{ij}(-\theta)$, the transition density is symmetric, and the Markov chain induced on the set of multidimensional rotations is reversible with respect to Haar measure. Problem 5 asks the reader to demonstrate that this Markov chain is irreducible. ■

25.3 Hidden Markov Chains

Hidden Markov chains incorporate both observed data and missing data. The missing data are the sequence of states visited by a Markov chain; the observed data provide partial information about this sequence of states. Denote the sequence of visited states by $Z_1, \ldots, Z_n$ and the observation taken at epoch i when the chain is in state Z_i by $Y_i = y_i$. Baum's algorithms recursively compute the likelihood of the observed data [1, 5, 27]

$$P = \Pr(Y_1 = y_1, \ldots, Y_n = y_n) \qquad (25.5)$$

without actually enumerating all possible realizations $Z_1, \ldots, Z_n$. Baum's algorithms can be adapted to perform an EM search as discussed in the next section. The references [8, 22, 27] discuss several concrete examples of hidden Markov chains.

The likelihood (25.5) is constructed from three ingredients: (a) the initial distribution π at the first epoch of the chain, (b) the epoch-dependent transition probabilities $p_{ijk} = \Pr(Z_{i+1} = k \mid Z_i = j)$, and (c) the conditional densities $\phi_i(y_i \mid j) = \Pr(Y_i = y_i \mid Z_i = j)$. The dependence of the transition probability p_{ijk} on i makes the chain inhomogeneous over time and allows greater flexibility in modeling. If the chain is homogeneous, then π is often taken as the equilibrium distribution. Implicit in the definition of $\phi_i(y_i \mid j)$ are the assumptions that $Y_1, \ldots, Y_n$ are independent given $Z_1, \ldots, Z_n$ and that Y_i depends only on Z_i. Finally, with obvious changes in notation, the observed data can be continuously rather than discretely distributed.

Baum's forward algorithm is based on recursively evaluating the joint probabilities

$$\alpha_i(j) = \Pr(Y_1 = y_1, \ldots, Y_{i-1} = y_{i-1}, Z_i = j).$$

At the first epoch, $\alpha_1(j) = \pi_j$ by definition; the obvious update to $\alpha_i(j)$ is

$$\alpha_{i+1}(k) = \sum_j \alpha_i(j)\phi_i(y_i \mid j)p_{ijk}. \qquad (25.6)$$

The likelihood (25.5) can be recovered by computing $\sum_j \alpha_n(j)\phi_n(y_n \mid j)$ at the final epoch n.

In Baum's backward algorithm, we recursively evaluate the conditional probabilities

$$\beta_i(k) = \Pr(Y_{i+1} = y_{i+1}, \ldots, Y_n = y_n \mid Z_i = k),$$

starting by convention at $\beta_n(k) = 1$ for all k. The required update is clearly

$$\beta_i(j) = \sum_k p_{ijk}\phi_{i+1}(y_{i+1} \mid k)\beta_{i+1}(k). \qquad (25.7)$$

In this instance, the likelihood is recovered at the first epoch by forming the sum $\sum_j \pi_j\phi_1(y_1 \mid j)\beta_1(j)$.

Baum's algorithms (25.6) and (25.7) are extremely efficient, particularly if the observations y_i strongly limit the number of compatible states at each epoch i. In statistical practice, maximization of the likelihood with respect to model parameters is usually an issue. Most maximum likelihood algorithms require the score in addition to the likelihood. These partial derivatives can often be computed quickly in parallel with other quantities in Baum's forward and backward algorithms. For example, suppose that a parameter θ impacts only the transition probabilities p_{ijk} for a specific epoch i [23]. Since we can write the likelihood as

$$P = \sum_j \sum_k \alpha_i(j)\phi_i(y_i \mid j)p_{ijk}\phi_{i+1}(y_{i+1} \mid k)\beta_{i+1}(k),$$

it follows that

$$\frac{\partial}{\partial\theta}P = \sum_j \sum_k \alpha_i(j)\phi_i(y_i \mid j)\Big[\frac{\partial}{\partial\theta}p_{ijk}\Big]\phi_{i+1}(y_{i+1} \mid k)\beta_{i+1}(k). \quad (25.8)$$

Similarly, if θ only enters the conditional density $\phi_i(y_i \mid j)$ for a given i, then the representation

$$P = \sum_j \alpha_i(j)\phi_i(y_i \mid j)\beta_i(j)$$

leads to the partial derivative formula

$$\frac{\partial}{\partial\theta}P = \sum_j \alpha_i(j)\Big[\frac{\partial}{\partial\theta}\phi_i(y_i \mid j)\Big]\beta_i(j). \quad (25.9)$$

Finally, if θ enters only into the initial distribution π, then

$$\frac{\partial}{\partial\theta}P = \sum_j \Big[\frac{\partial}{\partial\theta}\pi_j\Big]\phi_1(y_1 \mid j)\beta_1(j). \quad (25.10)$$

These formulas suggest that an efficient evaluation of P and its partial derivatives can be orchestrated by carrying out the backward algorithm first, saving all resulting $\beta_i(j)$, and then carrying out the forward algorithm while simultaneously computing all partial derivatives. Note that if a parameter θ enters into several of the factors defining P, then by virtue of the product rule of differentiation, we can express $\frac{\partial}{\partial\theta}P$ as a sum of the corresponding right-hand sides of equations (25.8), (25.9), and (25.10). Given a partial derivative $\frac{\partial}{\partial\theta}P$ of the likelihood P, we compute the corresponding entry in the score vector by taking the quotient $\frac{\partial}{\partial\theta}P/P$.

Besides evaluating and maximizing the likelihood, statisticians are often interested in finding a most probable sequence of states of the hidden Markov chain given the observed data. The Viterbi algorithm solves this problem by dynamic programming [10]. We proceed by solving the intermediate problems

$$\gamma_k(z_k) = \max_{z_1,\ldots,z_{k-1}} \Pr(Z_1 = z_1, \ldots, Z_k = z_k, Y_1 = y_1, \ldots, Y_k = y_k)$$

for each $k = 1, \ldots, n$, beginning with $\gamma_1(z_1) = \pi_{z_1}\phi_1(y_1 \mid z_1)$. When we reach $k = n$, then $\max_{z_n} \gamma_n(z_n)$ yields the largest joint probability

$$\Pr(Z_1 = z_1, \ldots, Z_n = z_n, Y_1 = y_1, \ldots, Y_n = y_n)$$

and consequently the largest conditional probability

$$\Pr(Z_1 = z_1, \ldots, Z_n = z_n \mid Y_1 = y_1, \ldots, Y_n = y_n)$$

as well. If we have kept track of one solution sequence $z_1(z_n), \ldots, z_{n-1}(z_n)$ for each $\gamma_n(z_n)$, then obviously we can construct a best overall sequence by taking the best z_n and appending to it $z_1(z_n), \ldots, z_{n-1}(z_n)$. To understand better the recursive phase of the algorithm, let

$$\delta_k(z_1, \ldots, z_k) \quad = \quad \Pr(Z_1 = z_1, \ldots, Z_k = z_k, Y_1 = y_1, \ldots, Y_k = y_k).$$

In this notation, we express $\gamma_{k+1}(z_{k+1})$ as

$$\begin{aligned}
\gamma_{k+1}(z_{k+1}) &= \max_{z_1,\ldots,z_k} \delta_{k+1}(z_1, \ldots, z_{k+1}) \\
&= \max_{z_1,\ldots,z_k} \delta_k(z_1, \ldots, z_k) p_{k,z_k,z_{k+1}} \phi_{k+1}(y_{k+1} \mid z_{k+1}) \\
&= \max_{z_k} p_{k,z_k,z_{k+1}} \phi_{k+1}(y_{k+1} \mid z_{k+1}) \max_{z_1,\ldots,z_{k-1}} \delta_k(z_1, \ldots, z_k) \\
&= \max_{z_k} p_{k,z_k,z_{k+1}} \phi_{k+1}(y_{k+1} \mid z_{k+1}) \gamma_k(z_k)
\end{aligned}$$

and create a maximizing sequence $z_1(z_{k+1}), \ldots, z_k(z_{k+1})$ for each z_{k+1} from the corresponding best z_k and its recorded sequence $z_1(z_k), \ldots, z_{k-1}(z_k)$.

25.4 Connections to the EM Algorithm

Baum's algorithms also interdigitate beautifully with the E step of the EM algorithm. It is natural to summarize the missing data by a collection of indicator random variables X_{ij}. If the chain occupies state j at epoch i, then we take $X_{ij} = 1$. Otherwise, we take $X_{ij} = 0$. In this notation, the complete data loglikelihood can be written as

$$\begin{aligned}
L_{\text{com}}(\theta) &= \sum_j X_{1j} \ln \pi_j + \sum_{i=1}^{n} \sum_j X_{ij} \ln \phi_i(Y_i \mid j) \\
&\quad + \sum_{i=1}^{n-1} \sum_j \sum_k X_{ij} X_{i+1,k} \ln p_{ijk}.
\end{aligned}$$

Execution of the E step amounts to calculation of the conditional expectations

$$\begin{aligned}
\mathrm{E}(X_{ij} X_{i+1,k} \mid Y, \theta_m) &= \frac{\alpha_i(j)\phi_i(y_i \mid j) p_{ijk} \phi_{i+1}(y_{i+1} \mid k) \beta_{i+1}(k)}{P} \Big|_{\theta=\theta_m} \\
\mathrm{E}(X_{ij} \mid Y, \theta_m) &= \frac{\alpha_i(j)\phi_i(y_i \mid j)\beta_i(j)}{P} \Big|_{\theta=\theta_m},
\end{aligned}$$

where Y is the observed data, P is the likelihood of the observed data, and θ_m is the current parameter vector. It is no accident that these are reminiscent of the sandwich formulas (25.8), (25.9), and (25.10). Problems 17, 19, and 20 of Chapter 13 explore the connections between the score vector and EM updates.

The M step may or may not be exactly solvable. If it is not, then one can always revert to the MM gradient algorithm discussed in Section 14.8. In the case of hidden multinomial trials, it is possible to carry out the M step analytically. Hidden multinomial trials may govern (a) the choice of the initial state, (b) the choice of an observed outcome Y_i at the ith epoch given the hidden state j of the chain at that epoch, or (c) the choice of the next state j given the current state i in a time-homogeneous chain. In the first case, the multinomial parameters are the π_i; in the last case, they are the common transition probabilities p_{ij}.

As a concrete example, consider estimation of the initial distribution π at the first epoch of the chain. For estimation to be accurate, there must be multiple independent runs of the chain. Let r index the various runs. The surrogate function delivered by the E step equals

$$Q(\pi \mid \pi = \pi_m) = \sum_r \sum_j \mathrm{E}(X_{1j}^r \mid Y^r = y^r, \pi = \pi_m) \ln \pi_j$$

up to an additive constant. Maximizing $Q(\pi \mid \pi = \pi_m)$ subject to the constraints $\sum_j \pi_j = 1$ and $\pi_j \geq 0$ for all j is done as in Example 11.3.1. The resulting EM updates

$$\pi_{m+1,j} = \frac{\sum_r \mathrm{E}(X_{1j}^r \mid Y^r = y^r, \pi = \pi_m)}{R}$$

for R runs can be interpreted as multinomial proportions with fractional category counts. Problem 9 asks the reader to derive the EM algorithm for estimating common transition probabilities.

25.5 Continuous-Time Markov Chains

Continuous-time Markov chains are often more realistic than discrete-time Markov chains. Just as in the discrete case, the behavior of a chain is described by an indexed family Z_t of random variables giving the state occupied by the chain at each time t. However, now the index t ranges over real numbers rather than integers. Of fundamental theoretical importance are the probabilities $p_{ij}(t) = \Pr(Z_t = j \mid Z_0 = i)$. For a chain having a finite number of states, these probabilities can be found by solving a matrix differential equation. To derive this equation, we use the short-time approximation

$$p_{ij}(t) = \lambda_{ij} t + o(t) \tag{25.11}$$

for $i \neq j$, where λ_{ij} is the transition rate (or infinitesimal transition probability) from state i to state j. Equation (25.11) implies the further short-time approximation

$$p_{ii}(t) = 1 - \lambda_i t + o(t), \tag{25.12}$$

where $\lambda_i = \sum_{j \neq i} \lambda_{ij}$.

The alternative perspective of competing risks sharpens our intuitive understanding of equations (25.11) and (25.12). Imagine that a particle executes the Markov chain by moving from state to state. If the particle is currently in state i, then each neighboring state independently beckons the particle to switch positions. The intensity of the temptation exerted by state j is the constant λ_{ij}. In the absence of competing temptations, the particle waits an exponential length of time T_{ij} with intensity λ_{ij} before moving to state j. Taking into account the competing temptations, the particle moves at the moment $\min_j T_{ij}$, which is exponentially distributed with intensity λ_i. Once the particle decides to move, it moves to state j with probability λ_{ij}/λ_i. Equations (25.11) and (25.12) now follow from the approximations

$$\begin{aligned} \left(1 - e^{-\lambda_i t}\right)\frac{\lambda_{ij}}{\lambda_i} &= \lambda_{ij} t + o(t) \\ e^{-\lambda_i t} &= 1 - \lambda_i t + o(t). \end{aligned}$$

Next consider the Chapman-Kolmogorov relation

$$p_{ij}(t+h) = p_{ij}(t)p_{jj}(h) + \sum_{k \neq j} p_{ik}(t)p_{kj}(h), \tag{25.13}$$

which simply says the chain must pass through some intermediate state k at time t en route to state j at time $t+h$. Substituting the approximations (25.11) and (25.12) in (25.13) yields

$$p_{ij}(t+h) = p_{ij}(t)(1 - \lambda_j h) + \sum_{k \neq j} p_{ik}(t)\lambda_{kj} h + o(h).$$

Sending h to 0 in the difference quotient

$$\frac{p_{ij}(t+h) - p_{ij}(t)}{h} = -p_{ij}(t)\lambda_j + \sum_{k \neq j} p_{ik}(t)\lambda_{kj} + \frac{o(h)}{h}$$

produces the forward differential equation

$$p'_{ij}(t) = -p_{ij}(t)\lambda_j + \sum_{k \neq j} p_{ik}(t)\lambda_{kj}. \tag{25.14}$$

The system of differential equations (25.14) can be summarized in matrix notation by introducing the matrices $P(t) = [p_{ij}(t)]$ and $\Lambda = (\Lambda_{ij})$, where

$\Lambda_{ij} = \lambda_{ij}$ for $i \neq j$ and $\Lambda_{ii} = -\lambda_i$. The forward equations in this notation become

$$\begin{aligned} P'(t) &= P(t)\Lambda \\ P(0) &= I, \end{aligned} \tag{25.15}$$

where I is the identity matrix. It is easy to check that the solution of the initial value problem (25.15) is furnished by the matrix exponential [17, 21]

$$\begin{aligned} P(t) &= e^{t\Lambda} \\ &= \sum_{k=0}^{\infty} \frac{1}{k!}(t\Lambda)^k. \end{aligned} \tag{25.16}$$

Probabilists call Λ the infinitesimal generator or infinitesimal transition matrix of the process. The infinite series (25.16) converges because its partial sums form a Cauchy sequence. This fact follows directly from the inequality

$$\left\| \sum_{k=0}^{m} \frac{1}{k!}(t\Lambda)^k - \sum_{k=0}^{m+n} \frac{1}{k!}(t\Lambda)^k \right\| \leq \sum_{k=m+1}^{m+n} \frac{1}{k!}|t|^k \|\Lambda\|^k.$$

A probability distribution $\pi = (\pi_i)$ on the states of a continuous-time Markov chain is a row vector whose components satisfy $\pi_i \geq 0$ for all i and $\sum_i \pi_i = 1$. If

$$\pi P(t) = \pi \tag{25.17}$$

holds for all $t \geq 0$, then π is said to be an equilibrium distribution for the chain. Written in components, the eigenvector equation (25.17) reduces to $\sum_i \pi_i p_{ij}(t) = \pi_j$. Again this is completely analogous to the discrete-time theory. For small t equation (25.17) can be rewritten as

$$\pi(I + t\Lambda) + o(t) = \pi.$$

This approximate form makes it obvious that $\pi\Lambda = \mathbf{0}$ is a necessary condition for π to be an equilibrium distribution. Multiplying (25.16) on the left by π shows that $\pi\Lambda = \mathbf{0}$ is also a sufficient condition for π to be an equilibrium distribution. In components this necessary and sufficient condition amounts to

$$\sum_{j \neq i} \pi_j \lambda_{ji} = \pi_i \sum_{j \neq i} \lambda_{ij} \tag{25.18}$$

for all i. If all the states of a Markov chain communicate, then there is one and only one equilibrium distribution π. Furthermore, each of the rows of $P(t)$ approaches π as t approaches ∞. Lamperti [21] provides a clear exposition of these facts.

Fortunately, the annoying feature of periodicity present in discrete-time theory disappears in the continuous-time theory. The definition and properties of reversible chains carry over directly from discrete time to continuous time provided we substitute infinitesimal transition probabilities for transition probabilities. For instance, the detailed balance condition becomes

$$\pi_i \lambda_{ij} = \pi_j \lambda_{ji}$$

for all pairs $i \neq j$. Kolmogorov's circulation criterion for reversibility continues to hold; when it is true, the equilibrium distribution is constructed from the infinitesimal transition probabilities exactly as in discrete time.

Example 25.5.1 *Oxygen Attachment to Hemoglobin*

A hemoglobin molecule has four possible sites to which oxygen (O_2) can attach. If the surrounding concentration s_o of O_2 is sufficiently high, then we can model the number of sites occupied on a hemoglobin molecule as a continuous-time Markov chain [28]. Figure 25.1 depicts the model; in the figure, each arc is labeled by an infinitesimal transition probability and each state by the number of O_2 molecules attached to the hemoglobin molecule. The forward rates $s_o k_{+j}$ incorporate the concentration of O_2. Because this chain is reversible, we can calculate its equilibrium distribution starting from the reference state 0 as $\pi_i = \pi_0 s_o^i \prod_{j=1}^{i} k_{+j}/k_{-j}$. ■

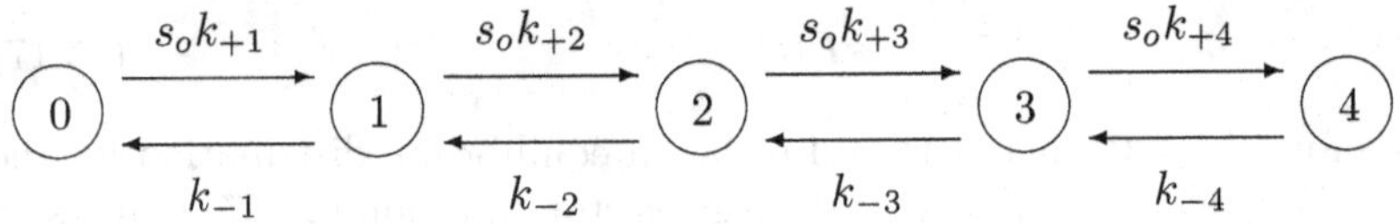

FIGURE 25.1. A Markov Chain Model for Oxygen Attachment to Hemoglobin

Example 25.5.2 *Continuous-Time Multitype Branching Processes*

Matrix exponentials also appear in the theory of continuous-time branching processes. In such a process one follows a finite number of independently acting particles that reproduce and die. In a multitype branching process, each particle is classified in one of n possible categories. A type i particle lives an exponentially distributed length of time with a death intensity of λ_i. At the end of its life, a type i particle reproduces both particles of its own type and particles of other types. Suppose that on average it produces f_{ij} particles of type j. We would like to calculate the average number of particles $m_{ij}(t)$ of type j at time $t \geq 0$ starting with a single particle of type i at time 0. Since particles of type j at time $t + h$ either arise from particles of type j at time t which do not die during $(t, t + h)$ or from particles of type k which die during $(t, t+h)$ and reproduce particles of type j, we find that

$$m_{ij}(t+h) = m_{ij}(t)(1-\lambda_j h) + \sum_k m_{ik}(t)\lambda_k f_{kj} h + o(h).$$

Forming the corresponding difference quotients and sending h to 0 yield the differential equations

$$m'_{ij}(t) = \sum_k m_{ik}(t)\lambda_k(f_{kj} - 1_{\{k=j\}}),$$

which we summarize as the matrix differential equation $M'(t) = M(t)\Omega$ for the $n \times n$ matrices $M(t) = [m_{ij}(t)]$ and $\Omega = [\lambda_i(f_{ij} - 1_{\{i=j\}})]$. Again the solution is provided by the matrix exponential $M(t) = e^{t\Omega}$ subject to the initial condition $M(0) = I$. ■

25.6 Calculation of Matrix Exponentials

From the definition of the matrix exponential e^A, it is easy to deduce that it is continuous in A and satisfies $e^{A+B} = e^A e^B$ whenever $AB = BA$. It is also straightforward to check the differentiability condition

$$\frac{d}{dt}e^{tA} = Ae^{tA} = e^{tA}A.$$

Of more practical importance is how one actually calculates e^{tA} [25]. In some cases it is possible to do so analytically. For instance, if u and v are column vectors with the same number of components, then

$$e^{suv^t} = \begin{cases} I + suv^t & \text{if } v^t u = 0 \\ I + \frac{e^{sv^t u} - 1}{v^t u} uv^t & \text{otherwise.} \end{cases}$$

This follows from the formula $(uv^t)^i = (v^t u)^{i-1} uv^t$. The special case having $u = (-\alpha, \beta)^t$ and $v = (1, -1)^t$ permits us to calculate the finite-time transition matrix

$$P(s) = \exp\left[s\begin{pmatrix} -\alpha & \alpha \\ \beta & -\beta \end{pmatrix}\right]$$

for a two-state Markov chain.

If A is a diagonalizable $n \times n$ matrix, then we can write $A = TDT^{-1}$ for D a diagonal matrix with ith diagonal entry ρ_i. In this situation note that $A^2 = TDT^{-1}TDT^{-1} = TD^2T^{-1}$ and in general that $A^i = TD^iT^{-1}$. Hence,

$$e^{tA} = \sum_{i=0}^{\infty} \frac{1}{i!}(tA)^i$$

$$\begin{aligned} &= \sum_{i=0}^{\infty} \frac{1}{i!} T(tD)^i T^{-1} \\ &= Te^{tD}T^{-1}, \end{aligned}$$

where

$$e^{tD} = \begin{pmatrix} e^{\rho_1 t} & \cdots & 0 \\ \vdots & \ddots & \vdots \\ 0 & \cdots & e^{\rho_n t} \end{pmatrix}.$$

Even if we cannot diagonalize A explicitly, we can usually do so numerically.

When $t > 0$ is small, another method is to truncate the series expansion for e^{tA} to $\sum_{i=0}^{n}(tA)^i/i!$ for n small. For larger t such truncation can lead to serious errors. If the truncated expansion is accurate for all $t \leq c$, then for arbitrary t one can exploit the property $e^{(s+t)A} = e^{sA}e^{tA}$ of the matrix exponential. Thus, if $t > c$, take the smallest positive integer k such that $2^{-k}t \leq c$ and approximate $e^{2^{-k}tA}$ by the truncated series. Applying the multiplicative property we can compute e^{tA} by squaring $e^{2^{-k}tA}$, squaring the result $e^{2^{-k+1}tA}$, squaring the result of this, and so forth, a total of k times. Problems 14 and 15 explore how the errors encountered in this procedure can be controlled. Padé approximations based on continued fractions work even better than truncated Taylor series [16].

25.7 Calculation of the Equilibrium Distribution

Finding the equilibrium distribution of a continuous-time Markov chain reduces to finding the equilibrium distribution of an associated discrete-time Markov chain. The latter task can be accomplished by the power method or a variety of other methods [30]. Consider a continuous-time chain with transition intensities λ_{ij} and infinitesimal generator Λ. If we collect the off-diagonal entries of Λ into a matrix Ω and the negative diagonal entries into a diagonal matrix D, then equation (25.18) describing the balance conditions satisfied by the equilibrium distribution π can be restated as

$$\pi D = \pi\Omega.$$

Close examination of the matrix $P = D^{-1}\Omega$ shows that its entries are nonnegative, its row sums are 1, and its diagonal entries are 0. Furthermore, P is sparse whenever Ω is sparse, and all states communicate under P when all states communicate under Λ. Nothing prevents the transition probability matrix P from being periodic, but aperiodicity is irrelevant in deciding whether a unique equilibrium distribution exists. Indeed, for any fixed constant $\alpha \in (0, 1)$, one can easily demonstrate that an equilibrium distribution of P is also an equilibrium distribution of the aperiodic transition probability matrix $Q = \alpha I + (1-\alpha)P$ and vice versa.

Suppose that we compute the equilibrium distribution ν of P by some method. Once ν is available, we set $\omega = \nu D^{-1}$. Because the two equations

$$\nu \;=\; \nu P, \qquad \omega D \;=\; \omega\Omega$$

are equivalent, ω coincides with π up to a multiplicative constant. In other words, Λ has equilibrium distribution $\pi = (\omega\mathbf{1})^{-1}\omega$. Hence, trivial adjustment of the equilibrium distribution for the associated discrete-time chain produces the equilibrium distribution of the original continuous-time chain.

25.8 Stochastic Simulation and Intensity Leaping

Many chemical and biological models depend on continuous-time Markov chains with a finite number of particle types [15]. The particles interact via a finite number of reaction channels, and each reaction destroys and/or creates particles in a predictable way. In this section, we consider the problem of simulating the behavior of such chains. Before we launch into simulation specifics, it is helpful to carefully define a typical process. If d denotes the number of types, then the chain follows the count vector X_t whose ith component X_{ti} is the number of particles of type i at time $t \geq 0$. We typically start the system at time 0 and let it evolve via a succession of random reactions. Let c denote the number of reaction channels. Channel j is characterized by an intensity function $r_j(x)$ depending on the current vector of counts x. In a small time interval of length s, we expect $r_j(x)s + o(s)$ reactions of type j to occur. Reaction j changes the count vector by a fixed integer vector v^j. Some components v_k^j of v^j may be positive, some 0, and some negative. From the wait and jump perspective of Markov chain theory, we wait an exponential length of time until the next reaction. If the chain is currently in state $X_t = x$, then the intensity of the waiting time is $r_0(x) = \sum_{j=1}^{c} r_j(x)$. Once the decision to jump is made, we jump to the neighboring state $x + v^j$ with probability $r_j(x)/r_0(x)$.

Table 25.1 lists typical reactions, their intensities $r(x)$, and increment vectors v. In the table, S_i denotes a single particle of type i. Only the nonzero increments v_i are shown. The reaction intensities invoke the law of mass action and depend on rate constants a_i. Each discipline has its own vocabulary. Chemists use the name propensity instead of the name intensity and call the increment vector a stoichiometric vector. Physicists prefer creation to immigration. Biologists speak of death and mutation rather than of decay and isomerization. Despite the variety of processes covered, the allowed chains form a subset of all continuous-time Markov chains. Chains with an infinite number of reaction channels or random increments are not allowed. For instance, most continuous-time branching

TABLE 25.1. Some Examples of Reaction Channels

Name	Reaction	$r(x)$	v
Immigration	$0 \to S_1$	a_1	$v_1 = 1$
Decay	$S_1 \to 0$	$a_2 x_1$	$v_1 = -1$
Dimerization	$S_1 + S_1 \to S_2$	$a_3 \binom{x_1}{2}$	$v_1 = -2,\ v_2 = 1$
Isomerization	$S_1 \to S_2$	$a_4 x_1$	$v_1 = -1,\ v_2 = 1$
Dissociation	$S_2 \to S_1 + S_1$	$a_5 x_2$	$v_1 = 2,\ v_2 = -1$
Budding	$S_1 \to S_1 + S_2$	$a_6 x_1$	$v_2 = 1$
Replacement	$S_1 + S_2 \to S_2 + S_2$	$a_7 x_1 x_2$	$v_1 = -1,\ v_2 = 1$
Complex Reaction	$S_1 + S_2 \to S_3 + S_4$	$a_8 x_1 x_2$	$v_1 = v_2 = -1$ $v_3 = v_4 = 1$

processes do not qualify. Branching processes that grow by budding serve as useful substitutes for more general branching processes.

The wait and jump mechanism constitutes a perfectly valid method of simulating one of these chains. Gillespie first recognized the practicality of this approach in chemical kinetics [11]. Although his stochastic simulation algorithm works well in some contexts, it can be excruciatingly slow in others. Unfortunately, reaction rates can vary by orders of magnitude, and the fastest reactions dominate computational expense in stochastic simulation. For the fast reactions, stochastic simulation takes far too many small steps. Our goal is to describe an alternative approximate algorithm that takes larger, less frequent steps. The alternative is predicated on the observation that reaction intensities change rather slowly in many models. Before describing how we can take advantage of this feature, it is worth mentioning the chemical master equations, which is just another name for the forward equations of Markov chain theory.

Let $p_{xy}(t)$ denote the finite-time transition probability of going from state x at time 0 to state y at time t. The usual reasoning leads to the expansions

$$p_{xy}(t+s) \quad = \quad p_{xy}(t)\left[1 - \sum_{j=1}^{c} r_j(y)s\right] + \sum_{j=1}^{c} p_{x,y-v^j}(t) r_j(y - v^j)s + o(s).$$

Forming the corresponding difference quotient and sending s to 0 produce the master equations

$$\frac{d}{dt}p_{xy}(t) \quad = \quad \sum_{j=1}^{c} \Big[p_{x,y-v^j}(t) r_j(y - v^j) - p_{xy}(t) r_j(y)\Big]$$

with initial conditions $p_{xy}(0) = 1_{\{x=y\}}$. Only in special cases can the master equations be solved. In deterministic models where particle counts are high, one is usually content to follow mean particle counts. The mean behavior

$\mu(t) = \mathrm{E}(X_t)$ is then roughly modeled by the system of ordinary differential equations

$$\frac{d}{dt}\mu(t) = \sum_{j=1}^{c} r_j[\mu(t)]v^j.$$

This approximation becomes more accurate as mean particle counts increase.

For the sake of argument, suppose all reaction intensities are constant. In the time interval $(t, t+s)$, reaction j occurs a Poisson number of times with mean $r_j s$. If we can sample from the Poisson distribution with an arbitrary mean, then we can run stochastic simulation accurately with s of any duration. If we start the process at $X_t = x$, then at time $t+s$ we have $X_{t+s} = x + \sum_{j=1}^{c} N_j v^j$, where the N_j are independent Poisson variates with means $r_j s$. The catch, of course, is that reaction intensities change as the particle count vector X_t changes. In the τ-leaping method of simulation, we restrict the time increment $\tau > 0$ to sufficiently small values such that each intensity $r_j(x)$ suffers little change over $(t, t+\tau)$ [4, 12].

Before we discuss exactly how to achieve this, let us pass to a more sophisticated update that anticipates how intensities change [29]. Assume that X_t is a deterministic process with a well-defined derivative. Over a short time interval $(t, t+\tau)$, the intensity $r_j(X_t)$ should then change by the approximate amount $\frac{d}{dt}r_j(X_t)\tau$. Reactions of type j now occur according to an inhomogeneous Poisson process with a linear intensity. Thus, we anticipate a Poisson number of reactions of type j with mean

$$\begin{aligned}\omega_j(t, t+\tau) &= \int_0^\tau \left[r_j(X_t) + \frac{d}{dt}r_j(X_t)s \right] ds \\ &= r_j(X_t)\tau + \frac{d}{dt}r_j(X_t)\frac{1}{2}\tau^2.\end{aligned}$$

At time $t+\tau$, we put $X_{t+\tau} = X_t + \sum_{j=1}^{c} Y_j v^j$, where the Y_j are independent Poisson variates with means $\omega_j(t, t+\tau)$. This is all to the good, but how do we compute the time derivatives of $r_j(X_t)$? The most natural approach is to invoke the chain rule

$$\frac{d}{dt}r_j(x) = \sum_{k=1}^{d} \frac{\partial}{\partial x_k} r_j(x) \frac{d}{dt}x_k,$$

and set

$$\frac{d}{dt}x_k = \sum_{j=1}^{c} r_j(x)v_k^j$$

as dictated by the approximate mean growth of the system. In most models the matrix $dr(x) = [\frac{\partial}{\partial x_k} r_j(x)]$ is sparse, with nontrivial entries that are constant or linear in x.

This exposition gives some insight into how we choose the increment τ in the τ-leaping method. It seems reasonable to take the largest τ such that

$$\left|\frac{d}{dt}r_j(x)\right|\tau \leq \epsilon r_j(x)$$

holds for all j, where $\epsilon > 0$ is a small constant. If $r_j(x) = 0$ is possible, then we might amend this to

$$\left|\frac{d}{dt}r_j(x)\right|\tau \leq \epsilon \max\{r_j(x), a_j\},$$

where a_j is the rate constant for reaction j. In each instance in Table 25.1, a_j is the smallest possible change in $r_j(x)$. In common with other τ-leaping strategies, we revert to the stochastic simulation update whenever the intensity $r_0(x)$ for leaving a state x falls below a certain threshold δ.

As a test case, we apply the above version of τ-leaping to Kendall's birth, death, and immigration process. In the time-homogeneous case, this Markov chain is governed by the birth rate α per particle, the death rate μ per particle, and the overall immigration rate ν. Starting with i particles at time 0, one can explicitly calculate the mean number of particles at time t as

$$m_i(t) = ie^{(\alpha-\mu)t} + \frac{\nu}{\alpha-\mu}\left[e^{(\alpha-\mu)t} - 1\right].$$

This exact expression permits us to evaluate the accuracy of τ-leaping. For the sake of illustration, we consider $t = 4$, $i = 5$, and average particle counts over 10,000 simulations. Table 25.2 lists the exact value of $m_5(4)$ and the average particle counts from τ-leaping for two methods. Method 1 ignores the derivative correction, and method 2 incorporates it. The table also gives for method 2 the time in seconds over all 10,000 runs and the fraction of steps attributable to stochastic simulation when the threshold constant $\delta = 100$. Although the table makes a clear case for the more accurate method 2, more testing is necessary. This is an active area of research, and given its practical importance, even more research is merited.

25.9 Problems

1. Demonstrate that a finite-state Markov chain is ergodic (irreducible and aperiodic) if and only if some power P^n of the transition matrix P has all entries positive. (Hints: For sufficiency, show that if some power P^n has all entries positive, then P^{n+1} has all entries positive. For necessity, note that $p_{ij}^{(r+s+t)} \geq p_{ik}^{(r)} p_{kk}^{(s)} p_{kj}^{(t)}$, and use the number theoretic fact that the set $\{s : p_{kk}^{(s)} > 0\}$ contains all sufficiently large positive integers s if k is aperiodic. See the appendix of [3] for the requisite number theory.)

TABLE 25.2. Mean Counts for $\alpha = 2$, $\mu = 1$, and $\nu = \frac{1}{2}$ in Kendall's Process.

epsilon	average 1	average 2	predicted	time	ssa fraction
1.0000	153.155	251.129	299.790	0.781	0.971
0.5000	195.280	279.999	299.790	0.875	0.950
0.2500	232.495	292.790	299.790	1.141	0.909
0.1250	261.197	297.003	299.790	1.625	0.839
0.0625	279.176	301.671	299.790	2.328	0.726
0.0313	286.901	298.565	299.790	3.422	0.575
0.0156	297.321	301.560	299.790	4.922	0.405
0.0078	294.487	300.818	299.790	7.484	0.256

2. Show that Kolmogorov's criterion (25.3) implies that definition (25.4) does not depend on the particular path chosen from i to j.

3. In the Bernoulli-Laplace model, we imagine two boxes with m particles each. Among the $2m$ particles there are b black particles and w white particles, where $b + w = 2m$ and $b \leq w$. At each epoch one particle is randomly selected from each box, and the two particles are exchanged. Let Z_n be the number of black particles in the first box. Is the corresponding chain irreducible, aperiodic, or reversible? Show that its equilibrium distribution is hypergeometric.

4. In Example 25.2.1, show that the chain is aperiodic if and only if the underlying graph is not bipartite.

5. Show that Hastings's Markov chain on multidimensional rotations is irreducible. (Hint: Prove that every multidimensional rotation R can be written as a finite product of matrices of the form $E_{ij}(\theta)$. Using a variation of Jacobi's method discussed in Chapter 8, argue inductively that you can zero out the off-diagonal entries of R in a finite number of multiplications by appropriate two-dimensional rotations $E_{ij}(\theta)$. The remaining diagonal entries all equal ± 1. There are an even number of -1 diagonal entries, and these can be converted to $+1$ diagonal entries in pairs.)

6. Let P be the transition matrix and π the equilibrium distribution of a reversible Markov chain with n states. Define an inner product $\langle u, v \rangle_\pi$ on complex column vectors u and v with n components by

$$\langle u, v \rangle_\pi = \sum_i u_i \pi_i v_i^*.$$

Verify that P satisfies the self-adjointness condition

$$\langle Pu, v \rangle_\pi = \langle u, Pv \rangle_\pi,$$

and conclude by standard arguments that P has only real eigenvalues. Formulate a similar result for a reversible continuous-time chain.

7. Let $Z_0, Z_1, Z_2, \ldots$ be a realization of an ergodic chain. If we sample every kth epoch, then show (a) that the sampled chain $Z_0, Z_k, Z_{2k}, \ldots$ is ergodic, (b) that it possesses the same equilibrium distribution as the original chain, and (c) that it is reversible if the original chain is. Thus, we can estimate theoretical means by sample averages using only every kth epoch of the original chain.

8. A Markov chain is said to be embedded in a base Markov chain if there exists a map $f : C^* \to C$ from the state space C^* of the base chain onto the state space C of the embedded chain [24]. This map partitions the states of C^* into equivalence classes under the equivalence relation $x \sim y$ when $f(x) = f(y)$. If $Q = (q_{uv})$ denotes the matrix of transition probabilities of the base chain, then it is natural to define the transition probabilities of the embedded chain by

$$p_{f(u)f(v)} = \sum_{w \sim v} q_{uw}.$$

For the embedding to be probabilistically consistent, it is necessary that

$$\sum_{w \sim v} q_{uw} = \sum_{w \sim v} q_{xw} \qquad (25.19)$$

for all $x \sim u$. A distribution ν on the base chain induces a distribution μ on the embedded chain according to

$$\mu_{f(u)} = \sum_{w \sim u} \nu_w. \qquad (25.20)$$

Mindful of these conventions, show that the embedded Markov chain is irreducible if the base Markov chain is irreducible and is aperiodic if the base chain is aperiodic. If the base chain is reversible with stationary distribution ν, then show that the embedded chain is reversible with induced stationary distribution μ given by (25.20).

9. In the hidden Markov chain model, suppose that the chain is time homogeneous with transition probabilities p_{ij}. Derive an EM algorithm for estimating the p_{ij} from one or more independent runs of the chain.

10. In the hidden Markov chain model, consider estimation of the parameters of the conditional densities $\phi_i(y_i \mid j)$ of the observed data

$y_1, \ldots, y_n$. When Y_i given $Z_i = j$ is Poisson distributed with mean μ_j, show that the EM algorithm updates μ_j by

$$\mu_{m+1,j} = \frac{\sum_{i=1}^n w_{mij} y_i}{\sum_{i=1}^n w_{mij}},$$

where the weight $w_{mij} = \mathrm{E}(X_{ij} \mid Y, \theta_m)$. Show that the same update applies when Y_i given $Z_i = i$ is exponentially distributed with mean μ_j or normally distributed with mean μ_j and common variance σ^2. In the latter setting, demonstrate that the EM update of σ^2 is

$$\sigma_{m+1}^2 = \frac{\sum_{i=1}^n \sum_j w_{mij}(y_i - \mu_{m+1,j})^2}{\sum_{i=1}^n \sum_j w_{mij}}.$$

11. Suppose that Λ is the infinitesimal transition matrix of a continuous-time Markov chain, and let $\mu \geq \max_i \lambda_i$. If $R = I + \mu^{-1}\Lambda$, then prove that R has nonnegative entries and that

$$S(t) = \sum_{i=0}^{\infty} e^{-\mu t} \frac{(\mu t)^i}{i!} R^i$$

coincides with $P(t)$. (Hint: Verify that $S(t)$ satisfies the same defining differential equation and the same initial condition as $P(t)$.)

12. Consider a continuous-time Markov chain with infinitesimal transition matrix Λ and equilibrium distribution π. If the chain is at equilibrium at time 0, then show that it experiences $t \sum_i \pi_i \lambda_i$ transitions on average during the time interval $[0, t]$, where $\lambda_i = \sum_{j \neq i} \lambda_{ij}$ and λ_{ij} denotes a typical off-diagonal entry of Λ.

13. Verify the inequalities

$$\begin{aligned} \|e^{tA}\| &\leq e^{|t| \cdot \|A\|} \\ \|e^{-tA}\| &\geq e^{-|t| \cdot \|A\|} \end{aligned}$$

for any square matrix A and matrix norm induced by a vector norm.

14. Derive the error estimate

$$\left\| e^{tA} - \sum_{i=0}^{n} \frac{1}{i!}(tA)^i \right\| \leq \frac{|t|^{n+1}\|A\|^{n+1}}{(n+1)!} \frac{1}{1 - \frac{|t| \cdot \|A\|}{n+2}}$$

for any square matrix A and matrix norm induced by a vector norm.

15. Consider the partial sums $S_n = \sum_{i=0}^n B^i/i!$ for some square matrix B. Show that $BS_n = S_nB$ and that for any $\epsilon > 0$

$$\begin{aligned} \|e^B - S_n\| &< \epsilon \\ \|S_n\| &\leq \|e^B\|\Big(1 + \frac{\epsilon}{\|e^B\|}\Big), \end{aligned}$$

provided n is large enough and the indicated matrix norm is induced by a vector norm. If n is chosen to satisfy the last two inequalities, then show that

$$\|e^{2^k B} - S_n^{2^k}\| \leq \epsilon \|e^B\|^{2^k-1}\Big(2 + \frac{\epsilon}{\|e^B\|}\Big)^{2^k-1}$$

for any positive integer k. In conjunction with Problem 14, conclude that we can approximate e^{tA} arbitrarily closely by $S_n^{2^k}$ for $B = 2^{-k}tA$ and n sufficiently large. Hint:

$$\|e^{2^k B} - S_n^{2^k}\| \leq \|e^{2^{k-1}B} + S_n^{2^{k-1}}\| \cdot \|e^{2^{k-1}B} - S_n^{2^{k-1}}\|.$$

16. Let A and B be the 2×2 real matrices

$$A = \begin{pmatrix} a & -b \\ b & a \end{pmatrix}, \qquad B = \begin{pmatrix} \lambda & 0 \\ 1 & \lambda \end{pmatrix}.$$

Show that

$$e^A = e^a \begin{pmatrix} \cos b & -\sin b \\ \sin b & \cos b \end{pmatrix}, \qquad e^B = e^\lambda \begin{pmatrix} 1 & 0 \\ 1 & 1 \end{pmatrix}.$$

(Hints: Note that 2×2 matrices of the form $\left(\begin{smallmatrix} a & -b \\ b & a \end{smallmatrix}\right)$ are isomorphic to the complex numbers under the correspondence $\left(\begin{smallmatrix} a & -b \\ b & a \end{smallmatrix}\right) \leftrightarrow a + bi$. For the second case write $B = \lambda I + C$.)

17. Prove that $\det(e^A) = e^{\text{tr}(A)}$, where tr is the trace function. (Hint: Since the diagonalizable matrices are dense in the set of matrices [17], by continuity you may assume that A is diagonalizable.)

18. In Moran's population genetics model, n genes evolve by substitution and mutation. Suppose each gene can be classified as one of d alleles, and let X_{ti} denote the number of alleles of type i at time t. The count process X_t moves from state to state by randomly selecting two genes, which may coincide. The first gene dies, and the second gene reproduces a replacement. If the second gene is of type i, then its daughter gene is of type j with probability p_{ij}. The replacement times are independent and exponentially distributed with intensity λ. Reformulate Moran's model to proceed by reaction channels. What are the intensity and the increment of each channel?

25.10 References

[1] Baum LE (1972) An inequality and associated maximization technique in statistical estimation for probabilistic functions of Markov processes. *Inequalities* 3:1–8

[2] Bhattacharya RN, Waymire EC (1990) *Stochastic Processes with Applications.* Wiley, New York

[3] Billingsley P (1986) *Probability and Measure,* 2nd ed. Wiley, New York

[4] Cao Y, Gillespie DT, Petzold LR (2006) Efficient leap-size selection for accelerated stochastic simulation. *J Phys Chem* 124:1–11

[5] Devijver PA (1985) Baum's forward-backward algorithm revisited. *Pattern Recognition Letters* 3:369–373

[6] Diaconis P (1988) *Group Representations in Probability and Statistics.* Institute of Mathematical Statistics, Hayward, CA

[7] Doyle PG, Snell JL (1984) *Random Walks and Electrical Networks.* The Mathematical Association of America

[8] Durbin R, Eddy S, Krogh A, Mitchison G (1998) *Biological Sequence Analysis: Probabilistic Models of Proteins and Nucleic Acids.* Cambridge University Press, Cambridge

[9] Feller W (1968) *An Introduction to Probability Theory and Its Applications, Volume 1,* 3rd ed. Wiley, New York

[10] Fredkin DR, Rice JA (1992) Bayesian restoration of single-channel patch clamp recordings. *Biometrics* 48:427–448

[11] Gillespie DT (1977) Exact stochastic simulation of coupled chemical reactions. *J Phys Chem* 81:2340–2361

[12] Gillespie DT (2001) Approximate accelerated stochastic simulation of chemically reacting systems. *J Chem Phys* 115:1716–1733

[13] Grimmett GR, Stirzaker DR (1992) *Probability and Random Processes,* 2nd ed. Oxford University Press, Oxford

[14] Hastings WK (1970) Monte Carlo sampling methods using Markov chains and their applications. *Biometrika* 57:97–109

[15] Higham DJ (2008) Modeling and simulating chemical reactions. *SIAM Review* 50:347–368

[16] Higham NJ (2009) The scaling and squaring method for matrix exponentiation. *SIAM Review* 51:747–764

[17] Hirsch MW, Smale S (1974) *Differential Equations, Dynamical Systems, and Linear Algebra.* Academic Press, New York

[18] Karlin S, Taylor HM (1975) *A First Course in Stochastic Processes,* 2nd ed. Academic Press, New York

[19] Karlin S, Taylor HM (1981) *A Second Course in Stochastic Processes.* Academic Press, New York

[20] Kelly FP (1979) *Reversibility and Stochastic Networks.* Wiley, New York

[21] Lamperti J (1977) *Stochastic Processes: A Survey of the Mathematical Theory.* Springer, New York

[22] Lange K (2002) *Mathematical and Statistical Methods for Genetic Analysis,* 2nd ed. Springer, New York

[23] Lange K, Boehnke M, Cox DR, Lunetta KL (1995) Statistical methods for polyploid radiation hybrid mapping. *Genome Res* 5:136–150

[24] Lazzeroni LC, Lange K (1997) Markov chains for Monte Carlo tests of genetic equilibrium in multidimensional contingency tables. *Ann Stat* 25:138–168

[25] Moler C, Van Loan C (1978) Nineteen dubious ways to compute the exponential of a matrix. *SIAM Review* 20:801–836

[26] Nachbin L (1965) *The Haar Integral.* Van Nostrand, Princeton, NJ

[27] Rabiner L (1989) A tutorial on hidden Markov models and selected applications in speech recognition. *Proc IEEE* 77:257–285

[28] Rubinow SI (1975) *Introduction to Mathematical Biology.* Wiley, New York

[29] Sehl ME, Alexseyenko AV, Lange KL (2009) Accurate stochastic simulation via the step anticipation (SAL) algorithm. *J Comp Biol* 16:1195–1208

[30] Stewart WJ (1994) *Introduction to the Numerical Solution of Markov Chains.* Princeton University Press, Princeton, NJ

[31] Waterman MS (1995) *Introduction to Computational Biology: Maps, Sequences, and Genomes.* Chapman & Hall, London

26

Markov Chain Monte Carlo

26.1 Introduction

The Markov chain Monte Carlo (MCMC) revolution sweeping statistics is drastically changing how statisticians perform integration and summation. In particular, the Metropolis algorithm and Gibbs sampling make it straightforward to construct a Markov chain that samples from a complicated conditional distribution. Once a sample is available, then any conditional expectation can be approximated by forming its corresponding sample average. The implications of this insight are profound for both classical and Bayesian statistics. As a bonus, trivial changes to the Metropolis algorithm yield simulated annealing, a general-purpose algorithm for solving difficult combinatorial optimization problems.

Our limited goal in this chapter is to introduce a few of the major MCMC themes, particularly Gibbs sampling. In describing the various methods we will use the notation of discrete-time Markov chains. Readers should bear in mind that most of the methods carry over to chains with continuous state spaces; our examples exploit this fact. One issue of paramount importance is how rapidly the underlying chains reach equilibrium. This is the Achilles heel of the whole business and not just a mathematical nicety. Our two numerical examples illustrate some strategies for accelerating convergence. We undertake a formal theoretical study of convergence in the next chapter.

Readers interested in pursuing MCMC methods and simulated annealing further will enjoy the pioneering articles [9, 12, 16, 19, 24]. The elementary surveys [3, 6] of Gibbs sampling and the Metropolis algorithm are quite readable, as are the books [8, 10, 15, 22, 28, 32]. The well-tested program WinBugs [29] is one of the best vehicles for Gibbs sampling. *Numerical Recipes* [27] provides a compact implementation of simulated annealing.

26.2 The Hastings-Metropolis Algorithm

The Hastings-Metropolis algorithm is a device for constructing a Markov chain with a prescribed equilibrium distribution π on a given state space [16, 24]. Each step of the chain is broken into two stages, a proposal stage and an acceptance stage. If the chain is currently in state i, then in the proposal stage a new destination state j is proposed according to a probability density $q_{ij} = q(j \mid i)$. In the subsequent acceptance stage, a random number is drawn uniformly from $[0, 1]$ to determine whether the proposed

K. Lange, *Numerical Analysis for Statisticians*, Statistics and Computing,
DOI 10.1007/978-1-4419-5945-4_26, © Springer Science+Business Media, LLC 2010

step is actually taken. If this number is less than the Hastings-Metropolis acceptance probability

$$a_{ij} = \min\Big\{\frac{\pi_j q_{ji}}{\pi_i q_{ij}}, 1\Big\}, \tag{26.1}$$

then the proposed step is taken. Otherwise, the proposed step is declined, and the chain remains in place.

A few comments about this strange procedure are in order. First, the resemblance of the Hastings-Metropolis algorithm to acceptance-rejection sampling should make the reader more comfortable. Second, like most good ideas, the algorithm has gone through successive stages of abstraction and generalization. For instance, Metropolis et al. [24] considered only symmetric proposal densities with $q_{ij} = q_{ji}$. In this case the acceptance probability reduces to

$$a_{ij} = \min\Big\{\frac{\pi_j}{\pi_i}, 1\Big\}. \tag{26.2}$$

In this simpler setting it is clear that any proposed destination j with $\pi_j > \pi_i$ is automatically accepted. Finally, in applying either formula (26.1) or formula (26.2), it is noteworthy that the π_i need only be known up to a multiplicative constant.

To prove that π is the equilibrium distribution of the chain constructed from the Hastings-Metropolis scheme (26.1), it suffices to check that detailed balance holds. If π puts positive weight on all points of the state space, it is clear that we must impose the requirement that the inequalities $q_{ij} > 0$ and $q_{ji} > 0$ are simultaneously true or simultaneously false. This requirement is also implicit in definition (26.1). Now suppose without loss of generality that the fraction

$$\frac{\pi_j q_{ji}}{\pi_i q_{ij}} \le 1$$

for some $j \neq i$. Then detailed balance follows immediately from

$$\begin{aligned} \pi_i q_{ij} a_{ij} &= \pi_i q_{ij} \frac{\pi_j q_{ji}}{\pi_i q_{ij}} \\ &= \pi_j q_{ji} \\ &= \pi_j q_{ji} a_{ji}. \end{aligned}$$

Besides checking that π is the equilibrium distribution, we should also be concerned about whether the Hastings-Metropolis chain is irreducible and aperiodic. Aperiodicity is the rule because the acceptance-rejection step allows the chain to remain in place. Problem 4 states a precise result and a counterexample. Irreducibility holds provided the entries of π are positive and the proposal matrix $Q = (q_{ij})$ is irreducible.

26.3 Gibbs Sampling

The Gibbs sampler is a special case of the Hastings-Metropolis algorithm for Cartesian product state spaces [9, 12, 15]. Suppose that each sample point $i = (i_1, \ldots, i_m)$ has m components. The Gibbs sampler updates one component of i at a time. If the component is chosen randomly and resampled conditional on the remaining components, then the acceptance probability is 1. To prove this assertion, let i_c be the uniformly chosen component, and denote the remaining components by $i_{-c} = (i_1, \ldots, i_{c-1}, i_{c+1}, \ldots, i_m)$. If j is a neighbor of i reachable by changing only component i_c, then $j_{-c} = i_{-c}$. For such a neighbor j the proposal probability

$$q_{ij} \quad = \quad \frac{1}{m} \cdot \frac{\pi_j}{\sum_{\{k:k_{-c}=i_{-c}\}} \pi_k}$$

satisfies $\pi_i q_{ij} = \pi_j q_{ji}$, and the ratio appearing in the acceptance probability (26.1) is 1.

In contrast to random sampling of components, we can repeatedly cycle through the components in some fixed order, say $1, 2, \ldots, m$. If the transition matrix for changing component c while leaving other components unaltered is $P^{(c)}$, then the transition matrices for random sampling and sequential (or cyclic) sampling are $R = \frac{1}{m}\sum_c P^{(c)}$ and $S = P^{(1)} \cdots P^{(m)}$, respectively. Because each $P^{(c)}$ satisfies $\pi P^{(c)} = \pi$, we have $\pi R = \pi$ and $\pi S = \pi$ as well. Thus, π is the unique equilibrium distribution for R or S if either is irreducible. However as pointed out in Problem 6, R satisfies detailed balance while S ordinarily does not.

Bayesian applications of Gibbs sampling rely heavily on conjugate distributions. A likelihood $p(x \mid \theta)$ and a prior density $p(\theta)$ are said to be conjugate provided the posterior density $p(\theta \mid x)$ has the same functional form as the prior density. Here θ is the parameter vector, and x is the data. Table 26.1 collects some of the more useful conjugate families. In the table the upper entry for the normal fixes the precision (reciprocal variance), and the lower entry for the normal fixes the mean. The table omits mentioning how the parameters of the prior distribution map to the parameters of the posterior distribution. As an example, the reader can check that

$$\begin{aligned} \theta \quad &\rightarrow \quad \frac{\omega\theta + \tau x}{\omega + \tau}, \qquad & \omega \quad &\rightarrow \quad \omega + \tau \\ \alpha \quad &\rightarrow \quad \alpha + \frac{1}{2}, \qquad & \beta \quad &\rightarrow \quad \beta + \frac{(x-\mu)^2}{2} \end{aligned}$$

for the two normal entries of the table. See the text [10] for a more extensive catalog of conjugate pairs.

TABLE 26.1. Conjugate Pairs

Likelihood	**Density**	**Prior**	**Density**
Binomial	$\binom{n}{x}p^x(1-p)^{n-x}$	Beta	$\frac{1}{B(\alpha,\beta)}p^{\alpha-1}(1-p)^{\beta-1}$
Poisson	$\frac{\lambda^x}{x!}e^{-\lambda}$	Gamma	$\frac{\beta^\alpha\lambda^{\alpha-1}}{\Gamma(\alpha)}e^{-\beta\lambda}$
Geometric	$(1-p)^x p$	Beta	$\frac{1}{B(\alpha,\beta)}p^{\alpha-1}(1-p)^{\beta-1}$
Multinomial	$\binom{n}{x_1\ldots x_k}\prod_{i=1}^k p_i^{x_i}$	Dirichlet	$\frac{\Gamma(\sum_{i=1}^k \alpha_i)}{\prod_{i=1}^k \Gamma(\alpha_i)}\prod_{i=1}^k p_i^{\alpha_i-1}$
Normal	$\sqrt{\frac{\tau}{2\pi}}e^{-\tau(x-\mu)^2/2}$	Normal	$\sqrt{\frac{\omega}{2\pi}}e^{-\omega(\mu-\theta)^2/2}$
Normal	$\sqrt{\frac{\tau}{2\pi}}e^{-\tau(x-\mu)^2/2}$	Gamma	$\frac{\beta^\alpha\tau^{\alpha-1}}{\Gamma(\alpha)}e^{-\beta\tau}$
Exponential	$\lambda e^{-\lambda x}$	Gamma	$\frac{\beta^\alpha\lambda^{\alpha-1}}{\Gamma(\alpha)}e^{-\beta\lambda}$

Example 26.3.1 *Ising Model*

Consider m elementary particles equally spaced around the boundary of the unit circle. Each particle c can be in one of two magnetic states—spin up with $i_c = 1$ or spin down with $i_c = -1$. The Gibbs distribution

$$\pi_i \quad \propto \quad e^{\beta\sum_d i_d i_{d+1}} \tag{26.3}$$

takes into account nearest-neighbor interactions in the sense that states like $(1,1,1,\ldots,1,1,1)$ are favored and states like $(1,-1,1,\ldots,1,-1,1)$ are shunned for $\beta > 0$. (Note that in (26.3) the index $m+1$ of i_{m+1} is interpreted as 1.) Specification of the normalizing constant (or partition function)

$$Z \quad = \quad \sum_i e^{\beta\sum_d i_d i_{d+1}}$$

is unnecessary to carry out Gibbs sampling. If we elect to resample component c, then the choices $j_c = -i_c$ and $j_c = i_c$ are made with respective probabilities

$$\frac{e^{\beta(-i_{c-1}i_c - i_c i_{c+1})}}{e^{\beta(i_{c-1}i_c + i_c i_{c+1})} + e^{\beta(-i_{c-1}i_c - i_c i_{c+1})}} \quad = \quad \frac{1}{e^{2\beta(i_{c-1}i_c + i_c i_{c+1})} + 1}$$

$$\frac{e^{\beta(i_{c-1}i_c + i_c i_{c+1})}}{e^{\beta(i_{c-1}i_c + i_c i_{c+1})} + e^{\beta(-i_{c-1}i_c - i_c i_{c+1})}} \quad = \quad \frac{1}{1 + e^{-2\beta(i_{c-1}i_c + i_c i_{c+1})}}.$$

When the number of particles m is even, the odd-numbered particles are independent given the even-numbered particles, and vice versa. This fact suggests alternating between resampling all odd-numbered particles and resampling all even-numbered particles. Such multi-particle updates take longer to execute but create more radical rearrangements than single-particle updates. ■

Example 26.3.2 *A Normal Random Sample with Conjugate Priors*

Consider a random sample $y = (y_1, \ldots, y_n)$ from a normal density with mean μ and precision τ. Suppose that μ is subject to a normal prior with mean 0 and precision ω and τ is subject to a gamma prior with shape parameter α and scale parameter β. Given that the two priors are independent, the joint density of data and parameters is

$$(2\pi)^{-\frac{n+1}{2}} \tau^{\frac{n}{2}} e^{-\frac{\tau}{2}\sum_{i=1}^{n}(y_i-\mu)^2} \omega^{\frac{1}{2}} e^{-\frac{\omega}{2}\mu^2} \frac{\tau^{\alpha-1}}{\Gamma(\alpha)\beta^{\alpha}} e^{-\frac{\tau}{\beta}}.$$

Gibbs sampling from the joint posterior distribution of μ and τ requires the conditional density of μ given y and τ and the conditional density of τ given y and μ. According to Table 26.1, the first of these two conditional densities is normally distributed with mean $n\tau\bar{y}/(\omega + n\tau)$ and precision $\omega + n\tau$, where $\bar{y}$ is the sample mean $\frac{1}{n}\sum_{i=1}^{n} y_i$. The second is gamma distributed with shape parameter $n/2+\alpha$ and scale parameter $1/(ns_n^2/2+1/\beta)$, where s_n^2 is the sample variance $\frac{1}{n}\sum_{i=1}^{n}(y_i - \mu)^2$. Choosing conjugate priors here eases the analysis as it does throughout Bayesian statistics. ■

Example 26.3.3 *Capture-Recapture Estimation*

Ecologists employ capture-recapture models to estimate the abundance of an animal species in a local habitat. Gibbs sampling offers a convenient way of implementing a Bayesian analysis [13]. For the sake of concreteness, suppose we are interested in estimating the number of fish f in a lake. We fish on t occasions, record the number of fish caught, and mark each fish caught. The data then consist of t count pairs (c_i, r_i), where c_i is the number of fish caught on trial i and r_i is the number of fish recaught. The number of new fish encountered on trial i is $c_i - r_i$. Over trials 1 through i, we encounter $u_i = \sum_{j=1}^{i}(c_i - r_i)$ unique fish.

Our first order of business in modeling is to construct a likelihood. The simplest and most widely used assumes independent binomial sampling with success probability p_i for trial i. These assumptions translate into the likelihood

$$\begin{aligned}
& \prod_{i=1}^{t} \binom{u_{i-1}}{r_i} p_i^{r_i}(1-p_i)^{u_{i-1}-r_i} \binom{f-u_{i-1}}{c_i-r_i} p_i^{c_i-r_i}(1-p_i)^{f-u_{i-1}-c_i+r_i} \\
= {} & \prod_{i=1}^{t} \binom{u_{i-1}}{r_i}\binom{f-u_{i-1}}{c_i-r_i} p_i^{c_i}(1-p_i)^{f-c_i} \\
= {} & \frac{f!}{(f-u_t)!}\prod_{i=1}^{t} \binom{u_{i-1}}{r_i} p_i^{c_i}(1-p_i)^{f-c_i},
\end{aligned}$$

where $u_0 = r_1 = 0$. It is mathematically convenient to put a Poisson prior on f and independent beta priors on the p_i. This yields the joint density

$$\begin{aligned}
& \frac{f!}{(f-u_t)!}\prod_{i=1}^{t}\binom{u_{i-1}}{r_i}p_i^{c_i}(1-p_i)^{f-c_i}\frac{\lambda^f e^{-\lambda}}{f!B(\alpha,\beta)^t}\prod_{i=1}^{t}p_i^{\alpha-1}(1-p_i)^{\beta-1} \\
= \; & \frac{\lambda^f e^{-\lambda}}{(f-u_t)!B(\alpha,\beta)^t}\prod_{i=1}^{t}\binom{u_{i-1}}{r_i}p_i^{\alpha+c_i-1}(1-p_i)^{\beta+f-c_i-1}.
\end{aligned}$$

To infer the conditional distribution of a parameter, we factor the joint density into a constant times a function of that parameter. Thus, the Gibbs update of parameter p_i is beta distributed with $\alpha + c_i$ replacing α and $\beta + f - c_i$ replacing β. The Gibbs update of f is more subtle, but close examination of the joint density implies that $f - u_t$ is Poisson distributed with mean $\lambda \prod_{i=1}^{t}(1-p_i)$.

As an example, consider the Gordy Lake sunfish data [4] recorded in Table 26.2. A simple frequentist analysis of these data constraining all p_i to be equal suggests the starting values $f = 457$ and $p_i = 0.02532$ for all i. We also take $\lambda = 457$ as a reasonable guess, $\alpha = \beta = 1$, and a burn-in period of 100 iterations. Figure 26.1 provides a histogram of the f values in the first 1000 Gibbs iterations after burn-in. Over these 1000 iterates the mean and median are 436.5 and 436, respectively. Although the iterates are dependent, they appear nearly normally distributed. We will comment later on our choice of priors. ■

TABLE 26.2. Capture-Recapture Counts for Gordy Lake Sunfish

Trial	c_i	r_i	Trial	c_i	r_i
1	10	0	8	15	1
2	27	0	9	9	5
3	17	0	10	18	5
4	7	0	11	16	4
5	1	0	12	5	2
6	5	0	13	7	2
7	6	2	14	19	3

Example 26.3.4 *Data Augmentation and Allele Frequency Estimation*

Data augmentation uses missing data in a slightly different fashion than the EM algorithm [32, 33]. In data augmentation we sample from the joint conditional distribution of the missing data and the parameters given the observed data. For example, consider the ABO allele frequency estimation problem of Chapter 13. If we put a Dirichlet prior with parameters $(\alpha_A, \alpha_B, \alpha_O)$ on the allele frequencies, then the joint density of the observed

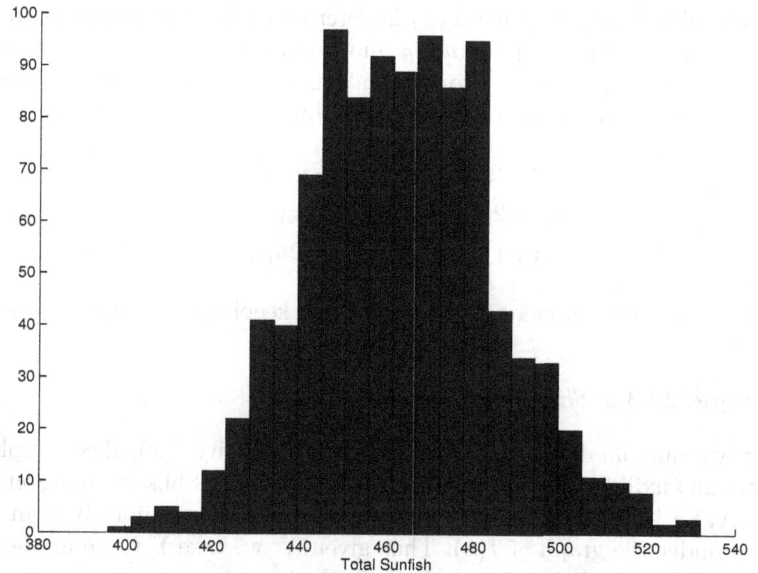

FIGURE 26.1. Histogram of the Number of Sunfish in 1000 Gibbs Iterations

data and the parameters amounts to

$$\binom{n}{n_A\ n_B\ n_{AB}\ n_O}(p_A^2 + 2p_Ap_O)^{n_A}(p_B^2 + 2p_Bp_O)^{n_B}(2p_Ap_B)^{n_{AB}}(p_O^2)^{n_O} \times \frac{\Gamma(\alpha_A+\alpha_B+\alpha_O)}{\Gamma(\alpha_A)\Gamma(\alpha_B)\Gamma(\alpha_O)}p_A^{\alpha_A-1}p_B^{\alpha_B-1}p_O^{\alpha_O-1}.$$

Extracting the posterior density of the allele frequencies p_A, p_B, and p_O appears formidable. However, sampling from the posterior distribution becomes straightforward if we augment the observed data by specifying the underlying counts $n_{A/A}$, $n_{A/O}$, $n_{B/B}$, and $n_{B/O}$ of individuals with genotypes A/A, A/O, B/B, and B/O, respectively.

Sequential Gibbs sampling alternates between sampling the complete data

$$(n_{A/A}, n_{A/O}, n_{B/B}, n_{B/O}, n_{AB}, n_O)$$

conditional on the observed data (n_A, n_B, n_{AB}, n_O) and the parameters (p_A, p_B, p_O) and sampling the parameters (p_A, p_B, p_O) conditional on the complete data. The marginal distribution of the parameters from a sequential Gibbs sample coincides with the posterior distribution. To sample $n_{A/A}$, we simply draw from a binomial distribution with n_A trials and success probability

$$\frac{p_A^2}{p_A^2 + 2p_Ap_O}.$$

The complementary variable $n_{A/O}$ is determined by the linear constraint $n_{A/A} + n_{A/O} = n_A$. Sampling $n_{B/B}$ and $n_{B/O}$ is done similarly. Sampling the parameters (p_A, p_B, p_O) conditional on the complete data is accomplished by sampling from a Dirichlet distribution with parameters

$$\begin{aligned} \alpha_A + 2n_{A/A} + n_{A/O} + n_{AB}, \\ \alpha_B + 2n_{B/B} + n_{B/O} + n_{AB}, \\ \alpha_O + n_{A/O} + n_{B/O} + 2n_O. \end{aligned}$$

Again choosing a conjugate prior is critical in keeping the sampling process simple. ■

Example 26.3.5 *Slice Sampling*

To create random deviates X with probability density $f(x)$, slice sampling creates an auxiliary random variable Y and conducts Gibbs sampling on the pair (X, Y) [1, 17, 26, 30]. The basic idea is to sample uniformly from the region under the graph of $f(x)$. Thus given X, we take Y to be uniformly distributed on the interval $[0, f(X)]$. The logic

$$\Pr(X \in A) = \int_A \frac{1}{f(x)} \int_0^{f(x)} dy\, f(x)\, dx = \int_A \int_0^{f(x)} dy\, dx$$

justifies a uniform distribution in the combined coordinates. If $cf(x)$ is a probability density for some constant $c \neq 1$, then the procedure is identical because choosing y uniformly from $[0, cf(x)]$ is the same as choosing y/c uniformly from $[0, f(x)]$ and choosing x uniformly from $\{x : f(x) \geq y/c\}$ is the same as choosing x uniformly from $\{x : cf(x) \geq y\}$. We will call any set $\{x : f(x) \geq y\}$ a top set.

For instance, suppose $f(x) = e^{-x^2/2}$ is the standard normal density up to a constant. Gibbs sampling alternates between choosing y uniformly over $[0, f(x)]$ and choosing x uniformly from the interval

$$\{x : f(x) \geq y\} = \Big[-\sqrt{-2 \ln y}, \sqrt{-2 \ln y} \Big].$$

The components of the correlated sample of $x_1, x_2, \ldots$ have a standard normal distribution at equilibrium. With very little change, one can simulate standard normal deviates conditioned to fall on a given interval $[a, b]$. All that is required is that in the second step of Gibbs sampling we choose x uniformly from the interval

$$\Big[-\sqrt{-2 \ln y}, \sqrt{-2 \ln y} \Big] \cap [a, b].$$

The biggest impediment to applying slice sampling is the complicated geometric nature of the top set $\{x : f(x) \geq y\}$. When $f(x)$ is concave or

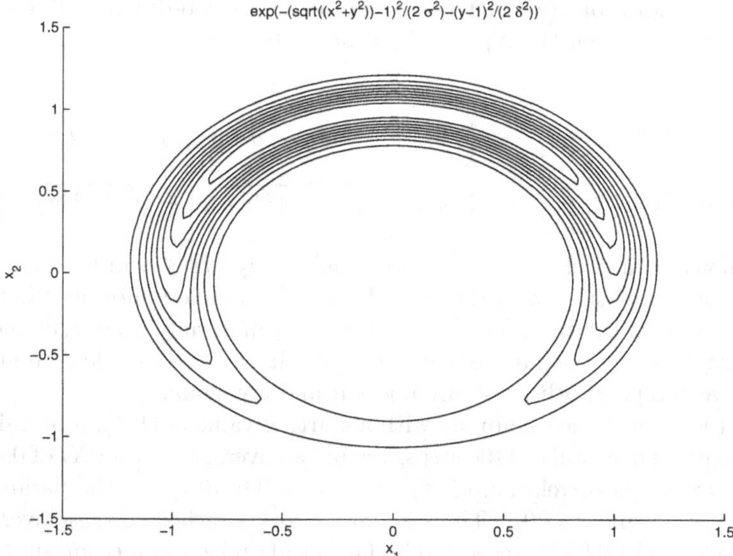

FIGURE 26.2. Contour Plot of $f(x) = \exp\left[-\frac{(\|x\|_2-1)^2}{2\sigma^2} - \frac{(x_2-1)^2}{2\delta^2}\right]$

log-concave, a top set is convex and therefore connected. Otherwise, it can be disconnected.

Slice sampling is very adaptable. Suppose we can write a probability density $f(x)$ as the product $f(x) = g(x)\prod_{i=1}^{j} f_j(x)$, where all of the displayed functions are nonnegative and have the same support. Furthermore, assume that $f(x)$ is a density relative to an arbitrary measure μ rather than Lebesgue measure. We now augment the underlying random variable X by j random variables $Y_1, \ldots, Y_j$. Given $X = x$, these are independently and uniformly distributed over the j intervals $[0, f_i(x)]$. Gibbs sampling alternates drawing the Y_i from these intervals with drawing X uniformly from the intersection of the top sets $\{x : f_i(x) \geq y_i\}$. The virtue of this more complicated procedure is that the intersection is often simpler geometrically than the single top set $\{x : f(x) \geq y\}$. The equation

$$\Pr(X \in A) = \int_A \prod_{i=1}^{j} \frac{1}{f_i(x)} \int_0^{f_i(x)} dy_i\, f(x)\, d\mu(x) = \int_A f(x)\, d\mu(x)$$

justifies the procedure.

For an example, consider slice sampling from the probability density proportional to the function

$$f(x) = e^{-\frac{(\|x\|_2-1)^2}{2\sigma^2}} e^{-\frac{(x_2-1)^2}{2\delta^2}} = f_1(x)f_2(x) \tag{26.4}$$

on R^2 [5]. Figure 26.2 plots the contours of this unusual function for the parameter values $\delta = 1$ and $\sigma = 0.1$. It is evident in the plot and from the functional form of $f(x)$ that the mean of the X_1 coordinate is 0 and the correlation between the X_1 and X_2 coordinates is also 0. Fortunately, the top sets

$$\begin{aligned}
\left\{x : e^{-\frac{(\|x\|_2-1)^2}{2\sigma^2}} \geq y_1\right\} &= \left\{\|x\|_2 \in \left[1-\sqrt{-2\sigma^2 \ln y_1}, 1+\sqrt{-2\sigma^2 \ln y_1}\right]\right\} \\
\left\{x : e^{-\frac{(x_2-1)^2}{2\delta^2}} \geq y_2\right\} &= \left\{x_2 \in \left[1-\sqrt{-2\delta^2 \ln y_2}, 1+\sqrt{-2\delta^2 \ln y_2}\right]\right\}
\end{aligned}$$

boil down to a circle $\|x\| \leq b$ or an annulus $a \leq \|x\| \leq b$ in the first case and a strip $c \leq x_2 \leq d$ in the second case. To sample uniformly from the intersection of the two sets, we can sample x_2 uniformly from $[c, d]$ and x_1 uniformly from $[-b, b]$. If the point (x_1, x_2) falls inside the circle or annulus, then we accept it; otherwise, we reject it and try again.

In 100 runs of slice sampling with a starting value of $(1, 1)$, a burn-in of 100 steps, and a total of 1100 steps, we find an average value of X_1 of 0.0026 and an average correlation of X_1 and X_2 of 0.0010 under the parameter values $\delta = 1$ and $\sigma = 0.1$. The standard errors attached to these averages are 0.0212 and 0.0258, respectively. Our crude rejection mechanism takes about six steps on average to generate a uniformly distributed point x in the intersection $\{x : f_1(x) \geq y_1\} \cap \{x : f_2(x) \geq y_2\}$. These results suggest that slice sampling is an excellent vehicle for simulation when the top sets are sufficiently simple. ■

26.4 Other Examples of Hastings-Metropolis Sampling

Although constructing good proposal densities is an art, two general techniques are worth mentioning [6, 34]. Our initial use of discrete notation should not obscure the fact that the methods have wider applicability.

Example 26.4.1 *Independence Sampler*

If the proposal density satisfies $q_{ij} = q_j$, then candidate points are drawn independently of the current point. To achieve quick convergence of the chain, q_i should mimic π_i for most i. This intuition is justified by introducing the importance ratios $w_i = \pi_i/q_i$ and rewriting the acceptance probability (26.1) as

$$a_{ij} = \min\left\{\frac{w_j}{w_i}, 1\right\}. \tag{26.5}$$

It is now obvious that it is difficult to exit any state i with a large importance ratio w_i.

In sampling a posterior density, the prior density can be a reasonable choice for the proposal density. In this situation the acceptance ratio (26.1) reduces to a likelihood ratio. Thus, a proposed move to a point with greater likelihood is always accepted. This generic scheme is often less efficient than more specialized schemes such as Gibbs sampling. ■

Example 26.4.2 *Random Walk*

Random walk sampling occurs when the proposal density $q_{ij} = q_{j-i}$ for some density q_k. This construction requires that the sample space be closed under subtraction. If $q_k = q_{-k}$, then the Metropolis acceptance probability (26.2) applies. In any case, normalizing constants are irrelevant in random walk sampling.

TABLE 26.3. Averages over 100 Runs of Random Walk Sampling

γ	$\bar{\mathbf{a}}$	$\bar{\mathbf{x}}_1$	std. err.	$\bar{\rho}$	std. err.
0.0156	0.9477	0.6848	0.2139	-0.4095	0.3065
0.0244	0.9213	0.6255	0.2709	-0.5078	0.3258
0.0381	0.8784	0.4777	0.4071	-0.3669	0.4956
0.0596	0.8130	0.3624	0.4835	-0.2271	0.5663
0.0931	0.7210	0.1558	0.4617	-0.0946	0.4773
0.1455	0.5960	0.0764	0.3875	-0.0497	0.4052
0.2274	0.4562	0.0416	0.2903	-0.0268	0.2550
0.3553	0.3240	0.0082	0.2208	-0.0137	0.1741
0.5551	0.2207	-0.0258	0.1495	0.0295	0.1294
0.8674	0.1481	-0.0009	0.1364	0.0090	0.1119
1.3553	0.0909	-0.0184	0.1190	0.0086	0.1227
2.1176	0.0475	0.0166	0.1448	-0.0335	0.1748
3.3087	0.0217	0.0008	0.2214	0.0167	0.2782
5.1699	0.0099	-0.0033	0.2718	0.0851	0.4025
8.0779	0.0044	0.0057	0.4491	0.0307	0.5777

As an example, consider random walk sampling from the probability density proportional to the function (26.4). Our random walk increments Y will be drawn from the symmetric normal distribution

$$g(y) \quad = \quad \frac{1}{2\pi\gamma^2} e^{-\frac{\|y\|_2^2}{2\gamma^2}}$$

for various values of γ^2 from the starting point $(1, 1)$. If on the one hand we take γ^2 too small, then the walk takes tiny steps and a long time to move to the left half-plane $x_1 < 0$. On the other hand, if we take γ^2 too large, almost all proposed steps are rejected.

Table 26.3 displays the average results of 100 runs of random walk at various values of γ. Each walk consists of 100 steps of burn-in followed by 1000 steps of sampling. The symbols $\bar{a}$, $\bar{x}_1$, and $\bar{\rho}$ denote the average

TABLE 26.4. Averages over 100 Runs of Leapfrog Random Walk Sampling

γ	$\bar{a}$	$\bar{x}_1$	std. err.	$\bar{\rho}$	std. err.
0.0156	0.9182	0.6134	0.2766	-0.5308	0.2902
0.0244	0.8769	0.4718	0.4316	-0.3200	0.5879
0.0381	0.8122	0.2510	0.5241	-0.0721	0.5906
0.0596	0.7174	0.1475	0.4391	-0.0289	0.4827
0.0931	0.6005	0.0717	0.3954	-0.0466	0.3693
0.1455	0.4668	0.0262	0.2826	-0.0371	0.2581
0.2274	0.3394	-0.0034	0.2070	-0.0134	0.1943
0.3553	0.2384	0.0177	0.1690	-0.0174	0.1276
0.5551	0.1572	0.0051	0.1412	-0.0018	0.1187
0.8674	0.0992	0.0049	0.1241	-0.0006	0.1255
1.3553	0.0548	-0.0075	0.1625	0.0100	0.1651
2.1176	0.0267	-0.0259	0.2203	0.0289	0.2220
3.3087	0.0122	0.0101	0.2745	-0.0396	0.3307
5.1699	0.0057	-0.0292	0.4085	-0.0050	0.5314
8.0779	0.0031	0.0417	0.4797	0.1886	0.6737

acceptance rate, the average x_1 coordinate, and the average correlation between the two coordinates. Standard errors are given for the latter two variables. It is evident from the table that as γ increases, the acceptance rate and the biases of x_1 and ρ all decrease. Standard errors show a more complex pattern. The best performance occurs for γ between about 0.25 and 2.0.

Various other measures can be taken to improve the efficiency of Monte Carlo sampling. One technique is to change the proposal density so that it allows a wider range of both small and large increments. For instance, one could draw increments from a bivariate t-distribution rather than a bivariate normal distribution. Here we feature a more generic device that takes a random number of proposal steps before checking for acceptance [7, 20, 22, 25]. Again suppose that the number of states is finite and that Q denotes the proposal transition matrix. Let p_k be the probability of taking k proposal steps before checking. Although finding the entries of the new transition matrix $R = \sum_k p_k Q^k$ may be challenging, it is clear that R is symmetric whenever Q is symmetric. This is all we need to run a Metropolis-driven random walk. The choice $p_k = (1-p)^{k-1}p$ of a geometric distribution is simple to implement and has fairly long tails. We will refer to randomizing the number of proposal steps as leapfrogging. Table 26.4 shows the results of leapfrog random walk under our earlier conditions for the choice $p = \frac{1}{3}$. In comparison with ordinary random walk, acceptance probabilities drop but overall performance improves for low values of γ. This accords with our intuition that occasional long excursions should be helpful. However, note that both ordinary and leapfrog random walk are inferior to slice sampling on this problem. ■

26.5 Some Practical Advice

When the first edition of this book was published, no commercial software existed for carrying out MCMC simulations. This has changed with the release of SAS PROC MCMC and open source additions to MATLAB. The freeware packages are too numerous to review here, but it is safe to say that WinBugs [29] still dominates the field. Because crafting good algorithms and writing software to implement them is time consuming and error prone, the long delays in implementation were probably inevitable.

The task of writing software is now less pressing than diagnosing convergence and interpreting output. Despite the limitations of MCMC, the range of problems it can solve is so impressive that more and more statisticians are willing to invest the time to perform a Bayesian analysis in novel applications. We therefore offer the following practical advice:

1. (a) Every chain must start somewhere. In sampling from posterior densities, there are four obvious possibilities. One can set initial parameter values equal to frequentist estimates, to sampled values from their corresponding priors, or to means or medians of their corresponding priors. Multiple random starts can give a feel for whether an MCMC algorithm is apt to be trapped by inferior local modes.

2. (b) In choosing priors for Gibbs sampling, it is clearly advantageous to select independent conjugate priors. This makes sampling from the various conditional densities straightforward. The introduction of hyperparameters can partially compensate for a poor choice of priors.

3. (c) Thoughtful reparameterization can make a chain converge faster. Ideally, parameters should be nearly independent under the posterior distribution. Centering predictors in regression problems is a good idea.

4. (d) As Example 26.3.4 shows, data augmentation can render sampling much easier. Whenever the EM algorithm works well in the frequentist version of a problem, data augmentation is apt to help in sampling from the posterior density in a Bayesian version of the same problem.

5. (e) Calculation of sample averages from a Markov chain should not commence immediately. Every chain needs a burn-in period to reach equilibrium.

6. (f) Not every epoch need be taken into account in forming a sample average of a complicated statistic. This is particularly true if the chain

reaches equilibrium slowly; in such circumstances, values from neighboring states are typically highly correlated. Problem 7 of Chapter 25 validates the procedure of sampling a statistic at every kth epoch of a chain.

7. (g) Just as with independent sampling, we can achieve variance reduction by replacing a sampled statistic by its conditional expectation. Suppose, for instance, in Example 26.3.4 that we want to find the posterior mean of the number $n_{A/A}$ of people of genotype A/A. If we run the chain m epochs after burn-in, then we can estimate the posterior mean by the sample average $\frac{1}{m}\sum_{i=1}^m n^i_{A/A}$. However, the estimator

$$\frac{1}{m}\sum_{i=1}^m n_A \frac{(p^i_A)^2}{(p^i_A)^2 + 2p^i_A p^i_O}$$

is apt to have smaller variance since we have eliminated the noise introduced by sampling $n^i_{A/A}$ at epoch i.

8. (h) Undiagnosed slow convergence can lead to grievous errors in statistical inference. A time series plot of each parameter can often suggest when equilibrium kicks in. Strong autocorrelation is diagnostic of slow convergence. Gelman and Rubin suggest some monitoring techniques that are helpful if multiple independent realizations of a chain are available [11]. Given the low cost of modern computing, there is little excuse for avoiding this precaution. Enough runs should be undertaken so that starting states are widely dispersed.

9. (i) If a chain is known to converge rapidly, multiple independent runs are no better than a single long chain in computing expectations.

10. (j) For chains that converge slowly, importance sampling and running parallel, coupled chains offer speedups. Problems 16 and 17 briefly explain these techniques.

To illustrate some of the above advice in action, let us return to Example 26.3.3. The most suspect assumption there was the imposition of beta priors on the catch probabilities p_i. The particular choice $\alpha = \beta = 1$ entails the prior mean $\mathrm{E}(p_i) = \frac{\alpha}{\alpha+\beta} = \frac{1}{2}$, which is hardly congruent with a low maximum likelihood estimate for the common value of the p_i. Here is where it is useful to introduce a hyperparameter. Instead of considering α and β fixed, we introduce an independent prior on each. The exact nature of the priors is less relevant than the fact that α and β are now allowed to migrate to values dictated by the data. For simplicity we assume exponential priors with intensity θ and mean θ^{-1}. Taking θ small yields a nearly flat prior.

Of course, imposing priors on α and β adds two more stages of Gibbs sampling. Ignoring the other parameters, the posterior density is now proportional to

$$B(\alpha,\beta)^{-t}\prod_{i=1}^{t} p_i^{\alpha}(1-p_i)^{\beta}e^{-\theta\alpha}e^{-\theta\beta}. \tag{26.6}$$

Fortunately, this complicated function is log-concave in the parameter vector (α, β); see Problem 18. Thus, one can sample from it fixing either component by the adaptive acceptance-rejection scheme sketched in Section 22.6. Table 26.5 shows the new simulation results under the earlier conditions except for replacement of the constraint $\alpha = \beta = 1$ by the more realistic hyperparameterization. The value μ in the table refers to the ratio $\alpha/(\alpha+\beta)$. As the exponential priors become flatter, the sample posterior means $\bar{f}$ and $\bar{\mu}$ approximate the maximum likelihood estimates of f and $\bar{p}$. However, the sample means $\bar{\alpha}$ and $\bar{\beta}$ continue to increase.

One can reasonably question the advantage of a Bayesian analysis over a frequentist analysis in this example. The generic answer, and a good one, is that MCMC reveals the entire posterior distribution of the parameters. This is predicated on choosing decent priors. In reality, the distribution of f, the most interesting quantity in the model, is more sensitive to its own prior than to the priors on α and β. Readers interested in this aspect of prior sensitivity should consult the original paper [13].

TABLE 26.5. Posterior Means for the Capture-Recapture Data

θ	$\bar{f}$	$\bar{\alpha}$	$\bar{\beta}$	$\bar{\mu}$
1.00000	452.51500	0.46502	5.74339	0.08186
0.50000	453.36600	0.61087	11.10304	0.05678
0.25000	452.94800	0.89373	21.75739	0.04177
0.12500	454.13300	1.29508	38.35684	0.03449
0.06250	454.71900	1.73865	58.33032	0.03008
0.03125	455.07700	2.44334	88.82707	0.02763
0.01563	454.18700	3.29712	122.88996	0.02663
0.00781	457.46700	3.63517	139.40565	0.02614
0.00391	456.67000	4.70653	181.60048	0.02555
0.00195	457.04700	4.27371	166.55931	0.02547
0.00098	455.68400	4.55649	175.54479	0.02581

Let us now return briefly to the random walk example. A time series plot of the x_1 component quickly demonstrates the long lag in reaching the half-plane $x_1 < 0$ starting from $(x_1, x_2) = (1, 1)$. Convergence diagnosis with less obvious examples often benefits from plotting the autocorrelation function of each component. Figure 26.3 plots the autocorrelation function of x_2 after burn-in under three scenarios: (a) $\gamma = 0.5$ and ordinary random walk as described, (b) $\gamma = 0.5$ and leapfrog random walk with $p = \frac{1}{3}$, and (c) slice sampling. In each case the autocorrelation function is averaged over 100 independent trials. Scenario (a) (left subplot) shows a very slow decline in the autocorrelation function. Matters improve under scenario (b) (middle subplot), but the decline is still fairly slow. Scenario (c) (right subplot) shows a rapid decline in the autocorrelation function, in agreement with our suggestion that slice sampling performs best. If we undertake a similar analysis with the x_1 component, the results are even more striking. The autocorrelation function under slice sampling is 0 for all nontrivial lags. This just reflects the symmetry of the sampling process in the x_1 component.

26.6 Simulated Annealing

In simulated annealing we are interested in finding the most probable state of a Markov chain [19, 27]. If this state is k, then $\pi_k > \pi_i$ for all $i \neq k$. To accentuate the weight given to state k, we can replace the equilibrium distribution π by a distribution putting probability

$$\pi_i^{(\tau)} = \frac{\pi_i^{\frac{1}{\tau}}}{\sum_j \pi_j^{\frac{1}{\tau}}}$$

on state i. Here τ is a small, positive parameter traditionally called temperature. With a symmetric proposal density, the distribution $\pi_i^{(\tau)}$ can be attained by running a chain with Metropolis acceptance probability

$$a_{ij} = \min\left\{\left(\frac{\pi_j}{\pi_i}\right)^{\frac{1}{\tau}}, 1\right\}. \tag{26.7}$$

In fact, what is done in simulated annealing is that the chain is run with τ gradually decreasing to 0. If τ starts out large, then in the early steps of simulated annealing, almost all proposed steps are accepted, and the chain broadly samples the state space. As τ declines, fewer unfavorable steps are taken, and the chain eventually settles on some nearly optimal state. With luck, this state is k or a state equivalent to k if several states are optimal. Simulated annealing is designed to mimic the gradual freezing of a substance into a crystalline state of perfect symmetry and hence minimum energy.

Example 26.6.1 *The Traveling Salesman Problem*

A salesman must visit n towns, starting and ending in his hometown. Given the distance d_{ij} between every pair of towns i and j, in what order should he

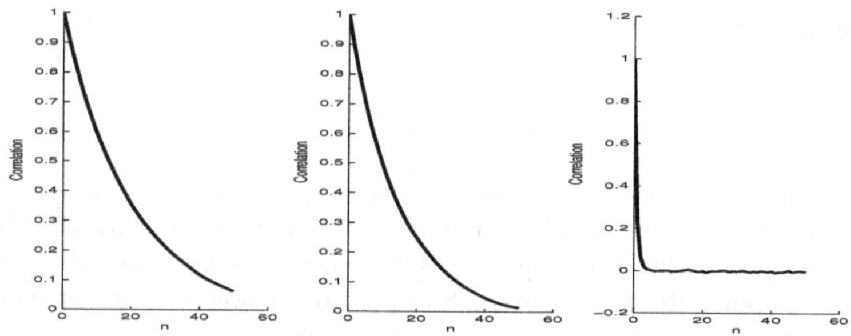

FIGURE 26.3. Plots of the Autocorrelation Function for the Random Walk.

visit the towns to minimize the length of his circuit? This problem belongs to the class of NP-complete problems; these have deterministic solutions that are conjectured to increase in complexity at an exponential rate in n.

In the simulated annealing approach to the traveling salesman problem, we assign to each permutation $\sigma = (\sigma_1, \ldots, \sigma_n)$ a cost $c_\sigma = \sum_{i=1}^n d_{\sigma_i, \sigma_{i+1}}$, where $\sigma_{n+1} = \sigma_1$. Defining $\pi_\sigma \propto e^{-c_\sigma}$ turns the problem of minimizing the cost into one of finding the most probable permutation σ. In the proposal stage of simulated annealing, we randomly select two indices $i \neq j$ and reverse the block of integers beginning at σ_i and ending at σ_j in the current permutation $(\sigma_1, \ldots, \sigma_n)$. This proposal is accepted with probability (26.7) depending on the temperature τ. In *Numerical Recipes'* [27] simulated annealing algorithm for the traveling salesman problem, τ is lowered in multiplicative decrements of 10 % after every $100n$ epochs or every $10n$ accepted steps, whichever comes first. ∎

Example 26.6.2 *Best Subset Regression*

Best subset regression can be tackled by simulated annealing [31]. The standard criterion for determining the best subset of predictors is Mallows' [23]

$$C_p = \frac{\|y - \hat{y}\|_2^2}{\hat{\sigma}^2} + \alpha p,$$

where p is the cardinality of the given subset, $y - \hat{y}$ is the residual vector under this set, $\hat{\sigma}^2$ is an unbiased estimator of the error variance, and $\alpha = 2$. If there are q potential predictors and n observations, and if $y - \tilde{y}$ is the residual vector under the full set of predictors, then the obvious choice of $\hat{\sigma}^2$ is $\|y - \tilde{y}\|_2^2/(n - q)$. To minimize C_p by simulated annealing, we must design a proposal mechanism to perturb the current subset. The simplest device is to choose a predictor at random. If it is in the subset, then we switch it out. If it is outside the subset, then we switch it in. Recall from Section 7.6 that regression is conducted by sweeping on the diagonal entries of the $(q+1) \times (q+1)$ matrix

$$\begin{pmatrix} X^tX & X^ty \\ y^tX & y^ty \end{pmatrix}. \tag{26.8}$$

Here X is the design matrix for the full set of predictors. If we sweep the p diagonal entries of this matrix corresponding to the current subset, then the lower-right entry $y^ty = \|y\|_2^2$ becomes $\|y - \hat{y}\|_2^2$. Let $M = (m_{ij})$ denote the partially swept matrix. Sweeping or unsweeping diagonal entry r sends entry $m_{q+1,q+1}$ into $m_{q+1,q+1} - m_{q+1,r}m_{r,r}^{-1}m_{r,q+1}$. Thus, it is trivial to check whether a proposed move should be taken. Actually taking the move involves performing the entire sweep, which is harder but still easily accomplished if q is not too large. ■

26.7 Problems

1. Implement a Metropolis-driven random walk to generate Poisson deviates with mean λ. If the random walk is in state x, then it should propose states $x-1$ and $x+1$ with equal probabilities. Check that the visited states have approximate mean and variance λ.

2. Implement a Metropolis-driven random walk to generate standard normal deviates. If the random walk is in state x, then it should propose to move to state $x + c(U - \frac{1}{2})$, where $c > 0$ and U is a uniform deviate from $[0, 1]$. Check that the walk increment has mean and variance 0 and $\frac{c^2}{12}$, respectively. What value of c would you suggest?

3. An acceptance function $a : [0, \infty] \mapsto [0, 1]$ satisfies the functional identity $a(x) = xa(1/x)$. Prove that the detailed balance condition

$$\pi_i q_{ij} a_{ij} \quad = \quad \pi_j q_{ji} a_{ji}$$

holds if the acceptance probability a_{ij} is defined by

$$a_{ij} \quad = \quad a\left(\frac{\pi_j q_{ji}}{\pi_i q_{ij}}\right)$$

in terms of an acceptance function $a(x)$. Check that the Barker function $a(x) = x/(1+x)$ qualifies as an acceptance function and that any acceptance function is dominated by the Metropolis acceptance function in the sense that $a(x) \leq \min\{x, 1\}$ for all x.

4. The Metropolis acceptance mechanism (26.2) ordinarily implies aperiodicity of the underlying Markov chain. Show that if the proposal distribution is symmetric and if some state i has a neighboring state j such that $\pi_i > \pi_j$, then the period of state i is 1, and the chain,

if irreducible, is aperiodic. For a counterexample, assign probability $\pi_i = 1/4$ to each vertex i of a square. If the two vertices adjacent to a given vertex i are each proposed with probability 1/2, then show that all proposed steps are accepted by the Metropolis criterion and that the chain is periodic with period 2.

5. Validate the conjugate pairs displayed in Table 26.1.

6. Consider the Cartesian product space $\{0,1\} \times \{0,1\}$ equipped with the probability distribution

$$(\pi_{00}, \pi_{01}, \pi_{10}, \pi_{11}) = \left(\frac{1}{2}, \frac{1}{4}, \frac{1}{8}, \frac{1}{8}\right).$$

Demonstrate that sequential Gibbs sampling does not satisfy detailed balance by showing that $\pi_{00}s_{00,11} \neq \pi_{11}s_{11,00}$, where $s_{00,11}$ and $s_{11,00}$ are entries of the matrix S for first resampling component 1 and then resampling component 2.

7. Let $f(x_1, x_2)$ denote the joint density of two random variables X_1 and X_2, and let $f(x_1 \mid x_2)$ and $f(x_2 \mid x_1)$ denote their corresponding conditional densities. Prove the formula

$$f(x_1, x_2) = f(x_1^0, x_2^0)\frac{f(x_1 \mid x_2)f(x_2 \mid x_1^0)}{f(x_1^0 \mid x_2)f(x_2^0 \mid x_1^0)}$$

showing how the joint density can be recovered up to a normalizing constant from the conditional densities.

8. Design and implement a Gibbs sampler for generating bivariate normal deviates.

9. In a Bayesian random effects model, m_i observations y_{ij} are drawn from a normal density with mean η_i and precision τ for $i = 1, \ldots, n$. Impose a normal prior on η_i with mean μ and precision ω, a normal prior on μ with mean 0 and precision δ, and gamma priors on τ and ω with shape parameters α_τ and α_ω and scale parameters β_τ and β_ω, respectively. Assume independence among the y_{ij} given all model parameters, independence of the η_i given the hyperparameters μ, τ, and ω, and independence among the hyperparameters themselves. Calculate the conditional densities necessary to conduct a Gibbs sample from the posterior distribution of the parameter vector $(\eta_1, \ldots, \eta_n, \mu, \tau, \omega)$.

10. Carry out the Gibbs sampling procedure for the ABO allele frequency model described in Example 26.3.4 using the duodenal ulcer data of Chapter 13. Estimate the posterior medians, means, variances, and covariances of the allele frequency parameters.

11. As a simple change-point problem, suppose we observe the independent random variables X_1 through X_m. The first k of these are Poisson distributed with mean λ_1, and the last $m-k$ are Poisson distributed with mean λ_2. We place independent priors on k, λ_1, and λ_2. The prior on k is uniform over $\{1, \ldots, m\}$. The prior on λ_i is gamma with shape parameter α_i and scale parameter β_i. The prior on the hyperparameter β_i is inverse gamma with shape parameter γ_i and scale parameter δ_i. The full prior can be factored as

$$\pi(k)\pi(\lambda_1 \mid \beta_1)\pi(\lambda_2 \mid \beta_2)\pi(\beta_1)\pi(\beta_2).$$

Derive the Gibbs sampling updates for the posterior density of the parameter vector $(k, \lambda_1, \lambda_2, \beta_1, \beta_2)$.

12. Let $\{x_i\}_{i=1}^n$ be observations from independent Poisson random variables with means $\{\lambda_i t_i\}_{i=1}^n$, where the t_i are known times and the λ_i are unknown intensities. Suppose that λ_i has a gamma prior with shape parameter α and scale parameter β and that β has an inverse gamma prior $\delta^\gamma e^{-\delta/\beta}/[\beta^{\gamma+1}\Gamma(\gamma)]$. Here α, δ, and γ are given, while the λ_i and β are parameters. If the priors on the λ_i are independent, what is the joint density of the data and the parameters? Design a sequential Gibbs scheme to sample from the posterior distribution of the parameters given the data.

13. If the component updated in the randomly sampled component version of Gibbs sampling depends probabilistically on the current state of the chain, how must the Hastings-Metropolis acceptance probability be modified to preserve detailed balance? Under the appropriate modification, the acceptance probability is no longer always 1.

14. Design and implement a slice sampler that generates deviates from the unnormalized density $f(x) = e^{-|x|^\alpha}$, with x real and $\alpha > 0$.

15. Write a program implementing the Markov chain of Example 25.2.4 of Chapter 25. Use the program to estimate $\mathrm{E}(\sum_{i=1}^n R_{ii}^2)$ for a random rotation R. This integral has exact value 1.

16. Importance sampling is one remedy when the states of a Markov chain communicate poorly [16]. Suppose that π is the equilibrium distribution of the chain. If we sample from a chain whose distribution is ν, then we can recover approximate expectations with respect to π by taking weighted averages. In this scheme, the state z is given weight $w_z = \pi_z/\nu_z$. If $Z_0, Z_1, Z_2 \ldots$ is a run from the chain with equilibrium distribution ν, then under the appropriate ergodic assumptions prove that

$$\lim_{n\to\infty} \frac{\sum_{i=0}^{n-1} w_{Z_i} f(Z_i)}{\sum_{i=0}^{n-1} w_{Z_i}} = \mathrm{E}_\pi[f(X)].$$

The choice $\nu_z \propto \pi_z^{1/\tau}$ for $\tau > 1$ lowers the peaks and raises the valleys of π [18]. Unfortunately, if ν differs too much from π, then the estimator

$$\frac{\sum_{i=0}^{n-1} w_{Z_i} f(Z_i)}{\sum_{i=0}^{n-1} w_{Z_i}}$$

of the expectation $\mathrm{E}_\pi[f(X)]$ will have a large variance for n of moderate size.

17. Another device to improve mixing of a Markov chain is to run several parallel chains on the same state space and occasionally swap their states [14]. If π is the target distribution, then let $\pi^{(1)} = \pi$, and define $m-1$ additional distributions $\pi^{(2)}, \ldots, \pi^{(m)}$. For instance, incremental heating can be achieved by taking

$$\pi_z^{(k)} \quad \propto \quad \pi_z^{\frac{1}{1+(k-1)\tau}}$$

for $\tau > 0$. At epoch n, we sample for each chain k a state Z_{nk} given the chain's previous state $Z_{n-1,k}$. We then randomly select chain i with probability $1/m$ and consider swapping states between it and chain $j = i+1$. (When $i = m$, no swap is performed.) Under appropriate ergodic assumptions on the m participating chains, show that if the acceptance probability for the proposed swap is

$$\min\left\{ \frac{\pi_{Z_{nj}}^{(i)} \pi_{Z_{ni}}^{(j)}}{\pi_{Z_{ni}}^{(i)} \pi_{Z_{nj}}^{(j)}}, 1 \right\},$$

then the product chain is ergodic with equilibrium distribution given by the product distribution $\pi^{(1)} \otimes \pi^{(2)} \otimes \cdots \otimes \pi^{(m)}$. The marginal distribution of this distribution for chain 1 is just π. Therefore, we can throw away the outcomes of chains 2 through m and estimate expectations with respect to π by forming sample averages from the embedded run of chain 1. (Hint: The fact that no swap is possible at each step allows the chains to run independently for an arbitrary number of steps.)

18. Show that the function (26.6) is log-concave in (α, β). (Hint: Based on the closure properties enumerated in Problem 7 of Chapter 22, argue that the normalizing function

$$\frac{\Gamma(\alpha)\Gamma(\beta)}{\Gamma(\alpha+\beta)} \quad = \quad \int_0^1 x^{\alpha-1}(1-x)^{\beta-1} dx$$

is log-convex.)

19. It is known that every planar graph can be colored by four colors [2]. Design, program, and test a simulated annealing algorithm to find a four coloring of any planar graph. (Suggestions: Represent the graph by a list of nodes and a list of edges. Assign to each node a color represented by a number between 1 and 4. The cost of a coloring is the number of edges with incident nodes of the same color. In the proposal stage of the simulated annealing solution, randomly choose a node, randomly reassign its color, and recalculate the cost. If successful, simulated annealing will find a coloring with the minimum cost of 0.)

20. A Sudoku puzzle is a 9-by-9 matrix, with some entries containing predefined digits. The goal is to completely fill in the matrix, using the digits 1 through 9, in such a way that each row, column, and symmetrically placed 3-by-3 submatrix displays each digit exactly once. In mathematical language, a completed Sudoku matrix is a Latin square subject to further constraints on the 3-by-3 submatrices. The initial partially filled in matrix is assumed to have a unique completion. Design, program, and test a simulated annealing algorithm to solve a Sudoku puzzle.

21. Consider a graph with $2n$ nodes. The graph bisection problem involves dividing the nodes into two disjoint subsets A and B of size n such that the number of edges extending between a node in A and a node in B is as small as possible. Design and implement a simulated annealing algorithm to find a best pair of subsets. This problem has implications for the design of microchips [19].

26.8 References

[1] Besag J, Green PJ (1993) Spatial statistics and Bayesian computation. *J Roy Stat Soc B* 55:25–37

[2] Brualdi RA (1977) *Introductory Combinatorics.* North-Holland, New York

[3] Casella G, George EI (1992) Explaining the Gibbs sampler. *Amer Statistician* 46:167–174

[4] Castledine B (1981) A Bayesian analysis of multiple-recapture sampling for a closed population. *Biometrika* 67:197–210

[5] Calvetti D, Somersalo E (2007) *Introduction to Bayesian Scientific Computing: Ten Lectures on Subjective Computing.* Springer, New York

[6] Chib S, Greenberg E (1995) Understanding the Metropolis-Hastings algorithm. *Amer Statistician* 49:327–335

[7] Duane S, Kennedy AD, Pendleton BJ, Roweth D (1987) Hybrid Monte Carlo. *Physics Letters B* 195:216–222

[8] Fishman GS (2006) *A First Course in Monte Carlo.* Duxbury Press, Belmont, CA

[9] Gelfand AE, Smith AFM (1990) Sampling-based approaches to calculating marginal densities. *J Amer Stat Assoc* 85:398–409

[10] Gelman A, Carlin JB, Stern HS, Rubin DB (2003) *Bayesian Data Analysis,* 2nd ed. Chapman & Hall, London

[11] Gelman A, Rubin DB (1992) Inference from iterative simulation using multiple sequences (with discussion). *Stat Sci* 7:457–511

[12] Geman S, Geman D (1984) Stochastic relaxation, Gibbs distributions and the Bayesian restoration of images. *IEEE Trans Pattern Anal Machine Intell* 6:721–741

[13] George EI, Robert CP (1992) Capture-recapture estimation via Gibbs sampling. *Biometrika* 79:677–683

[14] Geyer CJ (1991) Markov chain Monte Carlo maximum likelihood. *Computing Science and Statistics: Proceedings of the 23rd Symposium on the Interface,* Keramidas EM, editor, Interface Foundation, Fairfax, VA pp 156–163

[15] Gilks WR, Richardson S, Spiegelhalter DJ (1996) *Markov Chain Monte Carlo in Practice.* Chapman & Hall, London

[16] Hastings WK (1970) Monte Carlo sampling methods using Markov chains and their applications. *Biometrika* 57:97–109

[17] Higdon DM (1998) Auxiliary variable methods for Markov chain Monte Carlo with applications. *JASA* 93:585–595

[18] Jennison C (1993) Discussion on the meeting of the Gibbs sampler and other Markov chain Monte Carlo methods. *J Roy Stat Soc B* 55:54–56

[19] Kirkpatrick S, Gelatt CD, Vecchi MP (1983) Optimization by simulated annealing. *Science* 220:671–680

[20] Lange K, Sobel E (1991) A random walk method for computing genetic location scores. *Amer J Hum Genet* 49:1320–1334

[21] Liu JS (1996) Metropolized independent sampling with comparisons to rejection sampling and importance sampling. *Stat and Computing* 6:113–119

[22] Liu JS (2001) *Monte Carlo Strategies in Scientific Computing.* Springer, New York

[23] Mallows CL (1973) Some comments on Cp. *Technometrics* 15:661–675

[24] Metropolis N, Rosenbluth A, Rosenbluth M, Teller A, Teller E (1953) Equations of state calculations by fast computing machines. *J Chem Physics* 21:1087–1092

[25] Neal RM (1996) *Bayesian Learning for Neural Networks.* Springer, New York

[26] Neal RM (2003) Slice sampling. *Ann Stat* 31:705–767

[27] Press WH, Teukolsky SA, Vetterling WT, Flannery BP (1992) *Numerical Recipes in Fortran: The Art of Scientific Computing*, 2nd ed. Cambridge University Press, Cambridge

[28] Robert CP, Casella G (2004) *Monte Carlo Statistical Methods*, 2nd ed. Springer, New York

[29] Spiegelhalter D, Thomas D, Best N, Lunn D (2006) Winbugs Users Manual 1.4.1, Bayesian Inference Using Gibbs Sampling. MRC Biostatistics Unit, Institute of Public Health, Cambridge, Available: http://www.mrc-bsu.cam.ac.uk/bugs

[30] Swendsen RH, Wang JS (1987) Nonuniversal critical dynamics in Monte Carlo simulations. *Physical Review Letters* 58:86–88

[31] Suchard MA, Bailey JN, Elashoff DE, Sinsheimer JS (2001) SNPing away at candidate genes. *Genetic Epidemiology* 21:S643–S648

[32] Tanner MA (1993) *Tools for Statistical Inference: Methods for Exploration of Posterior Distributions and Likelihood Functions*, 2nd ed. Springer, New York

[33] Tanner M, Wong W (1987) The calculation of posterior distributions (with discussion). *J Amer Stat Assoc* 82:528–550

[34] Tierney L (1994) Markov chains for exploring posterior distributions (with discussion). *Ann Stat* 22:1701–1762

27

Advanced Topics in MCMC

27.1 Introduction

The pace of research on MCMC methods is so quick that any survey of advanced topics is immediately obsolete. The highly eclectic and decidedly biased coverage in our final chapter begins with a discussion of Markov random fields. Our limited aims here are to prove the Hammersley-Clifford theorem and introduce the Swendsen-Wang algorithm, a clever form of slice sampling. In the Ising model, the Swendsen-Wang algorithm is much more efficient than standard Gibbs sampling.

We then move on to reversible jump MCMC. Bayesian inference tends to perform better than frequentist inference in choosing between non-nested models. Green's [8] invention of reversible jump MCMC provides a practical path to implementing Bayesian model selection. In our opinion, reversible jump MCMC will have a larger impact on the field of statistics than the more glamorous device of perfect sampling.

The second half of the chapter is designed to help readers sort through the advanced literature on convergence rates of Markov chains. Our treatment of convergence commences with the introduction of the total variation and chi-square distances. Coupling is mentioned because it often yields surprisingly good total variation bounds with little effort. Convergence theory is much easier for finite-state Markov chains than it is for denumerable and continuous chains. For reversible finite-state chains, the full force of linear algebra can be brought to bear. This is a good setting to develop intuition. For the sake of comparison, we treat convergence of the independence sampler from the different perspectives of linear algebra and coupling.

The last two sections of the chapter emphasize Hilbert spaces, compact operators, our old friend the singular value decomposition, and orthogonal polynomials. The foundations of Hilbert and Banach spaces were laid in the nineteenth century and completed in the first half of the twentieth century. Hilbert spaces isolate and generalize the inner products that drive linear algebra. Linear operators such as expectations and conditional expectations are the infinite-dimensional analogs of matrices. Throughout our discussion, alert readers will notice our debt to the books [11, 13, 15] and the survey article [6].

The lovely interplay of different branches of mathematics yields enormous insight into toy models of MCMC. Unfortunately, the convergence behavior of more practical models is much more complicated. Statisticians are in the position of those whose practical reach exceeds their theoretical grasp.

K. Lange, *Numerical Analysis for Statisticians*, Statistics and Computing,
DOI 10.1007/978-1-4419-5945-4_27, © Springer Science+Business Media, LLC 2010

Perhaps a chapter reinforcing this humbling reality is a good place to close the present book. Certainly, there will be no lack of challenging theoretical problems to keep the next generation of mathematical statisticians busy.

Before turning to particular topics, a few words about vocabulary and notation are again timely. We will use the term kernel instead of the term transition probability matrix in discussing denumerable and continuous-state Markov chains. Integration signs will be replaced by summation signs whenever convenient. Most of the kernels one encounters in practice provide densities relative to a fixed measure such as counting measure or Lebesgue measure. Omitting mention of these measures eases the notational burden in complicated formulas. Sticklers for detail will want to mentally add the measures back in following our derivations.

27.2 Markov Random Fields

A Markov chain is indexed by points in time. A random spatial process is indexed by points in space. If we are willing to avoid continuous indices, we can abstract even further and consider random processes over finite graphs. The Ising process is a typical example. Suppose the graph G has g nodes (or sites). The state of node i is characterized by a random element drawn from a finite set $\mathcal{S}_i$. The state of the entire system is a random vector $\mathbf{N} = (N_1, \ldots, N_g)$ with $N_i \in \mathcal{S}_i$. In the Ising model $\mathcal{S}_i = \{0, 1\}$. In imaging problems, $\mathcal{S}_i$ might be a range of intensities. If color is important, then $\mathcal{S}_i$ could be the Cartesian product of three sets of intensities, one for each primary color. The structure of G is determined by the set of neighbors ∂i of each node i. Thus, i and $j \in \partial i$ are joined by an edge of the graph. The first g epochs of a Markov chain fit naturally within this framework. The nodes of the graph consist of the integers from 1 to g. The neighbors of i are $i-1$ and $i+1$ provided $1 < i < g$. This linear graph is the simplest connected graph. The sets $\mathcal{S}_i$ collapse to the possible states of the chain.

It is useful to adopt notation as compact as possible. Let $\pi(\mathbf{n})$ be the probability of the realization $\mathbf{n} = (n_1, \ldots, n_g)$ of $\mathbf{N}$. To circumvent certain logical traps, we will assume that all $\pi(\mathbf{n})$ are positive. Because they are probabilities, they sum to 1. In defining a Markov random field, we must deal with conditional probabilities. Let $\pi(\mathbf{n} \mid \mathbf{n}_A)$ denote the conditional distribution of $\mathbf{N}$ given $N_i = n_i$ for $i \in A$. The most important choices for the subset $A \subset G$ are $G - i = G \setminus \{i\}$ and ∂i. With this notation in hand, the Markov random field assumption is $\pi(n_i \mid \mathbf{n}_{G-i}) = \pi(n_i \mid n_{\partial i})$ for all nodes i. For instance, consider the first g epochs of a stationary Markov chain with transition probability matrix (p_{ij}) and equilibrium distribution ν. The calculation

$$\pi(n_i \mid \mathbf{n}_{G-i}) = \frac{\nu_{n_1} \prod_{j=1}^{g-1} p_{n_j, n_{j+1}}}{\nu_{n_1} \prod_{j \notin \{i-1, i\}} p_{n_j, n_{j+1}} \sum_m p_{n_{i-1}, m} p_{m, n_{i+1}}}$$

$$\begin{aligned} &= \frac{p_{n_{i-1},n_i} p_{n_i,n_{i+1}}}{\sum_m p_{n_{i-1},m} p_{m,n_{i+1}}} \\ &= \pi(n_i \mid \mathbf{n}_{\partial i}) \end{aligned}$$

demonstrates the Markov field property for $1 < i < g$. Verification when $i = 1$ and $i = g$ is equally easy.

Given a random field, Markovian or not, the conditional probabilities $\pi(n_i \mid \mathbf{n}_{G-i})$ determine the full distribution $\pi(\mathbf{n})$. To establish this fact, let T_i^m be the operator that changes the state of node i to m. The identity

$$\frac{\pi(n_i \mid \mathbf{n}_{G-i})}{\pi(m \mid \mathbf{n}_{G-i})} = \frac{\pi(\mathbf{n})}{\pi(T_i^m \mathbf{n})}$$

is the key to the proof. For each set $\mathcal{S}_i$ choose a distinguished state 0 and let $\mathbf{0} = (0, \ldots, 0)$. The equation

$$\frac{\pi(\mathbf{n})}{\pi(\mathbf{0})} = \frac{\pi(\mathbf{n})}{\pi(T_1^0 \mathbf{n})} \cdot \frac{\pi(T_1^0 \mathbf{n})}{\pi(T_2^0 T_1^0 \mathbf{n})} \cdots \frac{\pi(T_{g-1}^0 T_{g-2}^0 \cdots T_1^0 \mathbf{n})}{\pi(T_g^0 T_{g-1}^0 \cdots T_1^0 \mathbf{n})}$$

now determines $\pi(\mathbf{n})$ up to a normalizing constant.

It is worth emphasizing that merely postulating the conditional distributions $\pi(n_i \mid \mathbf{n}_{G-i})$ does not guarantee the existence of a random field. The conditional distributions have to be consistent. For instance, consider the conditional distributions

$$\Pr(N_1 = i \mid N_2 = j) = \begin{pmatrix} \frac{1}{2} & \frac{1}{4} \\ \frac{1}{2} & \frac{3}{4} \end{pmatrix}, \quad \Pr(N_2 = j \mid N_1 = i) = \begin{pmatrix} \frac{1}{2} & \frac{1}{2} \\ \frac{3}{4} & \frac{1}{4} \end{pmatrix}$$

for a two-node graph with the common state space $\{1, 2\}$ for each node. If we postulate the marginal distributions $(a, 1-a)$ for N_1 and $(b, 1-b)$ for N_2, then the joint distribution can be written in the two different ways

$$\Pr(N_1 = i, N_2 = j) = \begin{pmatrix} \frac{a}{2} & \frac{a}{2} \\ \frac{3(1-a)}{4} & \frac{(1-a)}{4} \end{pmatrix} = \begin{pmatrix} \frac{b}{2} & \frac{1-b}{4} \\ \frac{b}{2} & \frac{3(1-b)}{4} \end{pmatrix}.$$

The resulting four equations have no solution for the unknowns a and b.

Our next object is to characterize Markovian random fields. This exercise is intimately tied to the clique structure of the graph G. A subset $R \subset G$ is said to be a clique if all pairs i and j drawn from R are connected by an edge. Every singleton $\{i\}$ is a clique by default. Let $\mathcal{C}$ denote the collection of cliques. With this notation we prove the necessary and sufficient condition of Hammersley and Clifford for a random field to be Markovian [3].

Proposition 27.2.1 *A random field is Markovian if and only if its probability distribution $\pi(\mathbf{n})$ can be expressed as a product*

$$\pi(\mathbf{n}) = \prod_{C \in \mathcal{C}} \psi_C(\mathbf{n}_C) \tag{27.1}$$

over functions whose arguments depend only on the cliques of the underlying graph G.

Proof: Suppose the representation (27.1) holds. Then

$$\pi(n_i \mid \mathbf{n}_{G-i}) = \frac{\prod_{C\in\mathcal{C}:i\in C}\psi_C(\mathbf{n}_C)}{\sum_{m\in S_i}\prod_{C\in\mathcal{C}:i\in C}\psi_C(T_i^m\mathbf{n}_C)}$$

clearly depends only on the neighborhood ∂i of i.

To prove the converse, note that if G itself is a clique, then the asserted representation is obvious. For the general case, it is convenient to let $\mathbf{n}_R^0$ be the vector whose ith component is n_i when $i \in R$ and 0 when $i \notin R$. We define the clique functions of the representation (27.1) recursively via

$$\begin{aligned}
\psi_\emptyset(\mathbf{n}_\emptyset) &= \pi(\mathbf{0}) \\
\psi_C(\mathbf{n}_C) &= \begin{cases} \frac{\pi(\mathbf{n}_C^0)}{\prod_{B\subset C:B\neq C}\psi_B(\mathbf{n}_B)} & n_i \neq 0 \text{ for all } i \in C \\ 1 & \text{otherwise.} \end{cases}
\end{aligned}$$

Here $\emptyset$ is the empty set, and $C \neq \emptyset$. These definitions guarantee the validity of the representation (27.1) for all $\mathbf{n} = \mathbf{n}_C^0$ with C a clique. In particular, it holds when $\mathbf{n} = \mathbf{0}$. We therefore assume inductively that it holds when $\mathbf{n}$ has at most m nonzero components. Take $\mathbf{n} \neq \mathbf{n}_C^0$ for any clique C, and assume $\mathbf{n}$ has at most $m+1$ nonzero components. There clearly exist two nonneighboring nodes i and j with $n_i \neq 0$ and $n_j \neq 0$; otherwise, the set $C = \{i : n_i \neq 0\}$ is a clique with $n = n_C^0$. Since π is Markovian, the ratio $\pi(\mathbf{n})/\pi(T_j^0\mathbf{n})$ does not depend on n_i, and

$$\frac{\pi(\mathbf{n})}{\pi(T_j^0\mathbf{n})} = \frac{\pi(T_i^0\mathbf{n})}{\pi(T_i^0T_j^0\mathbf{n})}.$$

Because the state vectors $T_i^0\mathbf{n}$, $T_j^0\mathbf{n}$, and $T_i^0T_j^0$ have more components with value 0 than $\mathbf{n}$, the induction hypothesis implies

$$\begin{aligned}
&\pi(\mathbf{n}) \\
&= \frac{\pi(T_i^0\mathbf{n})\pi(T_j^0\mathbf{n})}{\pi(T_i^0T_j^0\mathbf{n})} \\
&= \frac{\prod_{i\in C}\psi_C(T_i^0\mathbf{n}_C)\prod_{i\notin C}\psi_C(T_i^0\mathbf{n}_C)\prod_{j\in C}\psi_C(T_j^0\mathbf{n}_C)\prod_{j\notin C}\psi_C(T_j^0\mathbf{n}_C)}{\prod_{i,j\notin C}\psi_C(T_i^0T_j^0\mathbf{n}_C)\prod_{i\in C}\psi_C(T_i^0T_j^0\mathbf{n}_C)\prod_{j\in C}\psi_C(T_i^0T_j^0\mathbf{n}_C)} \\
&= \frac{\prod_{i\in C}\psi_C(T_i^0\mathbf{n}_C)\prod_{i\notin C}\psi_C(\mathbf{n}_C)\prod_{j\in C}\psi_C(T_j^0\mathbf{n}_C)\prod_{j\notin C}\psi_C(\mathbf{n}_C)}{\prod_{i,j\notin C}\psi_C(\mathbf{n}_C)\prod_{i\in C}\psi_C(T_i^0\mathbf{n}_C)\prod_{j\in C}\psi_C(T_j^0\mathbf{n}_C)} \\
&= \frac{\prod_{i\notin C}\psi_C(\mathbf{n}_C)\prod_{j\notin C}\psi_C(\mathbf{n}_C)}{\prod_{i,j\notin C}\psi_C(\mathbf{n}_C)} \\
&= \prod_{i,j\notin C}\psi_C(\mathbf{n}_C)\prod_{i\in C}\psi_C(\mathbf{n}_C)\prod_{j\in C}\psi_C(\mathbf{n}_C),
\end{aligned}$$

which is just the representation (27.1) again. This advances the induction and proves the overall validity of the representation. ∎

As a simple illustration, suppose the field is binary (all $\mathcal{S}_i = \{0, 1\}$) and all cliques contain at most two nodes. The distribution $\pi(\mathbf{n})$ is determined by the ratios

$$\frac{\pi(n_i = 1 \mid \mathbf{n}_{\partial i})}{\pi(n_i = 0 \mid \mathbf{n}_{\partial i})} = \frac{\psi_i(1) \prod_{j \in \partial i} \psi_{\{i,j\}}(1, n_j)}{\psi_i(0) \prod_{j \in \partial i} \psi_{\{i,j\}}(0, n_j)}.$$

If we define

$$\begin{aligned} \alpha_i &= \frac{\psi_i(1) \prod_{j \in \partial i} \psi_{\{i,j\}}(1, 0)}{\psi_i(0) \prod_{j \in \partial i} \psi_{\{i,j\}}(0, 0)} \\ \beta_{\{i,j\}} &= \frac{\psi_{\{i,j\}}(0, 0)\psi_{\{i,j\}}(1, 1)}{\psi_{\{i,j\}}(1, 0)\psi_{\{i,j\}}(0, 1)}, \quad j \in \partial i, \end{aligned}$$

then this ratio becomes

$$\frac{\pi(n_i = 1 \mid \mathbf{n}_{\partial i})}{\pi(n_i = 0 \mid \mathbf{n}_{\partial i})} = \alpha_i \prod_{j \in \partial i} \beta_{\{i,j\}}^{n_j}.$$

Hence, $\pi(\mathbf{n})$ is determined by the node parameters α_i and the edge parameters $\beta_{\{i,j\}}$.

Example 27.2.1 *Swendsen-Wang Algorithm*

The most interesting special case of the model occurs when all $\beta_{ij} = \beta > 1$. In this setting it is convenient to reparameterize so that

$$\pi(\mathbf{n}) \propto e^{\sum_i \gamma_i n_i + \delta \sum_{\{ij\}} 1_{\{n_i = n_j\}}},$$

where the second sum extends over all edges and $\delta > 0$. Although Gibbs sampling is clearly possible, the slice sampler is much faster [7, 19]. To implement slice sampling, we must decompose $\pi(\mathbf{n})$ into a product of nonnegative functions with the same support. The auxiliary functions appearing in the product correspond to the edges $\{j, k\}$ of G and are defined as $f_{\{j,k\}}(\mathbf{n}) = e^{\gamma 1_{\{n_j = n_k\}}}$. For each auxiliary function, one creates an auxiliary random variable $U_{\{j,k\}}$ uniformly distributed on the interval $[0, f_{\{j,k\}}(\mathbf{n})]$. Because the $U_{\{j,k\}}$ are independent given $\mathbf{n}$, they are trivial to sample. Sampling from the intersection of the top sets $\{\mathbf{n} : f_{\{j,k\}}(\mathbf{n}) \geq u_{\{j,k\}}\}$ is more subtle.

The most fruitful way of thinking about sampling $\mathbf{n}$ is to separate the edges with $u_{\{j,k\}} > 1$ from those with $u_{\{j,k\}} \leq 1$. For those in the former category, the components n_j and n_k of a sampled $\mathbf{n}$ must be equal. For those in the latter category, no such restriction applies. Therefore, consider a new graph whose node set is G and whose edge set consists of the former category of edges. Divide the nodes in the new graph into connected components. Within a connected component, all nodes i must have the same value n_i. Nodes in different components have independent values for the n_i. Within a component A, we rely on Bayes' rule to set the common value of the n_i. This requires that 0 be chosen with probability

$$\frac{e^{\sum_{i \in A} \gamma_i 0}}{e^{\sum_{i \in A} \gamma_i 0} + e^{\sum_{i \in A} \gamma_i 1}} = \frac{1}{1 + e^{\sum_{i \in A} \gamma_i}}$$

and that 1 be chosen with the complementary probability. ■

27.3 Reversible Jump MCMC

One of the advantages of the Bayesian perspective is that it promotes graceful handling of non-nested models. Hypothesis testing by likelihood ratios requires the null model to be a smooth restriction of the alternative model. Many statistical problems involve selecting one model from a set of non-overlapping models or a hierarchy of non-nested models. For instance in biological taxonomy, several different evolutionary trees may plausibly trace the ancestry of a group of related species. The distinctions between trees are discrete and combinatorial rather than continuous.

Suppose x represents the data explained by a finite or countable number of competing models. Let $\mathcal{M}_m$ denote model m with prior ρ_m, parameter vector θ_m, prior $\pi_m(\theta_m)$, and likelihood $f_m(x \mid \theta_m)$. An MCMC sample from the joint density $f_m(x \mid \theta_m)\pi_m(\theta_m)\rho_m$ can be used to approximate the posterior probability

$$\Pr(\mathcal{M}_m \mid x) = \frac{\rho_m \int f_m(x \mid \theta_m)\pi_m(\theta_m)\, d\theta_m}{\sum_n \rho_n \int f_n(x \mid \theta_n)\pi_n(\theta_n)\, d\theta_n}.$$

Green's [8] reversible jump MCMC makes estimation of such posterior probabilities straightforward.

To implement Green's algorithm, one needs to specify the probability of jumping from $\mathcal{M}_m$ to $\mathcal{M}_n$. In many problems the models form a staircase, and choosing the jump probabilities j_{mn} as a nearest neighbor random walk is convenient. The diagonal terms j_{mm} should be substantial enough to permit exploration of the current model between jumps. If the dimensions of the parameter vectors θ_m and θ_n do not match, then the dimension of the less complex model is boosted by simulation. For instance, if the dimension of θ_n exceeds the dimension of θ_m, then a continuous random vector u_{mn} is simulated to fill out the missing dimensions of θ_m. Dimension matching is fully achieved by defining smooth invertible transformations $\theta_n = T_{mn}(\theta_m, u_{mn})$ and $(\theta_m, u_{mn}) = T_{nm}(\theta_n)$ that take subsets of positive Lebesgue measure into subsets of positive Lebesgue measure.

MCMC sampling is initiated by proposing the next model $\mathcal{M}_n$ according to the jump distribution $\{j_{mn}\}_n$. The motion of the Markov chain within the current model $\mathcal{M}_m$ should preserve its posterior distribution, which is proportional to $f_m(x \mid \theta_m)\pi_m(\theta_m)$. For instance, one can conduct random component Gibbs sampling. Jumps between models are governed by the Hastings-Metropolis mechanism to maintain reversibility. If model $\mathcal{M}_n$ is selected, then the boosting vector u_{mn} is sampled as needed from a fixed density $h_{mn}(u_{mn})$, often assumed uniform. The proposal step is completed by sending the vector θ_m into the vector $\theta_n = T_{mn}(\theta_m, u_{mn})$. Acceptance is more subtle. When the proposed model $\mathcal{M}_n$ differs from the current model $\mathcal{M}_m$, the acceptance probability is

$$\min\left\{\frac{f_n(x \mid \theta_n)\pi_n(\theta_n)\rho_n h_{nm}(u_{nm})j_{nm}}{f_m(x \mid \theta_m)\pi_m(\theta_m)\rho_m h_{mn}(u_{mn})j_{mn}} \cdot |dT_{mn}(\theta_m, u_{mn})|, 1\right\}. \quad (27.2)$$

Equation (27.2) differs from the standard Hastings-Metropolis acceptance probability by the inclusion of the jump probabilities, the boosting densities, and the dimension matching Jacobian $|dT_{mn}(\theta_m, u_{mn})|$.

The Jacobian plays the role of a volume magnification factor. If we think of the point (θ_m, u_{mn}) as surrounded by a small box of integration, then this box is transformed into a distorted box around the point (θ_n, u_{nm}). The Jacobian $|dT(\theta_m, u_{mn})|$ specifies how the volume of the transformed box relates to the volume of the original box. The boosted components u_{mn} and u_{nm}, one of which is empty, are ghosts that appear only when needed to facilitate the jumps between models. The whole apparatus requires the functions $f_m(x \mid \theta_m)$, $\pi_m(\theta_m)$, and $h_{mn}(u_{mn})$ to be densities relative to Lebesgue measure (ordinary volume). These densities need only be specified up to a constant, but the overall constants must be the same across models in order for the proper cancellations to occur in the acceptance probability.

Example 27.3.1 *Poisson versus Negative Binomial*

The negative binomial distribution (model 2) provides an alternative to the Poisson distribution (model 1) in a random sample $x = (x_1, \ldots, x_s)$ with potential over-dispersion. The two corresponding densities

$$\Pr(X = j) = \frac{\lambda^j}{j!}e^{-\lambda}, \quad \Pr(X = j) = \binom{n+j-1}{j}p^n(1-p)^j$$

are supported on the integers $\{0, 1, 2, \ldots\}$ and have means λ and $n(1-p)/p$ and variances λ and $n(1-p)/p^2$, respectively. The comparison is clearer if we reparameterize by setting $n(1-p)/p = \lambda$ and $n(1-p)/p^2 = \lambda(1+\kappa)$ for $\kappa > 0$. This entails taking $n = \lambda/\kappa$ and $p = 1/(1+\kappa)$. With this change in notation, the negative binomial density becomes

$$\Pr(X = j) = \frac{\Gamma(\lambda/\kappa + j)}{\Gamma(\lambda/\kappa)j!}\left(\frac{1}{1+\kappa}\right)^{\lambda/\kappa}\left(\frac{\kappa}{1+\kappa}\right)^j.$$

A reasonable prior on λ is gamma with shape parameter α_λ and scale parameter β_λ. Under the Poisson model a gamma prior is conjugate. The choice of a prior for the over-dispersion parameter κ is more problematic. One possibility is an independent gamma prior with shape parameter α_κ and scale parameter β_κ. The uninformative prior $\rho_1 = \rho_2 = \frac{1}{2}$ and jump probabilities $j_{mn} = \frac{1}{2}$ are both fairly natural.

To achieve dimension matching, one can simply sample an exponential deviate u with intensity δ and equate κ to u. The implied transformations $T_{12}(\lambda, u) = (\lambda, u)$ and $T_{21}(\lambda, \kappa) = (\lambda, \kappa)$ both have Jacobian 1. These choices fully define the acceptance probabilities and drive MCMC sampling. An obvious omission is how the chain moves within a model. Under the Poisson model, resampling the posterior distribution is straightforward. Under the negative binomial model, a good sampling procedure is less obvious. One possibility is to institute Hastings-Metropolis sampling by multiplying λ and κ by independent exponential deviates with mean 1. We encourage readers to play with the model and consult the expository paper [9] for further hints. For the sake of simplicity, we have modified some aspects of the model. ■

27.4 Metrics for Convergence

To discuss convergence rates rigorously, it is convenient to introduce two distance functions on probability distributions. For the sake of simplicity, suppose the common state space is the set of integers. The total variation distance between distributions π and ν is defined as [5]

$$\|\pi - \nu\|_{\mathrm{TV}} = \sup_A |\pi(A) - \nu(A)| = \frac{1}{2}\sum_i |\pi_i - \nu_i|, \qquad (27.3)$$

where A ranges over all subsets of the integers. Problem 5 asks the reader to check that these two definitions are equivalent. Another useful distance is the chi-square distance $\|\pi - \nu\|_{\chi^2} = \|\pi - \nu\|_{1/\nu}$. Its square equals

$$\|\pi - \nu\|_{\chi^2}^2 = \sum_i \frac{(\pi_i - \nu_i)^2}{\nu_i} = \mathrm{Var}_\nu\left(\frac{\pi_i}{\nu_i}\right).$$

If $\pi_i > 0$ and $\nu_i = 0$ for some i, then $\|\pi - \nu\|_{\chi^2}$ is infinite. The Cauchy-Schwarz inequality

$$\sum_i |\pi_i - \nu_i| \le \left[\sum_i \frac{(\pi_i - \nu_i)^2}{\nu_i}\right]^{1/2} \left(\sum_i \nu_i\right)^{1/2}$$

implies

$$\|\pi - \nu\|_{\mathrm{TV}} \le \frac{1}{2}\|\pi - \nu\|_{\chi^2}. \qquad (27.4)$$

Hence, chi-square distance is stronger than total variation distance. The same reasoning leads to the bound

$$|\,\mathrm{E}_\pi(Z) - \mathrm{E}_\nu(Z)| \;\; \leq \;\; \mathrm{E}_\nu(Z^2)^{1/2}\|\pi - \nu\|_{\chi^2}$$

for any random variable Z.

Coupling arguments depend on π and ν being attached to random variables X and Y defined on the same probability space. When this is the case, we have

$$\begin{aligned}
& |\Pr(X \in A) - \Pr(Y \in A)| \\
= \;& |\Pr(X \in A, X = Y) + \Pr(X \in A, X \neq Y) \\
& - \Pr(Y \in A, X = Y) - \Pr(Y \in A, X \neq Y)| \\
= \;& |\Pr(X \in A, X \neq Y) - \Pr(Y \in A, X \neq Y)| \\
\leq \;& \mathrm{E}(1_{\{X \neq Y\}}|1_A(X) - 1_A(Y)|) \\
\leq \;& \Pr(X \neq Y).
\end{aligned}$$

Taking the supremum over A yields

$$\|\pi - \nu\|_{\mathrm{TV}} \;\; \leq \;\; \Pr(X \neq Y). \qquad (27.5)$$

Example 27.4.1 *Distance between Poisson Random Variables*

Consider Poisson random variables X and Y with means μ and $\omega > \mu$, respectively. It is well known that one can construct a copy of Y by adding to X an independent Poisson random variable Z with mean $\omega - \mu$. In this case the bound (27.5) gives

$$\|\pi_Y - \pi_X\|_{\mathrm{TV}} \;\; \leq \;\; \Pr(X + Z \neq X) \;\; = \;\; \Pr(Z \neq 0) \;\; = \;\; 1 - e^{-(\omega-\mu)}$$

for the distributions π_X and π_Y of X and Y. This calculation succeeds because X and Z live on the same probability space. ■

In the next example, we concoct a stopping time $T < \infty$ connected with a Markov chain X_n. By definition the event $\{T = n\}$ depends only on the outcomes of $X_1, \ldots, X_n$. If at epoch T the chain achieves its equilibrium distribution π and X_T is independent of T, then T is said to be a strong stationary time. When π is uniform, T is a strong uniform time. In either circumstance, the inequality

$$\|\pi_{X_n} - \pi\|_{\mathrm{TV}} \;\; \leq \;\; \Pr(T > n) \qquad (27.6)$$

holds, where π_{X_n} is the distribution of X_n. The key elements

$$\begin{aligned}
& |\Pr(X_n \in A) - \pi(A)| \\
= \;& |\Pr(X_n \in A, T \leq n) + \Pr(X_n \in A, T > n) - \pi(A)| \\
= \;& |\pi(A)\Pr(T \leq n) + \Pr(X_n \in A, T > n) \\
& -\pi(A)\Pr(T \leq n) - \pi(A)\Pr(T > n)| \\
\leq \;& \Pr(T > n)
\end{aligned}$$

of the proof parallel the proof of the earlier coupling bound (27.5).

Example 27.4.2 *Riffle Shuffle*

As an example of a strong uniform time, we turn briefly to card shuffling. In the inverse shuffling method conceptualized by Reeds [5], at every shuffle we imagine that each of c cards is assigned independently and uniformly to a top pile or a bottom pile. Hence, each pile has a binomial number of cards with mean $\frac{c}{2}$. The order of the two subpiles is kept consistent with the order of the parent pile, and in preparation for the next shuffle, the top pile is placed above the bottom pile. To keep track of the process, one can mark each card with a 0 (top pile) or 1 (bottom pile). Thus, shuffling induces an infinite binary sequence on each card that serves to track its fate. Let T denote the epoch when the first n digits for each card are unique. At T the cards reach a completely random state where all c permutations are equally likely. Let π be the uniform distribution and π_{X_n} be the distribution of the cards after n shuffles. The probability $\Pr(T \leq n)$ is the same as the probability that c balls (digit strings) dropped independently and uniformly into 2^n boxes all wind up in different boxes. It follows from inequality (27.6) that

$$\|\pi_{X_n} - \pi\|_{\text{TV}} \leq \Pr(T > n) = 1 - \Pr(T \leq n) = 1 - \prod_{i=1}^{c-1}\Big(1 - \frac{i}{2^n}\Big).$$

This bound converges rapidly to 0 and shows that 11 or fewer shuffles suffice for $c = 52$ cards. ■

27.5 Convergence Rates for Finite Chains

Although an ergodic chain converges to equilibrium, finding its rate of convergence is challenging. Example 27.4.2 is instructive because it constructs an explicit and natural bound. Unfortunately, it is often impossible to identify a strong stationary time. The best estimates of the rate of convergence rely on understanding the eigenstructure of the transition probability matrix P [5, 16]. We now discuss this approach for a reversible ergodic chain with a finite number of states and equilibrium distribution π. The inner products

$$\langle u, v\rangle_{1/\pi} = \sum_i u_i v_i \frac{1}{\pi_i}, \qquad \langle u, v\rangle_{\pi} = \sum_i u_i v_i \pi_i$$

feature prominently in our discussion.

For the chain in question, detailed balance translates into the condition

$$\sqrt{\pi_i} p_{ij} \frac{1}{\sqrt{\pi_j}} = \sqrt{\pi_j} p_{ji} \frac{1}{\sqrt{\pi_i}}. \tag{27.7}$$

If D is the diagonal matrix with ith diagonal entry $\sqrt{\pi_i}$, then the validity of equation (27.7) for all pairs (i, j) is equivalent to the symmetry of the matrix $Q = DPD^{-1}$. Let $Q = U\Lambda U^t$ be its spectral decomposition, where U is orthogonal, and Λ is diagonal with ith diagonal entry λ_i. One can rewrite the spectral decomposition as the sum of outer products

$$Q = \sum_i \lambda_i u^i (u^i)^t$$

using the columns u^i of U. The formulas $(u^i)^t u^j = 1_{\{i=j\}}$ and

$$Q^k = \sum_i \lambda_i^k u^i (u^i)^t$$

follow immediately. The formula for Q^k in turn implies

$$P^k = \sum_i \lambda_i^k D^{-1} u^i (u^i)^t D = \sum_i \lambda_i^k w^i v^i, \tag{27.8}$$

where $v^i = (u^i)^t D$ and $w^i = D^{-1} u^i$.

Rearranging the identity $DPD^{-1} = Q = U\Lambda U^t$ yields $U^t DP = \Lambda U^t D$. Hence, the rows v^i of $V = U^t D$ are row eigenvectors of P. These vectors satisfy the orthogonality relations

$$\langle v^i, v^j \rangle_{1/\pi} = v^i D^{-2} (v^j)^t = (u^i)^t u^j = 1_{\{i=j\}}$$

and therefore form a basis of the inner product space $\ell^2_{1/\pi}$. The identity $PD^{-1}U = D^{-1}U\Lambda$ shows that the columns w^j of $W = D^{-1}U$ are column eigenvectors of P. These dual vectors satisfy the orthogonality relations

$$\langle w^i, w^j \rangle_\pi = (w^i)^t D^2 w^j = (u^i)^t u^j = 1_{\{i=j\}}$$

and therefore form a basis of the inner product space ℓ^2_π. Finally, we have the biorthogonality relations

$$v^i w^j = 1_{\{i=j\}}$$

under the ordinary inner product. The trivial rescalings $w^i = D^{-2}(v^i)^t$ and $(v^i)^t = D^2 w^i$ allow one to pass back and forth between row eigenvectors and column eigenvectors.

The distance from equilibrium in the $\ell^2_{1/\pi}$ norm bounds the total variation distance from equilibrium in the sense that

$$\|\mu - \pi\|_{\mathrm{TV}} \leq \frac{1}{2}\|\mu - \pi\|_{1/\pi}. \tag{27.9}$$

This is just a restatement of inequality (27.4). With the understanding that $\lambda_1 = 1$, $v^1 = \pi$, and $w^1 = \mathbf{1}$, the next proposition provides an even more basic bound.

Proposition 27.5.1 *An initial distribution μ for a reversible ergodic chain with m states satisfies*

$$\begin{aligned} \|\mu P^k - \pi\|_{\chi^2}^2 &= \sum_{j=2}^{m} \lambda_j^{2k}[(\mu-\pi)w^j]^2 &(27.10)\\ &\le \rho^{2k}\|\mu-\pi\|_{1/\pi}^2, &(27.11)\end{aligned}$$

where $\rho < 1$ is the absolute value of the second-largest eigenvalue in magnitude of the transition probability matrix P.

Proof: According to Sections 6.5.3 and 25.2, $\rho < 1$. In view of the identity $\pi P = \pi$, the expansion (27.8) gives

$$\begin{aligned} \|\mu P^k - \pi\|_{1/\pi}^2 &= \|(\mu-\pi)P^k\|_{1/\pi}^2 \\ &= (\mu-\pi)\sum_{i=1}^{m}\lambda_i^k w^i v^i D^{-2}\sum_{j=1}^{m}\lambda_j^k (v^j)^t (w^j)^t(\mu-\pi)^t \\ &= (\mu-\pi)\sum_{i=1}^{m}\lambda_i^{2k} w^i (w^i)^t(\mu-\pi)^t \\ &= \sum_{i=1}^{m}\lambda_i^{2k}\left[(\mu-\pi)w^i\right]^2. \end{aligned}$$

The two constraints $\sum_j \pi_j = \sum_j \mu_j = 1$ clearly imply $(\mu-\pi)w^1 = 0$. Equality (27.10) follows immediately. Because all remaining eigenvalues satisfy $|\lambda_j| \le \rho$, one can show by similar reasoning that

$$\begin{aligned} \sum_{j=2}^{m}\lambda_j^{2k}\left[(\mu-\pi)w^i\right]^2 &\le \rho^{2k}\sum_{j=1}^{m}\left[(\mu-\pi)w^i\right]^2 \\ &= \rho^{2k}\|\mu-\pi\|_{1/\pi}^2. \end{aligned}$$

This validates inequality (27.11). ■

If a nonreversible finite-state chain X_n is ergodic, then it still satisfies the bound

$$\|\pi^n - \pi\|_{\mathrm{TV}} \le c\rho^n$$

for some positive constant c. Many infinite-state chains also satisfy this condition; they are said to be uniformly ergodic. The central limit theorem holds for uniformly ergodic chains. Suppose $f(X_n)$ is square integrable under π with mean μ and variance σ_0. If we define the autocovariance sequence $\sigma_m = \mathrm{Cov}(X_n, X_{n+m})$, then the central limit theorem asserts that the adjusted sample averages

$$\frac{1}{\sqrt{n}}\sum_{m=0}^{n-1}(X_m - \mu)$$

tend in distribution to a normal random variable with mean 0 and variance $\sigma = \sigma_0 + 2\sum_{m=1}^{\infty} \sigma_m$. Further discussion of these results appear in the references [2, 10, 20].

27.6 Convergence of the Independence Sampler

For the independence sampler, it is possible to give a coupling bound on the rate of convergence to equilibrium [12]. Suppose that $X_0, X_1, \ldots$ represents the sequence of states visited by the independence sampler starting from $X_0 = x_0$. We couple this Markov chain to a second independence sampler $Y_0, Y_1, \ldots$ starting from the equilibrium distribution π. By definition, each Y_k has distribution π. The two chains are coupled by a common proposal stage and a common uniform deviate U sampled in deciding whether to accept the common proposed point. They differ in having different acceptance probabilities. If $X_n = Y_n$ for some n, then $X_k = Y_k$ for all $k \geq n$. Let T denote the random epoch when X_n first meets Y_n and the X chain attains equilibrium.

The importance ratios $w_j = \pi_j/q_j$ determine what proposed points are accepted. Without loss of generality, assume that the states of the chain are numbered $1, \ldots, m$ and that the importance ratios w_i are in decreasing order. If $X_n = x \neq y = Y_n$, then according to equation (26.1) the next proposed point is accepted by both chains with probability

$$\sum_{j=1}^{m} q_j \min\left\{\frac{w_j}{w_x}, \frac{w_j}{w_y}, 1\right\} = \sum_{j=1}^{m} \pi_j \min\left\{\frac{1}{w_x}, \frac{1}{w_y}, \frac{1}{w_j}\right\} \geq \frac{1}{w_1}.$$

In other words, at each trial the two chains meet with at least probability $1/w_1$. This translates into the tail probability bound

$$\Pr(T > n) \leq \left(1 - \frac{1}{w_1}\right)^n$$

and ultimately via inequality (27.6) into the bound

$$\|\pi^n - \pi\|_{\mathrm{TV}} \leq \left(1 - \frac{1}{w_1}\right)^n \qquad (27.12)$$

on the total variation distance of X_n from equilibrium.

It is interesting to compare this last bound with the bound entailed by Proposition 27.5.1. Based on our assumption that the importance ratios are decreasing, equation (26.5) shows that the transition probabilities of the independence sampler are

$$p_{ij} = \begin{cases} q_j & j < i \\ \pi_j/w_i & j > i\,. \end{cases}$$

In order for $\sum_j p_{ij} = 1$, we must set $p_{ii} = q_i + \lambda_i$, where

$$\lambda_i = \sum_{k=i}^{m}\left(q_k - \frac{\pi_k}{w_i}\right) = \sum_{k=i+1}^{m}\left(q_k - \frac{\pi_k}{w_i}\right).$$

With these formulas in mind, one can decompose the overall transition probability matrix as $P = U + \mathbf{1}q$, where $q = (q_1, \ldots, q_m)$ and U is the upper triangular matrix

$$U = \begin{pmatrix} \lambda_1 & \frac{q_2(w_2-w_1)}{w_1} & \cdots & \cdots & \frac{q_{m-1}(w_{m-1}-w_1)}{w_1} & \frac{q_m(w_m-w_1)}{w_1} \\ \vdots & \vdots & \vdots & \ddots & \vdots & \vdots \\ 0 & 0 & 0 & \cdots & \lambda_{m-1} & \frac{q_m(w_m-w_{m-1})}{w_{m-1}} \\ 0 & 0 & 0 & \cdots & 0 & \lambda_m \end{pmatrix}.$$

The eigenvalues of U are just its diagonal entries λ_1 through λ_m.

The reader can check that (a) $\lambda_1 = 1 - 1/w_1$, (b) the λ_i are decreasing, and (c) $\lambda_m = 0$. It turns out that P and U share most of their eigenvalues. They differ in the eigenvalue attached to the eigenvector $\mathbf{1}$ since $P\mathbf{1} = \mathbf{1}$ and $U\mathbf{1} = \mathbf{0}$. Suppose $Uv = \lambda_i v$ for some i between 1 and $m-1$. Let us construct a column eigenvector of P with the eigenvalue λ_i. As a trial eigenvector we take $v + c\mathbf{1}$ and calculate

$$(U + \mathbf{1}q)(v + c\mathbf{1}) = \lambda_i v + qv\mathbf{1} + c\mathbf{1} = \lambda_i v + (qv + c)\mathbf{1}.$$

This is consistent with $v + c\mathbf{1}$ being an eigenvector provided we choose the constant c to satisfy $qv + c = \lambda_i c$. Because $\lambda_i \neq 1$, it is always possible to do so. The combination of Proposition 27.5.1 and inequality (27.9) gives a bound that decays at the same geometric rate $\lambda_1 = 1 - w_1^{-1}$ as the coupling bound (27.12). Thus, the coupling bound is about as good as one could hope for. Problems 11 and 12 ask the reader to flesh out our convergence arguments.

27.7 Operators and Markov Chains

Classical functional analysis is the study of linear algebra on infinite-dimensional spaces [4, 17]. The topology of such a space is usually determined by a norm or an inner product giving rise to a norm. Assuming all Cauchy sequences converge, the space is called a Banach space in the former case and a Hilbert space in the latter case. Hilbert spaces have the richer structure and will be the object of study in this section. Chapters 6 and 17 summarize background material on norms, inner products, and Hilbert spaces. All Hilbert spaces in this section and the next will be real and separable. Recall that the norm associated with an inner product is defined by

$$\|x\| = \sqrt{\langle x, x\rangle} = \sup_{\|y\|=1} \langle x, y\rangle.$$

The second of these two definitions is a consequence of the Cauchy-Schwarz inequality and the choice $y = x/\|x\|$.

Linear operators and linear functionals are of paramount importance. Linear operators are the analogs of matrices; symmetric operators are the analogs of symmetric matrices. For the sake of simplicity, all of the operators we consider will be linear and continuous. A continuous linear operator is necessarily bounded and vice versa. The adjoint (transpose) T^* of an operator T satisfies $\langle Ty, x\rangle = \langle y, T^*x\rangle$ for all x and y. It follows that $T^{**} = T$ and that the kernel of T^* is perpendicular to the range of T. When $T^* = T$, the operator is symmetric. The norm of an operator T between two Hilbert spaces is defined via

$$\|T\| = \sup_{\|x\|=1} \|Tx\| = \sup_{\|x\|=1} \sup_{\|y\|=1} \langle Tx, y\rangle. \tag{27.13}$$

The second of these definitions makes it clear that $\|T^*\| = \|T\|$. The spectral radius of an operator T sending a Hilbert space into itself is the number $\rho = \lim_{n\to\infty} \|T^n\|^{1/n}$. If λ is an eigenvalue of T, then $\lambda \le \rho$. This follows from the inequality $|\lambda|^n\|v\| = \|T^n v\| \le \|T\|^n\|v\|$ for the associated eigenvector v. A symmetric operator has real eigenvalues and orthogonal eigenvectors corresponding to distinct eigenvalues.

We will deal almost exclusively with the Hilbert space $\mathrm{L}^2(\pi)$ of square integrable random variables relative to a probability distribution π. In many applications the subspace $\mathrm{L}_0^2(\pi)$ of random variables with zero means is more relevant. On this subspace, $\|h\|^2$ represents the variance of the random variable h, and $\langle g, h\rangle$ represents the covariance of the random variables g and h. Given a Markov chain X_n with transition probability matrix $P = (p_{ij})$ and equilibrium distribution π, one can define two operators summarizing the action of the chain. The forward and backward operators take a function $h \in \mathrm{L}^2(\pi)$ to the functions with components

$$F(h)_i = \sum_j h_j p_{ij}, \quad B(h)_j = \sum_i h_i p_{ij} \frac{\pi_i}{\pi_j},$$

respectively. Our discrete notation here should not obscure the fact that these operators are defined on arbitrary probability spaces. Despite their abstract definitions, the forward and backward operators are nothing more than the conditional expectations

$$(Fh)_i = \mathrm{E}[h(X_1) \mid X_0 = i], \quad (Bh)_j = \mathrm{E}[h(X_0) \mid X_1 = j].$$

Simple induction demonstrates that the powers of these operators

$$(F^n h)_i = \mathrm{E}[h(X_n) \mid X_0 = i], \quad (B^n h)_j = \mathrm{E}[h(X_0) \mid X_n = j]$$

are also conditional expectations.

These interpretations show that the forward and backward operators preserve the mean value $\sum_i h_i\pi_i$ of a function. Hence, both operators map $\mathrm{L}_0^2(\pi)$ into itself. The Cauchy-Schwarz inequalities

$$\begin{aligned}\sum_i (Fh)_i^2\pi_i &\le \sum_i\Big(\sum_j h_j^2 p_{ij}\pi_i\Big)\Big(\sum_j \frac{p_{ij}}{\pi_i}\Big)\pi_i = \sum_j h_j^2\pi_j \\ \sum_j (Bh)_j^2\pi_j &\le \sum_j\Big(\sum_i h_i^2\frac{p_{ij}\pi_i}{\pi_j}\Big)\Big(\sum_i \frac{p_{ij}\pi_i}{\pi_j}\Big)\pi_j = \sum_i h_i^2\pi_i\end{aligned}$$

prove that both operators are norm decreasing. The two operators are adjoint to one another because

$$\langle Fg, h\rangle_\pi = \sum_i\sum_j g_j p_{ij} h_i \pi_i = \sum_j g_j \sum_i h_i \frac{p_{ij}\pi_i}{\pi_j}\pi_j = \langle g, Bh\rangle_\pi.$$

When the chain satisfies the detailed balance condition $\pi_i p_{ij} = \pi_j p_{ji}$, the identity

$$\sum_j h_j p_{ij} = \sum_j h_j \frac{p_{ji}\pi_j}{\pi_i}$$

demonstrates that the operators coincide and define a single symmetric operator.

If a reversible Markov chain X_n is already at equilibrium, then the autocovariance sequence $\sigma_n = \mathrm{Cov}_\pi[h(X_0), h(X_n)]$ defined by a function h in $\mathrm{L}_0^2(\pi)$ is decreasing in n. To check this claim, suppose first that $n = 2m$. Then the equality $F = B$ yields

$$\begin{aligned}\mathrm{Cov}_\pi[h(X_0), h(X_n)] &= \mathrm{E}_\pi[h(X_0)h(X_n)] \\ &= \mathrm{E}_\pi\{\mathrm{E}[h(X_0)\mid X_m]\,\mathrm{E}[h(X_{2m})\mid X_m]\} \\ &= \langle B^m h, F^m h\rangle \\ &= \|F^m h\|^2 \\ &\le \|F\|^2\|F^{m-1}h\|^2,\end{aligned}$$

which is decreasing in m. On the other hand, when $n = 2m+1$, we have

$$\mathrm{Cov}_\pi[h(X_0), h(X_n)] = \langle F^m h, F^{m+1}h\rangle \le \|F\|\|F^m h\|^2$$

by virtue of definition (27.13).

To summarize, the forward and backward operators characterize the evolution of a Markov chain. The rate of convergence of the chain to equilibrium is determined by the common spectral radius of the two operators. Based on equation (27.13), this can be defined in terms of the maximal correlations

$$\gamma_n = \sup_{\mathrm{Var}_\pi(g)=1}\ \sup_{\mathrm{Var}_\pi(h)=1} \mathrm{Cov}_\pi[g(X_0), h(X_n)]$$

as $\lim_{n\to\infty} \gamma_n^{1/n}$. The chi-square inequality (27.4) offers a less precise but often more computable bound. In the chi-square inequality, we take ν to be the distribution π_{X_n} of X_n starting from an arbitrary initial distribution. Of course to be of any use, the chi-square distance must be finite.

27.8 Compact Operators and Gibbs Sampling

The spectral decomposition for symmetric matrices carries over to symmetric operators. The most natural generalization involves a discrete spectrum and hinges on the notion of a compact operator. Any such linear operator T is the limit of a sequence of operators T_n with finite-dimensional ranges. By limit we mean that $\lim_{n\to\infty} \|T - T_n\| = 0$ in the operator norm. The next proposition summarizes a few key facts.

Proposition 27.8.1 *Suppose T is a continuous linear operator from a Hilbert space $\mathcal{H}_1$ to a Hilbert space $\mathcal{H}_2$. Then the following properties hold:*

(a) If the operator $S : \mathcal{H}_0 \mapsto \mathcal{H}_1$ is compact, then TS is compact. If the operator $S : \mathcal{H}_2 \mapsto \mathcal{H}_3$ is compact, then ST is compact.

(b) If T_n is a sequence of compact operators with limit T, then T is compact as well.

(c) The adjoint operator T^* is compact whenever T is compact.

(d) The operator T is compact if and only if for every bounded sequence of vectors f_n from $\mathcal{H}_1$, there exists a subsequence f_{n_m} such that Tf_{n_m} converges to a vector g of $\mathcal{H}_2$.

Proof: Straightforward proofs can be found in the references [1, 17, 18]. ■

Property (d) is often taken as the definition of a compact operator. Based on these results, one can prove the spectral decomposition.

Proposition 27.8.2 *Suppose T is a compact symmetric operator on the Hilbert space $\mathcal{H}$. Then there exists an orthonormal basis $\{v^n\}_n$ consisting entirely of eigenvectors of T. The corresponding eigenvalues $\{\lambda_n\}_n$ are real and satisfy $\lim_{n\to\infty} \lambda_n = 0$. Conversely, if an operator T is diagonalizable in this sense, then it is compact and symmetric.*

Proof: See the references [1, 17, 18] for proofs. ■

The forward and backward operators are not the only operators of interest in MCMC. Gibbs sampling suggests two novel operators. Let r_{ij} denote the density of the bivariate random vector $X = (X_1, X_2)$. The corresponding marginal distributions are $\mu_i = \sum_j r_{ij}$ and $\nu_j = \sum_i r_{ij}$. The conditional distribution of X_2 given $X_1 = i$ is r_{ij}/μ_i; similarly, the conditional distribution of X_1 given $X_2 = j$ is r_{ij}/ν_j. Marginally, Gibbs sampling can be described by the two kernels

$$\begin{aligned}
\Pr(X_1^{n+1} = j \mid X_1^n = i) &= p_{ij} = \sum_k \frac{r_{ik} r_{jk}}{\mu_i \nu_k} \\
\Pr(X_2^{n+1} = j \mid X_2^n = i) &= q_{ij} = \sum_k \frac{r_{ki} r_{kj}}{\nu_i \mu_k}.
\end{aligned}$$

These kernels satisfy detailed balance because

$$\begin{aligned}
\mu_i p_{ij} &= \sum_k \frac{r_{ik} r_{jk}}{\nu_k} = \sum_k \frac{r_{jk} r_{ik}}{\nu_k} = \mu_j p_{ji} \\
\nu_i q_{ij} &= \sum_k \frac{r_{ki} r_{kj}}{\mu_k} = \sum_k \frac{r_{kj} r_{ki}}{\mu_k} = \nu_j q_{ji}.
\end{aligned}$$

One can define two associated operators

$$(Tf)_j = \sum_i f_i \frac{r_{ij}}{\nu_j} = \sum_i f_i \frac{r_{ij}}{\mu_i \nu_j} \mu_i \tag{27.14}$$

$$(T^* g)_i = \sum_j g_j \frac{r_{ij}}{\mu_i} = \sum_j g_j \frac{r_{ij}}{\mu_i \nu_j} \nu_j \tag{27.15}$$

that are formally adjoints of one another. This fact follows from the identities

$$\langle Tf, g \rangle_\nu = \sum_j \sum_i f_i \frac{r_{ij}}{\nu_j} g_j \nu_j = \sum_i f_i \sum_j g_j \frac{r_{ij}}{\mu_i} \mu_i = \langle f, T^* g \rangle_\mu.$$

In order for the maps to preserve square integrability, we must constrain the joint density r_{ij}. The Cauchy-Schwarz inequalities

$$\begin{aligned}
\sum_j \Big(\sum_i f_i \frac{r_{ij}}{\nu_j} \Big)^2 \nu_j &\le \|f\|_\mu^2 \sum_j \sum_i \frac{r_{ij}^2}{\mu_i \nu_j} \\
\sum_i \Big(\sum_j g_j \frac{r_{ij}}{\mu_i} \Big)^2 \mu_i &\le \|g\|_\nu^2 \sum_i \sum_j \frac{r_{ij}^2}{\mu_i \nu_j}
\end{aligned}$$

suggest a single sufficient condition, which can be restated as the trace requirement

$$\sum_i p_{ii} = \sum_i q_{ii} = \sum_i \sum_j \frac{r_{ij}^2}{\mu_i \nu_j} < \infty. \tag{27.16}$$

The trace condition is a consequence of either of the stronger conditions

$$\begin{aligned}
\sup_i \frac{p_{ii}}{\mu_i} &= \sup_i \sum_j \Big(\frac{r_{ij}}{\mu_i} \Big)^2 \frac{1}{\nu_j} < \infty \\
\sup_j \frac{q_{jj}}{\nu_j} &= \sup_j \sum_i \Big(\frac{r_{ij}}{\nu_j} \Big)^2 \frac{1}{\mu_i} < \infty.
\end{aligned}$$

In particular, if either marginal chain is finite, then the trace condition holds.

The operators T^*T and TT^* are

$$\begin{aligned}
(T^*Tf)_k &= \sum_j \sum_i f_i \frac{r_{ij}}{\nu_j} \frac{r_{kj}}{\mu_k} = \sum_i f_i \sum_j \frac{r_{ij}}{\mu_k} \frac{r_{kj}}{\nu_j} = \sum_i f_i p_{ik} \frac{\mu_i}{\mu_k} \\
(TT^*f)_k &= \sum_i \sum_j f_j \frac{r_{ij}}{\mu_i} \frac{r_{ik}}{\nu_k} = \sum_j f_j \sum_i \frac{r_{ik}}{\nu_k} \frac{r_{ij}}{\mu_i} = \sum_j q_{kj} f_j.
\end{aligned}$$

The first of these is the backward operator associated with $L^2(\mu)$, and the second is the forward operator associated with $L^2(\nu)$. In view of reversibility, each operator is symmetric and agrees with its forward or backward counterpart. Furthermore, each operator possesses only nonnegative eigenvalues. For example, if $TT^*f = \lambda f$, then this claim follows from the identity

$$\lambda \|f\|_\mu^2 = \langle TT^*f, f\rangle_\mu = \langle T^*f, T^*f\rangle_\nu = \|T^*f\|_\nu^2.$$

In view of parts (a) and (c) of Proposition 27.8.1, if we can show that T is compact, then the three operators T^*, T^*T, and TT^* are compact as well. It suffices to identify T as the limit of a sequence T_k of operators with finite-dimensional ranges. According to the trace criterion (27.16), the function $r_{ij}/(\mu_i \nu_j)$ belongs to $L^2(\mu \times \nu)$. If $\{u^m\}_m$ is an orthonormal basis of $L^2(\mu)$, and $\{v^n\}_n$ is an orthonormal basis of $L^2(\nu)$, then Problem 18 demonstrates that the product collection $\{u^m v^n\}_{mn}$ is an orthonormal basis of $L^2(\mu \times \nu)$. The expansion

$$\frac{r_{ij}}{\mu_i \nu_j} = \sum_m \sum_n c_{mn} u_i^m v_j^n$$

suggests that we define T_k via integration against the function

$$w_{ij}^k = \sum_{|m|+|n|\le k} c_{mn} u_i^m v_j^n.$$

The representation

$$(T_k f)_j = \sum_i f_i w_{ij}^k \mu_i = \sum_{|m|+|n|\le k} c_{mn} \Big(\sum_i f_i u_i^m \mu_i \Big) v_j^n$$

proves that the range of T_k is contained in the finite-dimensional subspace with basis $\{v^n\}_{|n|\le k}$. Furthermore, the Cauchy-Schwarz inequality

$$\begin{aligned}
\|Tf - T_k f\|_\nu^2 &= \sum_j \Big[\sum_i f_i \Big(\frac{r_{ij}}{\mu_i \nu_j} - w_{ij}^k \Big) \mu_i \Big]^2 \nu_j \\
&\le \|f\|_\mu^2 \sum_i \sum_j \Big(\frac{r_{ij}}{\mu_i \nu_j} - w_{ij}^k \Big)^2 \mu_i \nu_j.
\end{aligned}$$

proves that $\|T - T_k\|^2$ tends to 0. Hence, T and therefore T^*, T^*T, and TT^* are all compact operators.

Given the trace condition (27.16), Proposition 27.8.2 implies that T^*T and TT^* possess spectral decompositions. Suppose T^*T has orthonormal eigenvectors u^n with nontrivial associated eigenvalues β_n^2. The vectors Tu^n are orthogonal because

$$\langle Tu^m, Tu^n \rangle \quad = \quad \langle u^m, T^*Tu^n \rangle \quad = \quad \beta_n^2 \langle u^m, u^n \rangle.$$

Therefore, the vectors v^n defined by $Tu^n = \beta_n v^n$ are orthonormal. In addition, these vectors satisfy the identities $T^*v^n = \beta_n u^n$ and $TT^*v^n = \beta_n^2 v^n$. Eigenvectors u^n and v^n with eigenvalue 0 get mapped into null vectors under T and T^*. These are precisely the properties characterizing the singular value decomposition of linear algebra.

27.9 Convergence Rates for Gibbs Sampling

Before we tackle some concrete examples, it is helpful to comment on the relationships between the metrics defined for the joint chain and the marginal chains in Gibbs sampling. There are two versions of the joint chain, depending on whether we resample X_1 or X_2 first. These versions have the kernels

$$\begin{aligned} s_{i_1 j_1, i_2 j_2} &= \Pr(X_2^1 = j_2 \mid X_1^0 = i_1) \Pr(X_1^1 = i_2 \mid X_2^1 = j_2) = \frac{r_{i_1 j_2}}{\mu_{i_1}} \frac{r_{i_2 j_2}}{\nu_{j_2}}, \\ t_{i_1 j_1, i_2 j_2} &= \Pr(X_1^1 = i_2 \mid X_2^0 = j_1) \Pr(X_2^1 = j_2 \mid X_1^1 = i_2) = \frac{r_{i_2 j_1}}{\nu_{j_1}} \frac{r_{i_2 j_2}}{\mu_{i_2}}. \end{aligned}$$

We will write $s_{i_1 j_1, i_2 j_2}^n$ and $t_{i_1 j_1, i_2 j_2}^n$ for the corresponding n-step kernels. Let $\chi_p^2(n)$, $\chi_q^2(n)$, $\chi_s^2(n)$, and $\chi_t^2(n)$ denote the chi-square distances from equilibrium after n steps for the chains with kernels p, q, s, and t, respectively. The following bounds are hardly surprising since the marginal chains at step $n-1$ determine the joint chains at step n, which in turn determine the marginal chains at step n.

Proposition 27.9.1 *In the above notation we have the bounds*

$$\chi_p^2(n) \quad \leq \quad \chi_s^2(n) \quad \leq \quad \chi_p^2(n-1) \tag{27.17}$$
$$\chi_q^2(n) \quad \leq \quad \chi_t^2(n) \quad \leq \quad \chi_q^2(n-1). \tag{27.18}$$

Similar inequalities hold for the corresponding total variation distances.

Proof: First note that

$$\frac{t_{i_1 j_1, i_2 j_2}^n}{r_{i_2 j_2}} - 1 \quad = \quad \sum_k q_{j_1 k}^{n-1} \frac{r_{i_2 k}}{\nu_k} \frac{1}{\mu_{i_2}} - \sum_k \frac{r_{i_2 k}}{\mu_{i_2}} \quad = \quad \sum_k \frac{r_{i_2 k}}{\mu_{i_2}} \Big(\frac{q_{j_1 k}^{n-1}}{\nu_k} - 1 \Big).$$

Hence, Jensen's inequality implies that

$$\begin{aligned}
\sum_{i_2 j_2}\Big(\frac{t^n_{i_1j_1,i_2j_2}}{r_{i_2j_2}}-1\Big)^2 r_{i_2j_2} &= \sum_{i_2j_2}\Big[\sum_k \frac{r_{i_2k}}{\mu_{i_2}}\Big(\frac{q^{n-1}_{j_1k}}{\nu_k}-1\Big)\Big]^2 r_{i_2j_2} \\
&= \sum_{i_2}\Big[\sum_k \frac{r_{i_2k}}{\mu_{i_2}}\Big(\frac{q^{n-1}_{j_1k}}{\nu_k}-1\Big)\Big]^2 \mu_{i_2} \\
&\le \sum_{i_2}\sum_k \frac{r_{i_2k}}{\mu_{i_2}}\Big(\frac{q^{n-1}_{j_1k}}{\nu_k}-1\Big)^2 \mu_{i_2} \\
&= \sum_{i_2}\sum_k \frac{r_{i_2k}}{\nu_k}\Big(\frac{q^{n-1}_{j_1k}}{\nu_k}-1\Big)^2 \nu_k \\
&= \sum_k \Big(\frac{q^{n-1}_{j_1k}}{\nu_k}-1\Big)^2 \nu_k .
\end{aligned}$$

The first quantity in this string is $\chi^2_t(n)$, and the last quantity is $\chi^2_q(n-1)$.

For the lower bound, invoking Jensen's inequality yields

$$\begin{aligned}
\sum_{i_2 j_2}\Big(\frac{t^n_{i_1j_1,i_2j_2}}{r_{i_2j_2}}-1\Big)^2 r_{i_2j_2} &= \sum_{j_2}\sum_{i_2}\Big(\frac{t^n_{i_1j_1,i_2j_2}}{r_{i_2j_2}}-1\Big)^2 \frac{r_{i_2j_2}}{\nu_{j_2}}\nu_{j_2} \\
&\ge \sum_{j_2}\Big[\sum_{i_2}\Big(\frac{t^n_{i_1j_1,i_2j_2}}{r_{i_2j_2}}-1\Big)\frac{r_{i_2j_2}}{\nu_{j_2}}\Big]^2 \nu_{j_2} \\
&= \sum_{j_2}\Big(\frac{q^n_{j_1j_2}}{\nu_{j_2}}-1\Big)^2 \nu_{j_2} .
\end{aligned}$$

The first quantity in this string is again $\chi^2_t(n)$, and the last quantity is now $\chi^2_q(n)$. The inequalities (27.17) involving $\chi^2_p(n)$, $\chi^2_s(n)$, and $\chi^2_p(n-1)$ are handled in the same manner. ■

The combination of Propositions 27.5.1 and 27.9.1 with inequality (27.4) gives precise information on the rate of convergence of a Gibbs sampler to equilibrium. Unfortunately, diagonalizing the corresponding operators is impossible in most cases. In a few simple examples, the following proposition provides the key.

Proposition 27.9.2 *Suppose the marginal distributions in Gibbs sampling are one-dimensional and satisfy the conditions of Proposition 17.4.1. This condition guarantees the existence of two polynomial sequences* $\{p_n(x_1)\}_n$ *and* $\{q_n(x_2)\}_n$ *furnishing orthonormal bases of* $L^2(\mu)$ *and* $L^2(\nu)$. *In addition suppose the conditional expectations* $\mathrm{E}[f(X_1) \mid X_2]$ *and* $\mathrm{E}[g(X_2) \mid X_1]$ *send polynomial functions* $f(x_1)$ *and* $g(x_2)$ *of degree* n *into polynomial functions of degree* n *in the dual variables* x_2 *and* x_1. *Then there exist constants*

γ_n and δ_n such that

$$\begin{aligned} \mathrm{E}[p_n(X_1) \mid X_2] &= \gamma_n q_n(X_2) & (27.19) \\ \mathrm{E}[q_n(X_2) \mid X_1] &= \delta_n p_n(X_1). & (27.20) \end{aligned}$$

As a consequence, the X_1 chain has eigenvectors $p_n(x_1)$ and corresponding nonnegative eigenvalues $\theta_n = \gamma_n\delta_n$. Likewise, the X_2 chain has eigenvectors $q_n(x_2)$ and the same eigenvalues.

Proof: The conditional expectations $\mathrm{E}[f(X_1) \mid X_2]$ and $\mathrm{E}[g(X_2) \mid X_1]$ correspond to the operators Tf and T^*g in equations (27.14) and (27.15). The identities (27.19) and (27.20) show that $\{p_n(x_1)\}_n$ and $\{q_n(x_2)\}_n$ diagonalize the symmetric operators T^*T and TT^*. Hence, it suffices to verify these two identities. Because $p_0(x_1) = q_0(x_2) = 1$, they are true for the choices $\gamma_0 = \delta_0 = 1$ when $n = 0$. For $0 \le m < n$, we have the full expectation

$$\mathrm{E}[X_1^m q_n(X_2)] = \mathrm{E}\{\mathrm{E}[X_1^m \mid X_2] q_n(X_2)\} = \mathrm{E}[f_m(X_2) q_n(X_2)],$$

where $f_m(x_2)$ is a polynomial of degree m. In view of the orthogonality of the sequence $q_n(x_2)$, the expectation $\mathrm{E}[X_1^m q_n(X_2)]$ vanishes. We can reinterpret this result as

$$\mathrm{E}[X_1^m q_n(X_2)] = \mathrm{E}\{X_1^m \,\mathrm{E}[q_n(X_2) \mid X_1]\} = 0.$$

Because the conditional expectation $\mathrm{E}[q_n(X_2) \mid X_1]$ is both a polynomial of degree n and orthogonal to all polynomials of degree $m < n$, it reduces to a multiple of $p_n(X_1)$. Similar reasoning applies to the conditional expectation $\mathrm{E}[p_n(X_1) \mid X_2]$. ■

Example 27.9.1 *Poisson-Gamma*

Consider a Poisson likelihood with a gamma prior. Replace X_1 by Y and X_2 by λ in our earlier notation. The joint and marginal densities are

$$\begin{aligned} r_{y\lambda} &= \frac{\lambda^y}{y!} e^{-\lambda} \frac{\alpha(\alpha\lambda)^{\beta-1}}{\Gamma(\beta)} e^{-\alpha\lambda} \\ \nu_\lambda &= \frac{\alpha(\alpha\lambda)^{\beta-1}}{\Gamma(\beta)} e^{-\alpha\lambda} \\ \mu_y &= \frac{\alpha^\beta}{y!\Gamma(\beta)} \int_0^\infty \lambda^{\beta+y-1} e^{-(\alpha+1)\lambda} d\lambda \\ &= \frac{\Gamma(\beta+y)}{y!\Gamma(\beta)} \Big(\frac{\alpha}{\alpha+1}\Big)^\beta \Big(\frac{1}{\alpha+1}\Big)^y. \end{aligned}$$

The trace criterion for compactness is

$$\sum_{y=0}^\infty \int_0^\infty \frac{r_{y\lambda}^2}{\mu_y \nu_\lambda} d\lambda = \sum_{y=0}^\infty \frac{(\alpha+1)^{\beta+y}}{y!\Gamma(\beta+y)} \int_0^\infty \lambda^{\beta+2y-1} e^{-(\alpha+2)\lambda} d\lambda$$

$$= \sum_{y=0}^{\infty} \frac{\Gamma(\beta+2y)}{y!\Gamma(\beta+y)} \frac{(\alpha+1)^{\beta+y}}{(\alpha+2)^{\beta+2y}}.$$

If we denote the general term of this series by c_y, then the limiting ratio

$$\lim_{y\to\infty} \frac{c_{y+1}}{c_y} = \frac{4(\alpha+1)}{(\alpha+2)^2} < 1$$

for $\alpha > 0$ is enough to guarantee convergence.

The two operators $(Tf)(\lambda)$ and $(T^*g)(y)$ are simply the conditional expectations

$$\begin{aligned} \mathrm{E}[f(Y) \mid \lambda] &= \sum_{y=0}^{\infty} f(y)\frac{\lambda^y}{y!}e^{-\lambda} \\ \mathrm{E}[g(\lambda) \mid Y] &= \frac{(\alpha+1)^{\beta+y}}{\Gamma(\beta+y)} \int_0^\infty g(\lambda)\lambda^{\beta+y-1}e^{-(\alpha+1)\lambda}d\lambda. \end{aligned}$$

The choices $f(y) = y(y-1)\cdots(y-n+1)$ and $g(\lambda) = \lambda^n$ produce

$$\mathrm{E}[f(Y) \mid \lambda] = \lambda^n, \quad \mathrm{E}[g(\lambda) \mid Y] = \frac{\Gamma(\beta+Y+n)}{\Gamma(\beta+Y)(\alpha+1)^n}. \quad (27.21)$$

Hence, the two operators send polynomials of degree n into polynomials of degree n.

At this stage, it is worth noting that the proof of Proposition 27.9.2 does not actually require polynomials of unit norm. Only the orthogonality of the $p_n(x_1)$ and $q_n(x_2)$ is invoked. It is now convenient to assume that these polynomials are monic in the sense that their leading coefficients are 1. This change makes it possible to recover the factors γ_n and δ_n by taking the conditional expectation of any monic polynomial of degree n and extracting the leading coefficient of the result. For instance, the choices $f(y) = y(y-1)\cdots(y-n+1)$ and $g(\lambda) = \lambda^n$ in the identities (27.21) yield the conclusions $\gamma_n = 1$ and $\delta_n = (\alpha+1)^{-n}$. Hence, $\theta_n = (\alpha+1)^{-n}$, and the second-largest eigenvalue of either marginal chain is $\rho = (\alpha+1)^{-1}$. The inequalities (27.9) and (27.10) therefore imply that

$$\begin{aligned} \|\mu^n - \mu\|_{\mathrm{TV}} &\le \frac{1}{2}\|\mu^n - \mu\|_{\chi^2} \le \frac{1}{2(\alpha+1)^n}\|\mu^0 - \mu\|_{1/\mu} \\ \|\nu^n - \nu\|_{\mathrm{TV}} &\le \frac{1}{2}\|\nu^n - \nu\|_{\chi^2} \le \frac{1}{2(\alpha+1)^n}\|\nu^0 - \nu\|_{1/\nu}, \end{aligned}$$

where μ^n and ν^n are the distributions of X_1 and X_2 at epoch n of Gibbs sampling. Similar bounds are possible for the joint distribution if we substitute $n-1$ for n in the marginal chains. More definitive bounds are possible if we use the eigenvector part of the bound (27.10). The Laguerre and Meixner polynomials developed in Chapter 17 are pertinent. ■

Example 27.9.2 *Normal-Normal*

Consider a normal likelihood with a normal prior on the mean. The joint and marginal densities are

$$
\begin{aligned}
r_{x_1x_2} &= \frac{1}{\sqrt{2\pi\sigma^2}}e^{-\frac{(x_1-x_2)^2}{2\sigma^2}}\frac{1}{\sqrt{2\pi\delta^2}}e^{-\frac{(x_2-\gamma)^2}{2\delta^2}} \\
\nu_{x_2} &= \frac{1}{\sqrt{2\pi\delta^2}}e^{-\frac{(x_2-\gamma)^2}{2\delta^2}} \\
\mu_{x_1} &= \frac{1}{\sqrt{2\pi\sigma^2}}\frac{1}{\sqrt{2\pi\delta^2}}\int_{-\infty}^{\infty} e^{-\frac{(x_1-x_2)^2}{2\sigma^2}}e^{-\frac{(x_2-\gamma)^2}{2\delta^2}}\,dx_2 \\
&= \frac{1}{\sqrt{2\pi(\sigma^2+\delta^2)}}e^{-\frac{(x_1-\gamma)^2}{2(\sigma^2+\delta^2)}}.
\end{aligned}
$$

These assumptions imply that X_2 given X_1 is normally distributed with mean and variance

$$
\alpha X_1+\phi = \frac{\delta^2(X_1-\gamma)}{\sigma^2+\delta^2}+\gamma, \quad \eta = \frac{\sigma^2\delta^2}{\sigma^2+\delta^2}.
$$

The trace criterion for compactness

$$
\begin{aligned}
&\int_{-\infty}^{\infty}\int_{-\infty}^{\infty}\frac{r_{x_1x_2}^2}{\mu_{x_1}\nu_{x_2}}dx_1dx_2 \\
&= \frac{\sqrt{\delta^2(\sigma^2+\delta^2)}}{\sigma^2}\int_{-\infty}^{\infty}\int_{-\infty}^{\infty} e^{-\frac{(x_1-x_2)^2}{\sigma^2}}e^{-\frac{(x_2-\gamma)^2}{2\delta^2}}e^{\frac{(x_1-\gamma)^2}{2(\sigma^2+\delta^2)}}dx_1dx_2
\end{aligned}
$$

is finite because, as the reader can check, the integrand is proportional to e^{-x^tAx} for a positive definite matrix A.

The two operators $(Tf)(x_2)$ and $(T^*g)(x_1)$ are

$$
\begin{aligned}
\mathrm{E}[f(X_1)\mid X_2] &= \frac{1}{\sqrt{2\pi\sigma^2}}\int_{-\infty}^{\infty} f(x_1)e^{-\frac{(x_1-x_2)^2}{2\sigma^2}}dx_1 &\quad (27.22)\\
\mathrm{E}[g(X_2)\mid X_1] &= \frac{1}{\sqrt{2\pi\eta^2}}\int_{-\infty}^{\infty} g(x_2)e^{-\frac{(x_2-\alpha x_1-\phi)^2}{2\eta^2}}dx_2. &\quad (27.23)
\end{aligned}
$$

The choices $f(x_1) = x_1^n$ and $g(x_2) = x_2^n$ produce polynomials of degree n in x_2 and x_1, respectively. Hence, the two operators send polynomials of degree n into polynomials of degree n. If we assume the orthogonal polynomials are monic, then we can again extract the leading coefficient of X_2^n in equation (27.22) and the leading coefficient of X_1^n in equation (27.23). This action gives $\gamma_n = 1$, $\delta_n = \alpha^n$, and $\theta_n = \alpha^n$. The second-largest eigenvalue of either marginal chain is therefore $\rho = \alpha = \delta^2/(\sigma^2+\delta^2)$. The orthogonal polynomials for the two chains are related to the Hermite polynomials discussed in Chapter 17. ■

27.10 Problems

1. The Hammersley and Clifford theorem can fail if the distribution of the Markov field is not everywhere positive. As a counterexample, consider the graph corresponding to the four vertices of a square. Each vertex is assigned one of the two values 0 or 1, and each pair of neighboring vertices is connected by an edge. The eight realizations of (N_1, N_2, N_3, N_4)

$$\begin{array}{cc} (0\ 0\ 0\ 0) & (0\ 0\ 0\ 1) \\ (1\ 0\ 0\ 0) & (0\ 0\ 1\ 1) \\ (1\ 1\ 0\ 0) & (0\ 1\ 1\ 1) \\ (1\ 1\ 1\ 0) & (1\ 1\ 1\ 1) \end{array}$$

each have probability $\frac{1}{8}$. The remaining eight realizations have probability 0. Prove that the consistency conditions

$$\pi(n_i \mid \mathbf{n}_{G-i}) = \pi(n_i \mid n_{\partial i})$$

hold, yet Proposition 27.2.1 is false.

2. The Swendsen-Wang algorithm depends on finding the connected components of a graph. Design and test a quick algorithm for this purpose.

3. Implement the reversible jump chain described by Green [8] for a Poisson process with multiple change points.

4. Implement the reversible jump chain described by Richardson and Green [14] for analyzing a mixture of univariate normals with an unknown number of components.

5. Demonstrate that the two definitions (27.3) of total variation distance are equivalent. (Hint: Consider the set $A = \{i : \pi_i \geq \nu_i\}$ and the identities $\sum_i \pi_i = \sum_i \nu_i = 1$.)

6. Let X have a Bernoulli distribution with success probability p and Y a Poisson distribution with mean p. Prove the total variation inequality

$$\|\pi_X - \pi_Y\|_{\mathrm{TV}} \leq p^2$$

involving the distributions π_X and π_Y of X and Y.

7. Let Y be a Poisson random variable with mean λ. Demonstrate that $\Pr(Y \geq k)$ is increasing in λ for k fixed.

8. Suppose Y follows a negative binomial distribution that counts the number of failures until n successes. Demonstrate by a coupling argument that $\Pr(Y \geq k)$ is decreasing in the success probability p for k fixed.

9. Let X_1 follow a beta distribution with parameters α_1 and β_1 and X_2 follow a beta distribution with parameters α_2 and β_2. If $\alpha_1 \leq \alpha_2$ and $\alpha_1 + \beta_1 = \alpha_2 + \beta_2$, then demonstrate that $\Pr(X_1 \geq x) \leq \Pr(X_2 \geq x)$ for all $x \in [0, 1]$. How does this result carry over to the beta-binomial distribution? (Hint: Construct X_1 and X_2 from gamma-distributed random variables.)

10. Suppose P is the transition probability matrix of a reversible Markov chain with equilibrium distribution π. Verify that P satisfies the symmetry condition

$$\langle Pu, v \rangle_\pi \quad = \quad \langle u, Pv \rangle_\pi,$$

which yields a direct proof that P has only real eigenvalues.

11. In our analysis of convergence of the independence sampler, we asserted that the eigenvalues $\lambda_1, \ldots, \lambda_m$ satisfied the properties: (a) $\lambda_1 = 1 - 1/w_1$, (b) the λ_i are decreasing, and (c) $\lambda_m = 0$. Verify these properties.

12. Find the row and column eigenvectors of the transition probability matrix P for the independence sampler. Show that they are orthogonal in the appropriate inner products.

13. Suppose T is a linear operator from a Hilbert space into itself. Prove that

$$\|TT^*\| \quad = \quad \|T^*T\| \quad = \quad \|T\|^2 \quad = \quad \|T^*\|^2.$$

14. An orthogonal projection P onto a closed subspace S of a Hilbert space satisfies $Pf = f$ for $f \in S$ and $Pf = \mathbf{0}$ for $f \in S^\perp$. Show that the properties $P^2 = P$ and $P^* = P$ are necessary and sufficient for a continuous linear operator to be an orthogonal projection. Use the obvious orthogonal projection to prove that a closed subspace S of a separable Hilbert space is separable.

15. A continuous linear operator O from a Hilbert space to itself is said to be orthogonal if and only if $OO^* = O^*O = I$. It is said to be an isometry if and only if $\|Ov\| = \|v\|$ for every vector v. Prove that an orthogonal operator is an isometry and that an isometry with full range is orthogonal. Produce an infinite-dimensional example of an isometry that is not orthogonal.

16. A diagonalizable operator T sends a vector $f = \sum_n c_n v_n$ into the vector $Tf = \sum_n \lambda_n c_n v_n$. Assuming the basis $\{v_n\}_n$ is orthonormal, prove the following assertions:

    (a) T is continuous if and only if the λ_n are bounded. In this case $\|T\| = \sup_n |\lambda_n|$.

    (b) T is symmetric if and only if all λ_n are real.

    (c) T is an orthogonal operator if and only if all λ_n have absolute value 1.

    (d) T is an orthogonal projection if and only if all λ_n equal 0 or 1.

    (e) T is compact if and only if $\lim_{n\to\infty} \lambda_n = 0$.

    See the previous two problems for definitions.

17. Prove that the identity operator cannot be compact on an infinite-dimensional Hilbert space. Use this fact to demonstrate that a compact operator cannot have a continuous inverse in the same circumstances. (Hint: Apply Proposition 27.8.1.)

18. If $\{u^m\}_m$ is an orthonormal basis of $\mathrm{L}^2(\mu)$, and $\{v^n\}_n$ is a basis of $\mathrm{L}^2(\nu)$, then demonstrate that the product collection $\{u^m v^n\}_{mn}$ is an orthonormal basis of $\mathrm{L}^2(\mu \times \nu)$. (Hint: Use Fubini's theorem to show that the vectors $u^m v^n$ are orthogonal unit vectors. If f_{ij} satisfies

$$\sum_i \sum_j f_{ij} u_i^m v_j^n \mu_i \nu_j = 0$$

for all m and n, then $g_j^m = \sum_i f_{ij} u_i^m \mu_i$ satisfies $\sum_j g_j^m v_j^n \nu_j = 0$ for all n.)

19. Prove the analogs of the chi-square bounds (27.17) and (27.18) for total variation distance.

20. Let X be binomially distributed with n trials and success probability P. Assume that P follows a beta distribution with parameters (α, β). If one alternates Gibbs sampling of X and P, then

    (a) Derive the transition probabilities and stationary distributions for the X and P marginal chains.

    (b) Show that the associated Markov chain operators are compact and send polynomials of degree k into polynomials of degree k.

    (c) Show that the eigenvalues of the marginal chains are

$$\theta_k = \frac{n(n-1)\cdots(n-k+1)}{(n+\alpha+\beta)(n+\alpha+\beta+1)\cdots(n+\alpha+\beta+k-1)}$$

    for $k = 0, 1, \ldots, n$.

(Hints: The marginal moments are

$$\begin{aligned} E(X^k \mid P) &= n \cdots (n-k+1)P^k \\ E(P^k \mid X) &= \frac{(X+\alpha)\cdots(X+\alpha+k-1)}{(n+\alpha+\beta)\cdots(n+\alpha+\beta+k-1)}. \end{aligned}$$

The relevant eigenvectors are Hahn and Jacobi polynomials, but these are not explicitly needed.)

21. In a binomial location model, X is distributed as $Y + Z$, where Z is binomial with n_1 trials and success probability p, Y is binomial with n_2 trials and success probability p, and Y and Z are independent. If one alternates Gibbs sampling of X and Y, then

    (a) Derive the transition probabilities and stationary distributions of the X and Y marginal chains.

    (b) Show that the associated Markov operators are compact and send polynomials of degree k into polynomials of degree k.

    (c) Show that the eigenvalues of the marginal chains are

    $$\theta_k = \frac{n_1 \cdots (n_1-k+1)}{(n_1+n_2)\cdots(n_1+n_2-k+1)}$$

    for $k = 0, 1, \ldots, n_1 + n_2$.

    (Hints: The marginal moments are

    $$\begin{aligned} E(X^k \mid Y) &= \sum_{j=0}^{k} \binom{k}{j} Y^j \, E(Z^{k-j}) \\ E[Y \cdots (Y-k+1) \mid X] &= \frac{n_1 \cdots (n_1-k+1) X \cdots (X-k+1)}{(n_1+n_2)\cdots(n_1+n_2-k+1)}. \end{aligned}$$

    The relevant eigenvectors are the Krawtchouk polynomials, but these are not explicitly needed.)

27.11 References

[1] Akhiezer NI, Glazman IM (1993) *Theory of Linear Operators in Hilbert Space.* Dover, New York

[2] Brémaud P (1999) *Markov Chains: Gibbs Fields, Monte Carlo Simulation, and Queues.* Springer

[3] Brook D (1964) On the distinction between the conditional and the joint probability approaches in the specification of nearest-neighbor systems. *Biometrika* 51:481–483

[4] Conway JB (1985) *A Course on Functional Analysis.* Springer, New York

[5] Diaconis P (1988) *Group Representations in Probability and Statistics.* Institute of Mathematical Statistics, Hayward, CA

[6] Diaconis P, Khare K, Saloff-Coste L (2008) Gibbs sampling, exponential families and orthogonal polynomials. *Stat Science* 23:151–178

[7] Edwards RG, Sokal AD (1988) Generalizations of the Fortuin-Kasteleyn-Swendsen-Wang representation and Monte Carlo algorithm. *Physical Review D* 38:2009–2012

[8] Green PJ (1995) Reversible jump Markov chain Monte Carlo computation and Bayesian model determination. *Biometrika* 82:711–732

[9] Hastie D, Green PJ (2009) Reversible jump MCMC. (unpublished lecture notes)

[10] Jones GL (2004) On the Markov chain central limit theorem. *Prob Surveys* 1:299–320

[11] Levin DA, Peres Y, Wilmer EL (2008) *Markov Chains and Mixing Times.* Amer Math Soc, Providence, RI

[12] Liu JS (1996) Metropolized independent sampling with comparisons to rejection sampling and importance sampling. *Stat and Computing* 6:113–119

[13] Liu JS (2001) *Monte Carlo Strategies in Scientific Computing.* Springer, New York

[14] Richardson S, Green PJ (1997) On Bayesian analysis of mixtures with an unknown number of components. *J Royal Stat Soc B* 59:731–792

[15] Robert CP, Casella G (2004) *Monte Carlo Statistical Methods*, 2nd ed. Springer, New York

[16] Rosenthal JS (1995) Convergence rates of Markov chains. *SIAM Review* 37:387–405

[17] Rynne BP, Youngson MA (2008) *Linear Functional Analysis.* Springer, New York

[18] Stein EM, Shakarchi R (2005) *Real Analysis: Measure Theory, Integration, and Hilbert Spaces.* Princeton University Press, Princeton, NJ

[19] Swendsen RH, Wang JS (1987) Nonuniversal critical dynamics in Monte Carlo simulations. *Physical Review Letters* 58:86–88

[20] Tierney L (1994) Markov chains for exploring posterior distributions (with discussion). *Ann Stat* 22:1701–1762

Appendix: The Multivariate Normal Distribution

In dealing with multivariate distributions such as the multivariate normal, it is convenient to extend the expectation and variance operators to random vectors. The expectation of a random vector $X = (X_1, \ldots, X_n)^t$ is defined componentwise by

$$\mathrm{E}(X) = \begin{pmatrix} \mathrm{E}[X_1] \\ \vdots \\ \mathrm{E}[X_n] \end{pmatrix}.$$

Linearity carries over from the scalar case in the sense that

$$\begin{aligned} \mathrm{E}(X+Y) &= \mathrm{E}(X) + \mathrm{E}(Y) \\ \mathrm{E}(MX) &= M\,\mathrm{E}(X) \end{aligned}$$

for a compatible random vector Y and a compatible matrix M. The same componentwise conventions hold for the expectation of a random matrix and the variances and covariances of a random vector. Thus, we can express the variance-covariance matrix of a random vector X as

$$\mathrm{Var}(X) = \mathrm{E}\{[X - \mathrm{E}(X)][X - \mathrm{E}(X)]^t\} = \mathrm{E}(XX^t) - \mathrm{E}(X)\,\mathrm{E}(X)^t.$$

These notational choices produce many other compact formulas. For instance, the random quadratic form X^tMX has expectation

$$\mathrm{E}(X^tMX) = \mathrm{tr}[M\,\mathrm{Var}(X)] + \mathrm{E}(X)^t M\,\mathrm{E}(X). \tag{A.1}$$

To verify this assertion, observe that

$$\begin{aligned} \mathrm{E}(X^tMX) &= \mathrm{E}\Big(\sum_i\sum_j X_i m_{ij} X_j\Big) \\ &= \sum_i\sum_j m_{ij}\,\mathrm{E}(X_iX_j) \\ &= \sum_i\sum_j m_{ij}[\mathrm{Cov}(X_i, X_j) + \mathrm{E}(X_i)\,\mathrm{E}(X_j)] \\ &= \mathrm{tr}[M\,\mathrm{Var}(X)] + \mathrm{E}(X)^t M\,\mathrm{E}(X). \end{aligned}$$

Among the many possible definitions of the multivariate normal distribution, we adopt the one most widely used in stochastic simulation. Our point of departure will be random vectors with independent standard normal components. If such a random vector X has n components, then its density is

$$\prod_{j=1}^{n} \frac{1}{\sqrt{2\pi}} e^{-x_j^2/2} = \left(\frac{1}{2\pi}\right)^{n/2} e^{-x^t x/2}.$$

Because the standard normal distribution has mean 0, variance 1, and characteristic function $e^{-s^2/2}$, it follows that X has mean vector $\mathbf{0}$, variance matrix I, and characteristic function

$$\mathrm{E}(e^{is^t X}) = \prod_{j=1}^{n} e^{-s_j^2/2} = e^{-s^t s/2}.$$

We now define any affine transformation $Y = AX + \mu$ of X to be multivariate normal [1, 2]. This definition has several practical consequences. First, it is clear that $\mathrm{E}(Y) = \mu$ and $\mathrm{Var}(Y) = A\,\mathrm{Var}(X)A^t = AA^t = \Omega$. Second, any affine transformation $BY + \nu = BAX + B\mu + \nu$ of Y is also multivariate normal. Third, any subvector of Y is multivariate normal. Fourth, the characteristic function of Y is

$$\mathrm{E}(e^{is^t Y}) = e^{is^t\mu}\,\mathrm{E}(e^{is^t AX}) = e^{is^t\mu - s^t AA^t s/2} = e^{is^t\mu - s^t\Omega s/2}.$$

Fifth, the sum of two independent multivariate normal random vectors is multivariate normal. Indeed, if $Z = BU + \nu$ is suitably dimensioned and X is independent of U, then we can represent the sum

$$Y + Z = (A \quad B)\begin{pmatrix} X \\ U \end{pmatrix} + \mu + \nu$$

in the required form.

This enumeration omits two more subtle issues. One is whether Y possesses a density. Observe that Y lives in an affine subspace of dimension equal to or less than the rank of A. Thus, if Y has m components, then $n \geq m$ must hold in order for Y to possess a density. A second issue is the existence and nature of the conditional density of a set of components of Y given the remaining components. We can clarify both of these issues by making canonical choices of X and A based on the QR decomposition of a matrix.

Assuming that $n \geq m$, we can write

$$A^t = Q\begin{pmatrix} R \\ \mathbf{0} \end{pmatrix},$$

where Q is an $n \times n$ orthogonal matrix and $R = L^t$ is an $m \times m$ upper-triangular matrix with nonnegative diagonal entries. (If $n = m$, we omit the zero matrix in the QR decomposition.) It follows that

$$AX = (L \quad \mathbf{0}^t)\,Q^t X = (L \quad \mathbf{0}^t)\,Z.$$

In view of the usual change-of-variables formula for probability densities and the facts that the orthogonal matrix Q^t preserves inner products and has determinant ± 1, the random vector Z has n independent standard normal components and serves as a substitute for X. Not only is this true, but we can dispense with the last $n - m$ components of Z because they are multiplied by the matrix $\mathbf{0}^t$. Thus, we can safely assume $n = m$ and calculate the density of $Y = LZ + \mu$ when L is invertible. The change-of-variables formula then shows that Y has density

$$\begin{aligned} f(y) &= \left(\frac{1}{2\pi}\right)^{n/2} |\det L^{-1}| e^{-(y-\mu)^t (L^{-1})^t L^{-1}(y-\mu)/2} \\ &= \left(\frac{1}{2\pi}\right)^{n/2} |\det \Omega|^{-1/2} e^{-(y-\mu)^t \Omega^{-1}(y-\mu)/2}, \end{aligned}$$

where $\Omega = LL^t$ is the variance matrix of Y. By definition LL^t is the Cholesky decomposition of Ω.

To address the issue of conditional densities, consider the compatibly partitioned vectors $Y^t = (Y_1^t, Y_2^t)$, $X^t = (X_1^t, X_2^t)$, and $\mu^t = (\mu_1^t, \mu_2^t)$ and matrices

$$L = \begin{pmatrix} L_{11} & \mathbf{0} \\ L_{21} & L_{22} \end{pmatrix}, \qquad \Omega = \begin{pmatrix} \Omega_{11} & \Omega_{12} \\ \Omega_{21} & \Omega_{22} \end{pmatrix}.$$

Now suppose that X is standard normal, that $Y = LX + \mu$, and that L_{11} has full rank. For $Y_1 = y_1$ fixed, the equation $y_1 = L_{11}X_1 + \mu_1$ shows that X_1 is fixed at the value $x_1 = L_{11}^{-1}(y_1 - \mu_1)$. Because no restrictions apply to X_2, we have

$$Y_2 = L_{22}X_2 + L_{21}L_{11}^{-1}(y_1 - \mu_1) + \mu_2.$$

Thus, Y_2 given Y_1 is normal with mean $L_{21}L_{11}^{-1}(y_1 - \mu_1) + \mu_2$ and variance $L_{22}L_{22}^t$. To express these in terms of the blocks of $\Omega = LL^t$, observe that

$$\begin{aligned} \Omega_{11} &= L_{11}L_{11}^t \\ \Omega_{21} &= L_{21}L_{11}^t \\ \Omega_{22} &= L_{21}L_{21}^t + L_{22}L_{22}^t. \end{aligned}$$

The first two of these equations imply that $L_{21}L_{11}^{-1} = \Omega_{21}\Omega_{11}^{-1}$. The last equation then gives

$$\begin{aligned} L_{22}L_{22}^t &= \Omega_{22} - L_{21}L_{21}^t \\ &= \Omega_{22} - \Omega_{21}(L_{11}^t)^{-1}L_{11}^{-1}\Omega_{12} \\ &= \Omega_{22} - \Omega_{21}\Omega_{11}^{-1}\Omega_{12}. \end{aligned}$$

These calculations do not require that Y_2 possess a density. In summary, the conditional distribution of Y_2 given Y_1 is normal with mean and variance

$$\begin{aligned} \mathrm{E}(Y_2 \mid Y_1) &= \Omega_{21}\Omega_{11}^{-1}(Y_1 - \mu_1) + \mu_2 \\ \mathrm{Var}(Y_2 \mid Y_1) &= \Omega_{22} - \Omega_{21}\Omega_{11}^{-1}\Omega_{12}. \end{aligned} \qquad \text{(A.2)}$$

A.1 References

[1] Rao CR (1973) *Linear Statistical Inference and Its Applications*, 2nd ed. Wiley, New York

[2] Severini TA (2005) *Elements of Distribution Theory.* Cambridge University Press, Cambridge

Index